Artificial Intelligence for Scientific Discoveries

Raban Iten

Artificial Intelligence for Scientific Discoveries

Extracting Physical Concepts from Experimental Data Using Deep Learning

Springer

Raban Iten
ETH Zürich
Zürich, Switzerland

ISBN 978-3-031-27018-5 ISBN 978-3-031-27019-2 (eBook)
https://doi.org/10.1007/978-3-031-27019-2

This Springer imprint is published by the registered company Springer Nature Switzerland AG
The registered company address is: Gewerbestrasse 11, 6330 Cham, Switzerland

Dedicated to Theres and Aschi Iten

Preface

This book is written for scientists interested in using machine learning to discover physical concepts. No prior knowledge of machine learning is required, but a first degree in a science subject is assumed.

The motivation to write this book is rooted in my Ph.D. research, which focused on the long-term goal of using machine learning to make progress on fundamental questions in modern physics. At the beginning of the Ph.D., I had only a very basic knowledge of machine learning, so I started reading a lot of papers in the field of machine learning as well as papers that used machine learning for physics. Most of the work using machine learning for physics focuses on finding model parameters for specific physical systems, and I could only find a few papers that use machine learning in a way to gain insights about the model itself rather than just its parameters. These papers come from a variety of research communities, and it took me some effort to get an overview of the relevant machine learning methods for automating parts of a physicist's discovery process. Therefore, I thought that providing an overview over this field could be helpful to other physicists who are interested in learning about how to use machines to get insights from experimental data. I would be very interested in any feedback on the book or suggestions for work to be covered in a second version. If you have any comments, please just email me at itenr@ethz.ch.

Zurich, Switzerland
January 2023

Raban Iten

Acknowledgements

This book is based on my Ph.D. and would not exist in this form without the close collaboration with Tony Metger and Hendrik P. Nautrup. I would like to warmly thank these brilliant friends for the fruitful and very enjoyable collaboration and for reviewing parts of this book.

In addition, I would like to thank Angela Lahee, Ashok Arumairaj and their team from Springer for their enthusiasm for the topic covered in this book and their great support while writing it.

I would also like to sincerely thank my Ph.D. supervisor Renato Renner for interesting discussions and for supporting me in writing this book. His precision in scientific thinking is highly inspiring and clearly sharpened my mind. Renato gave me a lot of freedom to work on topics of my interest and was always open to discuss any questions.

Furthermore, I would like to thank Giuseppe Carleo and Matthias Troyer for reviewing my work as co-examiners for my doctoral thesis.

I am also very grateful to Manfred Sigrist for interesting discussions and the occasional chats in the office or during a jogging round, and last but not least for creating a very friendly atmosphere on the K floor in the HIT building of ETH Zurich.

During the last couple of years, I was extremely lucky to collaborate with various brilliant people including Mario Berta, Hans J. Briegel, Matthias Christandl, Roger Colbeck, Emanuel Malvetti, Hendrik P. Nautrup, Tony Metger, Romain Moyard, Navneeth Ramakrishnan, Joseph M. Renes, Renato Renner, Henrik Wilming and Stefan Woerner. I would like to thank all of them for the interesting collaborations.

Hans J. Briegel, Mario Krenn, Alexey Melnikov, Tony Metger, Hendrik P. Nautrup, Henrik Wilming and Monika Zimmermann helped me to improve this book, and I thank them very much. I would also like to thank Mario Krenn for very interesting and extremely motivating discussions about how to use machine learning to discover conceptual information.

In addition, I would like to thank all the former and current members of the quantum information theory group at ETH Zurich for many interesting and stimulating discussions. In particular, I would like to thank my office mates Rotem Arnon-Friedman and Mischa Woods for many interesting discussions and their scientific advice. Further, I would like to thank Christa Zoufal for helping me to organize an internship at IBM Research in Zürich with Stefan Woerner during my Ph.D., which was a very enriching experience and a truly wonderful time, not only because of the delicious food at the IBM canteen.

Furthermore, I have to thank my girlfriends along the Ph.D. for setting me challenges outside of the physics world and thus contributing to my personal learning experience. They taught me complex lessons (sometimes painful to learn) and provided me with various experiences, each of them in their unique way. This contributed greatly to making my Ph.D. time an instructive period of my life. In particular, I would like to thank Monika for creating a steady, supportive and stimulating social environment which led to a productive time towards the end of my Ph.D.

Finally and most importantly, I would like to thank my two sisters and my parents for their constant support and encouragement.

Zurich, Switzerland
January 2023

Raban Iten

Contents

Chapter 1
Introduction

There has been amazing progress in the field of artificial intelligence (AI) and machine learning in recent years. Not long ago, robots could not walk, now they can even dance (as for example demonstrated by robots from Boston Dynamics in 2020). Computer vision and natural language processing have reached a level of performance that make them indispensable in many areas today, such as face recognition, speech interpretation and self-driving cars. While most AI is developed based on domain knowledge, i.e., specific knowledge about the data under consideration or the environment of interest, there are also impressive results where AI learns from scratch. A breakthrough in this direction was achieved with MuZero by DeepMind [1]. MuZero is a machine learning system that learns to play Go, chess and shogi from scratch without knowing anything about the rules of the games or what it would take to win. You may wonder how it is possible to learn a game without watching a human play the game and learning from him or her. The strategy used by DeepMind is to let machines play against each other and tell them when they have won a game. Surprisingly, this information (together with the information about which moves are allowed) is sufficient for the machine to build a model of the game and its rules and develop a winning strategy without learning from a human player. Impressively, the trained machine has been able to beat the best human players in the world in all three games, Go, chess and shogi. Having algorithms that can learn models of different games without any prior knowledge about the games, nothing prevents us from applying such algorithms to scientific problems. Indeed, the methods used for MuZero have already been successfully applied to a range of complex problems in chemistry [2] and quantum physics [3]. However, in science, we are not only interested in the prediction, but also in how we make the prediction. The main goal of physicists is to find the simplest models that describe nature. Having such a model at hand makes it possible to predict the behavior of previously unobserved systems. Unfortunately, understanding the model learned by MuZero remains very challenging and hence we do not get much insight into how MuZero makes its predictions. This is a fundamental problem often faced when applying machine learning

R. Iten, *Artificial Intelligence for Scientific Discoveries*,
https://doi.org/10.1007/978-3-031-27019-2_1

to scientific problems. As scientists, we are particularly interested in understanding *how* the machine learning system makes its predictions and hence what the underlying model looks like. In this book we focus on this question, i.e., on how we can extract conceptual information from physical systems using machine learning. This question is of interest for several reasons. On one hand, it is an important step for the long term goal of building AI physicists [4], on the other hand, it may help us to gain insight into fundamental problems in modern physics. The latter is explained in more detail in the following section.

1.1 Motivation

History has shown that physical theories often only provide accurate predictions under certain restrictions of the environment under consideration. For example, Newtonian physics works extremely well for describing the behavior of objects we know from our daily life like a table or a football. However, the theory does not apply to extremely small objects like molecules. Rather, physical phenomena on the length scale of atoms are described more precisely by quantum theory. New physical theories usually contain certain elements of previously known theories. This is expected since the new theory should be more fundamental than the previous one and hence, we expect to find the previous theory as a limit case of the new theory. In the mentioned case, we recover Newtonian physics from quantum physics by considering the limit of vanishing Planck constant. However, transferring too many elements from the previous theory to a new theory may not lead to the simplest model to explain experimental data, but rather to the one that most naturally followed from the previous theory.

For example, before the discovery of special relativity, the constant speed of light that appears in Maxwell's equations of electromagnetism was explained by a medium called "aether" in which light can propagate. The aether served as an absolute and unique frame in which Maxwell's equations hold. The idea of aether goes back to the seventeenth century. Robert Boyle considered two sort of aethers, one serving as a medium that explains mechanical interactions between separated bodies, and another one explaining phenomenas like magnetism [5]. Therefore, the idea of the aether was already strongly based on prior knowledge of physics at the time of its introduction, namely the idea that objects can only interact with each other through other particles. Although the idea of the aether became more magical over time and the theory came into conflict with several experiments, the idea survived until the twentieth century, when Einstein came up with the special theory of relativity. The essential idea of special relativity is that light propagates with the same speed with respect to every inertial system. A conclusion of this statement is that time and space cannot be considered separately: two events that occur at the same time for one observer, may occur at different times for another observer. Special relativity required a complete rethink of our physical intuition and we had to abandon the idea of absolute time and an aether as a medium for the propagation of light. This historical example shows that prior knowledge of physics can prevent us from finding the simplest explanations to describe observational data.

Nowadays, it could be that we are in a similar situation: although we can use quantum theory—which is built on classical mechanics—to predict measurement outcomes with very high accuracy in the realm of the very small, we still face several conceptual problems. One such problem, related to the measurement problem of quantum physics, is the question if an observer of a quantum system can itself be described by quantum mechanics. This problem was raised by Wigner in [6], deepened in [7] and recently sharpened in [8, 9]. Experimentally, it is investigated whether we can imitate observers with autonomous agents [10, 11]. A serious problem in modern physics, which may be related to the conceptual problems encountered in quantum physics, is that general relativity—impressively successful in the restricted regime of the very large—is fundamentally incompatible with quantum mechanics, as evidenced by paradoxes such as the black hole information loss [12, 13]. This raises an interesting question: are the laws of quantum physics (or other physical theories) the most natural ones to explain data from experiments if we assume no prior knowledge of physics? While this question is unlikely to be answered in the near future, the recent advances described in this book make a first step in this direction.

1.2 Physicist's Discovery Process

Building machines that can autonomously learn physical principles by interacting with an environment is an extremely challenging task. Roughly, the cyclic process a physicist goes through to find laws in nature can be divided into the following parts.

- **Choose a physical subsystem**: As humans, we collect a lot of data with our sensory organs every second. A physicist focuses on a particular part of this data. Usually, the data on which is focused is related to one or several physical subsystems, such as the data of a pendulum together with the air in its environment.
- **Create an experimental setup**: Create devices that are able to collect relevant data from physical processes.
- **Observations**: The physicist collects data about the chosen physical systems by using devices of an experimental setup.
- **Create a model**: Analyze observed data to build a (or expand an already existing) model describing the behavior of the underlying physical system. A useful model should make predictions that are experimentally falsifiable.
- **Test the model**: Test predictions made by a model. If the model is falsified, the physicist should create another model (or at least analyze the restricted scope of validity of the model). If it is not falsified, the model can be further tested and used for applications.

Each of these parts can be very challenging, and there are different approaches from machine learning trying to automize them. We provide an overview over such approaches in Part II. In Part III of this book we focus on one particular step of the model building process, namely the determination of the relevant degrees of freedom of a physical system from experimental data. Recent advances in this field have made it possible to discover compact representations of physical systems without requiring any prior knowledge of the considered physical system other than the given experimental data. Minimizing the prior knowledge is not only interesting for the long-term goal of finding an alternative description of quantum mechanics, but also an essential requirement for building an AI physicist [4]. Indeed, otherwise it could not be claimed that the AI physicist discovered the found laws completely by itself. The basic idea of the approach presented in Part III is explained in Sect. 1.3.

1.3 Extracting Relevant Parameters from Experimental Data

The goal of using machines to discover physical laws underlying experimental data has been pursued in several contexts (see Part II for a detailed discussion and [14–17] for recent reviews). A lot of early work focused on finding mathematical expressions describing a given dataset (see e.g. [18–20] and Sect. 6.2 for a detailed discussion of these works). For example, in [19] an algorithm recovers the laws of motion of mechanical systems, like a double pendulum, by searching over a space of mathematical expressions on given input variables (searching the form as well as the parameters of mathematical expressions is known under the term *symbolic regression* [21]). However, such approaches require prior knowledge in the form of knowing what the relevant variables are. In certain situations one might not have such prior knowledge or does not want to impose it to allow the machine to find entirely different representations of the physical system. Defining properties of such representations and describing how they can be found using machine learning is the main focus of this book (Part III) and is regarded as a first step towards discovering physical concepts with machine learning without using prior knowledge about physics and maths. In fact, in order to write down a physical law, we must first know which parameters occur in the law and thus which degrees of freedom the physical system has. In the following, the basic idea for finding the relevant degrees of freedom is presented.

Finding a minimal set of parameters that is sufficient to describe a given data set (without using any prior knowledge about the system that created the data set) is a complex task, but essential for the process of building physical models. Let us consider a very simple toy example, where we are given time series data $(t_i, x(t_i))_{i\in\{1,\dots,N\}}$ of a particle moving along the x-direction with constant speed (for some $3 \leq N \in \mathbb{N}$). The goal of a physicist could be to build a model that allows to predict the x-coordinate of the position of the particle for any time t'. The standard approach is to extract the speed $v = (x(t_1) - x(t_0))/(t_1 - t_0)$ and the start position $x_0 = x(t_0) - vt_0$ from the data and then calculate $x(t') = x_0 + vt'$. Hence, the models that a physicist builds do not deal with the observations directly, but rather with a representation of the underlying physical state of the observed system, e.g., the two parameters *initial*

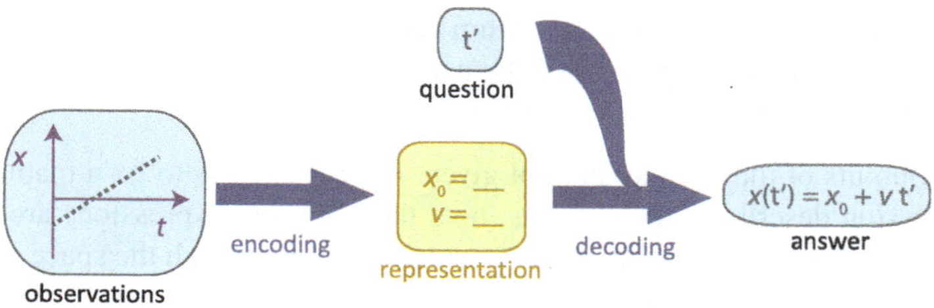

Fig. 1.1 Human learning of physical representations. (Figure reproduced from Iten and Metger et al., *Physical Review Letters*, 2020 [22]). A physicist compresses experimental observations into a simple representation (*encoding*). When later asked any question about the physical setting, the physicist should be able to produce a correct answer using only the representation and not the original data. We call the process of producing the answer from the representation *decoding*. For example, the observations may be the first few seconds of the trajectory of a particle moving with constant speed; the representation could be the parameters "speed v" and "initial position x_0" and the question could be "where will the particle be at a later time t'?"

position and *speed*, (x_0, v). Which parameters are used is an important part of the model. In Part III of this book we follow the ideas presented in [22, 23] and formalize what we understand by a "natural" representation and describe a machine learning system that finds such representations without using any prior knowledge about the specific physical system that is considered. One natural criteria for the representation is that it is sufficient, hence that it contains all relevant information to predict the system's (future) behavior. To know which parameters are required for a sufficient representation, we also have to model the prediction process of a physicist.

We hence split the human modeling process into two interdependent tasks. Firstly, finding the relevant parameters in experimental data (e.g., the *initial position* and *speed*, (x_0, v)). And, secondly, building a physical model that specifies how to make predictions (i.e., answer questions) based on the knowledge of the relevant parameters of the system (e.g. "where is the particle at time t'?"). Since our goal is to build a machine that solves these tasks, it is helpful to formalize them. Formally, the physical modelling process can be regarded as an "encoder" $E : \mathcal{O} \to \mathcal{R}$ mapping the set of possible observations $\mathcal{O}$ to representations $\mathcal{R}$, followed by a "decoder" $D : \mathcal{R} \times \mathcal{Q} \to \mathcal{A}$ mapping the sets of all possible representations $\mathcal{R}$ and questions $\mathcal{Q}$ to answers $\mathcal{A}$ (Fig. 1.1). Throughout this book, we refer to this structure as *SciNet* for simplicity. From a machine learning perspective, the structure of *SciNet* is closely related to autoencoders, a method from representation learning for finding compressed representations of high dimensional data (see Chap. 4 for a non-technical introduction to autoencoders). We would like to allow for arbitrary encoder and decoder functions, and let a computer learn these by itself from a given data set containing triples of the form $[o, q, a^\star(o, q)]$, where $a^\star(o, q)$ describes the correct answer to question q given observation o. Similar to [19], one could try to approach this task by using symbolic regression (discussed in Sect. 6.2.1) to find functions from input-output pairs for the encoder and the decoder synchronously. Thereby, one uses a certain kind of search algorithm based on evolutionary computation [24, 25], to search through the space of mathematical expressions fitting the input-output pairs. However, such an approach faces several challenges:

1. The outputs of the encoder are not given. Hence, one has to fix a mathematical expression describing the encoder (note that different expressions are allowed to have different output-dimensions) and then search trough the space of mathematical expressions representing the decoder to see if there exists a suitable one that provides the correct answers. Such a search is computationally extremely costly.
2. For common physical problems, the decoder function that calculates the predictions from the (small number of) relevant physical parameters is usually quite simple (in the example considered above, it has to implement the function $x(t') = x_0 + vt'$ with input (x_0, v, t')). However, the encoder function might be of a more complex form, and hence difficult to find with symbolic regression.
3. The input to the encoder can be very high dimensional, and a lot of the input parameters might not be relevant for answering the questions. Finding the relevant inputs by looping trough all possible subsets of input parameters is computationally extremely costly.
4. The search over functions is biased towards simple mathematical expressions. However, such a bias might not be desirable.

Artificial neural networks, introduced in Chap. 3, on the other hand can approximate any (continuous) function (Sect. 3.4), are unbiased towards simple mathematical expressions, perform very well in selecting only the relevant features from input data, and, probably most importantly, implementing the encoder and decoder with neural networks allows to learn both functions simultaneously. Crucially, having no restrictions on the encoder and decoder functions E and D, the proposed scheme does not use any prior knowledge about a specific physical system or mathematics. These are the main reasons why the methods discussed in Part III for finding the relevant parameters of physical systems are based on neural networks. A difficulty that arises due to approximating functions with neural networks is that it can be difficult for humans to interpret the found encoding mapping. We discuss some methods to overcome this difficulty, but a general solution is not yet known and is being actively researched in the field of machine learning.

1.4 Outline

The book is structured as follows. First, we provide a brief introduction into machine learning in Part I, which contains the computer science background for a scientist that is necessary to understand the following parts of the book. Readers who are familiar with the concepts of machine learning and artificial neural networks can safely skip this part.

The following two parts, Parts II and III, are self-contained and the reader is encouraged to directly focus on the part he is mostly interested in.

In Part II, we describe a selection of work that uses machine learning for automatizing different steps of a physicist's discovery process. Often, machine learning is applied for specific physical systems and prior knowledge about the system at hand is built into the machine learning structure. In contrast, the chapters in Part II focus on work that is applicable to a broad range of systems and uses a minimal amount of prior knowledge. The chapters in Part II are self-contained and the reader may pick the ones he or she is particularly interested in. In Chap. 5, we discuss how experimental setups can be created using reinforcement learning. In Chap. 6, we discuss model creation and optimization under the assumption that the relevant parameters of the considered systems are known. And finally, in Chap. 7, we investigate how to test a model that makes probabilistic predictions and how to find clusters in the experimental data that deviates the most from the model predictions.

Part III focuses on a particular part of model creation, namely extracting the relevant degrees of freedom from experimental data using minimal prior knowledge about physics or maths. The main idea to achieve this is to map a simplified version of the physicist's reasoning process (Fig. 1.1) to an artificial neural network structure. First, in Chap. 8, we formalize the setting and provide a mathematical formulation of the simplified physical reasoning process as well as of the properties of "natural" representations. In Chap. 9, we then describe machine learning methods that allow to find such representations. In Chap. 10, we apply these methods to various toy examples from classical and quantum physics, demonstrate that they allow to find the same variables that are used by physics textbooks and explain how conceptual insights can be extracted from the found representations. Finally, we conclude with some open questions and directions for future work on representation learning for physics in Chap. 11.

In the last part of the book (Part IV), we discuss future perspectives for automating the discovery process of physicists and the use of machine learning for the foundation of physics.

Part I
Machine Learning Background

In this part, we give a rough overview of machine learning whose aim is to create software that can learn by itself from given data or by interacting with an environment. We discuss the three important subfields of *supervised*, *unsupervised* and *reinforcement* learning. While the first two subfields consider systems that learn from given data sets, in reinforcement learning an agent interacts with an environment and hence, the input data is created continuously and may depend on the actions chosen by the artificial agent. While the data for supervised learning has to be labeled with the expected outputs, in unsupervised learning no such labels are provided.

After discussing the basics of machine learning, we introduce the reader to artificial neural networks, the (currently) most important tool for machine learning and essential for Part III of this book. Roughly, neural networks can represent and implement functions on real input data. We discuss the basic structure of neural networks, which kind of functions they can approximate and how to train them to learn a function from input–output pairs.

Finally, we introduce the reader to *autoencoders*, a tool for unsupervised learning or more concretely for *representation learning*. Representation learning aims to find "natural" and low-dimensional representations of high-dimensional data and builds the basis for Part III of this book.

Chapter 2
Machine Learning in a Nutshell

Machine learning (ML) has started to gain traction over the past years and found a lot of applications in science and industry. The main idea is to create algorithms that can learn from data themselves. Traditionally, we can divide ML into *supervised, unsupervised* and *reinforcement* learning. The idea behind machine learning is simplest explained with supervised learning, which is covered in the following section.

2.1 Supervised Learning

One celebrated application of ML is the recognition of objects on images. Let us assume that you are asked to write a program that takes images of cats and dogs as an input, and the program should output which of these animals is shown on the image. Writing an algorithm for this task would be extremely complex, since there are all sorts of cats and dogs, they might not be fully visible on the image, the image might be of bad quality and so on. Instead we could let us inspire from the human learning process. How do we train a child to distinguish cats and dogs? We show him or her a lot of examples of cats and dogs and the child will learn to distinguish them by him- or herself. But it would be pretty impossible to teach the child a recipe (which could be translated into an algorithm) of how to distinguish cats and dogs. This leads us to the main idea behind ML: we provide a lot of examples of labeled images (where we have the two labels "cat" and "dog"), the so called *training data*, to a computer algorithm, whose task is to output a (probably extremely complex) function F that takes an image as an input and provides the correct label as an output (see Fig. 2.1). The difficult task of writing directly a program implementing the function F is now shifted to designing an algorithm that is able to learn a good approximation of the function F from labeled images. This is exactly the goal of *supervised* machine learning. Clearly, this is a very difficult task, since the algorithm has to create a function that recognizes the relevant features that distinguish cats from dogs from a set of images. Otherwise, if the algorithm just created a function that memorizes the correct labels for the provided images in the training set, it could

R. Iten, *Artificial Intelligence for Scientific Discoveries*,
https://doi.org/10.1007/978-3-031-27019-2_2

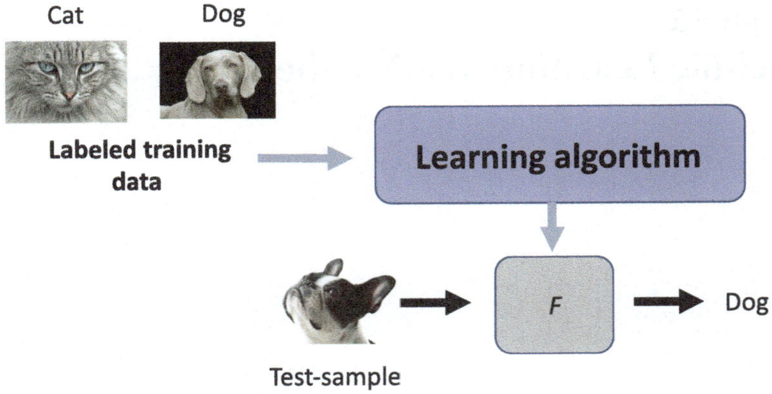

Fig. 2.1 Image classification. An common example for supervised machine learning is image classification. A learning algorithm is provided with a data set of labeled images (the so called *training data set*). For example, this could be a set of images of cats and dogs together with the correct label "cat" or "dog". The task of the learning algorithm is to infer a function F (which might be given as a blackbox), which is able to correctly label images. If the learning algorithm performs well, the inferred function F should also generalize, i.e., it should also label images correctly that were not seen during the training

not generalize to images not seen during the training. One option to create such an algorithm is to use artificial neural networks. For the present purpose, one may just think of a neural network as a parametrized family of functions $\{F_\theta\}_\theta$ from the space of images to the output labels (see Chap. 3 for the details). A training algorithm is then used to optimize over the choice of the parameters θ (determinig the function F_θ) using the labeled images with the goal that $F_{\theta^\star}$ gets close to the function F for some parameters $\theta^\star$. Interestingly, neural networks perform very well for this task and can generalize to images never seen during the training and without hardcoding specific information about cats and dogs into the network structure or the learning algorithm (see for example [26]).

2.2 Unsupervised Learning

Whereas in supervised learning one is given data samples together with the desired output for all the given samples, in *unsupervised* learning, one tries to learn some features about the given data without knowing the correct outputs for the training data. For example, given again images of cats and dogs, one might want to compress the images, i.e., to reduce the dimensionality of the original image, still maintaining all essential information about the image. Here, we may not know how such a representation should look like and we would like the algorithm to find the optimal one. The probably most widely known method for dimensionality reduction is *Principal Component Analysis* (PCA). Again, the goal of PCA is to reduce the dimension of

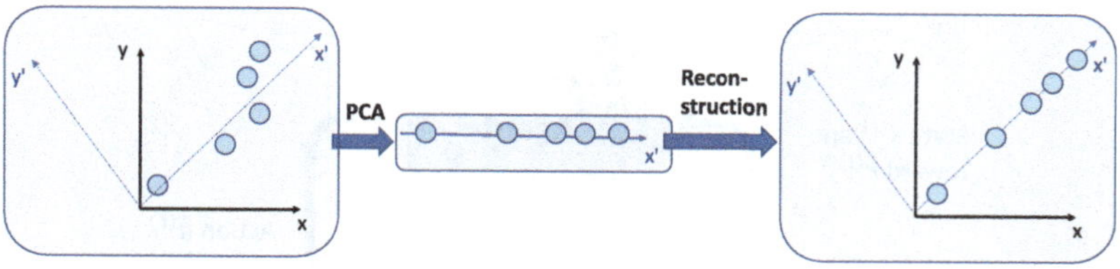

Fig. 2.2 Principal Component Analysis (PCA). A well known method for unsupervised learning is PCA, which exploits linear correlations in a data set to lower its dimension without loosing much information. The figure shows some data points in two dimensions (x, y) with a positive correlation of the x- and the y-coordinates. PCA transforms the coordinate system to (x', y'), such that the data points approximately lie on the x'-axis. In other words, the distribution of the y'-coordinate is approximately independent of the x'-coordinate. The y'-coordinate can be ignored without loosing too much information. Hence, PCA (approximately) reduced a two-dimensional data set to a one-dimensional one. The original distribution of the data can then be approximately recovered from the one-dimensional representation

the input data, but one restricts the allowed transformations from the input data to the new representation to be linear (Fig. 2.2). Finding compressed representations of data is the focus of *representation learning*, a subfield of unsupervised learning (see [27] for a recent review).

2.3 Reinforcement Learning

In *reinforcement learning*, the training data is not given, but is collected by an artificial agent from an environment. Hence, reinforcement learning is closely related to the objective of artificial intelligence (AI) of creating an agent that can perform actions in an environment to achieve a certain goal. The "environment" is considered to be everything of interest that is not part of the agent itself. For example, the environment of an agent playing chess against another agent could be the chess board and the other agent. The basic model for reinforcement learning is shown in Fig. 2.3. At a (time) step t, an agent gets the state $s^{(t)}$ of an environment (encoding some information of the environment like positions of objects) and a reward $r^{(t)}$ quantifying its performance in achieving the desired goal as an input. Based on the input, the agent chooses an action $a^{(t)}$ of a set of possible actions. The action transforms the environment from a state $s^{(t)}$ to a state $s^{(t+1)}$. In the example of plying chess, the action could for example be the movement of a game figure, and the new state $s^{(t+1)}$ could correspond to an encoding of all the positions of the game figures after the movement. Further a reward $r^{(t+1)}$ is returned to the agent, providing information about how useful the action $a^{(t)}$ was in getting closer to achieving the agent's goal. Typical examples for reinforcement learning environments are games like chess or Go, where the goal of the agent is simply to win the game. The actions of the agent are restricted to the ones allowed by the rules of the game and the reward could be zero as long as the game is not finished, and -1 or 1 if the agent looses or wins the game, respectively.

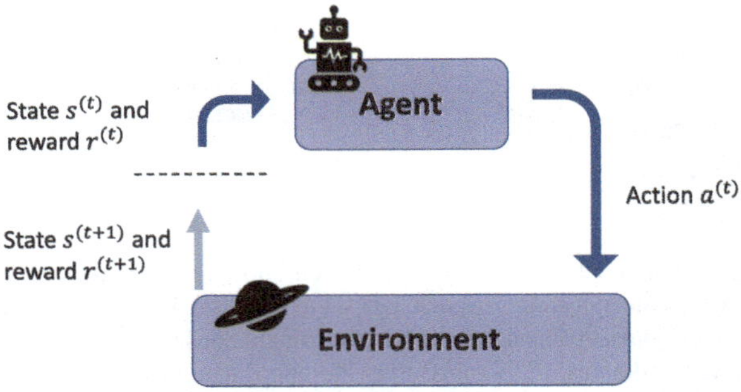

Fig. 2.3 Basic setting for reinforcement learning. An agent has to achieve a given goal in an environment. In the t-th interaction step, the agent gets an encoding $s^{(t)}$ of the current state of the environment and a reward $r^{(t)}$ quantifying its current performance as an input. Based on this input, the agent decides which action $a^{(t)}$ to perform next. The action transforms the environment from a state $s^{(t)}$ to a state $s^{(t+1)}$, and returns the reward $r^{(t+1)}$. This cycle is then repeated until the agent achieves its goal

2.4 Bias-Variance Tradeoff

The main goal of machine learning is to learn a model from a training data set (or a training environment) that generalizes well to new data samples (or new environments) not seen during training. The difficulty thereby is to handle the tradeoff between two kind of errors, the so called *bias* and *variance* of a machine learning algorithm. Roughly, bias is the tendency of an algorithm to consistently learn the same wrong thing, where variance is the tendency to learn random things not related to the real data.

Let us explain this with a simple example that should be familiar to most readers: fitting a polynomial trough 5 given data points (x_i, y_i) with $i \in \{1, \ldots, 5\}$ (see Fig. 2.4). The points are generated by a linear model with some Gaussian noise, i.e., $y_i = cx_i + \varepsilon z_i$ for constants $c \in \mathbb{R}$, $\varepsilon > 0$ and $z_i \sim \mathcal{N}(0, 1)$ sampled from the normal distribution. Assume now that we do not know exactly the underlying model for the data creation, but our guess is that it is a polynomial model (with some noise). One may for example try a linear fit to the data and one with a polynomial of degree four. A linear model is more restrictive and is described with less parameters (just a constant shift in the y-direction and the slope of the line). From a machine learning perspective, we expect such a model to lead to a larger error due to the low expressivity of the model, i.e., a such a model has a higher bias than a polynomial of degree four (for a mathematical discussion of the bias and variance we refer to [28]). In the example demonstrated in Fig. 2.4, we find that the linear model can not predict the data point perfectly, but actually quite well for such a simple model. On the other hand, fitting a polynomial of degree four clearly can fit the data point perfectly and has no error due to the simplicity of the model. Unfortunately, this comes in tradeoff

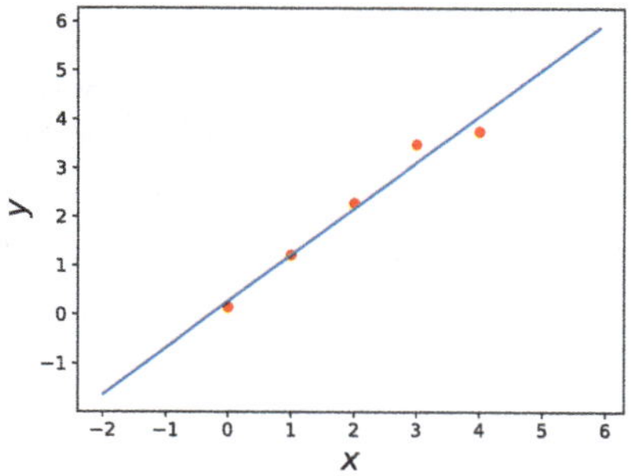

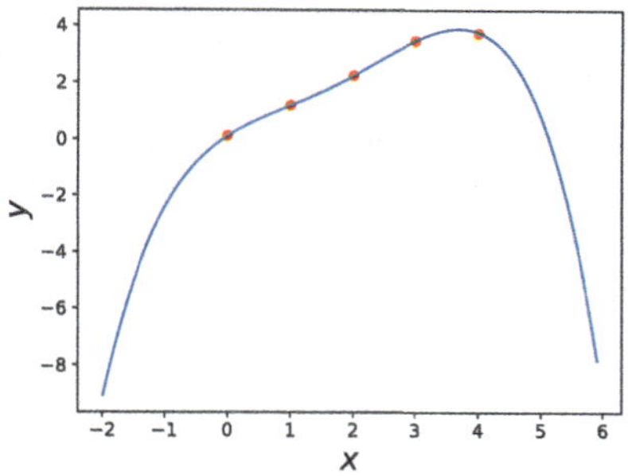

Fig. 2.4 Bias-variance tradeoff for polynomial regression. The data points are given by (x_i, y_i) with $x_i = i - 1$ and $y_i = x_i + 0.5z_i$ with randomly sampled $z_i \sim \mathcal{N}(0, 1)$ and $i \in \{1, \dots, 5\}$. **a** Linear least squares fit of the data. **b** Polynomial least squares fit of the data with a polynomial of degree four. Although this polynomial perfectly fits the data points, it does not reflect the underlying linear model. Sampling a new data point (x, y) with $x = 6$, the prediction of this model would be around -8, which is far away of the expected value of $y = 6$. This demonstrated the tradeoff between fitting the training data and generalizing well to the test data. The tradeoff between making prediction errors due to too simplistic models and overfitting the training set with too expressive models is called the *bias-variance tradeoff* in machine learning

with a higher variance, meaning that the fit depends strongly on the samples in the training set, i.e., in particular on the random samples $z_i \sim \mathcal{N}(0, 1)$. The problem with models that are too expressiv do not directly manifest itself on the training data (we have zero prediction error there!), but will lead to bad generalization behavior. Indeed, in the example in Fig. 2.4, predicting the value of y for a test sample with $x = 6$ would be close to $y = -8$, where the expected value is $y = 6$. We say that we *overfitted* the training data by choosing a model with a too high variance.

In the example above we knew the underlying model, and hence, it was clear that the linear fit will have better generalization power. However, in general, the underlying model is not known. It could be that the data samples are indeed created by a four dimensional polynomial with zero noise. In this case, choosing a model with high bias such as a liner fit, would fit and generalize badly. This demonstrates a fundamental problem of machine learning, the so called *bias-variance tradeoff* stating that we can not lower both errors, the variance and the bias, at the same time. The goal of most machine learning strategies is to find the optimal tradeoff of the variance and the bias to achieve good generalization properties.

Remark 2.1 (Cross-validation) In the example discussed above, it remains unclear how to systematically test the generalization power of a machine learning model. One option is *cross-validation*, which partitions the training set into two complementary subsets and trains the model on one of the subsets. The other subset is then used to test the generalization power of the trained model. This is done with different partitions of the training data. If the model performs well on the training subsets, but badly generalizes to the complementary subset, this points towards the fact that we have a high variance and hence that our model has too high expressivity (or is not regularized properly). On the other hand, if our model has high prediction error on the training subsets as well as on the complementary subsets, this points towards a model with too high bias, and hence we may want to replace the model with one with higher expressivity.

Chapter 3
Artificial Neural Networks

Artificial neural networks have become the state of the art for tackling machine learning problems and to build AI-agents. Furthermore, they are considered for gaining insights into how the human brain develops physical intuition from observations [29–35]. For example, in [36] neural network were shown to be able to predict whether a tower consisting of wooden blocks is stable or will fall. Such experiments suggest that the processing of neural networks may have similarities to the intuitive reasoning of humans. In this section, we provide a short overview of the basics of neural networks. We still do not have a deep understanding of the training process of such networks and why they often generalize well to data not seen during the training. Hence, training neural networks requires some experience and knowledge about several tricks, and can not be taught easily. The book by Michael Nielsen on neural networks and deep learning [37] provides an overview of several such tricks. Understanding how to train neural networks is a subject of current research. For example, an information theoretic approach was suggested in [38]. Here we focus on a basic understanding of what neural networks are and how the training method works in principle.

3.1 Single Artificial Neuron

The building blocks of neural networks are single neurons (Fig. 3.1a). We can think of a neuron as a map that takes several real inputs $x_1, \ldots, x_n$ and provides an output $\sigma(\sum_i w_i x_i + b)$, according to an *activation function* $\sigma : \mathbb{R} \to \mathbb{R}$, where the weights $w_i \in \mathbb{R}$ and the bias $b \in \mathbb{R}$ are tunable parameters. The output of the neuron is itself sometimes denoted by *activation*, and there are different possible choices for the activation function.

R. Iten, *Artificial Intelligence for Scientific Discoveries*,
https://doi.org/10.1007/978-3-031-27019-2_3

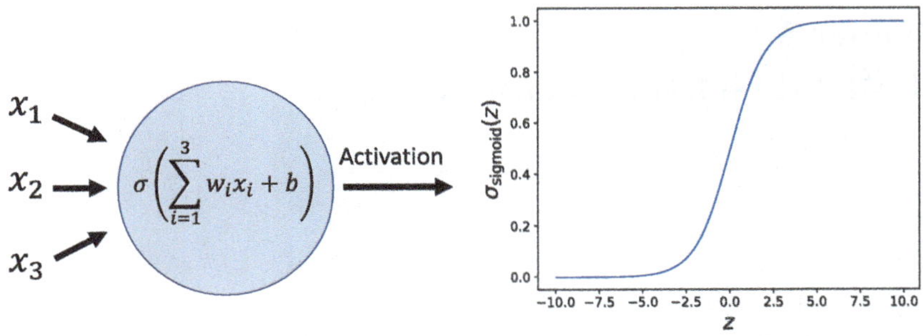

(a) (b)

Fig. 3.1 Sigmoid neuron. **a** Single artificial neuron with weights w_i, bias b and activation function σ. The inputs to the neuron are denoted by x_1, x_2 and x_3. The output of the neuron is called activation and is given by $\sigma(\sum_i w_i x_i + b) \in \mathbb{R}$. **b** Sigmoid activation function given in (3.2). The sigmoid activation function can be seen as a smooth approximation of the step function given in (3.1). Hence, the neuron provides an output that is approximately equal to one if the input is significantly bigger than zero, and it provides an output close to zero if the input is significantly smaller than zero

For example, consider a step function

$$\sigma(z) = \begin{cases} 1 & \text{if } z > 0 \\ 0 & \text{if } z \leq 0 \end{cases} . \tag{3.1}$$

Such a neuron is called perceptron and was developed in the 1950s and 1960s by Frank Rosenblatt, inspired by earlier work by Warren McCulloch and Walter Pitts. The choice of such an activation function is (loosely) motivated by the functioning of a biological neuron, which only "fires" (i.e., produces an output) if the sum of the weighted input signals is higher than a given threshold. One can consider a perceptron as a simple model for decision making by weighting up evidence. For example, your decision to go surfing may mainly depend on two factors, the strength of the wind (corresponding to input x_1) and how sunny it is (corresponding to input x_2). The weights represent your personal weighting between the two factors; choosing the weight $w_1 \gg w_2$ means that it is more important to have strong wind than to have sunshine. The bias b then sets your decision boundary and the outcome 1 corresponds to the decision to go surfing, whereas the outcome zero means that you will not go surfing. Clearly, this is a very simple decision model, but we will show in Sect. 3.3 that building networks of many neurons allows to represent complex models.

3.2 Activation Functions

The artificial neural networks currently used in practice use different activation functions for their neurons. Unfortunately, there is no theory that tells us which activation function works best for a certain machine learning task. However, as we will see in

Sect. 3.5, to be able to train the neural network, it is important that we choose a (piecewise) smooth activation function. Further, choosing a smooth activation function also allows a network to output real numbers instead of discrete ones, which is required for regression tasks in machine learning. Instead of using the discontinuous step function of a perceptron, we may use a smooth approximation of it, which is given by the sigmoid activation function (Fig. 3.1b), defined by

$$\sigma(z) = \frac{1}{1 + e^{-z}} . \tag{3.2}$$

In practice, activation functions that do not saturate, i.e., that do not converge to a constant value for large inputs, usually perform better. Two commonly used activation functions are the Rectified Linear Unit (ReLU) [39, 40] and the Exponential Linear Unit (ELU) activation function [41], given by

$$\sigma_{\mathrm{ReLU}}(z) = \begin{cases} z & \text{for } z > 0 , \\ 0 & \text{for } z \leq 0 , \end{cases} \tag{3.3}$$

and, for some $\alpha > 0$

$$\sigma_{\mathrm{ELU}}(z) = \begin{cases} z & \text{for } z > 0 , \\ \alpha \, (e^{z} - 1) & \text{for } z \leq 0 . \end{cases} \tag{3.4}$$

The two functions are plotted in Fig. 3.2. For the implementation of the examples discussed in Part III of this book, we use the ELU activation function, apart from the example discussed in Sect. 10.3, which uses the ReLU activation function.

3.3 Neural Networks

A (feed-forward) neural network is created by arranging neurons in layers and forwarding the outcomes of the neurons in the ith layer to neurons in the $(i + 1)$th layer (see Fig. 3.3). The network as a whole can represent a class of functions $\{F_\theta\}_\theta$, parametrized by θ, which contains the weights and biases of all the neurons in the network. The inputs $x_1, \ldots, x_n$ of a function $F_\theta : \mathbb{R}^n \to \mathbb{R}^m$ in this class correspond to the activations of the neurons in the first layer (which is called the *input layer*). The activations of the input layer form the input for the second layer, which is a *hidden layer* (since it is neither an input nor an output layer). In the case of a fully connected network, each neuron in the $(i + 1)$th layer receives the activations of all neurons in the ith layer as input. The activations of the m neurons in the last layer, which is called *output layer*, are then interpreted as the output of the function F_θ. Clearly, such

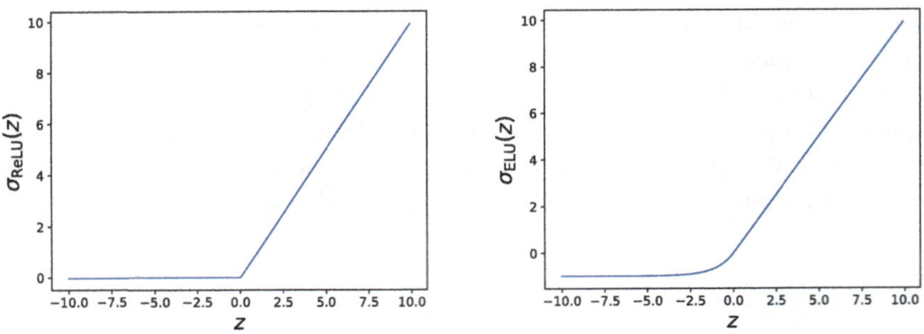

(a) (b)

Fig. 3.2 ReLU and ELU activation functions. **a** Rectified Linear Unit (ReLU) [39, 40] given in (3.3). The ReLU activation function is linear for an input $z > 0$ and constantly zero otherwise. **b** Exponential Linear Unit (ELU) [41] given in (3.4) for $\alpha = 1$. The ELU activation function can be seen as a smooth version of the ReLU activation function (vertically shifted by –1)

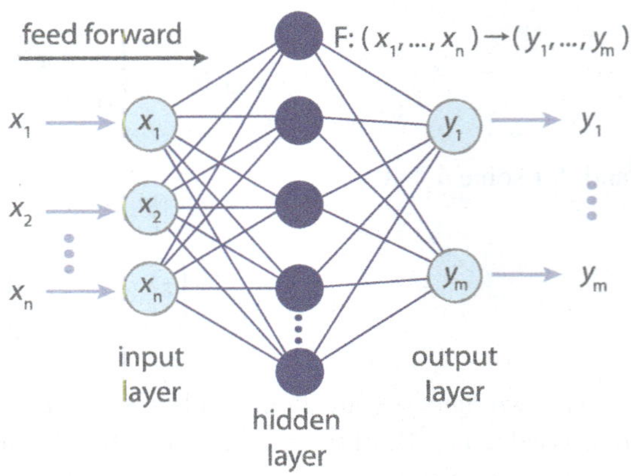

Fig. 3.3 Feedforward neural network. (Figure reproduced from Iten and Metger et al., *Physical Review Letters*, 2020 [22]). Fully connected (feed-forward) neural network with 3 layers. The network as a whole can be thought of as a function mapping the inputs $(x_1, \ldots, x_n)$ to the output $(y_1, \ldots, y_m)$. The layers consists of a series of artificial neurons (see Fig. 3.1)

networks can represent much more complex models than a single neuron. In fact, as discussed in the next section, for certain classes of activation functions, neural networks are *universal* in the sense that any continuous function can be approximated arbitrarily well by a sufficiently large feedforward network for a certain choice of the the weights and biases of the neurons in the network.

3.4 Universality Theorem

In this section, we investigate the expressiveness of neural networks, i.e., we investigate the class of functions that can be represented by neural networks of variable size. There are two classes of universality theorems. The first class considers the expressivity of neural networks for a fixed depth, i.e., a fixed number of hidden layers.

The second class considers the expressivity for fixed width, i.e. for a fixed number of neurons in each hidden layer. For this book, only the first class of universality theorems is relevant, and we refer to [42] for the second class.

A general result for the expressivity of neural networks with only one hidden layer was developed in [43, 44]. A version of the general universality theorem given in [44], which is particularly relevant for this book, is stated in the following theorem.

Theorem 3.1 *For any continuous, bounded and non-constant activation function, let us denote the class of functions that can be represented by a feed-forward neural network with one hidden layer and* p *hidden neurons by* $\{F_\theta^p : \mathbb{R}^n \to \mathbb{R}^m\}$. *Then, for any continuous function* $G : \mathbb{R}^n \to \mathbb{R}^m$ *and compact subset* $X \subset \mathbb{R}^n$ *and* $\epsilon > 0$, *there exists* $p \in \mathbb{N}$ *and a function* F_θ^p *such that* $\sup_{x \in X} \left| F_\theta^p(x) - G(x) \right| < \varepsilon$.

Hence, since the sigmoid activation function is continuous, bounded and non-constant, any continuous function can be approximates arbitrary well by a feed-forward network with a sufficiently large hidden layer containing sigmoid neurons. We refer to [44] for a proof of the statement and to [37] for an intuitive proof using visualization. The theorem does not apply to networks using the ReLU or ELU activation function, since they are not bounded. Nonetheless, a result established in 2017 [45] shows that neural networks with non-bounded and non-polynomial activation functions are also universal.

3.5 Training of Neural Networks

Even if we would know the target function that a neural network should implement, tuning the weights and the biases of all the neurons in the network by hand, such that the network well approximates the target function, would be a tedious task. Furthermore, using neural networks as a tool for machine learning, we would like them to learn a function themselves from training samples, i.e., known input-output pairs (x, y) with $x \in \mathbb{R}^n$ and $y \in \mathbb{R}^m$. Hence, the goal is to find a training algorithm that takes a training set $\mathcal{S}$ consisting of input-output pairs as an input and adapts the biases and weights, summarized in a variable θ, of the neural network such that the function F_θ implemented by the network satisfies:

1. **Fitting**: $F_\theta(x) \approx y$ for all training input-output-pairs (x, y),
2. **Generalization**: $F_\theta(x') \approx y'$ for all test input-output-pairs (x', y') that has not been observed during training.

You may recall the example discussed in Sect. 2.1, where the machine learning task is to classify images of dogs and cats. The learning algorithm is thereby fed with input-output-pairs (x, y) consisting of images of cats and dogs (corresponding to x) and the correct label "cat" or "dog" (corresponding to y). More concretely, the input x to the neural network could be the values of the pixels of the images, and the output $F_\theta(x) \in [0, 1]$ could be the activation of a sigmoid neuron, which can be interpreted as the probability of having a dog on the image.[1] For example, if we get an outcome of 0.9 from the network, it is pretty sure that there is a dog on the image. The classification can then be done by classifying the image as "dog" if $F_\theta(x) \geq 0.5$ and as a cat otherwise. The optimal output would be $F_\theta(x) = y$, where we set $y = 1$ if the label is "dog" and $y = 0$ if the label is "cat". The test data then checks how well images that have not been seen during training are classified.

Let us first focus on a training algorithm to achieve goal 1. To train a neural network, we need some performance measure of the network that smoothly depends on the network parameters θ (i.e., the biases and weights of the network). In the example above, such a measure could be chosen as $C((x, y), \theta) = (F_\theta(x) - y)^2$ for a training sample (x, y). We call $C((x, y), \theta) \geq 0$ the *cost function*. The name is motivated by the fact that we would like to minimize the cost, i.e., we would like to have $C((x, y), \theta) \approx 0$ for all training examples (x, y), which corresponds to goal 1. The performance of the network over the whole training set is then measured by the average cost defined by

$$\bar{C}(\theta, \mathcal{S}) = \frac{1}{|\mathcal{S}|} \sum_{(x,y)\in\mathcal{S}} C((x, y), \theta) , \tag{3.5}$$

where $|\mathcal{S}|$ denotes the number of samples (x, y) in the training set. Note that the averaged cost function is a sum of smooth functions in θ, and hence, depends itself smoothly on θ.

3.5.1 Stochastic Gradient Descent

The goal of the training algorithm for neural networks is to find a choice for the weights and biases in the network, such that the average cost given in (3.5) is minimized. The basic idea is to start from randomly chosen weights and biases and then stepwise adapt them to lower the average cost. In every step, we change each of the weights and biases a little bit, i.e., $w_i \rightarrow w_i + \delta w_i$ and $b_i \rightarrow b_i + \delta b_i$ for some small $\delta w_i > 0$ and $\delta b_i > 0$, and observe the effect of the change on the average cost. Then, we update the weights and biases for which the cost decreases by $w_i \rightarrow w_i + \delta w_i$ and $b_i \rightarrow b_i + \delta b_i$, and the ones for which the cost increases by $w_j \rightarrow w_j - \delta w_j$ and $b_j \rightarrow b_j - \delta b_j$. For small enough δw_j and δb_j, the average cost is decreased by these updates.

In practice, instead of working with small changes, one calculates the gradient of the cost function with respect to all weights and biases, which indicates the direction

[1] The hidden structure of the networks that are commonly used for image classification are not fully connected, but are so called *convolutional neural networks* (see for example [37] for an introduction), which are optimized for processing visual data. However, this is irrelevant for the conceptual discussion here.

in which the cost function maximally increases (in an infinitesimal neighborhood). Moreover, it is usually computationally too expensive to calculate the change of the average cost function, since one has to calculate the change separately for all training samples and usually, the training set is quite large. Hence, instead one estimates the change of the cost function by averaging only over a part of the training set, called a *mini-batch*. More concretely, the training algorithm works as follows (using again the vector θ containing all weights and biases of the network as elements to simplify the notation):

1. Initialize the weights and biases randomly.
2. For $t \in \{1, 2, \ldots, T\}$, where T denotes the number of training steps, repeat the steps 3 and 4.
3. Randomly choose a subset (called mini-batch) $\mathcal{M}_t \subset \mathcal{S}$ consisting of $k \in \mathbb{N}$ samples.
4. Update the network parameters by

$$\theta_t \to \theta_{t+1} = \theta_t - \eta \nabla_\theta \bar{C} \left(\theta_t, \mathcal{M}_t\right) , \tag{3.6}$$

for some learning rate $\eta > 0$.

This learning algorithm is called *stochastic gradient descent*. The name comes from the fact that the mini-batch is chosen randomly leading to a stochastic estimate of the gradient. The gradients can be efficiently calculated by the *backpropagation algorithm* that essentially calculates the gradients of the parameters of the neurons at the output layer, and then subsequently applies the chain rule for derivatives to calculate the gradients for earlier layers until ending up with the input layer (the interested reader can find more details in [37]). The mini-batch size k and the learning rate η are so called *hyperparameters*. Hence, they are not adapted by the algorithm, but have to be chosen by hand before running the algorithm. The bigger we choose the mini-batch size, the better the estimate of the gradient gets in average in tradeoff with larger computation time for each update step. The learning rate determines the size of $\delta\theta$, i.e., how much we update the network parameters in each step.

3.5.2 Convergence and Choice of Hyperparameters

Generally, given an optimization problem and some hyperparameters, the goal is to choose the hyperparameters in a way that maximizes the convergence speed. However, for cost functions that are non-convex in θ, which is commonly the case for neural networks, we have no guarantee that stochastic gradient descent converges to the global minimum of the cost function. Indeed, usually it will get stuck in a local minima, and it depends on the random initialization of the network parameters, in which local minima we get stuck. Nonetheless, in practice, training a network several times with random initialized parameters θ and then choosing the network with the best performance works well for a lot of problems. Unfortunately, it is not well understood yet why this is the case and it is an active field of research to better understand the training of neural networks. Furthermore, we do also not have a theory that tells us the optimal choice of the hyperparameters to optimize the convergence speed or alternatively to optimize the probability to end up in a local minima that achieve prediction accuracy close to the optimal minima.

Therefore, choosing good hyperparameters and random initialization strategies are currently based on experience and some intuition and tricks. A comprehensible overview of the current state of the art is given in [37], but it is beyond the scope of this introduction into neural networks to provide an extensive overview here. In the following, we just describe the essential difficulty in choosing good hyperparameters. A priori, one may think that choosing a large mini-batch size and a small learning rate will improve the performance of the trained network in tradeoff with computational time. Unfortunately, things are more complicated with neural networks. Basically, the learning rate and the mini-batch size should be chosen in a way to achieve a good tradeoff between (i) converging fast to a local minima, and (ii) missing out local minima with bad performance. Indeed, choosing a small mini-batch size introduces some stochasticity in the updates of the network parameters, which may at first hand seem like something one would like to avoid since it temporarily may increase the average cost if the mini-batch is too small to be representative. However, it turns out that a certain amount of such stochasticity can help to jump out of local minima with bad performance, and hence to converge to another local minima with better performance. Similarly, choosing a small learning rate ensures that we converge to the closest local minima. On the other hand, choosing a large learning rate may lead to an "overshooting" of a local minima, i.e., the average cost over the full training set may actually increase in such a training step. Again, this could help to avoid bad local minima. On the other hand, choosing a too small mini-batch size and a too large learning rate will essentially lead to a random walk in the cost landscape and the network may not be able to improve its performance at all. To conclude, training neural networks and choosing the optimal hyperparameters is currently more based on experience than rigorous insights and requires some patience in trying out different choices of hyperparameters until a good choice is found.

3.5.3 Generalization

A neural network of moderate size consists of a large amount of parameters. For example, a fully connected network with 10 input- and output-neurons, and a hidden layer consisting of 100 neurons, contains 120 biases and $2 \cdot 10 \cdot 100$ weights, which are over 2000 parameters in total. According to the bias-variance tradeoff discussed in Sect. 2.4, we would expect such a complex model to overfit and generalize badly due to high variance. In other words, we would expect that the training data can be fitted well, but that the generalization power of a neural network is very bad.

Surprisingly, using some tricks for regularization, neural networks show amazing generalization power for tasks like image classifications.[2] However, to achieve good generalization power, the choice of the regularization technique is essential and, again, we do not have a deep understanding yet of which regularization methods should be used for a given setting. One regularization method is for example to add a term $\|\theta\|_1$ to the cost function to motivate the network keeping parameters small and setting unnecessary parameters equal to zero. Intuitively, motivating the network to reduce the model complexity by setting unnecessary parameters to zero lowers the variance of the model and hence may increase the generalization power. For a further discussion on state of the art techniques for regularizing neural networks we again refer to [37].

3.6 Deep Learning

Very likely, you have already heard about deep learning. So what is it exactly? Essentially, training any neural network with at least one hidden layer can be considered to be deep learning. In practice, the term is often used for networks containing many hidden layers (quite common are numbers of around five to ten hidden layers). Recalling from Sect. 3.4 that a neural network with one hidden layer can approximate any continuous function with sufficiently many hidden neurons, one may wonder why we should even consider networks with more than one hidden layer. Just because something is possible does not mean it is efficient, in the sense that with multiple layers, fewer neurons may be needed overall to achieve the same expressiveness as a network with one hidden layer (see e.g. [46] for a mathematical comparison of the expressiveness of shallow and deep ReLU networks). Furthermore, in principle, it could be the case that deep neural networks have better generalization properties than shallow ones.

Intuitively, there is a compositional argument why deep learning works well in praxis. Recall again the task of classifying images of cats and dogs considered in Sect. 2.1. As demonstrated in [47, 48], neural networks that are trained to classify images store gradually more complex features in later layers. While the first few layers may represent simple features like edges of different orientation, later layers can build on these features to form more complex part-of-object shapes like typical eyes of dogs or cats. The part-of-object shapes can then be combined further to make a decision on the classification task and output the label "cat" or "dog".

In praxis, deep neural networks often outperform shallow networks for tasks like image recognition. But unfortunately, deep networks tend to be difficult to train in praxis. While deep networks are also trained using gradient descent (and the

[2] For image classifications, one does not use fully connected neural networks, but so called *convolutional neural networks*. These networks use some prior knowledge about "locality" of information in images and translation invariance of visual features. Nonetheless, the number of parameters in such networks is still significantly larger compared to more conventional machine learning models.

backpropagation algorithm is used for efficiently calculating the gradients), the size of the gradients turn out to behave unstable in such networks. Hence, the gradients in the earlier layers of such networks tend to explode (*exploding gradient problem*) or to nearly vanish (*vanishing gradient problem*). We refer again to [37] for a discussion why this problem appears and how it is handled in praxis.

Sometimes one may not only be interested in the classification of pictures of dogs and cats, but also in extracting relevant features of cats and dogs from images. Finding a compact representation of data by extracting high level features is exactly the goal of representation learning discussed on the basis of autoencoders in Chap. 4. Deep neural networks are a powerful tool for representation learning, since they quite naturally learn high level features in deeper layers.

Chapter 4
Autoencoders

Autoencoders are a tool for representation learning, which is a subfield of unsupervised machine learning and deals with feature detection in raw data. A well known example for representation learning is PCA, discussed in Sect. 2.2. The most methods that are currently used for representation learning are based on artificial neural networks. While, in principle, all deep neural network architectures learn some representation within their hidden layers (see also Sect. 3.6), most work in representation learning is dedicated to defining and finding *good* representations [27]. The main focus of Part III of this book is the formalization of what a *good* representation is for physical systems and designing neural network structures that are able to find such representations. While there are a variety of tools for representation learning [27], the one most relevant for this book is the autoencoder.

In the following, we provide a non-technical introduction to autoencoders. A mathematical discussion of a specific kind of autoencoders (so called *variational autoencoders*) is given in Appendix B. An autoencoder is a tool to find a compact representation of samples in a data set. For concreteness, think about a data set containing images of circles of different radius and centered at different points in the s-t-plane (see Fig. 4.1). As discussed in Sect. 3.6, deep neural networks are able to extract abstract features from data. However, so far, we have only considered neural networks for supervised learning. Here, we only have given images of circles, but we do not have any labels, so what should the output of the neural network be? As we search the hidden layers of the network for representations that still contain enough information to reconstruct the image, we should ask the network to reproduce the input image again as the output of the network, i.e., for an image x and a network F_θ we want $F_\theta(x) \approx x$. But this is just the identity function, so what is the point of training a network to learn the identity function? The network becomes interesting by introducing a dimension "bottleneck" in a hidden layer, i.e., by introducing a hidden layer with a small number of neurons (called *latent layer*). This forces the network to compress the input to a few relevant parameters and does not allow it to just successively pass the original image from one layer to the next. More concretely, we can split the network into two parts, the *encoder* function $E(x) = r$ mapping an input x to a latent representation $r \in \mathbb{R}^p$ with p latent neurons and where the elements

R. Iten, *Artificial Intelligence for Scientific Discoveries*,
https://doi.org/10.1007/978-3-031-27019-2_4

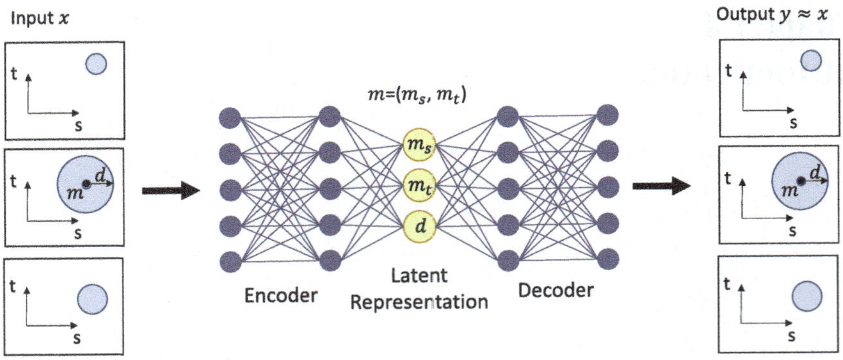

Fig. 4.1 Autoencoder. The figure shows an autoencoder used to compress images of circles. Three possible input images are shown. The position as well as the radius of the circle varies over the training data. An encoder mapping is used to compress the image data to a small dimensional latent representation. The high dimensional input to the encoder may corresponds to the values of the pixels of the image, and the output of the encoder is given by the activations of the neurons in the latent layer. Note that the number of neurons depicted for the encoder (and decoder) is not representative. An efficient encoder recognizes that it is sufficient to "store" the coordinates $m = (m_s, m_t)$ of the center of the circle together with its radius d. We would like to stress that these values are not parameters of the latent neurons, but correspond to the real valued output of theses neurons (given an image as an input to the encoder). A decoder is then used to reconstruct the image using only the information from the latent representation

of r correspond to the activation of the latent neurons for the given input x. The latent representation r is then mapped by the second half of the network, the *decoder* function $D(r) = y$, to the output y of the network (which should be approximately equal to the input image, i.e., $x \approx y$). In the considered example (Fig. 4.1), knowing the centre (m_s, m_t) of the circle and its radius d is sufficient to reconstruct the image. Hence, the input image, which might be quite high dimensional, e.g., a vector with 10,000 entries if the image dimension is 100×100 pixels, can be compressed to a three-dimensional latent representation storing m_s, m_t and d. Note that features that stay the same over all images, such as the shape (in this case a circle), are "stored" in the decoder. The latent representation only has to store the information that may change from one image to another.

Where it seems most natural as a human to store m_s, m_t and d in a latent representation consisting of three neurons, there is no a priori reason for the neural network not to store e.g. $a = m_s + m_t$, $b = m_s - m_t$ and $c = d$, since the decoding function can easily recover $m_s = (a + b)/2$, $m_t = (a - b)/2$ and $d = c$ from this representation. Therefore, in resulting representations, different parameters stored in the latent neurons are often highly correlated and do not have a straightforward interpretation. A lot of work in representation learning has recently been devoted to *disentangling* such representations, i.e., to seperate the parameters in the representation in a meaningful way (see e.g. [49–53]). In particular, these works introduce criteria, also referred to as *priors* in representation learning, by which we can disentangle representations. Where several work focusses on a prior of having statistically independent latent neurons, a recently proposed prior is the so called *consciousness prior* [54]. It suggests to disentangle the latent representation via an attention mechanism by assuming that, at any given time, only a small fraction of the features or concepts stored in the representation are sufficient to make a useful statement about reality. In Part III of this book, we discuss criteria to disentangle representations of physical system in a natural way.

Part II
Overview of Using Machine Learning for Scientific Discoveries

In this part, we give an overview of different directions in which machine learning can be applied to automate different aspects of the scientific discovery process. In recent years, machine learning has become a common tool for science, and it is beyond the scope of this part of the book to discuss all related results. Rather, we focus on a selection of results with the following characteristics: (i) they are based on machine learning methods that can be applied to a wide range of physical systems, i.e., the methods are not specialized to a particular class of systems; (ii) they are representative of a research direction (and hence, they provide the reader with the necessary machine learning background to read the latest results in the corresponding research field); and (iii) they allow the extraction of conceptual information from the physical system. Each chapter in this part of the book is self-contained, and hence, the reader is free to pick any chapters he or she is particularly interested in, and skip others if they are less relevant to his or her field of study.

The most methods discussed in this part require the knowledge about which physical parameters are relevant to describe the system at hand. For example, it may be assumed that the relevant parameters to describe the time evolution of a charged particle in a homogeneous electric field are the charge and the mass of the particle, as well as the field strength. Recent advances in deep learning and representation learning allow to get rid of this prior knowledge and to let the machine discover the relevant parameters itself. This is discussed in Part III of this book.

Chapter 5
Creating Experimental Setups

As discussed in Sect. 1.2, creating experimental setups is a fundamental step in a physicist's discovery process. Creating experimental setups for quantum systems is particularly challenging, since the behavior of such systems is often unintuitive. In the last view years, automated search techniques and reinforcement-learning based schemes have been used to generate new experimental setups for quantum optics [55, 56] (see [57] for a review of computer inspired experiments). This research has lead to the solution of several previously unsolved questions [55, 58], experimental implementation of computer inspired experiments [59–64] and last but not least, to the discovery of new scientific ideas and concepts [65–67]. Recently, a new design algorithm named THESEUS was introduced in [68] specifically aiming at conceptual insights. THESEUS uses graph representations fo quantum experiments [66], which significantly speeds up the runtime compared to the automated search algorithm MELVIN [55]. Instead of completely automatizing the generation of experiments, one may alternatively also use machine learning to generate hypotheses which can help to improve a human expert's intuition about the system [69].

In this chapter, we focus on the results described in [56], which are based on a physically motivated kind of reinforcement learning, called projective simulation [70]. Projective simulation is particularly well suited for reinforcement learning problems where one wants to gain some insight into the thinking process of the agent. Indeed, projective simulation allows to discover combinations (or *gadgets*) of experimental tools that turn out to be useful outside of the given task of the agent. Hence, the following discussion can be considered in a broader context as providing an example of how to gain conceptual insights into the interaction of an artificial physicist with its environment using reinforcement learning.

R. Iten, *Artificial Intelligence for Scientific Discoveries*,
https://doi.org/10.1007/978-3-031-27019-2_5

5.1 Problem Setting from Quantum Optics

In this section, we provide the background required to understand the quantum optics setup discussed in the following sections (and based on [56]). It is not necessary to have an understanding of quantum mechanics to follow this section. Enough background knowledge is provided so that the considered goal of generating high-dimensional entangled quantum states can be regarded as a purely mathematical optimisation problem.

A (pure) state of a d-dimensional quantum system can be represented as a complex vector $|\psi\rangle \in \mathbb{C}^d$ that is normalized, i.e., with unit Euclidean norm $\| |\psi\rangle \|_2 = 1$, and where two states $|\psi\rangle$ and $|\psi'\rangle$ are identified if and only if they differ by a global phase factor, i.e., if there exists $\phi \in \mathbb{R}$ such that $|\psi\rangle = e^{i\phi}|\psi'\rangle$. We work with the standard inner product on the complex vector space, which is given by $\langle v|w\rangle = \sum_{i=1}^{d} v_i w_i^*$, for two d-dimensional vectors $v_i, w_i \in \mathbb{C}^d$ and where the $*$-symbol denotes the complex conjugate. The vector space $\mathbb{C}^d$ together with the standard inner product is then called a *Hilbert space*.[1]

In the quantum optics setup used in [56], the position and the orbital angular momentum (OAM) of photons are used as the degrees of freedom to store and manipulate quantum states. We work with four different paths, labeled by a, b, c and d where the photons can travel, and with $2M + 1$ angular momentum states of each photon, labeled with $-M, \ldots, 0, \ldots, M$, and where $M \in \mathbb{N}$ is some finite upper bound on the angular momentum number m of a photon. Hence, the state of a photon lives in a d-dimensional complex Hilbert space with $d = 4 \times (2M + 1)$ and orthogonal basis vectors $|m\rangle_p$ with $m \in \{-M, \ldots, 0, \ldots, M\}$ and $p \in \{a, b, c, d\}$.

The goal considered in [56] is to create a special kind of photon states, so called *entangled* states. Before we (mathematically) describe the experimental tools that are available to achieve this goal, let us discuss the definition of entanglement. Entanglement plays a fundamental role in various fields of research that are based on quantum mechanics. It describes an intrinsic property of quantum systems that forms the basis for the conceptually challenging behavior of such systems [71] and finds many applications in quantum information theory (see [72] for an overview) and in condensed matter theory, where a strong connection between phase transitions in complex systems and entanglement was discovered [73]. Let us first define entanglement for bipartite states, and consider the more evolved case of many parties afterwards.

5.1.1 *Entanglement of Bipartite Systems*

Consider two quantum systems A and B whose states live in Hilbert spaces $\mathcal{H}_A$ and $\mathcal{H}_B$ of dimensions d_A and d_B, respectively. The state of the composite system

[1] A Hilbert space is a real or complex inner product space that is a complete metric space with respect to the distance function induced by the inner product. The distance function, induced by an inner product $\langle \cdot | \cdot \rangle$, between two vectors x and y is defined by $\langle x - y | x - y \rangle$.

then lives in the tensor product of the Hilbert spaces $\mathcal{H}_{AB} = \mathcal{H}_A \otimes \mathcal{H}_B$ (see [72] for a detailed introduction to tensor products in the context of quantum mechanics). For example, if the system A and B are in states $|\psi_A\rangle$ and $|\phi_B\rangle$, respectively, the composite state is given by $|\psi_A\rangle \otimes |\phi_B\rangle$. States that can be written in tensor product form are called *separable* or *product states*. Maybe surprisingly if one is not already familiar with quantum mechanics, not all states in $\mathcal{H}_{AB}$ are product states, hence, there are states that can not be described by two local states of each of the systems. Rather, if the system is in a non-separable state (also called *entangled state*), the two systems form an inseparable whole. An example for a state of two photons at path a and b that is (maximally) entangled in the OAM basis is give by

$$|\psi_{AB}\rangle = \frac{1}{\sqrt{3}} \sum_{m=-1}^{1} |m\rangle_a |-m\rangle_b\,. \tag{5.1}$$

Note that the factor $1/\sqrt{3}$ is required, such that $\langle\psi_{AB}|\psi_{AB}\rangle = 1$ is normalized.[2] Such states can be experimentally generated by a double spontaneous parametric down-conversion process in two nonlinear crystals [55, 59]. A way to characterize entanglement is based on the Schmidt decomposition.

Theorem 1 (Schmidt decomposition [72]) *Let $|\psi_{AB}\rangle$ be a composite state of two systems A and B. Then, there exists orthonormal states $|\psi_i\rangle_A$ for system A, and orthonormal states $|\phi_i\rangle_B$ for system B, such that*

$$|\psi_{AB}\rangle = \sum_i \lambda_i |\psi_i\rangle_A |\phi_i\rangle_B\,, \tag{5.2}$$

where λ_i are non-negative real numbers satisfying $\sum_i \lambda_i^2 = 1$ and known as the Schmidt co-efficients.

The proof of the decomposition is based on the singular value decomposition and is given in [72]. Apart from many other applications of the Schmidt decomposition in quantum information theory, the minimal number of Schmidt coefficients, called the *Schmidt rank*, can be used to characterize the entanglement by the minimal number of required degrees of freedom of the local systems that are entangled with each other. A product state requires only one Schmidt coefficient and hence has a Schmidt rank equal to one. This corresponds to the interpretation that the system is "entangled" over one degree of freedom, which is equivalent to say that each local system is independent of the other one and hence the state is separable. On the other hand, any state with Schmidt rank bigger than one is entangled. For example, one can show that the state given in (5.1) has Schmidt rank three, hence it is entangled in

---

[2] The notation $\langle\psi_{AB}|\psi_{AB}\rangle$ abbreviates the expression $\langle(|\psi_{AB}\rangle)|(|\psi_{AB}\rangle)\rangle$ and is actually one of the underlying reasons why it is common to bracket a quantum state by $|\cdot\rangle$, which is called the "ket"-notation. Together with the "bra"-notation $\langle\psi_{AB}|$, denoting the complex conjugate row vector of the column vector ψ_{AB}, the "braket"-notation $\langle\psi_{AB}|\psi_{AB}\rangle$ can be read as the matrix multiplication of $\langle\psi_{AB}|$ with $|\psi_{AB}\rangle$.

all three OAM dimensions. Note that this characterization does not depend on the size of the Schmidt coefficients, and hence is not robust against noise: an arbitrary small disturbance of the state may change the Schmidt rank e.g. from one to two and hence the characterization from "separable" to "entangled". Nevertheless, this characterization works well in the context considered in the following sections, in particular since the quantum systems used in [56] are simulated on a computer with arbitrary accuracy.

5.1.2 Entanglement of Multipartite Systems

Characterizing the entanglement of multipartite systems becomes much more involved than for bipartite systems. To demonstrate this, let us consider a tripartite system ABC. We can characterize the entanglement between the system $A - BC$, $B - AC$, and $C - AB$ separately by the Schmidt rank. However, in general, the three Schmidt ranks are not equal, and hence we have not a single number characterizing the entanglement, but three numbers (which are not fully independent, but also not determined by each other [74]). Hence, we may define the Schmidt rank vector (SRV) of a tripartite system as a three dimensional vector with integer entries corresponding to the Schmidt ranks of the different bipartite splittings of the system. The entries of the SRV can be considered as the *entanglement dimensions* between the different splittings of the system. The goal considered in [56] is to create high-dimensional entangled tripartite photon states with SRV (3, 3, 2) and (3, 3, 3). Such states are not only of fundamental interest, but also because of the many applications in quantum communication and computation [75–80].

5.1.3 Preparation of Photon States

Initially, two photon pairs are created that are maximally entangled in the OAM basis, i.e., we start from a state of four photons given by

$$|\psi\rangle = \frac{1}{3}\left(\sum_{m=-1}^{1} |m\rangle_a |-m\rangle_b\right) \otimes \left(\sum_{m=-1}^{1} |m\rangle_c |-m\rangle_d\right). \tag{5.3}$$

Then, we can apply different transformation to this state corresponding to well known tools from quantum optics with the goal of creating high-dimensional entanglement between three of the photons. The available tools together with their mathematical description are listed below.

1. $\mathrm{BS}_{a,b}$: Nonpolarizing symmetric 50/50 beamsplitter for photons on path a and b:

$$|m\rangle_a \rightarrow \frac{1}{\sqrt{2}}\left(i|-m\rangle_a + |m\rangle_b\right), \tag{5.4}$$

$$|m\rangle_b \rightarrow \frac{1}{\sqrt{2}}\left(|m\rangle_a + i|-m\rangle_b\right). \tag{5.5}$$

2. $\text{Holo}_{a,k}$: Hologram in path a with a OAM shift $k \in \{\pm 1, \pm 2\}$:

$$|m\rangle_a \rightarrow |m+k\rangle_a. \tag{5.6}$$

3. DP_a: Dove prism on path a:

$$|m\rangle_a \rightarrow ie^{i\pi m}|-m\rangle_a. \tag{5.7}$$

4. Refl_a: Mirror on path a:

$$|m\rangle_a \rightarrow |-m\rangle_a. \tag{5.8}$$

The goal is, starting from the state $|\psi\rangle$ given in (5.3), to apply a sequence of the transformations listed in 1 to 4, such that we end up with a high-dimensional entangled state on three photons. After applying all transformations, the photon in path a is post-selected in the OAM basis meaning that it is measured in the OAM basis and that for all possible measurement outcomes, the states of the remaining three photons are investigated. For example, if we would not apply any transformation on the initial state $|\psi\rangle$, postselecting in the OAM basis would lead to three non-trivial state corresponding to postselection on $|m_0\rangle_a$ with $m_0 \in \{-1, 0, 1\}$. Dependent on the measurement result m_0 of the orbital angular momentum of the photon in path a, we would obtain the following state on the remaining three photons

$$\frac{1}{\sqrt{3}}|-m_0\rangle_b \otimes \left(\sum_{m=-1}^{1} |m\rangle_c|-m\rangle_d\right). \tag{5.9}$$

This state has a SRV of $(1, 3, 3)$, hence, the photon in path b is not entangled at all. The goal is to find a sequence of transformations to end up with a three photon state with SRV $(3, 3, 2)$ or $(3, 3, 3)$. This is a quite challenging task and it is discussed in the next section how reinforcement learning can be used to tackle it.

5.2 Creating Experimental Setups Using Projective Simulation

Projective simulation (PS) is a specific method for reinforcement learning introduced in 2012 in [70]. PS has shown to perform well for standard reinforcement learning problems [81–84], advanced robotics problems [85, 86], quantum error correction [87, 88], quantum communication [89], setting up experimental Bell tests [90], modelling collective behavior based on the principle of agency [91–93], and even turned out to allow for quantum enhancements for reinforcement learning [94–100]. Furthermore, some interesting philosophical challenges of agency were considered in the context of projective simulation [101–103]. Recall from Sect. 2.3 that the goal of an agent in reinforcement learning is to process an input encoding the state of the environment to decide which action it should perform next to achieve a goal. One of the fundamental ideas behind PS is that the agent does the processing of an input by projecting itself, based on previous experiences and variations thereof, to potential future situations. Based on the associated situations (and the current state of the environment given as input), the agent chooses the next action to perform on the environment. This idea is related to episodic memory whose role for planning and prediction was already discussed in psychology in the 1970s [104, 105].

Let us informally explain the idea on the basis of a simple example (discussed formally in [70]). Consider an agent that can decide to choose actions a_r or a_l in each time step, meaning that he moves to the right or left, respectively. The input to the agent is a red arrow pointing to the left s_l^{red} or to the right s_r^{red}, and the correct movement of the agent should be according to the shown arrow. The PS agent learns to associate the shown arrow with the correct actions, i.e., it associate the action a_r with s_r and a_l with s_l. Now, let's assume that the arrow shown to the agent is sometimes colored blue (and the other times it is still colored red), hence there are two additional potential inputs to the agent denoted by s_l^{blue} and s_r^{blue} corresponding to the blue arrow pointing left and right, respectively. Instead of learning to associate the blue errors again with the correct actions, the PS agent could also learn to associate the blue and red arrow pointing in the same direction with each other, hence recognizing that the "meaning" of the arrow stays the same independent of the color. The memory episode would then look like this, for example given an input s_l^{blue}, this input would activate s_l^{red}, which in turn would lead to choosing the correct action a_l. The reflection is hence not realized as a sophisticated computational process, but it can be seen as a structural-dynamical feature of the memory itself. In particular, this allow to generalize the idea of PS to the quantum world as discussed in [70]. The elementary excitations of the episodic memory are called *clips*, which are given by s_l^{blue}, s_l^{red}, s_r^{blue}, s_r^{red}, a_l and a_r in the considered example.

The discussed example is very simplistic and ignores several aspect of the general framework of PS [70]. In the general framework the associations between different clips are probabilistic. PS allows to reason about various potential situations by performing a random walk in the clip network without applying any action that is exited during this process to the environment. Hence, it is only reasoned about the actions in the first place, but they are not directly implemented in "reality". Finally, dependent on the whole reasoning process, a "real" action is chosen to perform on the environment. Importantly, clips can not only consist of one state or action, but also of a composition. PS is able to learn such compositions on its own, which is a key feature helping to extract conceptual information from the trained PS agent. This idea will be explained in detail in the concrete example from quantum optics (Sect. 5.1) in the following.

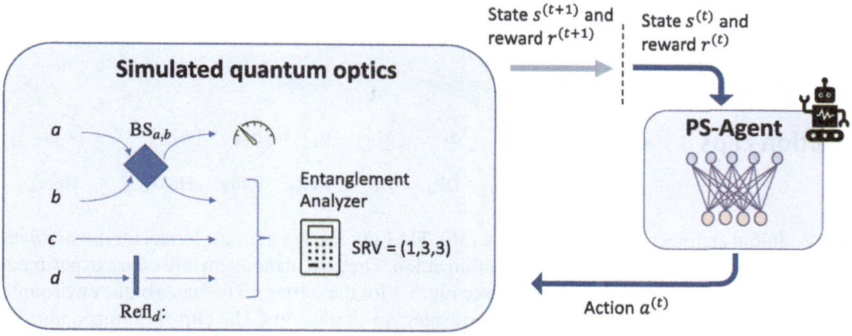

Fig. 5.1 Environment for projective simulation agent [56]. A PS agent places in each step t (dependent on the current optical setup $s^{(t)}$) additional optical tools at some position on a (virtual) optical table. The action $a^{(t)}$ determines the number of tools and which tools are added at what place on the optical table. For example, the action $a^{(t-1)}$ could have been set to a beam splitter $BS_{a,b}$ on path a and b (which is shown in current state $s^{(t)}$ in the image). The evolution of four photons propagating on four paths a, b, c and d is simulated by a computer and the output state on path b, c, d (postselected in the OAM basis of the photon on path a) is calculated. Then, the Schmidt-Rank vector (SRV) of the output state is determined as an entanglement measure. The agent obtains a positive reward $r^{(t+1)}$ if the SRV of the three photon state is (3, 3, 2) or (3, 3, 3). If the reward is positive or if the maximal number L of optical elements is achieved, the experiment ends. Otherwise, the agent choses the next elements to place on the optical table dependent on state $s^{(t+1)}$

5.2.1 Architecture for PS Agent

The PS agent used in [56] has a simple architecture, where each remembered state of the optical setup and each action corresponds to a single clip. The PS agent builds up the optical setup stepwise, where in each step it places a new optical element on the optical table (Fig. 5.1). Hence, the set of $K \in \mathbb{N}$ possible actions $\{a_i\}_{i\in\{1,\dots,K\}}$ correspond to the possible transformations listed in 1–4. We have 6 possible ways to place a beam splitter (choosing two paths out of four), 16 different variants of Holograms, and 4 choices for Dove prisms as well as for reflections summing up to $K = 30$ possible actions in total. The clips corresponding to the optical setups are added to the architecture on the go, i.e., starting from an architecture with state clips $\{s_i\}_{i\in\{1,\dots,N\}}$, each created setup that is not already contained as a clip s_i in the architecture is added as as an additional clip s_{N+1}.

We are interested in finite experiments, so the maximal number of optical elements in one experiment is limited to $L \in \mathbb{N}$. This is not solely a theoretical restriction, but rather of practical interest: since each optical element adds some noise to the quantum state, e.g. through misalignment and interferometric instability, we expect a decreasing fidelity of the resulting state for longer experiments. The number of possible states of the optical setup with a fixed number l of optical elements is given by K^l. And hence, the set of all possible states $\{s_i\}_{i\in\{1,\dots,N_{\max}\}}$ consists of $N_{\max} = \sum_{l=0}^{L} K^l$ elements. Note that thereby, s_1 represents the empty optical table.

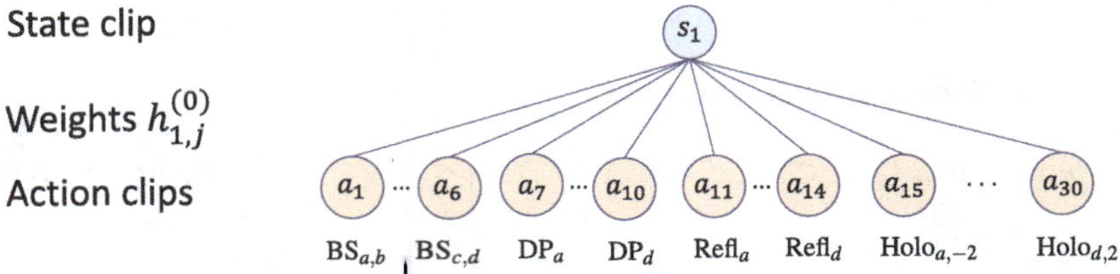

Fig. 5.2 Initial architecture for the PS agent [56]. The figure shows a simple model for the simulation of the episodic memory of the PS agent at initialization. The goal of the agent is to create experimental setups that create entangled photon states (see Fig. 5.1 for the setting). The states of the environment and the actions that the agent can choose correspond to single clips. The clip s_1 corresponds to the empty optical table. (Further clips corresponding to states of the agent are memorized on the go during the training of the agent.) The action clips $a_1, \ldots, a_{30}$ correspond to placing one of the possible tools listed in 1 to 4 on the optical table. The state s_1 is connected to all actions a_j via weighted edges $(1, j)$. The weights $h_{1,j}^{(0)} = 1$ are all initialized with one, and hence, the first action is chosen uniformly at random

Working with at most $L = 6$ optical elements, this already corresponds to a maximal number of clips of more than half a billion.

The initial architecture (which may be changed during training) for the PS agent then just consists of the clip s_1 corresponding to the empty optical table and of clips $a_1, \ldots, a_K$ corresponding to actions placing a single optical tool listed in 1 to 4 on the table (Fig. 5.2). The state s_1 is connected to every action a_j via a weighted edge $(1, j)$ with weight $h_{1,j}^{(0)}$, where the superscript denotes the number of passed training steps. An edge (i, j) represents the possibility of taking an action a_j in a situation s_i. More concretely, the probability $p_{i,j}^{(0)}$ of choosing action a_j given state s_i is proportional to the weight $h_{i,j}^{(0)}$, i.e. we have

$$p_{i,j}^{(0)} = \frac{h_{i,j}^{(0)}}{\sum_{k=1}^{K} h_{i,k}^{(0)}}. \tag{5.10}$$

Initially, all weights are set equal to one [56], and hence, the probability of performing action a_j starting from the empty optical table s_0 is uniform, i.e., $p_{1,j} = 1/K$ for every $j \in \{1, \ldots, K\}$. Similarly, if an additional state clip s_i is memorized during the training process, the weights for the edges (i, j) are initialized with one.

5.2.2 *Training of the PS Agent*

Starting from randomly generating optical setups by letting the PS agent choose which optical element to place next, we would like to train the PS agent to learn producing setups that lead to high-dimensional entangled states with high probability. In contrast to supervised learning, we have no knowledge about which actions the PS agent should choose to achieve this goal. The only feedback on its performance that the PS agent gets is a positive reward in the case he created an optical setup leading to a high-dimensional entangled state, or no reward if the maximal number of elements on the optical table is reached. In both cases, we reset the optical table afterwards to the empty state and start a new experiment where the PS agent again places optical tools on the table. From such rewards, the agent has to learn which elements to place next, given an optical table with some elements already placed. Three different kind of adaptions during the training allow the agent to adapt his behavior [56]:

1. Adapting the weights of the PS architecture, and hence, the probabilities to choose an action a_j given a state s_i of the optical table.
2. Action composition, i.e. adding compositions of several actions that have proven useful in achieving the goal as an action clip to the architecture.
3. Clip deletion, i.e., deleting states and actions that are likely not required to achieve the goal.

We discuss all three learning techniques separately in the following (readers not interested in the details how the learning works may safely skip Sects. 5.2.2.1 and 5.2.2.3). Action composition (Sect. 5.2.2.2) is in particularly interesting since it can help to get some conceptual insights into which gadgets, i.e., combination of experimental tools, are especially useful in combination.

5.2.2.1 Training of the Weights

The goal is to update the weights during the training of the PS agent to optimize the (average) probability of success, i.e., the chances of receiving a positive reward. The basic idea to achieve this is the following: each time the PS agent receives a reward, all the weights of the edges that correspond to the choices of actions made during building up the optical setup are increased. For example, assume that, starting from the empty optical table state s_1, the PS agent first performs an action a_{14}, leading to a state s_3 corresponding to an optical table with a single mirror placed in path d (Fig. 5.3). The state s_3 in turn samples an action a_{30} and the resulting optical state leads to a reward. Then, the weights of the edges $(1, 14)$ and $(3, 30)$ should be increased, to raise the probability to choose action a_{14} or a_{30} when the agent observes the states s_1 or s_3, respectively.

Let us formalize this idea and mathematically describe how the weights are updated. We denote the number of interaction circles (see Fig. 5.1) passed during training by t, and the weights after t cycles by $h_{i,j}^{(t)}$. The reward $r^{(t)} \in \{0, \lambda\}$ after t cycles is equal to $\lambda > 0$ if and only if the agent produced a state with desired SVR of $(3, 3, 2)$ or $(3, 3, 3)$. Then, the update rule for the weights used in [56] is

$$h^{(t+1)} = h^{(t)} + r^{(t)} g^{(t+1)}, \tag{5.11}$$

where $h^{(t)}$ is the weight matrix with entries $h_{i,j}^{(t)}$ and the *glow matrix* $g^{(t+1)}$ allows to redistribute rewards such that decisions that were made in the past are rewarded less than the more recent decisions. The glow matrix $g^{(0)} = \mathbb{O}$ is initialized with the zero matrix and is updated in each step t by setting the entry $g_{i,j} = 1$ if the edge (i, j) was traversed in the last decision process in the PS architecture. Furthermore, the g matrix is updated after each interaction with the environment to lower the extent to which actions far in the past are rewarded by

$$g^{(t+1)} = (1 - \eta)g^{(t)}, \tag{5.12}$$

where $\eta \in [0, 1]$ is the *glow parameter* of the PS model. The hyperparameter η can be chosen heuristically, or also learned by PS itself [106]. From (5.11) together with (5.12) one can then see that a positive reward $r^{(t)}$ increases the weights corresponding to recent transactions from states to actions.

5.2.2.2 Action Composition

Action composition [70] enables the PS agent with some primitive kind of creativity [102] by coming up with its own actions. In [56], a concrete version of this idea is implemented to learn actions that lead to a positive reward. Each time the PS agent achieves a positive reward, the sequence of actions that lead to this reward is added as an additional action clip to the PS model (Fig. 5.3). This allows the agent

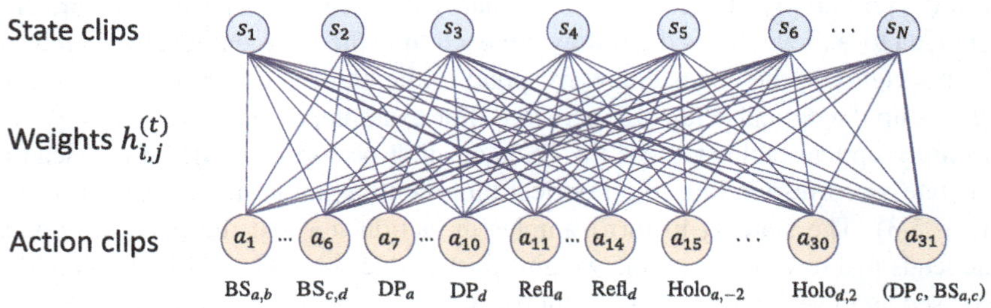

Fig. 5.3 Trained architecture for the PS agent [56]. The figure demonstrates how the architecture for the PS agent shown in Fig. 5.2 may have changed after t steps of training. State clips $s_2, \ldots, s_N$ that were explored during a rewarded experiment have been added during training. The probability that action a_j is chosen for an input state s_i is related to the weight $h_{i,j}^{(t)}$ by (5.13). The weight is increased if choosing action a_j in situation s_i leads to a positive reward in a situation experienced during training (high weights are depicted with solid lines in the figure). Hence, the probability to choose actions that will lead to rewarded experiments is increased during the training process. Furthermore, action composition changes the structure of the PS architecture. For example, the PS agent may realize during training that the combination of a Dove prism on path c together with a beam splitter on path a and c is particularly useful to achieve its goal. Hence, the composite action clip $(DP_c, BS_{a,c})$ is added, which allows the agent to place both elements in one training step

to directly place several optical tools that turned out to be useful in past experiences together in one step. Hence, the number of available actions varies over the number of training steps, and we denote this number after t training steps by $K^{(t)}$. The weights connecting the new action to all the state clips are initialized with value one and the probability to choose action a_j given state s_i is then given by

$$p_{i,j}^{(t)} = \frac{h_{i,j}^{(t)}}{\sum_{k=1}^{K^{(t)}} h_{i,k}^{(t)}}. \tag{5.13}$$

Rescaling of the tool box size $K^{(t)}$ leads to a smaller change of the probability $p_{i,j}^{(t)}$ with respect to a change of a weight. To compensate for this, we also rescale the reward $r^{(t)} \in \{0, \lambda\}$ by rescaling λ dependent on t proportionally to the size of the toolbox as

$$\lambda^{(t)} = \lambda^{(0)} \frac{K^{(t)}}{K^{(0)}}, \tag{5.14}$$

where $\lambda^{(0)}$ corresponds to the initial reward λ, and where a positive reward $r^{(t)} \in \{0, \lambda^{(t)}\}$ after t steps is rewarded with $\lambda^{(t)}$.

5.2.2.3 Clip Deletion

The number of all possible states of the optical setup grows with $\mathcal{O}\left(K^L\right)$, where k is the number of actions and L the maximal number of elements placed on the optical setup. Hence, adding all states that are built by the PS agent as a clip to the PS model may lead to a huge number of state clips, and hence, to a huge number of weights that must be trained. To avoid this, we delete state clips that are only used infrequently. More concretely, if, after placing a maximal number L of elements on the optical table, no positive reward was obtained, all state clips and edges that were created (and added to the PS architecture) are deleted.

Similarly, we also use a deletion mechanism to keep the total number of actions $K^{(t)}$ reasonable small. The idea is to only keep composite actions if they turn out to be useful in many situations. Technically, we check if after some training steps, the sum of the weights of the incoming edges to the added action clip is large enough to justify its existence. If this is not the case, we remove the action again. (For a full description of a method that implements this idea we refer to the Supplemental material of [56].)

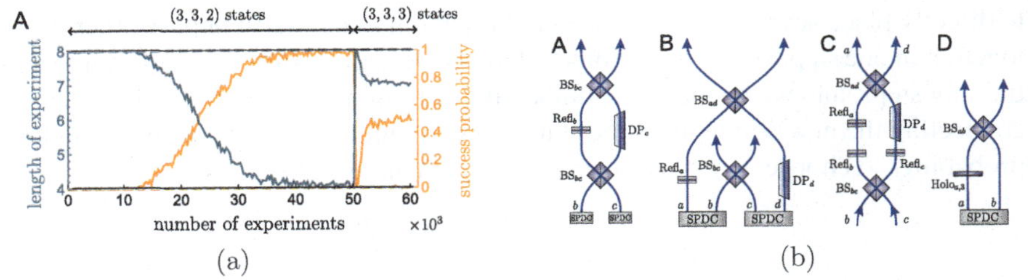

Fig. 5.4 Results for creating optical setups using PS (Figures taken from Melnikov and Nautrup et al., *Proceedings of the National Academy of Sciences* 2018 [56]). **a** The plot shows the evolution of the success probability as well as the average length (i.e., the number of optical elements) of the experiments during the training process. The PS agent is trained for 5×10^4 experiments with the goal of creating states with SRV (3, 3, 2), and subsequently for 10^4 experiments to create states with SRV (3, 3, 3). The trained agent successfully creates states with SRV (3, 3, 3) with probability of around 50%. The dashed lines corresponds to an agent that is trained with the goal of creating states with SRV (3, 3, 3) from the beginning. Such an agent does essentially not make any learning progress. This suggest that the techniques to create states with SRV (3, 3, 2) are also useful for the creation of states with SRV (3, 3, 3). **b** Experimental setups frequently used by the PS agent. (A) Local parity sorter. (B) Nonlocal parity sorter. (C) Nonlocal parity sorter in the Klyshko wave front picture [107], in which the paths a and d are identical to the paths b and c, respectively. (D) Setup to increase the dimensionality of the entanglement of photons

5.2.3 Results

The PS agent is trained for 5×10^4 experiments with the goal of learning photon states with SRV (3, 3, 2). The glow parameter is $\eta = 1/16$, the maximal number of optical elements is $L = 8$ and the initial size of the reward is $\lambda^{(0)} = 1$. The PS agent learns to create experimental setups that prepare photon states with SRV (3, 3, 2) with high probability and with a decreasing number of optical elements in the training process [56] (Fig. 5.4a). Then, the same PS agent is trained further with 10^4 experiments to learn creating photon states with SRV (3, 3, 3). Within this short additional period of training, the PS learns quickly to create such states with 50% success probability. This suggests that either knowing how to create states with SRV (3, 3, 2) helps the agent to construct states with SRV (3, 3, 3), or alternatively, that constructing states with SRV (3, 3, 3) is simpler than constructing states with SRV (3, 3, 2). To check this, a newly initialized PS agent is trained for 6×10^4 experiments with the goal of creating states with SRV (3, 3, 3), but the success probability stays near zero (Fig. 5.4a). This demonstrates that the agent trained to produce states with SRV (3, 3, 2) discovered some structures that also help in producing states with SRV (3, 3, 3). Some insight into how the PS agent builds the setups that lead to the desired states is discussed in Sect. 5.3.

5.3 Conceptual Insights from Action Composition

In Sect. 5.2, we discussed how PS can be used to create experimental setups for a specific task of preparing entangled states. As scientists, however, we are not only interested in knowing experimental setups that produce such states, but we also seek some understanding of the principles used to build the setups. Indeed, the connection between experimental setups and the entanglement structure of the created state is not well understood [55]. Action decomposition, which allows a PS agent to compose new actions during the training process that turned out to be particularly useful, can help to get some insight into this relation. Identifying gadgets often appearing in successful experimental setups can help to get some understanding of which building blocks are particularly useful for creating highly entangled states.

The added composed actions of the PS agent that was trained as described in Sect. 5.2 are analyzed in [56]. The importance of an action a_j is estimated according to the sum $\sum_k h_{k,j}$ of the weights of all incoming edges (which is related to the probability that the action a_j is chosen). One finds that the PS agent extensively uses a combination of actions corresponding to an optical interferometer (shown in Fig. 5.4b A). Usually, such interferometers are used to sort OAM modes with different parities [108], hence, the PS agent essentially (re-)discovered an optical setup that is well known and useful for various tasks.

To discover other relevant gadgets, one may train the PS agent in a way such that only novel experimental setups are rewarded. Hence, if the PS agent finds a setup that achieves an SRV (3, 3, 2) (or (3, 3, 3)), a reward is only given if this setup was not already found earlier in the training process. Analyzing again the composed actions found by such an agent, leads to several new interesting gadgets [56] (see Fig. 5.4b B,C, and D). The gadget shown in Fig. 5.4b B, a nonlocal interferometer, has recently been analyzed and motivated the work in [65]. Further, it can be shown that the gadget shown in Fig. 5.4b D allows to create high-dimensional entangled states beyond the initial state dimension of three in the OAM basis.

Chapter 6
Model Creation

In this chapter, we discuss how to optimize and build physical models using machine learning. There is an extensive literature on the use of machine learning to solve scientific (optimization) problems. Work in this direction has many success stories and could be used to solve many challenging problems in science. Often, a model for the physical system is assumed and the aim of the machine is mainly to optimize the parameters of the models. Therefore, despite the great practical usefulness of such approaches, they do not usually lead to new scientific knowledge beyond the specific solution to the given problem. In Sect. 6.1, we briefly review some recent work in this area. However, since the focus of this book is on discovering new models with machine learning, we keep this section short and provide several references for further reading. In Sect. 6.2, we discuss how to find mathematical formulas describing the time evolution of physical systems. Finding mathematical formulas is an essential part of a physicist's discovery process. The bias that the underlying model is described by "simple" mathematical formulas has led to models with great generalization power in history.

6.1 Optimizing Model Parameters

A lot of literature considers building machine learning systems modeling specific physical systems. Then, the parameters of the model are optimized to solve a specific problem. Such methods have proven very successful in solving challenging scientific problems, e.g. in the design of materials and molecules through accelerated simulations of quantum systems (see [109] for an extensive review using machine learning for quantum physics). Neural networks have become the most important machine learning tool for solving scientific problems in recent years. For example, in condensed matter physics and generally in many-body settings, neural networks have proven particularly useful to characterize phase transitions (see [14] and references

R. Iten, *Artificial Intelligence for Scientific Discoveries*,
https://doi.org/10.1007/978-3-031-27019-2_6

therein) and to learn local symmetries [110]. In quantum chemistry, neural networks and graph based networks show impressive results for synthesis of molecules and predicting their properties [111–113] (see [114, 115] for recent reviews and further references). A more extensive review on the applications of machine learning in physics can be found in [17].

The incorporation of prior knowledge about a particular physical model into machine learning systems fundamentally limits what can be "discovered" by the machine. In fact, we cannot claim a machine to discover a physical model itself if some knowledge about the model is already hardcoded in the machine learning system. Another approach, discussed in full detail in Part III of this book, is to model the physicist's reasoning process with the machine learning system. This approach has the potential to discover physical models on its own in tradeoff with the performance of solving specific physical problems. The tradeoff between the generality of a model and its performance is fundamental, in the sense that discovering something from scratch is always harder than starting from some prior knowledge about the system.

6.2 Discovering Physical Laws

Discovering the physical laws underlying experimental data is one of the main goals of a physicist. Automatizing this process was considered in an early study in [18] in 1987. In 2007 an algorithm was introduced in [116] that searches a space of mathematical expressions on given input variables to recover the differential equations underlying time-series data. Two years later, it was demonstrated in [19] that searching for conserved quantities in time-series data can be used to find the laws of motion of various mechanical systems, such as a double pendulum. In 2011, the same algorithm was applied to infer analytical models for metabolic networks [20]. In the last few years, significant progress was made in extracting dynamical equations from experimental data (see e.g. [117–120]), which is known to be an NP-hard problem [121]. The methods in [118], for example, have been successfully applied to complex physical systems such as water flowing through a pipe (defined by the Navier-Stokes equations). Recently, physical prior knowledge was incorporated into deep neural networks leading to so called *physics-imformed* deep neural networks [122]. Physics-informed neural networks have been used for the data-driven discovery of partial differential equations by exploiting the a priori knowledge that physical dynamics should belong to a class of partial differential equations [122–125]. Last but not least, tensor networks were used in [126] to learn non-linear dynamical laws. Building physical principles, such as the locality of interactions, into the tensor network naturally ensures scalability of the approach.

In the following sections, we will focus on the methods introduced in [19], because of three reasons: (i) the methods described in [19] cover the basic ideas for extracting mathematical laws from data, (ii) a minimal amount of prior knowledge on the expected form of the formulas is required, and (iii) the work in [19] provides a basis to understand more recent papers in this field. For readers interested in using the method that achieves the best performance in discovering physical laws from a given data set, please refer to the more recent references above.

All the work on extracting physical laws mentioned above assumes prior knowledge in the form of knowledge of the relevant variables describing the physical system. In Part III of this book we explain how to train a neural network to learn the relevant parameters and the differential equation describing the time evolution of the physical system in parallel (see Sects. 8.5, 9.6 and 10.3).

6.2.1 Symbolic Regression

Symbolic regression [21] is a form of regression in which the goal is to find a mathematical expression (the form as well as the parameters) that accurately describes the data. The accuracy and simplicity of the solution are equally essential. In the history of physics, the bias towards preferring "simple" mathematical models over complex formulas (which can be seen as a special case of Occam's razor, i.e., preferring simple explanations over complex ones) often lead to models with great generalization power. More concretely, given data samples (x, y) with $x \in \mathbb{R}^n$ for some $n \in \mathbb{N}$ and, without loss of generality, $y \in \mathbb{R}$, symbolic regression searches for a simple mathematical expression $f : \mathbb{R}^n \to \mathbb{R}$ such that $f(x) \approx y$ for all samples (x, y). For example, given data points $(0, 0)$, $(1, 1)$, $(2, 4)$, we may expect to find formulas like $f(x) = x^2$. Of course, we could also fit the data with a third-degree polynomial, but we consider higher-degree polynomials to be more "complex" than lower-degree ones and therefore prefer expressions of low-degree polynomials. In praxis, the data samples are usually noisy and there is a tradeoff between accuracy and simplicity of the solutions.

The space of mathematical equations is infinitely large and it is a hard task to efficiently search for the simplest expression describing the samples in this space. Furthermore, to automatize the search of simple expressions, we have to define the "complexity" of an expression. Mostly, *genetic algorithms* [127] are used as a search method. Genetic programming is a machine learning technique that is often used for searching an element in a discrete space. In principle, it can also be applied to continuous spaces, however, often we have more efficient methods at hand in this case. The idea of genetic algorithms is based on evolutionary theory that attempts to solve a given task by developing potential solutions over generations, discarding the least fit contestants and mutating the fittest ones. This algorithm can theoretically be used to solve a wide range of problems, as long as a way to determine the fitness and to genetically modify a potential solution exists. In the following, we describe how to use genetic algorithms for symbolic regressions, hence where the *genotype* corresponds to encodings of symbolic expressions. The aim of the following discussion is to introduce the reader to the basic concepts, and not to provide all details required for an efficient implementation. For readers interested in implementing or using symbolic regression, we refer to [19] and to [128] for a discussion of symbolic regression with a focus for applications in physics.

Let us consider a set of samples $\mathcal{S} = \{(x^i, y^i)\}_i$, and represent mathematical expression with graphs where each node corresponds to a mathematical building block and each leave to a system variable or parameter (examples are shown in Fig. 6.1). Crucially, the choice of the set from which the mathematical building blocks are chosen, introduces a bias into the search over expressions. The genetic algorithm starts by randomly choosing a set $\mathcal{P}_0$, called *population*, of mathematical expressions using the system variables $(x_1, \ldots, x_n)$ corresponding to the coordinates of a sample $x^i \in \mathbb{R}^n$. The *fitness* $F(f, \mathcal{S})$ of an expression f may then be defined by the negative averaged mean square prediction error, i.e.,

$$F(f, \mathcal{S}) = -\frac{1}{|\mathcal{S}|} \sum_{i \in \{1,\ldots,|\mathcal{S}|\}} \| f(x^i) - y^i \|_2^2 \,. \tag{6.1}$$

Hence, the fitness is a measure of how well an expression approximates the given data samples. The complexity $C(f)$ of the expression f can be measured, for example, by the number of terms in the expression. To include the simplicity of the expression into the fitness, one may define

$$F'(f, \mathcal{S}) = -C(f) - \lambda \frac{1}{|\mathcal{S}|} \sum_{i \in \{1,\ldots,|\mathcal{S}|\}} \| f(x^i) - y^i \|_2^2 \,, \tag{6.2}$$

where the hyperparameter λ regulates the tradeoff between finding accurate and simple expressions.

Inspired by evolutionary theory, we calculate the fitness of each expression in the population $\mathcal{P}_0$ and further consider the expressions with the highest fitness. The fittest individuals are then "slightly" modified to generate a new population $\mathcal{P}_1$ (see Fig. 6.1 for an overview). Again motivated by evolutionary theory, the fittest expressions in the population are modified by *mutations* and *crossovers*. A biological mutation is a random change in the genes of an individual, and corresponds to replacing a small part of the graph representing the expression with a random graph in the considered case here (Fig. 6.1). On the other hand, crossover operations are analogous to the crossover that happens during sexual reproduction in biology. Here, they may be represented as cutting out a subgraph of a representation of an expression (with high fitness) and insert it into another expression (Fig. 6.1). The process of choosing the fittest individuals and generating a new population is then repeated $m \in \mathbb{N}$ times until a satisfying expression is found in the mth population $\mathcal{P}_m$.

Based on evolutionary theory, genetic search algorithms are expected to be more efficient than brute-force search provided that parts of fit candidates can be combined and mutated in a reasonable way, i.e., in a way such that combining two parts of fit individuals likely leads to another fit or maybe even fitter individual. Nontheless, the runtime of such algorithms often prohibits searching for complex expressions in many variables. Indeed, in [19], an exponential scaling in the complexity of the target

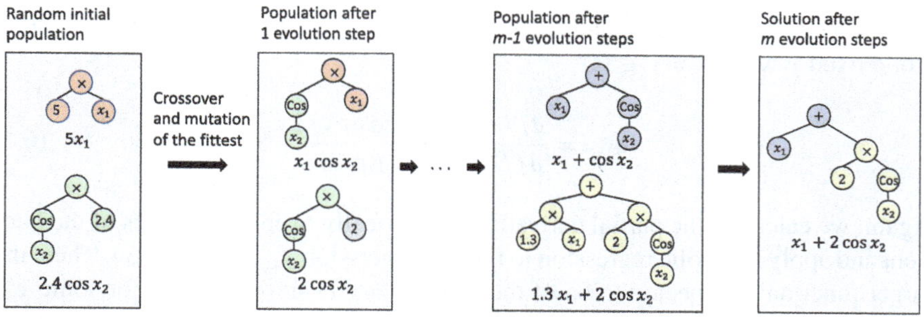

Fig. 6.1 Symbolic regression. The goal of symbolic regression is to find a simple mathematical expression f fitting with a given set $\mathcal{S} = \{(x^i, y^i)\}_i$ of input-output pairs, i.e., we search for a simple symbolic expression with $f(x^i) \approx y^i$ for all $i \in 1, \dots, |\mathcal{S}|$ with $x^i = (x_1^i, \dots, x_n^i) \in \mathbb{R}^n$ and $y \in \mathbb{R}$. For concreteness, here we consider $n = 2$ and a data set with $y^i \approx x_1^i + 2\cos x_2^i$, i.e., the target expression we would like to find is $f(x) = x_1 + 2\cos x_2$. Often, genetic algorithms are used to find the fittest expression, i.e., the simplest expression that fits the data well. The genetic algorithm starts from a random initial set of several symbolic expressions represented as graphs in the system variables x_1 and x_2 (the number of expressions depicted in the figure is not representative). Then, the fittest expressions are selected and changed by mutations and crossover operations. The upper graph in the initial population in the figure is combined with the lower graph by a cross over operation, corresponding to combining genes during sexual reproduction in biology. The lower graph is changed by a mutation of the leaf storing the parameter 2.4 by replacing it with a random number. We repeat this procedure $m \in \mathbb{N}$ times, until we end up with a simple expression fitting the data

expression was observed. A promising approach to lower the runtime by creating symbolic expressions in a step-wise fashion was proposed in [129] (see Remark 6.1).

Remark 6.1 (Stepwise symbolic regression) In [129], an approach is suggested to find symbolic expressions by step-wise adding system variables. The approach has similarities to the one in [19] and has not been fully automatized yet, but its idea is pretty simple and explained on the basis of an example in the following. Let us consider a sample set with pairs $((x, y, z), f(x, y, z))$ with an expression $f = z\sin(xe^y)$. The goal is to find the symbolic expression from the sample set. Assuming that we have numerical access to partial derivatives, we may numerically create samples for the following quotient

$$q_{x,y} = \frac{\partial f/\partial x}{\partial f/\partial y} = \frac{z\cos(xe^y)e^y}{z\cos(xe^y)xe^y} = \frac{1}{x}. \tag{6.3}$$

We may use (standard) symbolic regression to obtain an expression $f_{x,y}$ for the samples $((x, y), q_{x,y}(x, y))$ such that the quotient $\frac{\partial f_{x,y}/\partial x}{\partial f_{x,y}/\partial y}$ approximates the numerical values of $q_{x,y}(x, y)$. A simple expression that symbolic regression may find is $f_{x,y} = xe^y$, hence we recovered the relation between two of the system variables in the target expression. The form of the target expression is then expected to be given as $f = g(xe^y, z)$ for some function $g : \mathbb{R}^2 \to \mathbb{R}$.

In a second step, we may consider the relation between x and z by investigating (for a fixed y with $e^y = c \in \mathbb{R}$)

$$q_{x,z} = \frac{\partial f/\partial x}{\partial f/\partial z} = \frac{cz\cos(cx)}{\sin(cx)}. \tag{6.4}$$

Again, we calculate the partial derivatives numerically to create samples of the quotient and apply symbolic regression to find an expression $f_{x,z} = z\ \sin(cx)$. Then, the target function is expected to be of the form $f = g'(z\sin(c(y)x), y)$ for some $g' : \mathbb{R}^2 \to \mathbb{R}$ and $c : \mathbb{R} \to \mathbb{R}$. From the requirement $f = g(xe^y, z)$, we see that setting $g(a, z) = z\sin(a)$ provides the same expression as given by $f = g'(z\sin(c(y)x), y)$ setting $g'(a, y) = a$ and $c(y) = e^y$. Hence, we discovered the target expression $f = z\ \sin(xe^y)$ by applying twice symbolic regression to find simpler expressions in only two system variables, which can be done more efficiently than directly searching for the full expression. However, combining the expression may not always be easy and has not been automatized so far. It would be interesting to consider this for future work.

6.2.2 *Extracting Physical Laws from Data*

In this section, we describe the ideas used in [19] to discover physical laws from time-series data. The approach in [19] was successfully applied to find the equation of motion of various physical systems including the chaotic double-pendula without using any prior knowledge about physics, apart from the knowledge of the system variables (discovering the system variables with deep neural networks is discussed in Part III of this book). The conceptual basis of the approach in [19] is that nearly all physical laws in nature are based on mathematical symmetries and invariants [130], implying that the search for many natural laws is inextricably linked to the search for conserved quantities and invariant equations [131, 132]. The main difficulty in searching for conserved quantities using symbolic regression is to avoid finding trivial invariants, such as $\cos^2(x) + \sin^2(x)$ for a system variable $x \in \mathbb{R}$, which is constantly equal to one but does not contain any physical insights. In the following, we explain the approach taken in [19] to avoid such trivial invariants on the basis of an example.

Let us consider a pendulum, i.e. a non-linear oscillator whose position is described by an angle θ. The sample set $\mathcal{S} = \{(t^i, \theta(t^i), \omega(t^i))\}_i$ consists of triples containing the time t^i, as well as the corresponding position $\theta(t^i)$ and angular velocity $\omega(t^i)$. Note that we have no values provided for the conserved quantities we are searching for, hence, they must be learned in an unsupervised way. The secret to computationally detecting nontrivial conservation laws is that the candidate equations should predict relations between the dynamics of the system's subcomponents. More specifically, the conservation equation should be able to predict relationships between derivatives of groups of variables over time. Such relations can be captured by the quotient of partial derivatives (in a similar spirit as we discussed in Remark 6.1). In other words, the expression $f(\theta, \omega)$ for a conserved quantity is considered to be non-trivial if it satisfies

$$\frac{\partial\theta/\partial t}{\partial\omega/\partial t} = \frac{\partial\theta}{\partial\omega} = \frac{\partial f/\partial\omega}{\partial f/\partial\theta}\,. \tag{6.5}$$

We can numerically estimating the left hand side of (6.5) as

$$\frac{\partial\theta/\partial t}{\partial\omega/\partial t} \approx \frac{\Delta\theta}{\Delta\omega}\,, \tag{6.6}$$

where $\Delta\theta$ and $\Delta\omega$ denote the changes of the parameters θ and ω within a small time interval Δt. The fitness of an expression f can then be estimated by the *mean-log-error* [19]

$$F(f, \mathcal{S}) = -\frac{1}{|\mathcal{S}|}\sum_{i=1}^{|\mathcal{S}|} \log\left(1 + \left|\frac{\Delta\theta(t^i)}{\Delta\omega(t^i)} - \frac{\partial f/\partial\omega}{\partial f/\partial\theta}\bigg|_{t^i}\right|\right)\,, \tag{6.7}$$

where $\Delta\theta(t^i)$ and $\Delta\omega(t^i)$ estimate the time derivative of the variables θ and ω at time t^i (using the set $\mathcal{S}$) and where the partial derivative is calculated symbolically using the analytical expression for the candidate function f.

We can then use symbolic regression (as described in Sect. 6.2.1) to search for candidate functions f with the highest fitness (see Fig. 6.2 for an overview).[1] In [19], it was demonstrated, that this method recovers the invariant expression $f(\theta, \omega) = c_1\omega^2 + c_2\cos(\theta)$ for two parameters $c_1, c_2 > 0$ depending on the mass m and the length l of the pendulum.

The invariant expression corresponds to the Lagrangian of the pendulum, however we have left to find the dependence of the parameters c_1 and c_2 on the mass and length of the pendulum. Since the invariant f is scale invariant, we may divide it by c_2 leading to $\tilde{f}(\theta, \omega) = c_1/c_2\omega^2 + \cos(\theta)$ and search for a symbolic expression of c_1/c_2 in the system variables m and l. The expression $c_1/c_2 \equiv k(m, l)$ can for example be found as follows: One creates time-series data for pendulums with different masses m_j and of different length l_j. For each pendulum j, we determine the value of k_j such that the expression $k\omega^2 + \cos(\theta)$ optimally fits the time-series data, in the sense that it stays approximately constant. Then, we can again run symbolic regression with the system variables m and l and the sample set $\{([m_j, l_j], k_j)\}_j$, i.e., we are

[1] For simplicity, we do not go into the technical details here of how to manage the tradeoff between high fitness and low model complexity. One way to solve this problem is presented in [19]. An alternative method, presented in [133], uses a corpus of closed-form mathematical models from Wikipedia to obtain the posterior probability of each expression from basic probabilistic arguments and explicit approximations. The posterior probability leads to a prior over expressions that accounts for model complexity.

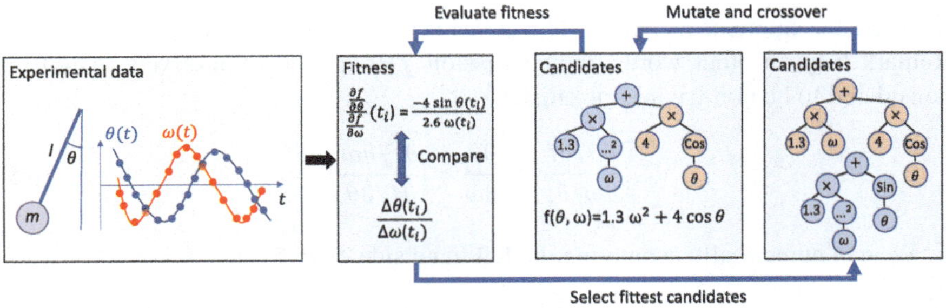

Fig. 6.2 Extracting invariant expressions from time-series data [19]. We consider experimental data collected from a pendulum whose position at a time t is described by an angle $\theta(t)$ and we denote the angular velocity by $\omega(t) := \dot{\theta}(t)$. Symbolic regression is used to extract time-invariant expressions, i.e., conserved quantities, in θ and ω from a data set $\mathcal{S} = \{(t^i, \theta(t^i), \omega(t^i))\}_i$ consisting of time series of the angle and velocity of the pendulum. Starting from randomly generated symbolic expressions (represented as graphs), we evaluate their fitness according to (6.7). The chosen expression (6.7) for the fitness is a measure of how well a candidate expression is able to predict relations between the dynamics of the system's subcomponents. Hence, a high fitness points towards an invariant expression that contains non-trivial physical information. Then, the fittest candidate expressions are chosen, and modified by mutations and crossover operations with the aim of creating even fitter individuals. This process is repeated until an invariant expression with fitness approximately equal to zero is found, which corresponds to an expression that can approximately recover the relation between the dynamics of the angle $\theta(t)$ and the velocity $\omega(t)$. In the considered example, the found expression is the Lagrangian of the pendulum given by $f(\theta, \omega) = c_1\omega^2 + c_2\cos(\theta)$ for two parameters $c_1, c_2 > 0$ depending on the mass m and the length l of the pendulum

searching symbolic expressions $k(m, l)$ with $k(m_j, l_j) \approx k_j$ for all j.[2] Finally, we find the form of the Lagrangian to be independent of the mass given by $L(\theta, \omega) = l/(2g)\,\omega^2 + \cos(\theta)$ with $g = 9.81\,m/s^2$ being the gravitational acceleration on the surface of Earth.

The equation of motion is then recovered by the Euler-Lagrange equation

$$\frac{\mathrm{d}}{\mathrm{d}t}\left(\frac{\mathrm{d}L}{\mathrm{d}\omega}\right) - \frac{\mathrm{d}L}{\mathrm{d}\theta} = \frac{l}{g}\ddot{\theta} + \sin\theta = 0\,, \tag{6.8}$$

and hence $\ddot{\theta} = -g/l\ \sin(\theta)$. Alternatively, we could search for invariant expressions in the system variables θ, ω and α, where $\alpha := \dot{\omega}$ denotes the acceleration of the pendulum. Doing so, we would directly recover the equation of motion [19]. This demonstrates on one hand the dependence of the approach on the chosen system variables, and on the other hand also that we discover useful invariants in settings using different system variables.

The runtime of the genetic search algorithm for finding invariant expressions takes from some minutes up to 30 h for the double pendulum parallelizing the search by

[2] Here we use the knowledge that the pendulum is fully described by its mass and length. Recovering these parameters automatically is discussed in Part III of this book.

using 32 cores [19]. On the other hand, finding the expressions for the parameters usually takes significantly less time. We conclude that when considering systems with only a few known system variables, the described approach provides a powerful method to find symbolic expressions for invariant quantities. The symbolic nature of the result can help a physicist to gain new conceptual insights into the system under consideration. On the other hand, the prior assumption that simple analytic expressions exist may prohibit us from recovering invariants that cannot be described as simple analytic formulas in the chosen system variables. For systems containing of many (probably unknown) system variables one way consider more recent methods in this field (see the introduction of Sect. 6.2 for relevant references) together with the methods introduced in Part III of this book.

Chapter 7
Model Testing

Testing physical models is an essential part of a physicist's discovery process (discussed in Sect. 1.2). Falsifying a model in a certain physical domain (such as in a high energy range) helps to understand the limits of a model and guides future research towards improved models with a broader application range. Testing new models is of particular interest in high-energy physics. Although the Standard Model (SM) of particle physics very successfully describes a large number of elementary particle processes with high accuracy, unexplained phenomena such as the neutrino masses or the appearance of dark matter in the universe point to the need to extend or adapt the SM. Searching for new phenomena in high-energy physics often relies on testing suggestions of new models by running experiments (e.g. with the Large Hadron Collider) and then comparing the collected data with predictions of a *reference model* (e.g. the SM of particle physics). Often, the considered models make probabilistic predictions, which makes testing particularly challenging. In several papers, a reference model was tested against another hypothetical model by constructing hypothesis tests using standard techniques (see [134] for a review for model testing in particle physics). Such an approach requires to come up with an alternative model and has the disadvantage that a statistical test designed for a particular alternative hypothesis is typically insensitive to data deviations of other kinds from the reference model.

How can we discover the limitations of a model from test data without having an alternative model at hand? The basic idea is to test the reference model against a model constructed by a machine based on the (experimental) test data. If the class of models is not restricted to a specific class of physical models, we call this a *model-independent* strategy (in a physical sense). If the machine constructs a model that deviates significantly from the reference model, one can conclude that the reference model does not fit the test data well. Furthermore, we can analyze the regions where the two models diverge the most, which can lead to conceptual insights into what problems the reference model has. Such model-independent searches for new physics are less commonly considered in the literature [135–139, 139–142] than model-dependent searches. The methods described in [141, 142] follow similar ideas

R. Iten, *Artificial Intelligence for Scientific Discoveries*,
https://doi.org/10.1007/978-3-031-27019-2_7

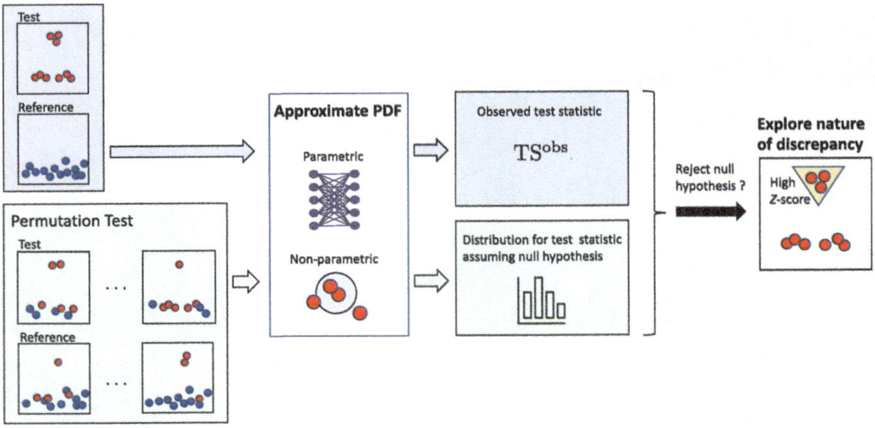

Fig. 7.1 Overview of checking model compatibility [141, 142]. The figure shows how to test the compatibility of a set $\mathcal{T}$ of test samples with a set of reference samples $\mathcal{R}$ in a model-independent way. The samples depicted in the figure live in a $d = 2$ dimensional feature space. The null hypothesis is that both data sets $\mathcal{T}$ and $\mathcal{R}$ are generated by the same model, i.e., the same probability density function (PDF). First, the test and the reference samples are provided as an input to a machine learning system whose goal is to estimate the PDFs underlying the test and the reference set and to provide the observed test statistic $\mathrm{TS}^{\mathrm{obs}}$ as an output (scenario marked in blue in the figure). The PDFs can be estimated with a parametrized model (such as neural networks [141]) or a non-parametrized method (such as the k-nearest neighbor estimation [142]). Here we use the test statistic given in (7.5), which approximates the KL-divergence between the true underlying test and reference PDF. To interpret the observed value $\mathrm{TS}^{\mathrm{obs}}$, we compare it with the probability distribution of the test statistic assuming that the null hypothesis is true. The distribution is obtained by the permutation test, i.e., by shuffling samples between the test and reference set and evaluating the test statistic for the obtained test and reference sets (scenario marked in green). Then, we can estimate the p-value, i.e., the probability that we observe $\mathrm{TS}^{\mathrm{obs}}$ or more extreme values of the test statistic assuming the null hypothesis. If the p value is lower than the desired significance level $\alpha \in [0, 1]$ of the test, the null hypothesis is rejected. Finally, we use the Z-score given in (7.16) to locate regions of high incompatibility in the feature space

for model-independent testing of a reference model. These ideas are described in the following (and summarized by Fig. 7.1) and can in principle be applied to test any physical model and to find regions where the collected data deviates significantly from the reference model. However, applying such techniques is praxis is still challenging due to reasons discussed in Sect. 7.5.

7.1 Statistical Setting

In this section, we describe the formal statistical setting for model-independent testing (following the lines in [142]). The setting considers a basic scenario of comparing two data sets with each other and can be applied to statistical problems beyond model testing. We consider two data sets with d-dimensional real data points

$$\mathcal{T} = \{x_i : x_i \in \mathbb{R}^d \,, 1 \leq i \leq n_T\} \quad \text{with } x_i \sim P_T \text{ and} \tag{7.1}$$

$$\mathcal{R} = \{x_i' : x_i' \in \mathbb{R}^d \,, 1 \leq i \leq n_R\} \quad \text{with } x_i \sim P_R \,, \tag{7.2}$$

where the observations $x_i \in \mathbb{R}^d$ and $x_i' \in \mathbb{R}^d$ are sampled i.i.d. from a test and a reference probability density function (PDF) P_T and P_R, respectively. The d-dimensional space in which the samples live is called the *feature space*. For the following considerations, neither P_T nor P_R have to be known analytically. In a typical practical scenario, the reference set $\mathcal{R}$ is created by running a simulation on a computer based on the reference model, while the test set $\mathcal{T}$ is obtained experimentally.

The goal is to determine wether the two data sets are (in)compatible, i.e., in a language of hypothesis testing, we would like to find out up to which significance level the null hypothesis $\{H_0 : P_T = P_R\}$ is rejected. A well-defined statistical test scenario requires an alternative hypothesis, which we define as follows $\{H_1 : P_T = \hat{P}_T\}$, where the PDF $\hat{P}_T$ depends on the test data set $\mathcal{T}$ and is chosen to approximate the true PDF P_T as accurately as possible. The choice of approximation method leads to different alternative hypotheses, whose choice is crucial for the sensitivity of the test. The goal considered here is to find approximation methods that lead to tests that are sensitive to a broad range of data deviations from the reference model. This is a subtle problem since every unjustified assumption on the class of PDFs used to approximate P_T may cause insensitivity to the data deviations occurring in the test sample.

As a measure of compatibility, we may consider the ratio λ of probability densities to observe the test data $\mathcal{T}$ under the PDF P_T (numerator) and under the PDF P_R (denominator) [142]

$$\lambda = \prod_{x_i \in \mathcal{T}} \frac{P_T(x_i)}{P_R(x_i)} \,. \tag{7.3}$$

A value of $\lambda \gg 1$ means that the data in the test sample $\mathcal{T}$ is much more likely to be sampled from the PDF P_T than from P_R. Since the PDFs are not known, we have to approximate them by $\hat{P}_\mathcal{T} \approx P_T$ and $\hat{P}_\mathcal{R} \approx P_R$ (see Sect. 7.2 for approximation strategies) to find an estimate on λ as

$$\hat{\lambda} = \prod_{x_i \in \mathcal{T}} \frac{\hat{P}_\mathcal{T}(x_i)}{\hat{P}_\mathcal{R}(x_i)} \,. \tag{7.4}$$

As a test statistic $\text{TS}(\mathcal{T}, \mathcal{R})$, we may use [142]

$$\mathrm{TS}(\mathcal{T},\mathcal{R}) = \log \hat{\lambda}^{\frac{1}{n_T}} = \frac{1}{n_T}\sum_{i=1}^{n_T} \log \frac{\hat{P}_{\mathcal{T}}(x_i)}{\hat{P}_{\mathcal{R}}(x_i)} . \tag{7.5}$$

The test statistic is used as a distance measure between the two PDFs $\hat{P}_{\mathcal{T}}$ and $\hat{P}_{\mathcal{R}}$ (Remark 7.1). A value of TS close to zero means that the two PDFs are essentially the same and hence, we may not reject the null hypothesis. On the other hand, a large value of TS suggests that the PDF $\hat{P}_{\mathcal{T}}$ underlying the test set is significantly different from the PDF $\hat{P}_{\mathcal{R}}$ underlying the reference set, and hence that we should reject the null hypothesis. How to determine the threshold value of TS above which we reject the null hypothesis for a desired test significance $\alpha \in [0, 1]$ is discussed in Sect. 7.3. First, we discuss how we can approximate the PDF with methods from machine learning in Sect. 7.2.

Remark 7.1 (Test statistic as an approximation of the Kullback-Leibler divergence [142]) The Kullback-Leibler (KL) divergence (see also Definition 8.4) of the PDFs P_T and P_R is defined as

$$D_{\mathrm{KL}}\left(P_T \| P_R\right) = \int P_T(x) \log\left(\frac{P_T(x)}{P_R(x)}\right) \mathrm{d}x , \tag{7.6}$$

where we integrate over the full feature space. The KL divergence is a fundamental measure in information theory that can be seen as a distance measure between two PDFs. Indeed, it is non negative and it equals zero if and only if $P_T = P_R$ (however, it is not a metric, since it is not symmetric). Replacing the integral with an empirical average over the samples in the test set, and P_T and P_R with the approximations $\hat{P}_{\mathcal{T}}$ and $\hat{P}_{\mathcal{R}}$, respectively, we see that the expression of the test statistic given in (7.5) approximates $D_{\mathrm{KL}}\left(P_T \| P_R\right)$.

7.2 Approximation of Probability Density Functions

In this section, we discuss how to obtain an estimate $\hat{P}_{\mathcal{R}}$ (or $\hat{P}_{\mathcal{T}}$) for the PDF P_R (or P_T) underlying a sample set $\mathcal{R}$ (or $\mathcal{T}$). More specifically, since we are ultimately interested in calculating the test statistic given in (7.5), we only need to estimate the quotient of the PDFs P_T and P_R at any point in the test set (a comprehensive review on density ratio estimators can be found in [143]). One can distinguish two types of approaches [141, 142]:

1. **Parametric modelling**: We consider a parametrized class of PDFs $\mathcal{C}_\Theta = \{\hat{P}_\theta\}_{\theta\in\Theta}$ from which $\hat{P}_{\mathcal{R}}$ is chosen from.
2. **Non-parametric modelling**: We do not restrict the class of PDFs and construct $\hat{P}_{\mathcal{R}}$ directly from $\mathcal{R}$.

The following sections present the general ideas behind both approaches on the basis of a specific machine learning strategy for leaning PDFs. Which method should be chosen to achieve the highest sensitive to new physics signals depends on the specific setting including the reference model and what kind of new signals are expected. For most methods, we have no exact understanding about the class of signals that can be detected with it with high significance. Hence, the choice of the model also depends on some experience of the user with various methods. The goal of the following sections is not to discuss many examples to gain such experience, but rather to introduce the reader to the general idea of how to approximate PDFs to search for unknown physical signals in data.

7.2.1 Parametric Modelling

In the parametric approach, first one has to choose the class of PDFs $\mathcal{C}_\Theta$. Instead of choosing a parametrized class of physical models, here we are interested in choosing an expressiv class that can approximate a wide range of PDFs. One option is to parametrize the class of PDFs by neural network [141], which are low-biased and universal function approximators (see Sect. 3.4). Furthermore, neural networks represent smooth functions, which is usually a reasonable bias for PDFs coming from physical models. Indeed, we would not expect a reasonable physical model to predict a discontinuous PDF. Last but not least, there is evidence that neural networks can break the curse of dimensionality. The number of events to approximate a PDF with a histogram grows exponentially with the dimension d of the feature space. While there is still no proof, the work in [144–146] suggests that neural network require fewer events to approximate PDFs on high-dimensional feature spaces. Since in practical applications, the dimension d is usually quite large, this is an extremely desirable property.

Having chosen a class of PDFs $\mathcal{C}_\Theta = \{\hat{P}_\theta\}_{\theta\in\Theta}$ (were the parameter θ contains all the weights and biases of the neural network), we can use the maximum likelihood method to approximate the PDFs P_T and P_R, i.e.,

$$\hat{P}_\mathcal{T} = \hat{P}_{\theta_\mathcal{T}} \text{ with } \theta_\mathcal{T} = \arg\max_{\theta\in\Theta} \prod_{x_i\in\mathcal{T}} \hat{P}_\theta(x_i) \text{ and} \tag{7.7}$$

$$\hat{P}_\mathcal{R} = \hat{P}_{\theta_\mathcal{R}} \text{ with } \theta_\mathcal{R} = \arg\max_{\theta\in\Theta} \prod_{x'_i\in\mathcal{R}} \hat{P}_\theta(x'_i)\,. \tag{7.8}$$

From these approximations, we can calculate the test statistic $\text{TS}(\mathcal{T}, \mathcal{R})$ given in (7.5). Instead of using two neural networks, and solving both optimization problems given in (7.7) and (7.8), one can also directly represent the (logarithm of the) quotient of the PDFs with a neural network and minimize the test statistic. This method is described in full detail in [141] and is demonstrated to have high sensitivity to various data deviations.

7.2.2 Non-parametric Modelling

In the non-parametric approach, the PDFs are approximated without restricting to a parametrized model class. Here, we discuss the *nearest neighbors* (NN) approach [147–153], described in the context of model-independent testing in [142]. The basic idea is to estimate a PDF at a point x in the feature space by calculating the volume of the sphere that contains the $k \in \mathbb{N}$ nearest neighbors of x. The density estimate is then proportional to the number of points k divided by the volume of the sphere. Let us formalize this on the basis of estimating the probability density $\hat{P}_{\mathcal{T}}$ on the test set $\mathcal{T}$. Consider a sample $x_i \in \mathcal{T}$ and calculate the euclidean distance to all other samples in $\mathcal{T} \setminus \{x_i\}$. Then, the k-nearest neighbors (i.e., the k samples in $\mathcal{T} \setminus \{x_i\}$ with the smallest distances to x_i) are determined, and the radius $r_{i,\mathcal{T}}$ is set equal to the radius of the kth nearest neighbor. Hence, the k nearest neighbors of x_i are contained in a sphere of volume $\omega_d r_{i,\mathcal{T}}^d$ with centre at x_i, where ω_d denotes the volume of the d-dimensional unit sphere, i.e.,

$$\omega_d = \frac{\pi^{d/2}}{\Gamma\left(\frac{d}{2}+1\right)}, \tag{7.9}$$

where Γ denotes the Gamma function, which is given by $\Gamma(n) = (n-1)!$ for positive integers n. Hence, normalizing with the total number of samples $(n_{\mathcal{T}} - 1)$ in $\mathcal{T} \setminus \{x_i\}$, the NN-estimation of the PDF $P_{\mathcal{T}}$ is given by

$$\hat{P}_{\mathcal{T}}(x_i) = \frac{k}{(n_{\mathcal{T}} - 1)} \frac{1}{\omega_d\, r_{i,\mathcal{T}}^d}, \tag{7.10}$$

for any $x_i \in \mathcal{T}$. To estimate the reference PDF P_R on samples $x_i \in \mathcal{T}$ in the test set, we calculate the distances from x_i to the k nearest neighbors in $\mathcal{R}$. The distance to the kth nearest neighbors is denoted by $r_{i,\mathcal{R}}$, and the NN-estimation of the reference PDF P_R is then given by

$$\hat{P}_{\mathcal{R}}(x_i) = \frac{k}{n_R} \frac{1}{\omega_d\, r_{i,\mathcal{R}}^d}, \tag{7.11}$$

for any $x_i \in \mathcal{T}$. The test statistic $\mathrm{TS}(\mathcal{T}, \mathcal{R})$ given in (7.5) then simply becomes

$$\mathrm{TS}(\mathcal{T}, \mathcal{R}) = \log \frac{n_R}{n_{\mathcal{T}} - 1} + \frac{d}{n_{\mathcal{T}}} \sum_{i=1}^{n_{\mathcal{T}}} \log \frac{r_{i,\mathcal{R}}}{r_{i,\mathcal{T}}}. \tag{7.12}$$

It can be shown that the expression in (7.12) is a consistent and asymptotically unbiased estimator of the KL-divergence [154–156], i.e., the test statistic (7.12) converges almost surely to the KL-divergence $D_{\mathrm{KL}}(P_T, P_R)$ between the true PDFs P_T and P_R in the large sample limit $n_R, n_T \to \infty$.

7.3 Statistical Hypothesis Testing

The ultimative goal of the statistical setting described in Sect. 7.1 is to decide wether the null hypothesis $\{H_0 : P_T = P_R\}$ can be rejected with a significance level $\alpha \in [0, 1]$. For example, rejecting the null hypothesis with significance $\alpha = 0.05$ means that the probability to observe the test data assuming the null hypothesis is below 5%. To interpret the observed value of the test statistic $\mathrm{TS}(\mathcal{T}, \mathcal{R})$ given in (7.5), we have to derive a probability distribution $P_{\mathrm{TS}|H_0}$ over the values of TS under the assumption that the null hypothesis is true, i.e., assuming that $P_T = P_R$. Then, we can compute the two-sided p value of the null hypothesis, i.e., the probability that the value of the test statistic is at least as far away from the expected value under the null hypothesis as the observed test statistic $\mathrm{TS}^{\mathrm{obs}}$.

As suggested in [142], we us a resampling method known as the permutation test [157, 158] to compute the distribution $P_{\mathrm{TS}|H_0}$. The basic idea is that we resample various test and reference sets under the null hypothesis that both sets are generated by the same PDF (see also Fig. 7.1). Then, we evaluate the test statistic for the resampled sets and finally, we derive a distribution over the test statistic from the various data points that we obtained.

More formally, first we merge the test and reference set into a pool of samples $\mathcal{U} = \mathcal{T} \cup \mathcal{R}$. Then we randomly choose n_T samples from $\mathcal{U}$ and assign them to a set $\tilde{\mathcal{T}}$.[1] The remaining elements are assigned to a resampled reference set $\tilde{\mathcal{R}}$. Then, the test statistic for the set $\tilde{\mathcal{T}}$ and $\tilde{\mathcal{R}}$ is calculated. Repeating this procedure n_{perm} times, leads to a set $\{\mathrm{TS}_1, \ldots, \mathrm{TS}_{n_{\mathrm{perm}}}\}$ of n_{perm} data points for the test statistic. Dependent on the size of n_{perm}, we may choose an interval size $\Delta\mathrm{TS}$ and discretize the distribution $P_{\mathrm{TS}|H_0}$ with a histogram by assigning the data points TS_j to the bins $[m\Delta\mathrm{TS}, (m+1)\Delta\mathrm{TS}]$ for $m \in \mathbb{Z}$. To decide wether the null hypothesis should be rejected with significance level α, we calculate the p-value defined by

$$p = \int_{-\infty}^{\hat{\mu}-\delta} P_{\mathrm{TS}|H_0}(t)\mathrm{d}t + \int_{\hat{\mu}+\delta}^{\infty} P_{\mathrm{TS}|H_0}(t)\mathrm{d}t\,, \tag{7.13}$$

[1] Alternatively, one may think of the elements in $\mathcal{U}$ to have a fixed order and then applying a random permutation on the list of elements (which is the reason for the name "permutation test"). Assigning the first n_T elements to $\tilde{\mathcal{T}}$ leads to the same result as randomly sampling these elements from the set $\mathcal{U}$.

where $\hat{\mu}$ denotes the estimated mean value of the test statistic, $\delta = |\mathrm{TS}^{\mathrm{obs}} - \hat{\mu}|$ and where the integration is actually implemented as a summation over the histogram that approximates the distribution $P_{\mathrm{TS}|H_0}$. Finally, the null hypothesis is rejected if the p value is smaller than the significance level α, since in such a case, the probability that we observe a value of the test statistic that deviates at least by δ of the expected value $\hat{\mu}$ is smaller than α assuming the null hypothesis.

Summarizing, the statistical procedure is as follows:

1. Approximate the distributions $P_{\mathcal{T}}$ and $P_{\mathcal{R}}$ with $\hat{P}_{\mathcal{T}}$ and $\hat{P}_{\mathcal{R}}$ (at data points $x_i \in \mathcal{T}$).
2. Calculate the observed test statistic $\mathrm{TS}^{\mathrm{obs}}(\mathcal{T}, \mathcal{R})$ using (7.5).
3. Estimate the distribution $P_{\mathrm{TS}|H_0}$ over the test statistic under the assumption that the null hypothesis is true using the permutation test.
4. Compute the two-sided p value of the null hypothesis.
5. If the two-sided p value is smaller than α, we reject the null hypothesis.

7.4 Identifying the Discrepant Regions

Let us assume that the statistical test described in Sect. 7.3 leads to a rejection of the null hypothesis. This means that the observed test statistic $\mathrm{TS}^{\mathrm{obs}}$ obtains a value that is far away from the expected value under the null hypothesis. To get some conceptual insight into why the reference model can not accurately explain the test data, we investigate the region in the feature space that contributes significantly to the deviation of the test statistic from the expected value. To find such regions, we follow the approaches in [141, 142] and assign a density field $U(x_i)$ to each sample $x_i \in \mathcal{T}$, given by

$$U(x_i) \equiv \log \frac{\hat{P}_{\mathcal{T}}(x_i)}{\hat{P}_{\mathcal{R}}(x_i)} . \tag{7.14}$$

The observed test statistic (7.5) can then be considered as the empirical average over the density field representing the expectation value

$$\mathrm{TS}^{\mathrm{obs}} = \mathbb{E}_{\mathcal{T}}[U] . \tag{7.15}$$

Furthermore, we normalize the density field to obtain a *Z-score*

$$Z(x_i) \equiv \frac{U(x_i) - \mathbb{E}_{\mathcal{T}}[U]}{\sqrt{\mathbb{V}_{\mathcal{T}}[U]}} , \tag{7.16}$$

where $\mathbb{V}_{\mathcal{T}}[U]$ denotes the variance of the density field U. To simplify the following discussion, let us assume that $\mathrm{TS}^{\mathrm{obs}} > \hat{\mu}$, where $\hat{\mu}$ is the expected value for the test distribution under the null hypothesis. The Z-score can be used to identify samples in $\mathcal{T}$ with a significantly larger density filed than the average $\mathbb{E}_{\mathcal{T}}[U]$. These samples contribute particularly strong to the high value of $\mathrm{TS}^{\mathrm{obs}}$, and hence to the deviation of the test data in relation to the reference model. We may therefore characterize samples with a Z-score higher than some threshold, e.g., samples $x_i \in \mathcal{T}$ with $Z(x_i) > 3$, as the samples that contribute most to the deviation of the test PDF from the reference PDF (see also Fig. 7.1). To get a better understanding in which regions of the feature space the reference model is not compatible with the test data, we may further use a clustering algorithm to identify regions in the feature space with a high density of samples x_i with $Z(x_i) > 3$ [142]. Investigating the high-discrepancy regions can lead to a better understanding of the causes of the deviation from the reference model.

7.5 Application for Model Testing

The statistical setting described in the previous sections is very general and can be used to check the compatibility of any two data sets with each other. Hence, the described testing strategy has a wide range of scientific and engineering applications. Here, we are interested in its application for model testing and the discovery of discrepant regions.

Both, the parametric method using neural networks and the non-parametric method based on k nearest neighbor estimations, have been applied to some simulated toy models [141, 142]. In particular, the non-parametric approach has been applied to detect a (simulated) black matter signal embedded in reference data generated according to the SM of particle physics [142]. The data set $\mathcal{R}$ contains the background events according to the SM, and the data set $\mathcal{T}$ contains also background events but with added black matter signals. It is then demonstrated in [142] that the test statistic based on nearest neighbor estimations is enough sensitive to detect the black matter signals. Furthermore, the same scenario together with added noise to the background simulation is considered in [142]. As expected, such noise lowers the sensitivity to the black matter signals and hence, dependent on the magnitude of the noise, the test may not be sensitive enough to reject the null hypothesis (which is in this case that the SM fully describes the test set). Nonetheless, the method to identify discrepant regions may still be applied and point towards regions in the feature space where the test data deviates the most from the reference data. These are likely the regions where the black matter signal occurs.

We conclude that the work in [141, 142] provides a proof-of-concept for model-independent tests. It demonstrates that the tests have a good sensitivity to various hypothetical new physics signals. Even if the test and the reference model do not deviate in large regions of the feature space (which is common for applications in particle physics), deviations in a small region of the feature space may still be detected by the model-independent tests. We should however keep in mind that real-world data is unlikely as clean as the simulated data and that the suggested methods may require some further improvements to be successfully applied to real data. Furthermore, parametrized test with neural networks tend to overfit for high-dimensional feature spaces with sparse data points, and regularization methods might be needed to avoid this problem [141].

Part III
Representation Learning for Physical Discoveries

In this part, we focus on an essential sub-process of model creation, namely extracting compact representations describing a physical system from experimental data using minimal prior knowledge about physics or maths. To achieve this, we map a simplified version of the physicist's reasoning process, which we refer to as *SciNet*, to an artificial neural network structure and train the network to learn "natural" representations for the given data.

In Chap. 8, we provide a mathematical formulation of the simplified physical reasoning process and introduce several mathematical criteria that a "natural" representation of physical data should satisfy (Sect. 8.4). These criteria are not tailored to a specific physical system, but apply to any physical system in principle. Furthermore, we discuss how building some mathematical or physical prior knowledge into *SciNet*'s structure can help to interpret the found representation in a mathematical framework (Sect. 8.5) or as a classification of physical properties (Sect. 8.6).

In Chap. 9, we describe the artificial neural network structures that allow to find representations of data with the desired mathematical properties. The structures used for this purpose resemble the simplified model we are using to describe a physicist's reasoning process, and we refer to all of these network structures as *SciNet* for simplicity.

We then describe applications of *SciNet* to different physical toy examples from classical as well as quantum mechanics in Chap. 10. The results show that *SciNet* finds the same parameters in its representation that are used by physics textbooks in the considered examples (Sects. 10.2 and 10.6). Furthermore, it is demonstrated that conceptual information can be extracted from the found representations. In Sect. 10.4, it is demonstrated for a collision experiment of two bodies how conservation laws are related to the parameters found in *SciNet*'s representation. In Sect. 10.5.1, it is shown how to use *SciNet* to decide if a set of measurements is sufficient to fully specify the state of a quantum system, and finally, in Sect. 10.7, *SciNet* points towards a heliocentric solar system by switching from angular data of planet orbits collected on Earth to a representation of heliocentric angles. In addition, Sect. 10.3 shows that *SciNet* can linearize the dynamics of the nonlinear pendulum using mathematical prior knowledge about Koopman operator theory. Finally, in Sect. 10.8, *SciNet* is applied to a physical system consisting of several objects to classify the interaction type between particles.

We conclude with some open questions and directions for future work on representation learning for physics in Chap. 11.

Chapter 8
Theory: Formalizing the Process of Human Model Building

In this chapter, the process of physical model creation is formalised. Physical models rely on representations of physical systems. Crucially, we would like to find "natural" representations, since this simplifies the process of building physical models based on the parameters of the representation. We start with introducing a very broad formalization of what we consider to be an experimental setting and observations. Then, we define *minimal representations* (with respect to a set of questions), which incorporate the natural requirement that a representation should be as compressed as possible but still contain the information about the considered physical system that is relevant to answer all questions that might be asked. Further, we introduce two criteria for separating the parameters in a minimal representation in a natural way, leading to the definition of *statistically independent* and *operationally meaningful* representations. The criteria for operationally meaningful representations derive from an operational requirement on minimizing the communication between different physicists to achieve their tasks. In addition, in physics we often prefer representations that evolve in time according to a simple update rule. For example, a linear time evolution of the position x of a particle moving with constant velocity v is described by a simple update rule $x \mapsto x + v\Delta t$ for a time difference Δt (where e.g. storing x^2 in each time step would require a more complex update rule). We formalize this requirement on a representation and also consider a special form of update rules motivated by Koopman operator theory in more detail. Finally, we describe how to find representations of systems consisting of several interacting objects.

8.1 Motivation

As discussed in the introduction, we group the research process of a physicist into five processes: (1) Choosing a physical subsystem; (2) Creating experimental setups and (3) performing measurements; (4) Creating models and (5) testing them. In this

R. Iten, *Artificial Intelligence for Scientific Discoveries*,
https://doi.org/10.1007/978-3-031-27019-2_8

work, we focus on one part of the creation of models from experimental data, namely, finding "natural" representations of physical systems. This step is crucial to make progress towards an interpretable artificial intelligence agent, because compact and "natural" representations of data build the basis to find the mathematical equations that describe the behavior of the system, and hence to make progress in understanding the laws of our physical environment.

8.2 Physicist's Reasoning Process

Modeling the complete reasoning process of a physicist is an extremely complex task and way beyond the scope of this book. However, one may split the process of generating models based on experimental data into three parts:

1. Finding the relevant physical parameters necessary to predict the behavior of a system,
2. finding the equations describing the system's behavior (using the relevant parameters) and
3. generalizing the equations and unify them with already known equations.

Here we focus on finding the relevant physical parameters. As discussed in Sect. 1, we again consider the simplified model *SciNet* for the physical reasoning process given in Fig. 1.1. Recall that an encoder $E : \mathcal{O} \rightarrow \mathcal{R}$ maps observations to representations $\mathcal{R}$, followed by a decoder $D : \mathcal{R} \times \mathcal{Q} \rightarrow \mathcal{A}$ that maps the representation together with a question to an answer. Crucially, we have left to formalize what we exactly understand to be a "natural" representation. We discuss this in the following sections.

8.3 Experimental Setting and Data Creation

Before we formalize what we understand under a "natural" representation of a physical system, let us explain what we consider to be an "experimental setting" and how it can be used to generate experimental data (following the definition given in [23]). We take a very general view, and represent a state of Nature by a set of hidden parameters[1] $\mathcal{H} \subset \mathbb{R}^n$ for some $n \in \mathbb{N}$. Note that any kind of information can be encoded into a set of real numbers. Then, we define:

- An *experiment* is a mapping S from a space of input parameters $\mathcal{X}$ (think of the setting described by the positions of the dials and buttons in an experiment) to a subset of the hidden parameters $\mathcal{H}' \subset \mathcal{H}$. An *experimental setting* is an instance of an experiment with specified parameters, i.e., the map S together with an $x \in \mathcal{X}$.

[1] The expression "hidden parameters" is not to be confused with the "hidden variables" appearing in Bell's theorem [159].

- An observation is a stochastic mapping $G : \mathcal{H} \mapsto \tilde{\mathcal{O}}$, where $\tilde{\mathcal{O}} \subset \mathbb{R}^m$ is the set of possible measurement outputs.

We assume that the value of the hidden parameters that do not lie in $\mathcal{H}'$ are constant for different experimental settings. Hence, for different experimental settings, we can consider an observation to be an stochastic mapping (which we again denote by G) from $\mathcal{H}'$ to $\tilde{\mathcal{O}}$. Further, since the hidden parameters $\mathcal{H}$ are not directly observable, it is more operational to consider the concatenated stochastic map $V = G \circ S : \mathcal{X} \mapsto \tilde{\mathcal{O}}$. Nonetheless, for illustrative reasons, we will sometimes use the concept of the hidden parameters in the following sections.

Example Let us again consider a particle moving with constant speed along the x-axis. In an experiment, a physicist has access to three dials that regulate the mass m, the the kinetic energy E and the initial position x_0 of the particle. Crucially, the physicist does not have to know which parameters are regulated by the dials. The hidden parameters would contain the information about all properties of the particle, like the mass, color, shape, and so on. An observer then measures the position $x(t_i)$ of the particle at N different times t_i. Hence, the map V for the considered example would be $V(m, E, x_0) = ((t_0, x(t_0)), \ldots, (t_N, x(t_N)))$ with $x(t_i) = x_0 + vt_i$ where $v = \sqrt{2E/m}$. One could then generate different data samples by choosing different experimental settings, i.e., different settings of the dials regulating m, E and x_0, and observe time series of the position of the particle for each setting.

Since we allow for a stochastic map V, this model also works for theories with probabilistic observations such as quantum mechanics. There, the hidden parameters contain the information of the wave function of a physical system, and the predictions are sampled according to the distribution predicted by the Born rule (see Sect. 10.5 for a detailed discussion). In such cases, one may want to perform a measurement several times for a fixed experimental setting x, since one observation may not provide enough information about the underlying distribution $P_V(\tilde{o} \in \tilde{\mathcal{O}}|x)$ that corresponds to the stochastic map V. Hence, an observation corresponding to an experiment with parameters x may be described by a parametrization of the distribution $P_V(\tilde{o} \in \tilde{\mathcal{O}}|x)$. For example, the distribution P_V of a stochastic map V with two possible outcomes $\tilde{O} = \{0, 1\}$ and $P_V(0|x) = p(x)$, $P_V(1|x) = 1 - p(x)$ for any function $p : \mathcal{X} \mapsto [0, 1]$ can be described by the single parameter $p(x)$. This parameter can be estimated by choosing the experimental setting x and performing the measurement n times. For large n, we then have $p(x) \approx n_0/n$, where n_0 is the number of times the output 0 occurred. We refer to the value set of such parametrizations by $\mathcal{O}$ (and denote it simply by the "set of observations"). In the case of a deterministic mapping V, the distribution $P_V(\tilde{o} \in \tilde{\mathcal{O}}|x)$ ist fully described by its output o with $P_V(o|x) = 1$. Hence, in this case we can set $\mathcal{O} = \tilde{\mathcal{O}}$.

8.4 Criteria for Operationally Meaningful Representations

Have you ever wondered why physicists work with variables like the mass of a particle or the frequency of a pendulum? Sure, we have mathematical formulas using these parameters to predict the systems behavior. But assuming no background in maths, are there more fundamental reasons why we work with these quantities?

In this section we formalize different desirable criteria for "natural" representations of physical systems, based on the ideas introduced in [22, 23]. Later, in Sect. 10, we will demonstrate with several toy examples that these criteria lead to representations commonly used in physical textbooks. Hence, we gain some more fundamental understanding why physicists work with certain variables.

8.4.1 *Minimal Representation*

One requirement for a representation of experimental data is that it contains sufficient information for answering all questions that are asked about the underlying physical system. Naturally, one would like to have just the necessary amount of information in the representation and remove redundandy. Apart from saving memory, removing redundancy often also simplifies the interpretation of the representation. In the picture described in Sect. 8.3, we would like to find the minimal amount of hidden parameters that are required to correctly reply to the asked questions. In this section, we formalize these criteria for the case of real valued data and parameters that are stored in the representation.

We consider real-valued data $(\mathcal{O}, \mathcal{Q}, a^\star)$, with a set of observations $\mathcal{O} \subseteq \mathbb{R}^r$, a set of questions $\mathcal{Q} \subseteq \mathbb{R}^s$ and as set of answers $\mathcal{A} \subseteq \mathbb{R}^t$, for some $r, s, t \in \mathbb{N}$. Note that the questions are encoded into a sequence of real parameters before being fed to the neural network. Thereby, the actual representation of a single question is irrelevant as long as it enables the network to distinguish questions that require different answers. We recall that the function $a^\star : \mathcal{O} \times \mathcal{Q} \to \mathcal{A}$ maps an observation $o \in \mathcal{O}$ and a question $q \in \mathcal{Q}$ to the correct answer $a \in \mathcal{A}$. Here, we assume that the observations contain enough information to correctly reply to all questions $q \in \mathcal{Q}$. If this would not be the case, the function $a^\star$ would not exist, and hence there could also not be a representation of the observations that contains sufficient information.[2]

[2] If the observations would not contain enough information to reply to all questions, the prediction accuracy of *SciNet* would be theoretically bounded. Hence, even for optimal encoder and decoder mappings E and D, *SciNet* could not always reply correctly to all the questions. Therefore, observing low prediction accuracy after training *SciNet*, one can conclude that the observational data is incomplete (see Sect. 10.5.1 for an example).

Definition 8.1 *Sufficient representation (with smooth decoder) [22]* A representation[3] $\mathcal{R}$ (defined by an encoder mapping $E : \mathcal{O} \to \mathcal{R} \subset \mathbb{R}^p$) for the data described by the triple $(\mathcal{O}, \mathcal{Q}, a^\star)$ is called a *sufficient representation* if there exists a smooth map $D : \mathcal{R} \times \mathcal{Q} \mapsto \mathcal{A}$, such that $D(E(o), q) = a^\star(o, q)$ for all observations $o \in \mathcal{O}$ and questions $q \in \mathcal{Q}$.

This definition asserts that the encoder map E encodes all information of the observation $o \in \mathcal{O}$ that is necessary to reply to all questions $q \in \mathcal{Q}$.

Definition 8.2 *Minimal representation [22]* We call a sufficient representation $\mathcal{R} \subset \mathbb{R}^p$ a *minimal representation* if there is no other sufficient representation $\mathcal{R}' \subset \mathbb{R}^{p'}$ with $p' < p$.

This formalizes what we consider to be a "maximally compressed" representation, i.e., a representation requiring the minimal number of real parameters, but that still encodes all relevant information. Without the assumption that the decoder is smooth, it would, in principle, always be sufficient to have a single latent variable, since a real number can store an infinite amount of information. Hence, methods from standard information theory, like the information bottleneck [38, 160, 161], are not the right tool to give the number of variables a formal meaning. In Appendix A, we use methods from differential geometry to show that the number of variables p in a minimal representation corresponds to the number of degrees of freedom (hence, the hidden parameters in the setting given in Sect. 8.3) in the observation data required to answer all possible questions.

8.4.2 Separation of Physical Parameters

The notion of a *minimal representation* introduced in Sect. 8.4.1 does only consider the number of parameters and the information contained in a representation, however, it remains unclear by which criterion the different parameters should be separated. For example, feeding the decoder time series $(t_i, x(t_i))_{i \in \{1,\dots,N\}}$ of particles moving with constant velocity, a minimal representation would encode the velocity v and the start position x_0 to be able to predict the future time evolution of the particle's position. A priori, we could store any two parameters a, b from which one can recover the parameters v and x_0. For example, we could store $a = x_0 + v$ and $b = x_0 - v$ in the representation (considering x_0 and v as real numbers with no physical units assigned). It is easy to see that such a representation would still be a minimal representation, since it still contains the full information in two parameters: we recover x_0 as $x_0 = (a + b)/2$ and v as $v = (a - b)/2$. However, for a physicist, it would be much more natural to store x_0 and v as separate parameters. Why do we consider x_0 and v as a "good" choice of variables?

[3] For simplicity, we abuse notation and use the symbol $\mathcal{R}$ to refer to the set of possible values for the representation, as well as to the representation of observations $\mathcal{O}$ that is defined by an encoder mapping $E : \mathcal{O} \to \mathcal{R}$.

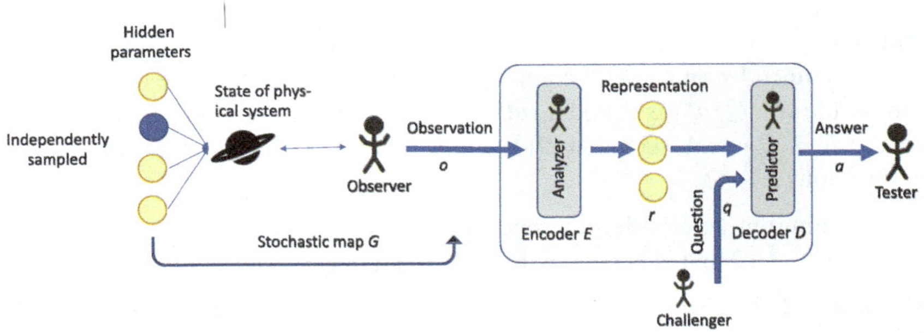

Fig. 8.1 Setting for disentangling statistically independent parameters. We assume that there is a small set of hidden parameters which are sampled independently and fully determine the state of a physical system. Note that the number of parameters (depicted as filled circles) in the figure is not representative. The map G from the hidden parameters to the measurement outputs of an observing agent is allowed to be stochastic. The observing agent then gives the observation o further to an analyzing agent (described by an encoding map E), whose task is to recover the relevant hidden parameters r that are necessary to reply to the questions asked by a challenging agent. The predicting agent that answers a question q based on the representation r is described by a decoder mapping D. Finally, a testing agent compares the answer a of the predicting agent with the correct answer obtained from performing measurements on the environment. Note that in practice, e.g. the analyzer, predictor and challenger could be represented by a single physicist

In the following sections, we introduce different requirements for a "natural" separation of the parameters in the representation. In the context of machine learning, one usually refers to the task of separating variables as *disentangling* latent variables, and to the requirements as *priors*. Later, in Sect. 10 , we will then apply these criteria to different examples and demonstrate that they lead to the separations commonly used by physicists.

8.4.2.1 Statistical Independence

A natural requirement on the parameters in the representation is statistical independence, which is also a common prior in representation learning.

> Different explanatory factors of the data tend to change independently of each other in the input distribution, (...) [27].

Although this criterium usually discovers quite "natural" features in various data structures (see for example [22, 50]), it has the disadvantage that it depends on the sampling method of the training data. Indeed, statistical independence over the training data ensures that knowing one parameter in the representation does not tell us anything about the others. But it remains unclear if this is only true under the specific method that was used to sample the training data or if the found parameters are more fundamentally "independent" features of Nature.

Here we notice that the process of an experimenter collecting data and the process of analyzing the data are mutually dependent on each other. The setting considered in this section is shown in Fig. 8.1, where one assumes that the hidden parameters are sampled independently when creating the observation data. We will consider an operational setting in conjunction with an alternative criterium for disentangling in Sect. 8.4.2.2. This alternative disentangles features based on an operational meaningful criterium that is independent of the sampling strategy used to generate the training data.

Let us now formalize the criterium of independent parameters. We consider real-valued data, which we think of being sampled from an unknown probability distribution. In other words, we assign random variables to the observations O, the questions Q, the latent representation R, and the answers A. We use the convention that a random variable $X = (X_1, \ldots, X_{|X|})$ takes samples in $\mathcal{X} \subset \mathbb{R}^{|X|}$, where $|X|$ denotes the dimension of the ambient space of X.

Definition 8.3 *(Statistically independent representations [22])* A representation R (defined by an encoder mapping $E : \mathcal{O} \to \mathcal{R}$) for the data described by the triple $(O, Q, a^\star)$ is called a *statistically independent representation* if the random variables $\{R_1, R_2, \ldots, R_{|R|}\}$ are mutually independent.[4]

In praxis, we will never find perfectly independent parameters in the representation. Hence, to quantify the independence of the parameters in the representation, we would like to use a correlation measure $C(R_1, R_2, \ldots, R_{|R|})$ with the following properties:

1. $C(R_1, R_2, \ldots, R_{|R|}) = 0$ if and only if $R_1, R_2, \ldots, R_{|R|}$ are statistically independent,
2. $C(R_1, R_2, \ldots, R_{|R|}) \geq 0$ for arbitrary distributions $p_{R_1, R_2, \ldots, R_{|R|}}$, and
3. $C(R_1, R_2, \ldots, R_{|R|})$ is continuous in the distribution $p_{R_1, R_2, \ldots, R_{|R|}}$.

A measure satisfying all of these properties is the total correlation. To define it, we first define the Kullback-Leibler divergence, which can be seen as a distance measure between the distributions of random variables (however, it is not a metric since it does not satisfy the triangular inequality).

Definition 8.4 *(Kullback-Leibler divergence)* For the distributions P_X and P_Y of two continuous random variables X and Y, where P_X is absolute continuous with respect to P_Y, the Kullback-Leibler divergence $D_{\mathrm{KL}}(P_X \| P_Y)$ is defined as

$$D_{\mathrm{KL}}(P_X \| P_Y) = \int_{-\infty}^{\infty} P_X(x) \log\left(\frac{P_X(x)}{P_Y(x)}\right) \mathrm{d}x \,. \tag{8.1}$$

Note that we use the convention that $0 \log 0 = 0$, which ensures continuity of the Kullback-Leiler divergence because $\lim_{x \to 0+} x \log x = 0$.

---

[4] The distribution for the random variables R_i is induced by the distribution over the observations and the encoder mapping E.

Definition 8.5 *(Total Correlation)* For a given set of random variables $\{R_1, R_2, \ldots, R_{|R|}\}$, the total correlation $C(R_1, R_2, \ldots, R_{|R|})$ is defined as

$$C(R_1, R_2, \ldots, R_{|R|}) = D_{\mathrm{KL}}\left(P_{R_1,R_2,\ldots,R_{|R|}} \| P_{R_1} P_{R_2} \ldots P_{R_{|R|}}\right). \tag{8.2}$$

Intuitively, the total correlation quantifies the dependency or redundancy among a set of several random variables. It measures the "distance" from the joint distribution of the random variables to the independently distributed random variables with marginal distributions.

The total correlation allows us to summarize the desired and practically applicable properties on a statistically disentangled representation in the following definition.

Definition 8.6 *(Minimally correlated representation)* A representation R for the data described by the triple $(O, Q, a^\star)$ is called a *minimally correlated representation* if and only if:

1. the representation R is minimal and
2. there is no other representation R' with

$$C(R'_1, R'_2, \ldots, R'_{|R|}) < C(R_1, R_2, \ldots, R_{|R|}).$$

Remark 8.1 *(Relation to Principal Component Analysis)* One of the most prominent methods to extract uncorrelated features from given data is Principal Component Analysis (PCA). The method performs a linear transformation on the data to extract uncorrelated features (see Chap. 2 and Fig. 2.2). Hence, we could understand PCA as an autoencoder whose encoder (and decoder) perform linear operations. For normal distributed input data, PCA is optimal, in a sense that it recovers the underlying statistically independent features. However, linearly uncorrelated features might still be statistically dependent in general. In such cases, non-linear transformations are required to find the statistically independent features. Since neural networks can in principle approximate any continuous function arbitrarily well (for large enough network sizes) [43, 44], implementing the encoder and decoder with neural networks allows for more complex transformations than PCA.

8.4.2.2 Efficient Communicable

In this section, we present an operational method to disentangle the parameters of a representation that is not biased by the sampling strategy of the data collection process [23]. To illustrate the idea, let us consider the following physical setting: There are two charged particles with masses m_1, m_2 and charges q_1, q_2, respectively. Agent B_1 can perform measurements on the first charge, and agent B_2 on the second charge. The task of agent B_1 is to predict the trajectory of the mass m_1, which depends on m_1, q_1 and q_2, but not on the mass m_2 (and vice versa for agent B_2).

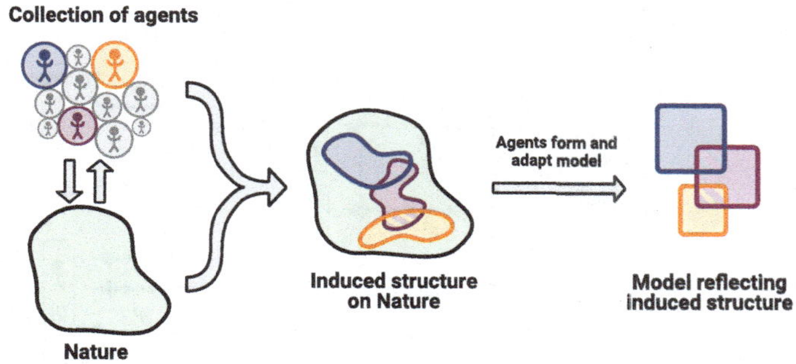

Fig. 8.2 Induced structure on Nature trough communicating agents (Figure taken from Nautrup, Metger and Iten et al., 2020 [23, 237]). Various agents interact with different aspects of Nature through experiments. Different agents may observe or interact with different subsystems, but might not have access to other parts of the environment. To solve certain tasks, agents may require information that can only be accessed by other agents. Therefore, the agents have to communicate. To this end, they encode their information into a representation that can be communicated efficiently to other agents, i.e., they have to find an efficient "language" to share relevant information.This induces structure on the description of Nature by the agents. When the agents build a model of Nature, i.e., learn to parameterize their experimental settings, we want this model to reflect the structure enforced by the requirement that agents compress and communicate parameters in the most efficient way

The agents have to communicate with each other and share information. Instead of sharing all the measurement data, we would like the agents to only share the relevant information. To do so, the agents first create a representation of their data and then share the relevant subset of their parameters that are stored in the representation. This is motivated by human interaction by communicating in an "abstract" language. To minimize the number of parameters the agent B_2 has to communicate to B_1, the charge q_2 should be stored separately from m_2. Indeed, if for example $p_1 = q_2 + m_2$ and $p_2 = q_2 - m_2$ would be stored in two parameters (in some fixed units) instead, the agent B_2 would have to communicate both parameters p_1 and p_2 to agent B_1, such that agent B_1 has enough information to predict the trajectory of the first charge. We conclude that motivating efficient communication in the context of given tasks for different agents poses operationally meaningful structure on the separation of physical parameters, as shown in Fig. 8.2.

We stress that the separation of the parameters depends on the chosen task. If agent B_1 would have to predict $q_2 + m_2$, then it would be optimal for agent B_2 to store the parameters $p_1 = q_2 + m_2$ and $p_2 = q_2 - m_2$. However, from a physical perspective, this task seems not be particularly useful. We define an "operationally meaningful parameter" as one that is useful in various tasks of interest. That is, while the parameters are biased by the tasks we would like to solve, they are not biased by how the measurement data was collected.

Let us formalize this approach. We consider several agents that interact with Nature (by performing experiments and collecting observations) and communicate with each other. The process of how agents could analyze a physical system together

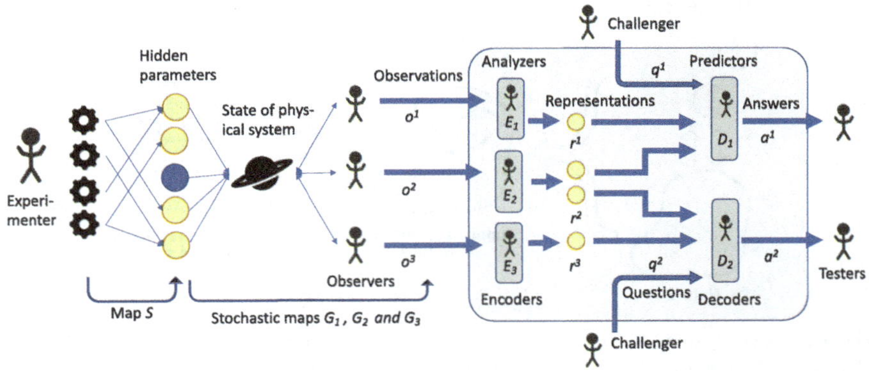

Fig. 8.3 Setting for disentangling operational meaningful parameters. One (or several) experimenters interact with the environment and choose an experimental setting (e.g. by choosing values for dials and buttons). A function S maps such a setting to a set of parameters describing the state of a physical system (the number of parameters depicted is not representative). Three observers have access to different parts of the physical system and based on the stochastic maps G_1, G_2 and G_3, the observations o^1, o^2 and o^3 are sampled. Then, analyzers (given by the encoding maps E_1, E_2, and E_3) produce compressed representations r^1, r^2 and r^3 of observations o^1, o^2 and o^3, respectively. The parameters in the representations are communicated to two predictors (described by the decoding maps D_1 and D_2), whose task is to reply to the questions q^1 and q^2, respectively. In general, some parameters of the representation may be shared with several decoders. The separation of the parameters in the representations is optimized in a way, such that a minimal amount of parameters is required by each of the agents. This criteria for separating parameters is motivated by minimizing the communication between different agents

is shown in Fig. 8.3. A first agent, the experimenter, has chosen an experiment described by the map $S : \mathcal{X} \mapsto \mathcal{H}$.[5] A number of m different observers $G_i : \mathcal{H} \mapsto \mathcal{O}^i$ then collect observations o^i of the physical system for an experimental settings x (and hence, hidden variables $S(x) \in \mathcal{H}$), and we denote the concatenated observations by $o = (o^1, \dots, o^m)$. The set of all concatenated observation is denoted by $\mathcal{O} \subset \mathcal{O}^1 \times \dots \times \mathcal{O}^m$. Note that in general $\mathcal{O}$ is a proper subset of $\mathcal{O}^1 \times \dots \times \mathcal{O}^m$, since the measurement results of different observers may be related to each other. Analyzing agents $E_i : \mathcal{O}^i \mapsto \mathcal{R}^i$ then encode the given experimental data o^i to compressed representations r^i. The analyzing agents then decide which parameters to communicate with whom of the predicting agents. We formalize the selection process as follows: For m analyzing agents, we concatenate all the representations r^i as $r = (r^1, \dots, r^m) \in \mathcal{R} \subset \mathcal{R}^1 \times \mathcal{R}^2 \times \dots \mathcal{R}^m$. Then, for each predicting agent D_j, we define a filter function $\phi_j : \mathcal{R} \mapsto \tilde{\mathcal{R}}^j \subset \mathbb{R}^{d_j}$ that selects a subset of d_j components from the vector $r \in \mathcal{R}$. The predicting agents $D_j : \tilde{\mathcal{R}}^j \times \mathcal{Q}^j \mapsto \mathcal{A}^j$ then get challenged with questions q^j chosen by challengers from sets $\mathcal{Q}^j$ and output answers $a^j \in \mathcal{A}^j$. Testing agents then perform experiments to provide the correct replies

[5] One could also consider several experimenters, however, this setting could be reduced again to the one having only one experimenter without changing the analysis in this work.

$(a^j)^\star (S(x), q^j)$ for hidden variables $S(x)$ and a question q^j. Intuitively, each filter should only chooses the part of the representation found by the analyzing agents that is required to reply to the questions asked by the challenging agents.

In the example considered at the start of this section, the agent B_1 and B_2 take the role of the observing, analyzing and predicting agents. Concretely, we have two observing agents G_1 and G_2, corresponding to the agents B_1 and B_2, that encode their observations (as analyzing agents E_1 and E_2) into the masses and charges m_1, q_1 and m_2, q_2, respectively. The analyzing agents then communicate m_1, q_1, q_2 and m_2, q_2, q_1 to the predicting agents D_1 and D_2, which again correspond to the agents B_1 and B_2, respectively.[6] In other words, defining the concatenated representation $r = (r_1, r_2, r_3, r_4) = (m_1, q_1, m_2, q_2)$, the two filters ϕ_1 and ϕ_2 map r to (r_1, r_2, r_4) and (r_2, r_3, r_4), respectively. The predicting agents, again corresponding to the agents B_1 and B_2, are then asked by challenging agents (which are different agents than B_1 and B_2) to predict the trajectory of the particles.

Remark 8.2 *(Attention mechanisms)* The formalization of the communication between agents given above could also be interpreted as an *attention mechanism*. Indeed, considering the extreme case where all the agents appearing in the communication scenario are the same, i.e., represent one agent only, the filters could be considered as an attention mechanism that focuses on the information required to answer a specific question. Such mechanisms have been investigated in depth for natural language processing [162–164]. In [162], an attention mechanism is introduced for machine translation. An encoder produces a sequence of annotations representing the information about an arbitrary input sentence. Then, an attention mechanism is used to filter (or weight) the annotations in accordance with their current relevance. The resulting context vector can be used by a decoder to produce a translation. This attention mechanism has been extended to multiple encoder-decoder models for multilingual neural machine translation [163] and summarization [164]. While the methods in these works have a strong focus on improving the outputs of the decoders, in this book, we are mainly interested in how such methods can bring further structure into the latent representation.

To define a *operationally meaningful representation*, we define the concatenation of n questions by $q = (q^1, \ldots, q^n) \in \mathcal{Q} \subset \mathcal{Q}^1 \times \cdots \times \mathcal{Q}^n$, and the concatenation of n answer by $a = (a^1, \ldots, a^n) \in \mathcal{A} \subset \mathcal{A}^1 \times \cdots \times \mathcal{A}^n$. Further, we note that under the assumption that there exists a sufficient representation (i.e., that the data collected by the observing agents is sufficient to reply to all questions), we can consider the function $a^\star : \mathcal{O} \times \mathcal{Q} \mapsto \mathcal{A}$ providing the (concatenated) correct answers a for the (concatenated) questions q as a function of the (concatenated) observations instead

---

[6] The parameters m_1, q_2 and m_2, q_2 are sent from agent B_1 and B_2, respectively, to themselves. Therefore, on the high-level picture considered at the start of this section, we do not only minimize the communication between B_1 and B_2, but also the "communication" between the agents themselves, i.e., we motivate to minimize the number of parameter that are used to solve the agent's tasks. In this specific case, where the analyzing agents are the same as the predicting agents, minimizing "communication" thus also corresponds to the motivation of the agents to concentrate on the relevant part of their representations (see Remark 8.2).

of the hidden variables. We denote the list $(\phi_1, \ldots, \phi_n)$ of filter functions $\phi_j : \mathcal{R} \mapsto \tilde{\mathcal{R}}^j \subset \mathbb{R}^{d_j}$ by ϕ, and the output dimension of ϕ_j by $\dim(\phi_j)$. Before we define an operationally meaningful representation, we need to extend the definition of a minimal representation (Definition 8.2) to multiple encoders. The generalization is straightforward, but it is required because using multiple encoders may increase the number of parameters in a minimal representation compared to the case where one uses one encoder (see Sect. 10.6.2 for an example).

Definition 8.7 *(Minimal representation with multiple encoders)* A representation $\mathcal{R} \subset \mathbb{R}^p$ (defined by encoders $E_1, \ldots, E_m$) for the data described by $(\mathcal{O}, \mathcal{Q}, a^\star)$ with $\mathcal{O} \subset \mathcal{O}^1 \times \cdots \times \mathcal{O}^m$ is called a *minimal representation (with m encoders)* if and only if

1. it is sufficient, i.e., there exists a smooth decoder D with $D(r = (r^1, \ldots, r^m), q) = a^\star(o = (o^1, \ldots, o^m), q)$ for all possible observations $o \in \mathcal{O}$, questions $q \in \mathcal{Q}$, and where we defined $r^i = E_i(o^i) \in \mathcal{R}^i$, and
2. there is no other sufficient representation $\mathcal{R}' \subset \mathbb{R}^{p'}$ with $p' < p$.

Definition 8.8 *(Operationally meaningful representation [23])* A representation $\mathcal{R}$ (defined by encoders $E_1, \ldots, E_m$) for the data and filters described by the quadruple $(\mathcal{O}, \mathcal{Q}, a^\star, \phi)$ is called a *operationally meaningful representation* if and only if:

1. the representation $\mathcal{R}$ is minimal and
2. there is no other minimal representation $\mathcal{R}'$ with a selection function ϕ', such that $\sum_j \dim((\phi')^j) < \sum_j \dim(\phi_j)$.

The first requirement ensures that the complete concatenated representation consists of a minimal number of parameters such that it still contains enough information to reply to all questions $q \in \mathcal{Q}$ that might be asked. The second requirement motivates a separation of the parameters in the representation in a way that minimizes the number of parameters sent to each of the decoders.

Remark 8.3 *(SciNet in the context of reinforcement learning)* In this book, we ask *SciNet* to predict different physical quantities based on given observation data collected from a physical environment. From a machine learning perspective, this task is categorized as representation learning. In general, there are tasks that require several interactions with the environment to be solved. For example, the task of moving a ball into a hole by hitting it with a stick (which may be a familiar task for readers paying golf), may require several steps. Further, if we do not hit the hole with the first stroke, it might be unclear how to quantify how "useful" the stroke was. In other words, we may not have a cost function that can be evaluated after each stroke, but we only get a reward if we finally hit the hole. This is a typical setting for reinforcement learning (see Chap. 2). Considering a reinforcement leaning environment and several agents with different tasks, we can combine *SciNet* with techniques from reinforcement learning to construct minimal representation of the state of the environment that is disentangled according to the agent's tasks: Each decoder of *SciNet* can be considered as a separate reinforcement learning agent that has to solve a certain task in its environment with help of additional information received from the representation found by encoders. Instead of learning the representation and the strategy in parallel, it might however be technically simpler to first learn the optimal strategy, and then train *SciNet* with a data set generated by letting a trained agent interact with the environment (see [23] for an example and further details). In other words, the trained agent provides the correct predictions for *SciNet*, and hence a cost function that can be evaluated for each input to *SciNet*.

8.4.2.3 Statistical Independence Versus Operationally Meaningful

In the previous sections, we have described two different methods for disentangling parameters in representation learning. Depending on the situation one approach may be preferable over the other. In operationally meaningful representations, the disentangling is operationally motivated by minimizing the number of parameters communicated between agents. However, to fully disentangle a representation consisting of k parameters, we would need at least k different "natural" questions about the physical system. If we are not aware of sufficiently many such questions, we may instead search for statistically independent parameters, which are biased by the data collection process. Alternatively, one may combine the methods by first disentangling according to questions, and then, disentangle further the resulting sub-representations by motivating statistical independence.

8.4.3 Simple Update Rules

Besides minimality and being disentangled, it is often desirable for physical representations that there are simple rules to obtain predictions from them. For example, it is often desirable that given a representation, there is a simple mathematical formula that evolves the representation in time. To formalize this, let us think about a set $\mathcal{U}$ of (simple) update rules $u : \mathcal{R} \mapsto \mathcal{R}$ acting on a representation $\mathcal{R}$. Further, we consider a specific form of questions $q = (t, \tilde{q})$, where $t \in \mathbb{N}_0$ denotes the number of times the update rule should be applied and the question $\tilde{q} \in \tilde{\mathcal{Q}}$ can be any question referring to the system after applying the updating rule t times.

Definition 8.9 *(Representation with simple update rule [22])* A representation $\mathcal{R}$ for the data described by the triple $(\mathcal{O}, \mathcal{Q} = \mathbb{N}_0 \times \tilde{\mathcal{Q}}, a^\star)$ (generated by an encoder $E : \mathcal{O} \mapsto \mathcal{R}$) is called a *simple representation with respect to a set of update rules* $\mathcal{U}$ if and only if:

1. the representation R is minimal and
2. there is an update rule $u \in \mathcal{U}$ and a decoder mapping $D : \mathcal{R} \mapsto \mathcal{A}$, such that

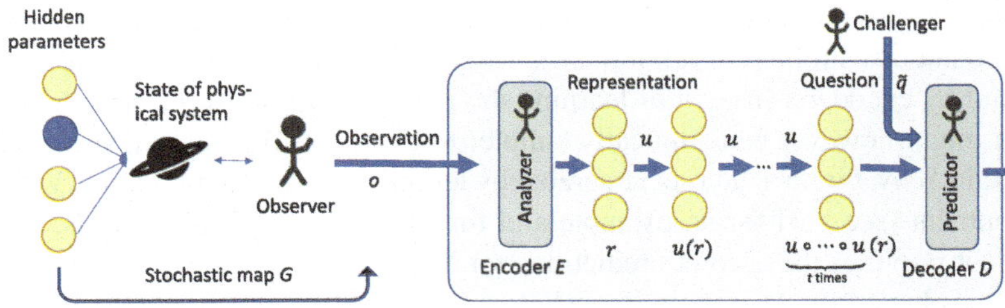

Fig. 8.4 Setting for searching parameters that evolve according to a simple update rule. A small set of hidden parameters fully determine the state of a physical system (the number of parameters depicted in the figure is not representative). The state of the system depends on a parameter $t \in \mathbb{N}_0$ (e.g., the time could correspond to $(\epsilon t + t_0)$s for some $\epsilon > 0$ and initial time t_0). We want to find a representation of the physical system that allows for a simple evolution $u : \mathcal{R} \mapsto \mathcal{R}$ mapping the representation of the system for a parameter t to the representation corresponding to a parameter $t + 1$. Hence, we consider questions of the form $q = (t, \tilde{q})$, i.e., we ask a question $\tilde{q}$ about the state of the system corresponding to the parameter t. To do so, an analyzer encodes an observation o into a representation r, that is then evolved t times by the simple update function u. Then, a decoding agent is asked the question $\tilde{q}$ and his answer a is checked by a testing agent

$$D(\underbrace{u \circ \cdots \circ u}_{t \text{ times}} \circ E(o), \tilde{q}) = a^\star(o, q)$$

for all observations $o \in \mathcal{O}$ and questions $q = (t, \tilde{q}) \in \mathcal{Q}$.

8.5 Criteria for Mathematically Meaningful Representations

In Sect. 8.4, we focused on "natural" requirements on representation of physical systems that do not use any mathematical or physical prior knowledge. However, in certain scenarios, we may want to use some mathematical prior knowledge to search for representations that can be interpreted within a mathematical framework. In this section we describe how to use deep learning together with Koopman operator theory [165, 166] to find mathematical descriptions of the dynamical behavior of non-linear dynamics [167–172]. This investigation serves as an example to demonstrate how mathematical knowledge can be used to describe mathematically meaningful representations.[7]

[7] An alternative method for discovering a representation of a system together with the equations describing its nonlinear dynamics is presented in [173]. The models used in [173] have the fewest terms necessary to describe the dynamics, balancing model complexity with descriptive ability, and thus promoting interpretability and generalizability. Furthermore, they also fit in the framework considered here, i.e., they can be considered as a method to find mathematically meaningful representations.

8.5.1 Koopman Operator Theory

Koopman operator theory was introduced in 1931 in [165, 166]. It is a leading candidate for a systematic approach to linearize nonlinear dynamics [174, 175]. There have been several recent advances in this field, in theory [174–178] as well as in related numerical methods such as dynamic mode decomposition [179–181]. Furthermore, an increasing amount of available measurement data supports applications of the data driven Koopman theory. In the following, we provide a short introduction into Koopman operator theory. For a more detailed recent review on this topic we refer to [182].

We consider a dynamical system whose state is described by a variable $o(t) \in \mathcal{O} \subset \mathbb{R}^n$ at time t and its dynamics by the vector field $f : \mathcal{O} \mapsto \mathbb{R}^n$ together with the differential equation

$$\dot{o}(t) = f(o(t)) . \tag{8.3}$$

The discrete time evolution for time Δt can be described by a function $F : \mathcal{O} \mapsto \mathcal{O}$, given by

$$o_{k+1} = F(o_k) , \tag{8.4}$$

where $o_k = o(k\Delta t)$ is the state of the system at time $k\Delta t$ with $k \in \mathbb{N}_0$. We are interested in finding the function F that describes the discrete-time dynamics of the considered system from given experimental data. A lot of physical systems of interest allow for a compact representation as differential equations in $o(t)$. However, in most cases apart from linear dynamics, the differential equations are analytically not trackable.

Koopman operator theory does not consider the evolution of the state $o(t)$ of the system, but rather the evolution of observables derived from $o(t)$ and hence described by real valued maps $g : \mathcal{O} \mapsto \mathbb{R}$ on the state space. Let us consider a box containing some ideal gas as an example. The state $o(t)$ of the system is fully described by all the positions and velocities of the particles in the gas. An example for an observable would be the pressure of the gas, which can be seen as a function of the sate $o(t)$. The *Koopman operator*, $\mathcal{K}$, is then the linear and infinite dimensional operator that advances measurement functions for a discrete time step Δt

$$\mathcal{K}g = g \circ F \implies \mathcal{K}g(o_k) = g(o_{k+1}) . \tag{8.5}$$

The linearity of the *Koopman operator* is a direct consequence of the linearity of the composition operation because for observables $g_1, g_2 : \mathcal{O} \mapsto \mathbb{R}$, we have

$$\mathcal{K}(g_1 + g_2) = (g_1 + g_2) \circ F = g_1 \circ F + g_2 \circ F = \mathcal{K}g_1 + \mathcal{K}g_2 . \tag{8.6}$$

So, the good news are that we replaced non linear dynamics with a linear operator $\mathcal{K}$; the bad news are that the operator $\mathcal{K}$ acts on a infinite dimensional space. To understand the dynamics described by $\mathcal{K}$, it can help to restrict the operator to finite dimensional subspaces that are invariant under the action of $\mathcal{K}$. Any subspace L (over the complex numbers) that is spanned by eigenfunctions of $\mathcal{K}$ is an invariant subspace, because for an n-dimensional subspace spanned by eigenfunctions $\phi_1, \ldots, \phi_l : \mathcal{O} \mapsto \mathbb{C}$ with eigenvalue $\lambda_1, \ldots, \lambda_l \in \mathbb{C}$ we have

$$\mathcal{K}(\alpha_1\phi_1 + \cdots + \alpha_l\phi_l) = \alpha_1\lambda_1\phi_1 + \cdots + \alpha_l\lambda_l\phi_l \in L\,, \tag{8.7}$$

for any complex coefficients $\alpha_1, \ldots, \alpha_l \in \mathbb{C}$. Hence, the Koopman operator restricted to a subspace spanned by eigenvectors allows for a linear finite dimensional representation of the dynamics on this subspace. Since we have a comprehensive theory for linear systems, such a restriction has the potential to enable advanced prediction and control of nonlinear systems.

8.5.2 Representation of Koopman Eigenfunctions

Obtaining eigenfunctions of the Koopman operator in practical applications has proven challenging. In this section we describe (based on [183]) how neural networks can be used to find these eigenfunctions. Searching for Koopman eigenfunctions can be considered as searching for transformations $\phi_1, \ldots, \phi_l : \mathcal{O} \mapsto \mathbb{C}$ of observations $o \in \mathcal{O}$, such that $r = (\phi_1(o), \ldots, \phi_l(o))$ evolves in time Δt as $r \mapsto (\mathcal{K}\phi_1(o), \ldots, \mathcal{K}\phi_l(o)) = (\lambda_1\phi_1(o), \ldots, \lambda_l\phi_l(o)) =: \mathcal{K}(\lambda)(\phi_1(o), \ldots, \phi_l(o))$, for some complex eigenvalues λ_i and where we defined $\mathcal{K}(\lambda) : \mathbb{C}^l \mapsto \mathbb{C}^l$ by $\mathcal{K}(\lambda)(\alpha_1, \ldots, \alpha_l) = (\lambda_1\alpha_1, \ldots, \lambda_l\alpha_l)$ for complex numbers $\alpha_1, \ldots, \alpha_l$. Defining the (complex valued) encoder $E : o \mapsto r = (\phi_1(o), \ldots, \phi_l(o))$, we can consider this setting as a special case of the one considered in Sect. 8.4.3, where we described representations that allow for a simple update rule u (see Fig. 8.4). Indeed, if we set $u = \mathcal{K}(\lambda)$ and choose the questions to be of the form $q = (k, \tilde{q})$, where $\tilde{q}$ is the (fixed) question "What is the state o_k at time $k\Delta t$ of the system?", Definition 8.9 (generalized for complex valued encoders) ensures that the encoder of the corresponding representation implements a minimal number of Koopman eigenfunctions $\phi_1, \ldots, \phi_l$, such that the resulting representation $r = (\phi_1(o), \ldots, \phi_l(o))$ is sufficient to recover the state o.

A wide range of physical systems requires a Koopman operator with continuous spectrum for its accurate description [166]. For example the frequency of a (non linearized) pendulum, whose angular displacement θ satisfies the differential equation $\ddot{\theta} = -\omega_0^2 \sin(\theta)$ for some $\omega_0 \in \mathbb{R}$, continuously decreases as the energy of the pendulum is increased. In such cases, approximations by a finite sum of a few eigenfunctions might not be accurate enough to be useful for a better understanding of the underlying system. One method that helps to handle continuous spectra and that is described in [183] is to allow the eigenvalues to depend on the representation r,

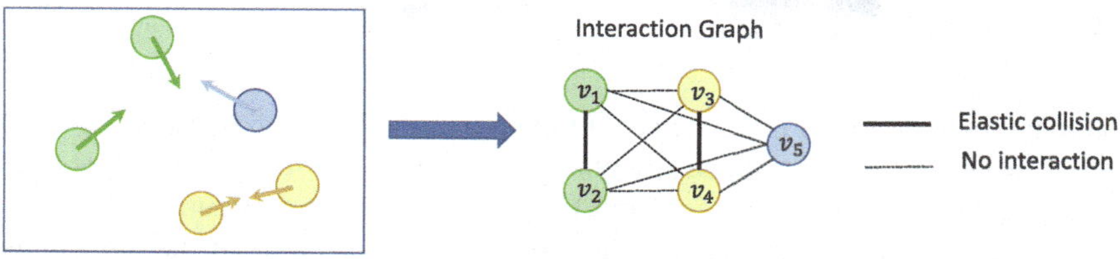

Fig. 8.5 Interaction graph for a toy model of moving balls. Five balls move within a box with different velocities (and collide elastically with the wall of the box). We consider the toy model of balls colliding with the same color elastically, whereas balls of different colors do not interact. From observing the time evolution of the system of the five balls, one can create an interaction graph that represents the information on how the five objects interact with each other. The vertices $v_1, \ldots, v_5$ of the graph represent the balls and the edges between the balls describe their interactions

i.e. we consider a Koopman operator $\mathcal{K}(\lambda)$ with $\lambda \equiv \lambda(r)$. In [183], this method is successfully applied to linearize the dynamics of the nonlinear pendulum (Sect. 10.3) and high dimensional nonlinear fluid flow.

8.6 Criteria for Physically Meaningful Representations

To simplify the interpretation of *SciNet*'s representation and to improve its generalization power, we may sometimes want to build some physical prior knowledge into the machine learning setup. There are a lot of options for the choice of prior knowledge like symmetries or pairwise interactions of subsystems, and also ways of implementation. We focus on an example that should serve as a demonstration of how such prior knowledge can be built into *SciNet* and how it can help to interpret the found representation. In this section we consider systems consisting of several objects (or more generally "parts") that interact with each other. We assume that our observation data is structured in a way that we know how it relates to the objects, in other words, we assume that an observation $o \in \mathcal{O}$ of k objects is of the form $o = (o_1, \ldots, o_k)$ for real valued vectors o_i. Different approaches have been considered to describe the behavior of such systems [184–188]. Here, we focus on the approach described in [184], which does require minimal prior knowledge about how objects interact with each other and on the form of the observation data o_i. Furthermore, it allows to find a description of the interaction structure given as an *interaction graph*. In the following sections, we describe what an interaction graph is and then how to extract a representation of it for dynamical systems.

8.6.1 Interaction Graph

Mathematically, a graph is defined as a tuple $(\mathcal{V}, \mathcal{E})$ of a set of vertices $v \in \mathcal{V}$ and edges $e = (v, v') \in \mathcal{E}$ that can be thought as connections between two vertices $v \in \mathcal{V}$ and $v' \in \mathcal{V}$ (see Fig. 8.5 for an example). Such a structure can naturally describe a set of interacting objects by assigning vertices $v_1, \ldots, v_n \in \mathcal{V}$ to the objects and edges $e_{i,j} = (v_i, v_j)$ to the action of the object assigned to v_i on the one assigned to v_j. However, to also describe the properties of the objects and kind of interactions, we work with embeddings of real vectors h_i and $h_{i,j}$ of the vertices and edges, respectively. For example, let us consider a toy model with n balls with positions x_i, velocities w_i and colors c_i. Hence, the embeddings of the vertices could simply be $h_i = (x_i, w_i, c_i)$. Let us assume that there are only two types of interactions between the balls: two balls with the same color collide fully elastically, whereas balls with different colors do not interact at all. Hence, the embedding of the edges could be $h_{i,j} = h_{j,i} = 1$, if $c_i = c_j$ and $h_{i,j} = h_{j,i} = 0$ if $c_i \neq c_j$. The interaction graph for such a system is shown in Fig. 8.5.

8.6.2 Representation of Interaction Graph

The goal of this section is to formalize the process of extracting the type of interactions, i.e., the embeddings $h_{i,j}$ of the edges of the interaction graph, from observation data of a dynamic system (following the approach in [184]). The number of objects may thereby vary between different observations. Let us consider time sequences $o_i = [x_i(t_l)]_{l \leq T} = x_i(t_1), \ldots, x_i(t_T)$ of length T of the d-dimensional states $x_i(t_l) \in \mathbb{R}^d$ of the object with label i at fixed times $t_1, \ldots, t_T$.

We would like to infer the structure of the interaction graph corresponding to the interactions between the objects described by the observations $[x_i(t_l)]_{l \leq T}$. This process can be considered to have the basic structure of *SciNet* as shown in Fig. 1.1. However, usually we consider real valued observations of fixed dimension, whereas here the set of observations $\mathcal{O}$ is the set of sequences of time sequences $[x_i(t_l)]_{l \leq T} \in \mathbb{R}^d \times \mathbb{R}^T$, hence $\mathcal{O} \subset \{[s_i]_{i \leq k} : s_i \in \mathbb{R}^d \times \mathbb{R}^T, k \in \mathbb{N}\}$. Now, an encoder should be able to map observation data of any number of objects to a representation that describes the interactions of the objects, and hence whose size scales quadratically in the number of objects k. More formally, the encoder should map an observation $o \in \mathcal{O}$ for k objects to a representation r that encodes all embeddings $h_{i,j}$ with $i, j \in \{1, \ldots, k\}$ and $i \neq j$. Assuming that the interactions can be fully characterized by an e-dimensional real vector $h_{i,j} \in \mathbb{R}^e$, we then have $r \in \mathbb{R}^{k(k-1)} \times \mathbb{R}^e$. For simplicity, we use double indices for the representation such that $r_{i,j} = h_{i,j}$. Note that such an encoder with variable input and output size can not trivially be implemented using a feedforward neural network. Instead, one can use *graph neural networks* (GNNs) to handle these difficulties, as described in Sect. 9.7.

We then ask *SciNet* to predict the future time evolution of all the objects, given an initial state $o(t_0) = o_1(t_0), \ldots, o_k(t_0)$ at some time t_0 and based on the found representation encoding the kind of interactions between all the objects. To be able to predict the time evolution with high accuracy far into the future (such that all the objects will interact once), *SciNet* must have inferred all the necessary information about the interactions and stored this information in the latent representation r. Hence, such a structure can be used to learn conceptual information about which objects interact with each other and which parameters describe these interactions. We describe in Sect. 9.7 how such representations can be found with GNNs [184].

Chapter 9
Methods: Using Neural Networks to Find Simple Representations

In the previous chapter, we have formalized what we consider to be a "simple" representation of physical data by introducing the notion of minimal *statistically independent* and *operpationally meaningful* representations, as well as representations allowing for a simple update rule and representation of systems consisting of several interacting objects. In this chapter, we discuss how neural networks can be used to find these kind of representations directly from a given data set (following the ideas described in [22, 23]). Apart from the given data and the theoretical requirements on representations found in the previous chapter, we focus on minimizing any further prior knowledge built into the machine learning system. Hence, the network architectures introduced in this chapter can be applied to a wide range of data.

9.1 Motivation

We would like to build software that is able to learn the different kinds of representations described in Sect. 8.4 directly from given data. To achieve this, the main difficulty is to find a function that encodes the observations into a representation that satisfies the properties formalized in Sects. 8.4, 8.5 and 8.6. One crucial property we require for any representation is that it is *sufficient*, i.e., that it contains enough information to reply to a set of questions. To check if a representation is sufficient, in general one has to search over decoding functions that map the representation and a question to an answer (Fig. 1.1). If one finds a decoder that outputs the correct answers for all questions, one can be sure that the found representation contains all necessary information. We conclude that we have to search for an encoder and decoder in parallel. This can be efficiently done by using artificial neural networks, which are a powerful tool to learn functions directly from data. Note that for many applications of neural networks, it is crucial that they can generalize to input samples that significantly differ from the samples seen during the training. However, for this work, neural networks are mainly used as a function-approximating tool.

R. Iten, *Artificial Intelligence for Scientific Discoveries*,
https://doi.org/10.1007/978-3-031-27019-2_9

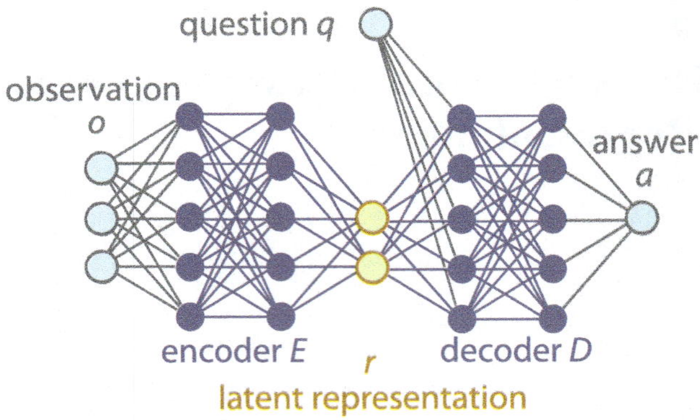

Fig. 9.1 Network structure for *SciNet* motivated by the human physical reasoning process (Figure taken from Iten and Metger et al., *Physical Review Letters*, 2020 [22]). Observations are encoded as real parameters fed to an encoder (a *feed-forward neural network*), which compresses the data into a representation (*latent representation*). The question is also encoded in a number of real parameters, which, together with the representation, are fed to the decoder network to produce an answer (The number of neurons depicted is not representative.)

9.2 General Network Structure to Learn Representations

The modelling process described in Sect. 8.2 can be translated directly into a neural network architecture (Fig. 9.1), which we refer to as *SciNet* in the following [22]. The encoder and decoder are both implemented as feed-forward neural networks. The resulting architecture, except for the question input, resembles an autoencoder in representation learning [27, 189], and more specifically the architecture in [190]. Note that if there is a finite set of questions, one could in principle replace the question input by asking the decoder to reply to all the questions in each run. However, when working with continuous variables describing the questions, such a strategy would not work.

During the training, we provide triples of the form $(o, q, a^\star(o, q))$ to the network, where $a^\star(o, q) \in \mathcal{A}$ is the correct reply to question $q \in \mathcal{Q}$ given the observation $o \in \mathcal{O}$. With this we train the encoder E_Θ and the decoder D_Φ, where Θ and Φ denote the collection of weights and biases of the encoder and the decoder, respectively. The training process of *SciNet* to learn minimal representations is summarized in Box 1. To find (an estimate of) the minimal number of parameters in the representation, one simply raises the number of parameters until the answer quality of *SciNet* does no longer increase significantly.[1] In this way, a representation consisting of a minimal

[1] One may want to use more efficient search algorithms over the number of parameters, in particular in the case where an initial estimate on the minimal number is given.

number of parameters can be found. As discussed later, there are more efficient methods to achieve this (see Remark 9.2). The learned parameterization (defined through the encoder E_Θ) is typically called latent representation [27, 189]. It is crucial that the encoder is completely free to choose a latent representation itself, instead of us imposing a specific one. Because neural networks with at least one hidden layer composed of sufficiently many neurons can approximate any continuous function arbitrarily well [191], the fact that the functions E and D are implemented as neural networks does not significantly restrict their generality. In other words, we consider significantly large networks E_Θ and D_Φ, so that there is always a choice of parameters Θ and Φ, such that E_Θ and D_Φ are close to the optimal functions for the encoder and the decoder. However, unlike in an autoencoder, the latent representation need not describe the observations completely; instead, it only needs to contain the information necessary to answer the questions posed.

Box 1: Train *SciNet* to learn minimal representations (Section 9.2)

Goal: Find a minimal representation for a given data set $\mathcal{D}$ consisting of triples of the form $(o, q, a^\star(o, q))$, where $a^\star(o, q) \in \mathcal{A}$ is the correct reply to question $q \in \mathcal{Q}$ given the observation $o \in \mathcal{O}$.

Network Structure: Shown in Figure 9.1. We denote the parametrized encoder mapping by $E_\Theta : \mathcal{O} \mapsto \mathcal{R}$, and the decoder mapping by $D_\Phi : \mathcal{R} \times \mathcal{Q} \mapsto \mathcal{A}$.

Cost Function: For a triple $(o, q, a^\star(o, q))$, the cost is given by $\mathrm{d}\left[D_\Phi(E_\Theta(o), q), a^\star(o, q)\right]$, for some (smooth) distance measure $\mathrm{d}[\cdot, \cdot]$.

Splitting of the data set: We split the data set $\mathcal{D}$ into a training, validation and a test set.

Training process: To find a minimal representation, we start with zero latent neurons and successively increase the number of latent neurons. For each number of latent neurons, we train the network with the training data and validate the precision of the predictions with the validation data. When the accuracy no longer increases significantly, the increase in the number of latent neurons is stopped. [A more efficient method to find the minimal number of required parameters is presented in Box 2.]

Testing: We test the performance of the trained *SciNet* using the test set.

Remark 9.1 In some cases, the neural network E_Θ implementing the encoder may have difficulty finding the perfect encoding. Either because it is limited in expressing the optimal encoding function (because the number of hidden neurons is not large enough or the encoding function is not continuous as demonstrated in Sect. 10.5.1) or because the training process does not converge to a solution that is close to the optimal encoding. In such cases, an alternative way to estimate the minimum number of parameters is to apply the Levina-Bickel algorithm [192].

Extracting information from the network. The architecture in Fig. 9.1 allows us to extract knowledge from the neural network: the relevant information is stored in the representation, and the size of this representation is small compared to the total number of degrees of freedom of the network. This helps to interpret the learnt representation. Specifically, we can compare *SciNet's* latent representation to a hypothesized parameterization to obtain a simple map from one to the other. If we do not even have any hypotheses about the system at hand, we may still gain some insights solely from the number of required parameters or from studying the change in the representation when manually changing the input, and the change in output when manually changing the representation (as in e.g. [49]).

9.3 Network Structures for Separating Parameters

A lot of work in representation learning focuses on methods to disentangle representations, i.e., to store "natural" features in separate neurons in the latent space (see e.g. [49, 193]). Usually, these works focus on the construction of a representation of the full observation input. In contrast, *SciNet* focuses on a representation that is sufficient to reply to all the questions that might be asked. Nonetheless, it is straightforward to use the methods introduced in the machine learning literature to separate parameters in *SciNet*'s representation. However, most of these methods depend on the distribution over the training data, and new methods were developed in [23] for operational disentangling, i.e., methods to disentangle parameters relevant for solving different physical tasks (see Sect. 8.4.2.2 for the details).

9.3.1 *Statistically Independent*

As described in Sect. 8.4.2.1, one may define a representation to be disentangled if the parameters in the representation are statistically independent over the given data set. To motivate *SciNet* to find minimal statistically independent representations, we could in principle use the same technique as described in Box 1 and add the total correlation $C(R_1, \ldots, R_{|R|})$ between the latent neurons as an additive term to the cost function. However, this training technique would be computationally extremely expensive. There are two reasons for this: Firstly, retraining *SciNet* for different number of latent neurons is costly. Secondly, and more importantly, estimating the total correlation during the training is computationally extremely expensive, since this measure depends on the distribution over the latent neurons, which in turn requires a lot of samples to be well approximated.

Box 2: Train *SciNet* to learn statistically independent representations (Section 9.3.1)

Goal: Find a statistically independent representation for a given data set $\mathcal{D}$ consisting of triples of the form $(o, q, a^\star(o, q))$, where $a^\star(o, q) \in \mathcal{A}$ is the correct reply to question $q \in \mathcal{Q}$ given the observation $o \in \mathcal{O}$.

Network Structure: Essentially the one shown in Figure 9.1 (see Figure B.1 in Appendix B for details). We denote the parametrized encoder mapping by $E_\Theta : \mathcal{O} \mapsto \mathcal{R}$, and the decoder mapping by $D_\Phi : \mathcal{R} \times \mathcal{Q} \mapsto \mathcal{A}$.

Desired Cost Function (computational infeasible): For $x = (o, q, a^\star(o, q))$, the desired cost function would be $\mathcal{L}(x) = \mathrm{d}\left[D_\Phi(E_\Theta(o), q), a^\star(o, q)\right] + \alpha\, C(R_1, \ldots, R_{|R|})$, for some (smooth) distance measure $\mathrm{d}[\cdot, \cdot]$, a hyperparameter α, and where $R_1, \ldots, R_{|R|}$ are the random variables for the latent neurons (determined by the encoder E_Θ and the distribution over the observations used as training data).

Computational Feasible Cost Function: β-VAE cost function given in (B.1) in Appendix B.

Splitting of the data set: We split the given data set $\mathcal{D}$ into a training, validation and test set.

Training process: To find a statistically independent representation, we choose a number of latent neurons that is expected to be larger or equal than the number requires for a minimal representation. We train *SciNet* with the β-VAE cost function, where we have to optimize the hyperparameter β. If we still have a high prediction error for the validation data, we may have chosen too few latent neurons in the beginning, and we repeat the process with a higher number of latent neurons.

Testing: We evaluate the prediction loss of the trained *SciNet* using the test set. Further, we would like to evaluate the total correlation of the latent neurons over the test set. This is computational feasible, because we only have to calculate it once after training. If the total correlation is still high, we could train *SciNet* again with a higher value of β to motivate the disentangling of the latent neurons more strongly.

Because of these difficulties, we follow the approach taken in [22] here: We use the techniques of beta-VAEs [49], i.e. of variational autoencoders (see Appendix B) with an adapted cost function that is meant to help disentangle the latent representation. In practice, beta-VAEs often find representations with statistically independent latent neurons. Further, if there are unnecessarily many latent neurons, superfluous neurons tend to be set to a constant zero-activation. However, since there is no guarantee that beta-VAEs always find statistically independent parameters, we verify this a posteriori by calculating the total correlation of the latent variables after finishing the training of *SciNet*. Since the details of how beta-VAEs work will not be relevant to understand this thesis, we explain them in Appendix B. The training process is summarized in Box 2.

9.3.2 *Operationally Meaningful*

In this section, we describe the approach introduced in [23] to find operationally meaningful representations (see Definition 8.8). Recalling the process given in

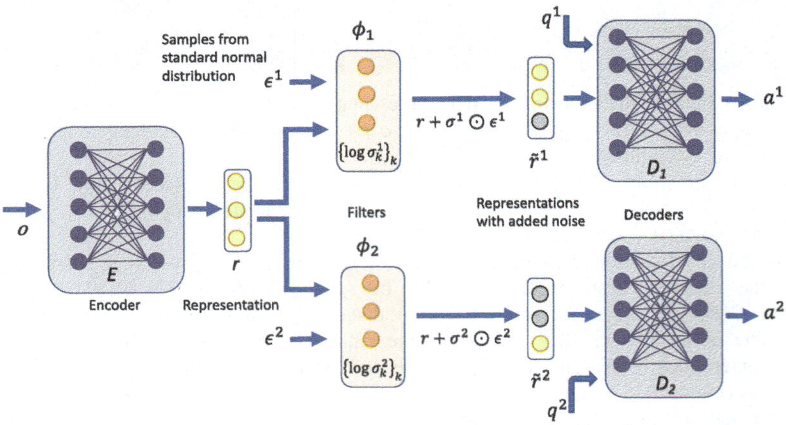

Fig. 9.2 Network architecture to learn operationally meaningful representations. An encoder gets an observation o and outputs a representation r. (In principle, one could consider several encoders, but for simplicity we only use one here). Then, for each decoder D_j, a filter ϕ_j is applied to the representation that adds Gaussian noise to the latent neurons r_k with a standard derivation σ_k^j. The sampling is done using the *reparameterization trick* [193], i.e., in each run of the network we sample $\epsilon_k^j \sim \mathcal{N}(0, 1)$ and provide these samples as an input to the filters. Then, the values $\tilde{r}_k^j = r_k + \sigma_k^j \epsilon_k^j$ are given to the decoder D_j, whose task is to output the response to the question q^j. The number of parameters and decoders in the figure is not representative. Technically, the filters are described by the trainable parameters $\{\log \sigma_k^j\}_k$. If a filter ϕ_j adds a lot of noise to a latent neuron k, there is essentially no information about the value r_k transmitted to the decoder D_j by $\tilde{r}_k^j$. Hence, such neurons (marked in grey in the figure) are considered to be filtered out by ϕ_j, i.e., the information contained in r_k^j is not accessible by the decoder D_j

Fig. 8.3, it is straightforward to replace the encoders and decoders with neural networks. The challenging part of the implementation are the filter functions. Indeed, these functions have to select which values of the latent neurons are sent to which of the decoders. Hence, they are of a binary nature, where for each latent neuron the filter can be "on" or "off" for a certain decoder D_j, meaning that the value of this latent neuron is or is not sent to the decoder D_j. To train such filters, we would like to "smooth" them, so that we can propagate gradients trough them while running the back-propagation training algorithm. We can smooth the filters based on the reparametrization trick [193], as described in the following. For simplicity, let us assume that we work with a single encoder. The implementation described below is summarized in Fig. 9.2.

Box 3: Train *SciNet* to learn operationally meaningful representations with a single encoder (Section 9.3.2)

Goal: Find operationally meaningful representations for a given data set $\mathcal{D}$ consisting of triples of the form $(o, (q^1, q^2, \ldots, q^n), (a_1^\star(o, q^1), a_2^\star(o, q^2), \ldots, a_n^\star(o, q^n))$, where $a_j^\star(o, q^j) \in \mathcal{A}^j$ is the correct response to question $q^j \in \mathcal{Q}$ given the observation $o \in \mathcal{O}$.

Network Structure: Shown in Figure 9.2. We denote the parametrized encoder mapping by $E_\Theta : \mathcal{O} \mapsto \mathcal{R}$, the filters by $\phi_{\sigma^j} : [0, 1]^n \times \mathcal{R} \mapsto \mathcal{R}$, where n denotes the number of latent neurons $\dim(\mathcal{R})$, and the decoder mappings by $D_{\Omega^j} : \tilde{\mathcal{R}}^j \times \mathcal{Q}^j \mapsto \mathcal{A}^j$.

Cost Function For a triple $x = (o, (q^1, \ldots, q^n), (a_1^\star(o, q^1), \ldots, a_n^\star(o, q^n))$, noise samples $\epsilon^j = (\epsilon_1^j, \ldots \epsilon_n^j)$ with $\epsilon_k^j \sim \mathcal{N}(0, 1)$, and outputs $a^j = (a_1^j, \ldots, a_{\dim(\mathcal{A}^j)}^j)$ for a decoder D_{Ω^j}, i.e., $a^j = D_{\Omega^j}\left[\phi_{\sigma^j}\left(\epsilon^j, E_\Theta(o)\right), q^j\right]$, the cost is given by

$$\mathcal{L}(x, \{\epsilon^j\}_j) = \sum_j \mathrm{d}\left[a^j, a_j^\star(o, q^j)\right] - \beta \sum_k \log(\sigma_k^j), \tag{9.1}$$

for some (smooth) distance measure $\mathrm{d}[\cdot, \cdot]$ and a hyper parameter β that regulates the tradeoff between good accuracy for the predictions and minimizing the number of latent neurons required by each decoder.

Splitting of the data set: We split the given data set $\mathcal{D}$ into a training, validation and test set.

Training process: To find a minimal representation, we start with zero latent neurons and successively increase the number of latent neurons. For each number of latent neurons, we train the network with the training data and the cost function given above, where we have to optimize the hyperparameter β, and validate the precision of the predictions with the validation data. If we achieve nearly perfect precision, we stop the process of increasing the number of latent neurons. (A more efficient method is described in Remark 9.2).

Testing: We evaluate the prediction loss of the trained *SciNet* using the test set.

Implementation of filters [23]. To train a filter in a smooth way, we need to be able to make a smooth transition from the "on" to the "off" state of a filter. Clearly, if the filter is "on" for a latent neuron, all its information is transferred to the decoder, where in the "off" state, no information about the parameter stored in the latent neuron is given to the decoder. If we add noise to the latent neuron, we can smoothly regulate how much information about the latent parameter is given to the decoder. Indeed, let us formally consider the latent neuron k to be sampled from a Gaussian distribution $\mathcal{N}(r_k, \sigma_k^j)$, where r_k is the value of the k-th latent neuron (output from the encoder) and σ_k^j determines the amount of noise added to the latent parameter that is subsequently given to the decoder D_j. For the decoder D_j, the value of σ_k^j has the following meaning:

1. A large value of σ_k^j (relative to the variation of the k-th latent neuron over the observation data) means that we add a lot of noise to the value r_k of the k-th latent neuron. Hence, the decoder D_j gets essentially no information about r_k and we consider this case to be a filter that does not select the k-th latent neuron.
2. A value of σ_k^j that is close to zero means that essentially all information about r_k is transferred to the decoder D_j, so the filter selects this latent neuron.

Although we can smoothly regulate the noise, we have to sample from the Gaussian distribution in each run of the network. Sampling is a non-differentiable process in the parameters in the set Θ, describing the weights and biases of the encoder. We can use the *reparameterization trick* introduced in [193] (for variational auto encoders) to solve this problem: we replace the sampling operations by using auxiliary random numbers $\epsilon_k^j \sim \mathcal{N}(0, 1)$ as inputs to the neural network. Then, the k-th latent variable with the filter applied for a decoder D_j and distributed with $\mathcal{N}(r_k, \sigma_k^j)$ can be generated by $r_k + \sigma_k^j \epsilon_k^j$. Using vector notation, we get the complete filter output for decoder D_j as a vector $r + \sigma^j \odot \epsilon^j$, where $r = (r_1, \ldots, r_m), \sigma^j = (\sigma_1^j, \ldots, \sigma_m^j)$ and $\epsilon^j = (\epsilon_1^j, \ldots, \epsilon_m^j)$ for m latent variables, and where we use the symbol "$\odot$" for component-wise multiplication. Sampling ϵ_k^j does not interfere with the gradient descent because ϵ_k^j is independent of the trainable parameters Θ.

Note that in our implementation (Fig. 9.2), we use $\log(\sigma_k^j)$ as trainable parameters of the filters instead of σ_k^j, because $\log(\sigma_k^j)$ can take on positive or negative real values, whereas σ_k^j must always be positive. Hence, this replacement simplifies the training, because we do not have to ensure positivity of these values in each training step. In addition, the values of σ_k^j can get very large during training and the logarithm maps these values into a reasonable range.

Cost function for minimizing the information transmitted by filters. As stated in Item 2 of Definition 8.8, we would like to minimize the number of "on"-states for the filters to minimize the information communicated between agents (see Sect. 8.4 for a detailed explanation of the motivation). With the implementation above, this translates into minimizing $-\sum_{j,k} \log(\sigma_k^j)$. Indeed, this cost term motivates the network to increase the noise added to the latent neurons and competes with the cost term motivating the decoders to output correct predictions. The full cost function and a description of the training process is given in Box 3.

Remark 9.2 (Minimizing the total number of latent neurons) If for a certain k, the value of σ_k^j is large for all j, the information of the k-th latent neuron is not transmitted to any of the decoders and hence could be ignored. Since we would like to find a minimal representation, we would like to maximize the number of latent neurons that are not used by any decoder under the condition that all the decoders are still able to output accurate predictions. The cost function term $-\sum_{k,j} \log(\sigma_k^j)$ does not necessarily motivate *SciNet* to minimize the total number of latent neurons. Indeed, assume that the representation has two latent neuron r_l and r_m that store the same information, and that two decoders D_i and D_j require this information. Then, the cost is the same wether the value of r_l (or r_m) is transferred to both decoders or r_m is transferred to D_i and r_l to D_j (or vice versa). To minimize the number of latent neurons, we could train *SciNet* with an increasing number of latent neurons (as described in Box 3), but this method is computationally expensive. Alternatively we could add a *global filter* parametrized by σ_k^{global} that selects a number of latent neurons before they are sent to all of the decoder specific filters ϕ_j. The cost term $-\sum_{k,j} \log(\sigma_k^j) - \gamma \sum_k \log(\sigma_k^{\text{global}})$ for some hyperparameter γ then motivates to minimize the total number of latent neurons (by motivating the global filter to select a minimal number of neurons that are sent further to the decoders). Although this method is quite efficient, it turns out to be less reliable than increasing the number of latent neurons one by one in numerical experiments, in the sense that the network gets harder to train and gets stuck more often in sub-optimal local minima. Nonetheless, the method is useful to get an estimate on the required number of latent neurons. Starting from this estimate, one can successively lower the number of latent neurons until the prediction accuracy of *SciNet* significantly decreases. When this happens, the current number of latent neurons plus one is likely the minimal number required.

9.4 Network Structure to Find Representations with Simple Update Rules

In this section, we describe how to find representations that evolve according to a simple update rule (see Definition 8.9), based on the ideas described in [22]. Recall that a decoder D is asked to reply to a question $\tilde{q} \in \tilde{\mathcal{Q}}$ after evolving the representation found by an encoder E for $t \in \mathbb{N}_0$ steps according to a simple update rule u chosen from a set $\mathcal{U}$ (Fig. 8.4). Again, it is straightforward to implement the encoder and the decoder appearing in this setting with neural networks. To speed up the training process, we ask the network to reply to a question $\tilde{q}$ after each application of the update rule. Hence, we ask the questions $(0, \tilde{q}), \ldots, (t, \tilde{q})$ in each training step (Fig. 9.3). This provides more feedback to the network and helps it to learn more quickly.

It remains to investigate how to implement the update rule. Let us assume that the functions $u \in \mathcal{U}$ are smooth. There are two technical cases that we have to distinguish:

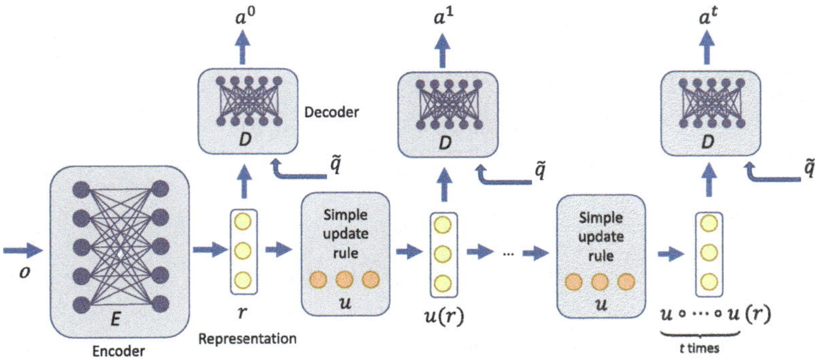

Fig. 9.3 Network architecture to learn representations that evolve according to a simple update rule. An encoder maps an observation o to a representation r that allows for a simple update rule u (think of evolving a representation of a physical system for a fixed time interval). A decoder should correctly respond to a question $\tilde{q}$ in each step. The set of questions is therefore described by the tuples $(0, \tilde{q}), \ldots, (t, \tilde{q})$, where the first element of the tuple determines the step at wich question $\tilde{q}$ is asked. The answers given by the decoder are labeled with the step number as $a^0, \ldots, a^t$. Crucially, the same decoder is used in each step. In the case where u evolves the system in time, and where $\tilde{q}$ asks to predict a physical quantity at a given time, using only one decoder means that the physical laws determining how to predict physical quantities do not change with time. The network architecture shown in the figure is similar to the one of recurrent neural networks [194]

1. The set of updates rules $\mathcal{U}$ can be smoothly parametrized by a set of parameters Γ, i.e., there is an open connected set of parameters $\Gamma \subset \mathbb{R}^l$ for some $l \in \mathbb{N}$, such that $u_\gamma(r) \in \mathcal{U}$ depends smoothly on $\gamma \in \Gamma$ for any $r \in \mathcal{R}$ and such that the sets $\{u_\gamma\}_{\gamma\in\Gamma}$ and $\mathcal{U}$ are in one-to-one correspondence.
2. The set of update rules cannot be smoothly parametrized.

In case 1, we can propagate gradients trough the update rule during the training of the neural network. Hence, the parametrized update rule u_γ can be considered as a part of the neural network, where the parameters γ are the trainable parameters analogous to the weights and biases of artificial neurons.

The training in case 2 is usually more costly and there are a lot of tricks in the machine learning literature to map it back to case 1. If, however, no such trick works for a given case, a brute-force strategy to tackle case 2 for a finite set $\mathcal{U}$ would be the following: For each $u \in \mathcal{U}$, train an encoder and a decoder to optimize *SciNet*'s prediction accuracy. If one finds a u, such that *SciNet* can predict nearly perfectly, one is done. However, if there are a lot of elements in $\mathcal{U}$, such a strategy gets computationally intractable. In the case where the set $\mathcal{U}$ can be partitioned into a small number of smoothly parametrizable subsets, one could handle these subsets separately by the strategy applied in case 1.

Box 4: Train *SciNet* to learn representations with simple update rules that can be smoothly parametrized (Section 9.4)

Goal: Find a representation for a given data set $\mathcal{D}$ consisting of triples of the form $(o, [(0,\tilde{q}), \ldots, (t,\tilde{q})], [a^\star(o,(0,\tilde{q})), \ldots, a^\star(o,(t,\tilde{q}))])$, where $a^\star(o,(s,\tilde{q}))$ is the correct reply to question $\tilde{q} \in \mathcal{Q}$ in step $s \in \mathbb{N}_0$ given the observation $o \in \mathcal{O}$.

Network Structure: Shown in Figure 9.3. We denote the parametrized encoder mapping by $E_\Theta : \mathcal{O} \mapsto \mathcal{R}$, the parametrized update rule by $u_\gamma : \mathcal{R} \mapsto \mathcal{R}$, and the decoder mapping by $D_\Omega : \mathcal{R} \times \tilde{\mathcal{Q}} \mapsto \mathcal{A}$.

Cost Function For a triple $x = (o, [(0,\tilde{q}), \ldots, (t,\tilde{q})], [a^\star(o,(0,\tilde{q})), \ldots, a^\star(o,(t,\tilde{q}))])$ and answers provided by the decoder D_Ω given as

$$a^s = D_\Omega(\underbrace{u_\gamma \circ \cdots \circ u_\gamma}_{s \text{ times}} \circ E_\Theta(o), \tilde{q}),$$

the cost is given by

$$\mathcal{L}(x) = \sum_s \mathrm{d}\left[a^s, a^\star(o,(s,\tilde{q}))\right]$$

for some (smooth) distance measure $\mathrm{d}[\cdot,\cdot]$.

Splitting of the data set: We split the given $\mathcal{D}$ into a training and test set.

Training process: To find a minimal representation, we start with zero latent neurons and successively increase the number of latent neurons. For each number of latent neurons, we train the network (including the update rule u_γ) with the training data and the cost function given above. If we achieve nearly perfect precision, we stop the process of increasing the number of latent neurons. Alternatively, one could try a more efficient method to find the minimal number of parameters, such as the ones used for beta-VAE [49] (see Appendix B).

Testing: We evaluate the prediction loss of the trained *SciNet* using the test set.

The resulting network structure shown in Fig. 8.4 is similar to standard *recurrent neural networks* [194] which are used to process sequential data in machine learning and have a wide range of applications such as machine translation [195–197].

9.5 Optimality Guarantees on the Representation

In the previous sections we introduced different methods to find "natural" representations. However, as common for methods based on neural networks, in general we do not have a guarantee that the suggested methods find the optimal representation. Nonetheless, some properties of a representation can be checked *a posteriori*:

- Sufficiency of the representation (by checking the prediction accuracy on the test data[2]).
- Statistical independence of the latent parameters (by calculating the total correlation of the latent parameters).
- The representation allows for a simple update rule (by checking the prediction accuracy after evolving the representation for several steps).

On the other hand, it is more challenging to test the following properties:

- Minimality of the representation.
- Optimality of the filters, i.e., minimality of the number of parameters that are communicated to the decoders in the setting for searching operationally meaningful representations (see Sect. 9.3.2).

These properties are defined trough the absence of any other sufficient representation that requires less parameters. Hence, to check minimality in a straightforward way, one would have to compare the found representation with all possible representations, which are infinitely many. Instead, one may for example reduce the number of latent neurons and train *SciNet* several times. If the prediction accuracy stays low, this points towards the fact that the number of latent neurons we started with is minimal. On the other hand, investigating the correlation between the latent neurons as well as the effect of varying their value on the predictions may help to better understand the found representation.

9.6 Network Structure to Find Koopman Eigenfunctions

As an example for mathematically meaningful representations, we considered representations that allow for linear dynamics in Sect. 8.5. In the case of a discrete spectrum of the Koopman operator, the setting considered there can be seen as a

---

[2] If the observations are very noisy, it might be more complex to check sufficiency of a representation.

special case of searching for representation with simple update rules. Hence, we can use the same methods as introduced in Sect. 9.4, which correspond to the methods used in [183] in this case. The only difference to Sect. 9.4 is that we are searching for complex valued representations. Since we can only represent real values with (standard) artificial neurons, we represent the complex valued latent variables $\tilde{r}_i \in \mathbb{C}$ by two real variables $r_{2i-1} \in \mathbb{R}$ and $r_{2i} \in \mathbb{R}$ corresponding to the real and imaginary part of $\tilde{r}_i$. Recall from Sect. 8.5 that each such pair corresponds to the output $\phi_i(o)$ of an eigenfunction ϕ_i of the Koopman operator for an observation input o. Further, we may write the eigenvalue for the eigenfunction ϕ_i as $\lambda_i = \exp[(\mu_i + \omega_i i)\,\Delta t]$ for coefficients $\mu_i, \omega_i \in \mathbb{R}$ and a discrete (fixed) time step $\Delta t > 0$, and denote the stacked coefficients as $\mu = (\mu_1, \ldots, \mu_m)$ and $\omega = (\omega_1, \ldots, \omega_m)$, respectively. Then, the Koopman operator $\mathcal{K}(\lambda) \equiv \mathcal{K}(\mu, \omega)$ acting on the (real) latent representation $r = (r_1, \ldots, r_{2m})$ consists of Jordan blocks $\mathcal{B}(\mu_i, \omega_i)$ of the form

$$\mathcal{B}(\mu_i, \omega_i) = e^{\mu_i \Delta t} \begin{bmatrix} \cos \omega_i \Delta t & -\sin \omega_i \Delta t \\ \sin \omega_i \Delta t & \cos \omega_i \Delta t \end{bmatrix}. \tag{9.2}$$

Hence, we have found a smooth parametrization of the set of update rules $\{u_\gamma\}_{\gamma \in \Gamma} = \{\mathcal{K}(\mu, \omega)\}_{\mu \in \mathbb{R}^m, \omega \in [0, 2\pi)^m}$. Instead of considering μ and ω as coefficients, one may allow them to depend on the state of the representation r before it gets updated. As discussed in Sect. 8.5.2, this dependence is investigated in [183] to describe the dynamics of nonlinear system with a continuous spectrum of the corresponding Koopman operator. The dependence of $\mu(r)$ and $\omega(r)$ on r can be simply integrated into the network structure shown in Fig. 9.3 by adding a neural network that gets the representation r as an input and provides μ and ω as an output. To enforce circular symmetry in the eigenfunction coordinates, one may provide $\|r\|_2^2$ as an input to the additional network instead of r itself [183].

9.7 Network Structure to Find Interaction Graphs

In Sect. 8.6, we described interaction graphs of dynamical systems. Here, we investigate how to find such graphs with neural networks following the approach in [184]. As noted in Sect. 8.6, we have to implement an encoder with a variable input and output size. Such an encoder has to respect the object structure of the considered system, otherwise it could not generalize from a system consisting of k objects to one consisting of $k + 1$ objects. One class of networks that are well adapted to inputs of a graph structure, are *graph neural networks* (GNNs) [111, 198, 199], which will be introduced in the next section.

9.7.1 Graph Neural Networks

Graph neural networks (GNNs) [111, 198, 199] allow to handle graph structures as an input and have found a wide range of applications, e.g. for predicting chemical properties [111], for control and inference for physical systems [184–186, 200–205] and for investigating social networks [206, 207]. Their expressive power was investigated in [208, 209]. In this section, we give a brief introduction to graph neural networks. There is a lot of flexibility in the structural details of graph neural networks, and here we focus on the structure used in [184]. The described structure can be used to infer the interaction structure of dynamical systems consisting of several objects under the assumption that there is a finite set of e interactions, i.e., working with discrete embeddings $h_{i,j}$ of the interactions between the objects with label i and j.

We consider a graph with embeddings h_i of its vertices. The basic idea for graph neural networks is to update these embeddings several times according to rules that depend on the graph structure, i.e., the update rule for the embedding h_i depends on how the corresponding vertex is connected to other vertices within the graph. Such update rules can be considered as message passing: the neighboring vertices of h_i send their information to h_i, which updates its values according to some rule dependent on the received information. Importantly, such a rule should be commutative, in the sense that the updated value does not depend on the order of how messages are proccessed. After L such updating steps, the updated embeddings h_i^L do not only contain information about the vertices, but also about the graph structure. If we would for example want to predict a binary property of the graph (such as ful connectedness), the embeddings h_i^L can be fed to a neural network, which is asked to classify the graph. Importantly, the neural network can only successfully classify if the update rule applied to the embeddings h_i are of a form, such that the resulting embeddings h_i^L will contain the necessary information about the graph. For a more thorough introduction to graph neural networks, we refer to [210].

Here, we are interested in inferring the type of interactions between objects from observation data of their time evolution. Hence, we also consider embeddings $h_{i,j}$ of the edges between vertices with labels i and j. The update rules are then split into two steps: i) $v \mapsto e$: updating the embeddings $h_{i,j}$ of the edges according to the values of the neighboring vertices, and ii) $e \mapsto v$: updating the embeddings h_i of the vertices according to the incoming edges $\mathcal{N}_i = \{h_{j,i}\}_j$. More concretely, we consider the update rules [210]

$$v \mapsto e: \quad h_{i,j}^l = f_e^l([h_i^l, h_j^l, y_{i,j}]) , \tag{9.3}$$

$$e \mapsto v: \quad h_i^{l+1} = f_v^l([\sum_{j \in \mathcal{N}_i} h_{j,i}^l, y_i]) , \tag{9.4}$$

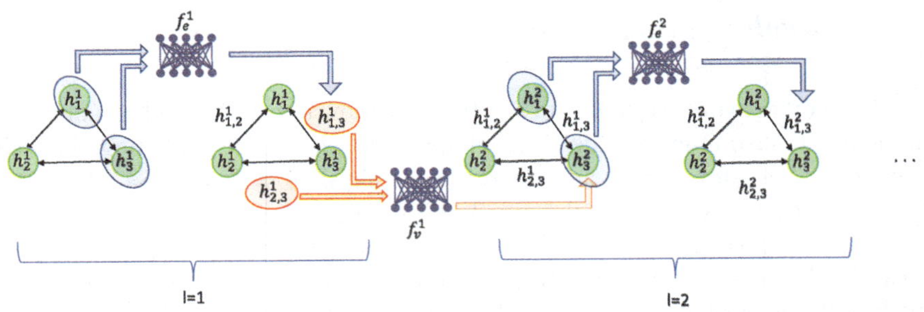

Fig. 9.4 Graph neural network. The figure shows a fully connected graph consisting of three vertices with embeddings h_1^1, h_2^1 and h_3^1. The symbol "↔" labels two directed edges pointing in opposite direction. The update rule given in (9.3) is applied on the first graph to get the embeddings $h_{i,j}^1$ of the edges with $i, j \in \{1, 2, 3\}$ and $i \neq j$. Note that we only show the edges $h_{i,j}^l$ with $i < j$ for simplicity. The update rule (9.3) takes the vertices h_1^1 and h_3^1 and auxiliary vertex features (which are not shown in the figure) as an input. Then, it applies a neural network representing the function f_e^1 to the inputs to obtain the edge embedding $h_{1,3}^1$. The other edge embeddings are found analogously (not shown in the figure). Then, the update rule (9.4) is implemented by applying a neural network f_v^1 to the vertex-embeddings $h_{2,3}^1$ and $h_{1,3}^1$ as well as to some auxiliary edge features to obtain the updated vertex embedding h_3^2. Again, the other vertices are updated analogously. This process is then repeated for $L \in \mathbb{N}$ times

where $h_{i,j}^l$ and h_i^{l+1} are the embeddings of the edges and vertices after l update steps and where $y_{i,j}$ and y_i encode initial or auxiliary edge and vertex features as for example node inputs and edge type. A GNN for these update rules is illustrated in Fig. 9.4. In the next section we describe how such GNNs can be used to learn the interaction types between objects from time series data.

9.7.2 Network Structure to Learn Interaction Graphs

In this section, we describe how to implement a network structure to learn the interaction types between objects as described in Sect. 8.6 and based on [184]. We start by constructing an encoder using GNNs that maps an observation $o = [o_1, \dots, o_k]$ of k objects with $o_i = [x_i(t_l)]_{l \leq T}$ to discrete variables $r_{i,j}$ representing the interaction type. For any labels i and j, the variable $r_{i,j}$ is a one-hot encoding of one of the e distinct interaction types.

Encoder: We use a GNN with the updates rule given in (9.3) and (9.4) for the encoder, where we start from a fully connected graph (see Fig. 9.5). A feedforward neural network f_{emb} is used to map the observations o_i to the vertex embeddings, i.e., $h_i^1 = f_{\text{emb}}(o_i)$. Then, we use the update rules (9.3) and (9.4) (with trivial auxiliary features $y_{i,j}$ and y_i) to evolve the GNN for L steps and obtain $h_{i,j}^L$ as an output.

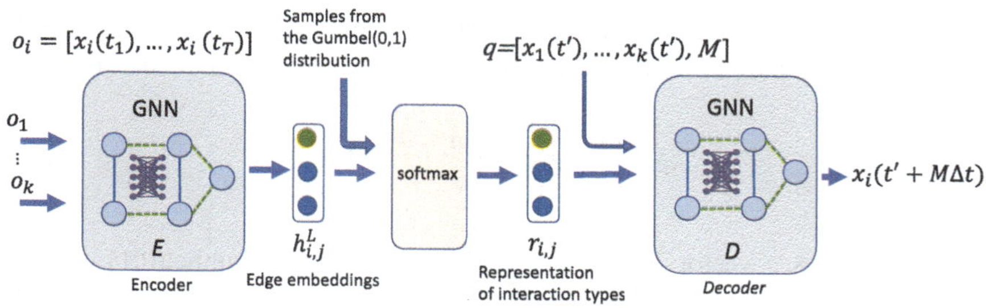

Fig. 9.5 Network structure to find interaction types. The encoder is a GNN with update rules given in (9.3) and (9.4). It takes time sequences of k objects as an input, where k may vary between different samples o, and provides edge embeddings $h^L_{i,j}$ as an output. The edge embeddings can be considered to parametrize a distribution, from which we sample continuous approximations of one-hot embedding $r_{i,j}$ of the interaction types. The stochastic input for this reparametrization trick is thereby provided by i.i.d. samples from the Gumbel(0,1) distribution. A decoder is then asked to evolve the state of all the objects starting from initial states $x_i(t')$ and based on the interaction types encoded in the representation r. The decoder is thereby implemented as a GNN with update rules given in (9.7) and (9.8)

Let us denote all parameters describing the GNN, and hence the encoding map, by Θ. To interpret the continuous output variables $h^L_{i,j}$ as discrete variables, we use a common trick in machine learning, namely considering $\mathrm{softmax}(h^L_{i,j})$ as a distribution $q_\Theta(r_{i,j}|o)$ over the discrete variable $r_{i,j}$, where the i-th component of the vector valued function $\mathrm{softmax}(a)$ of an n-dimensional real vector a is defined as

$$\mathrm{softmax}(a)_i = \frac{e^{a_i}}{\sum_{j=1}^{n} e^{a_i}} . \tag{9.5}$$

Since $\sum_{i=1}^{n} \mathrm{softmax}(a)_i = 1$, we can indeed consider the output of $\mathrm{softmax}(a)$ as a distribution over a discrete variable with n distinct values.

Sampling: Since we obtain a distribution over the representation as an output of the encoder, we have to sample from it to provide an input to the decoder. Unfortunately, sampling is a non-differentiable process in the parameters of the set Θ, and hence disallows the training of the GNNs with stochastic gradient descent. In Sect. 9.3.2, we encountered the same problem and solved it with the *reparameterization trick* [193]. The essence of this trick is to replace the sampling process by using a differentiable function of the output variables of the encoder and a random variable with fixed distribution. The samples for the random variable with fixed distribution are then provided as an input to the network. However, such a refactoring does not work for discrete random variables due to the discontinuous nature of discrete states. A recent approach to solve this problem was considered in [211], where it is suggested to approximate the discrete distribution with a continuous one and then use the reparameterization trick. More concretely, we draw samples from the distribution

$$r_{i,j} = \text{softmax}\left(\frac{h^L_{i,j} + g}{\tau}\right), \tag{9.6}$$

where $g \in \mathbb{R}^e$ is a vector of i.i.d. samples drawn from a Gumbel(0,1) distribution and the softmax temperature τ controls the strength of the smoothening. The Gambel(0,1) distribution is defined by the probability density function $f(x) = \exp(-x - \exp(-x))$. Note that, for $\tau \to 0$, the distribution converges to one-hot samples of our categorial distribution.

Decoder: The decoder takes the representation $r_{i,j}$ of the interactions as an input and a question $q = [x(t'), M]$ where $x(t') = (x_1(t'), \ldots, x_k(t'))$ describes the states of all objects at a time t' and where M is the number of time steps for which the decoder should make the state evolve for a fixed time difference Δt. Let us denote the future states predicted by the decoder for times t', $t' + \Delta t$, ..., $t' + M\Delta t$ by $x_i^0, \ldots, x_i^M$ for the i-th object, and let us initialize $x_i^0 = x_i(t')$ for all objects. We use again a GNN for the decoder with a different neural network for each type of interaction. Hence, we slightly adapt the update rules [210]:

$$v \mapsto e: \ \tilde{h}^c_{i,j} = \sum_{i,j,s} r_{i,j,s} \tilde{f}^s_e([x^c_i, x^c_j]), \tag{9.7}$$

$$e \mapsto v: \ x^{c+1}_i = x^c_i + \tilde{f}_v(\sum_{i \neq j} \tilde{h}^c_{j,i}), \tag{9.8}$$

where $r_{i,j,s}$ denotes the s-th element of $r_{i,j}$. The first update rule determines the interaction between the two objects with states x_i^c and x_j^c. In the case where the representation $r_{i,j}$ is a one-hot encoding whose a-th entry is equal to one (i.e., $r_{i,j,a} = 1$), only one decoder is applied to the input states and we have $\tilde{h}^c_{i,j} = \tilde{f}^a_e([x^c_i, x^c_j])$. Hence, there is indeed one decoder for each interaction type. The second update rule takes in a description of the interaction $\tilde{h}^c_{j,i}$ between the objects with labels i and j at the c-th time step and should provide the states x_i^{c+1} of the system as an output. Hence, applying these update rules M times should approximate the time evolution of the objects for time $M\Delta t$.

Training: The complete network shown in Fig. 9.5 can be trained with stochastic gradient descent. The loss function consists of two terms. A prediction loss given by

$$\sum_{i=1}^{k} \sum_{c=1}^{T} \left\| x_i(t' + c\Delta t) - x^c_i \right\|^2, \tag{9.9}$$

and a loss term motivated by the KL term appearing in the cost function of variational autoencoders (Appendix B) given by a sum of entropies [184]

$$\sum_{i \neq j} H(q_{\Theta}(r_{ij}|o)) \tag{9.10}$$

Chapter 10
Applications: Physical Toy Examples

In this chapter, we apply the network structures of Chap. 9 (to which we refer as *SciNet* for simplicity) to find representations of (simulated) observation data of six physical toy systems. The results demonstrate that *SciNet* can help to recover concepts in physics by providing the relevant physical parameters, both in quantum- and classical-mechanical settings. We also show that the results are robust against noise in the experimental data.

In summary, without providing any prior knowledge about the considered physical systems to *SciNet* (apart from the observational data), we find: (i) given a time series of the positions of a damped pendulum, *SciNet* can predict future positions with high accuracy and it uses the relevant parameters, namely frequency and damping factor (which are sampled independently to generate the training data), separately in two of the latent neurons (and sets the activation of unnecessary latent neurons to zero); (ii) given a time series of the angular displacement and velocity of a non-linear pendulum, *SciNet* switches to a representation that evolves linearly in time. (iii) *SciNet* finds and exploits conservation laws: it uses the total angular momentum to predict the motion of two colliding particles; (iv) given measurement data from a simple quantum experiment, *SciNet* can be used to determine the dimension of the underlying unknown quantum system and to decide whether a set of measurements is tomographically complete, i.e., whether it provides full information about the quantum state. Furthermore, operationally meaningful representations of mixed quantum states can separate local degrees of freedom in the representation from non-local ones; (v) considering observations coming from different experiments with charged particles, *SciNet* separates the charges and masses of the particles in its operationally meaningful representation; (vi) given a time series of the positions of the Sun and Mars as observed from Earth, *SciNet* switches to a heliocentric representation—that is, it encodes the data into the mean anomalies of the two planets with respect to the Sun; (vii) given time series of the positions and velocities of several particles in a box, where some of them are connected by a spring, *SciNet* classifies the interaction between the particles, i.e., *SciNet*'s representation encodes wether two particles are connected by a spring or not. The runtime to train the networks for all examples takes at most a few hours on a standard laptop.

R. Iten, *Artificial Intelligence for Scientific Discoveries*,
https://doi.org/10.1007/978-3-031-27019-2_10

10.1 Motivation

In Chap. 8, we have formalized what we consider to be a "simple" representation of a physical system and in Chap. 9, we described how to find such representations with neural networks. In this chapter, we apply these methods to several examples and analyze the extracted representations. The results show that *SciNet* finds, without having been given any prior information about the specific physical systems, the same quantities that we use in physics textbooks to describe the different settings. Hence, our formalization of "simple" representations corresponds well to what representations physicists prefer to describe data.

In all the examples, the training data we use is operational and could be generated from experiments, i.e., the correct answer is the one observed experimentally.[1] Here, we use simulations instead because we only deal with classical and quantum mechanics, theories whose predictions are experimentally well tested in the relevant regimes. One might think that using simulated data would restrict *SciNet* to rediscovering the theory used for data generation. However, in particular for quantum mechanics, we are interested in finding conceptually different formulations of the theory with the same predictions.

The specific examples considered in this chapter were chosen according to the following criteria:

- The total number of parameters that needs to be stored in the representation is small (i.e., around 5 real parameters).
- The examples should demonstrate the "usefulness" of the different criteria for representations defined in Sect. 8.4.
- Since our longterm goal is to apply the methods from Chap. 9 to better understand quantum mechanics, we also consider examples including quantum systems.
- We focus on examples where the representation contains conceptually interesting information.

The first criterium is purely technical and ensures that the interpretation of the representation does not get too cumbersome and that the training of the neural networks does not take too long on a standard laptop. Further examples, e.g. using video recordings of fire flames as observations, can be found in [192].

From a machine learning perspective, all the considered examples are similarly challenging. The Tensorflow library [212] makes setting up the network structures very simple. Usually, we work with encoders and decoders consisting of two hidden layers with around 200 neurons per layer (see Table C.1 for the details). Networks of this size were able to approximate the encoding and decoding mappings well for all the following examples.[2] The remaining difficulty was to find a "good" choice

[1] For example, one could consider a temporal sequence of positions as non-operational data, but use video recordings instead, since such observations are similar to those that humans perceive with their eyes. Although this is a valid argument, using video data as observations conceptually adds nothing to the examples considered in this chapter, but would complicate neural network training.

[2] A simple method to choose the network size is the following: Start with a small network size for *SciNet* (with enough latent neurons) and train it with the only target of minimizing the prediction

for the hyperparameters such as the parameter β appearing in the cost function 9.1 to learn operationally meaningful representations. Importantly, one does not have to know a minimal representation of the considered physical system to choose the hyperparameters (see Boxes 2 and 3). Although, as discussed in Sect. 9.5, the methods introduced in Chap. 9 are not guaranteed to find the "optimal" representation, they always find them in the following examples (without post-selecting examples, i.e., we did not post-select examples based on whether *SciNet* worked or not).

10.2 Damped Pendulum

Following the work in [22], we consider a simple example from classical physics, the damped pendulum, described in Box 5. The time evolution of the system is given by the differential equation $-\kappa x - b\dot{x} = m\ddot{x}$, where κ is the *spring constant*, which determines the frequency of the oscillation, and b is the *damping factor*. We keep the mass m constant (it is a scaling factor that could be absorbed by defining $\kappa' = \kappa/m$ and $b' = b/m$), such that κ and b are the only variable parameters. We consider the case of weak damping here, where the solution to the equation of motion is given in Box 5.

We generate the training data by uniformly sample the spring constant κ and the damping factor b. Hence, we would expect that a statistically independent representation (Definition 8.3) recovers these parameters. We choose a network structure for *SciNet* with 3 latent neurons. As an input, we provide a time series of positions of the pendulum and we ask *SciNet* to predict the position at a future time (see Box 5 for details). The training process is described in Box 1 and based on beta-VAE [49]. The accuracy of the predictions given by *SciNet* after training is illustrated in Fig. 10.1a.

Without being given any physical concepts, *SciNet* learns to extract the two relevant physical parameters from (simulated) time series data for the x-coordinate of the pendulum and to store them in the latent representation. As shown in Fig. 10.1b, the first latent neuron depends nearly linearly on b and is almost independent of κ, and the second latent neuron depends only on κ, again almost linearly. Indeed, more quantitatively, using the methods introduced in [213–215] and implemented in [216] to estimate the total correlation $C(R_1, R_2)$ (which corresponds to the mutual information in this case) between the two latent neurons, we find $C(R_1, R_2) = 0.107$, where in comparison we have $C(R_1, R_1) = 10.95$ for fully correlated latent neurons. We conclude that *SciNet* has recovered the same time-independent parameters b and κ that are used by physicists. The third latent neuron is nearly constant and does not provide any additional information—in other words, *SciNet* recognized that two parameters suffice to encode this situation.

loss. If the prediction accuracy stays low, increase the size of the network. Successively go on like this until good prediction accuracy can be achieved. (If the prediction accuracy stays low even for large networks, the input data might not contain sufficient information).

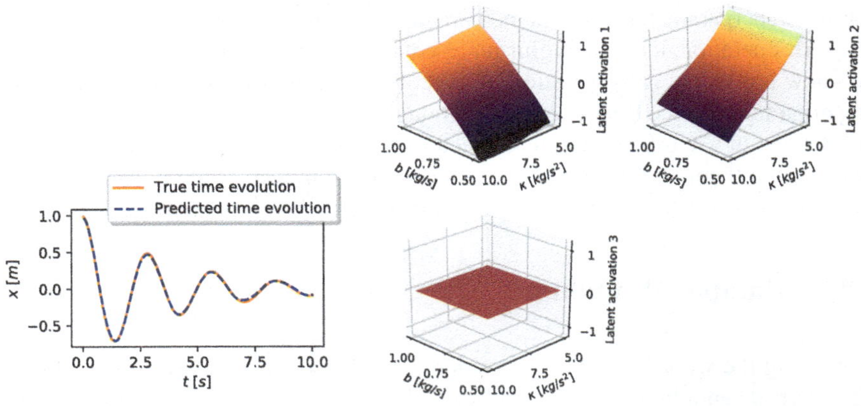

(a) (b)

Fig. 10.1 Damped pendulum (Figures taken from Iten and Metger et al., *Physical Review Letters*, 2020 [22]). *SciNet* is fed a time series of the trajectory of a damped pendulum. It learns to store the two relevant physical parameters, frequency and damping, in the representation, and makes correct predictions about the pendulum's future position. **a** Trajectory prediction of *SciNet*. Here, the spring constant is $\kappa = 5$ kg/s^2 and the damping factor is $b = 0.5$ kg/s. *SciNet's* prediction is in excellent agreement with the true time evolution. **b** Representation learned by *SciNet*. The plots show the activations of the three latent neurons of *SciNet* as a function of the spring constant κ and the damping factor b. The first two neurons store the damping factor and spring constant, respectively. The activation of the third neuron is close to zero, suggesting that only two physical variables are required. On an abstract level, learning that one activation can be set to a constant is encouraged by searching for unctorrelated latent variables, i.e., by minimizing the common information of the latent neurons during training

Box 5: Time evolution of a damped pendulum [22] (Sect. 10.2)

Problem: Predict the position of a one-dimensional damped pendulum at different times.

Physical model: Equation of motion: $m\ddot{x} = -\kappa x - b\dot{x}$.

Solution: $x(t) = A_0 e^{-\frac{b}{2m}t}\cos(\omega t + \delta_0)$, with $\omega = \sqrt{\frac{\kappa}{m}}\sqrt{1 - \frac{b^2}{4m\kappa}}$.

Observation: Time series of positions: $o = \left[x(t_i)\right]_{i \in \{1,\ldots,50\}} \in \mathbb{R}^{50}$, with equally spaced $t_i \in [0, 5]s$. Mass $m = 1$kg, amplitude $A_0 = 1$m and phase $\delta_0 = 0$ are fixed; spring constant $\kappa \in [5, 10]$ kg/s^2 and damping factor $b \in [0.5, 1]$ kg/s are independently and uniformly sampled during data creation.

Question: Prediction times: $q = t_{\text{pred}} \in [0, 10]s$.

Correct answer: Position at time t_{pred}: $a^\star = x(t_{\text{pred}}) \in \mathbb{R}$.

Network structure: Essentially, the network depicted in Fig. 9.1 with three latent neurons (see Appendix B for the detailed description of the network structure). The network size can be found in Table C.1.

Training: Described in Box 2 with parameters given in Table C.2.

Key findings:

- *SciNet* predicts the positions $x(t_{\text{pred}})$ with a root mean square error below 2% (with respect to the amplitude $A_0 = 1$m) (Fig. 10.1a).
- *SciNet* stores κ and b in two of the latent neurons, and does not store any information in the third latent neuron (Fig. 10.1b). Hence, it finds a statistically independent representation.

10.3 Dynamics of the Nonlinear Pendulum

In this section we demonstrate how mathematical prior knowledge integrated into *SciNet*'s network structure can help to investigate the dynamical behaviour of a nonlinear pendulum. The angular displacement θ of a nonlinear pendulum satisfies the differential equation $\ddot{\theta} = -\omega_0^2 \sin(\theta)$ with $\omega_0 = \sqrt{g/l}$, where g is the acceleration due to gravity and l denotes the length of the pendulum. This differential equation has complex analytic solutions that were described in terms of Jacobi elliptic functions in 2007 [217]. Koopman operator theory takes a different approach by searching for a coordinate transformation of $(\theta, \dot{\theta})$, such that the transformed coordinates allow for a linear time evolution (see Sect. 8.5). Here, we describe the results from [183], demonstrating how *SciNet* can be used to find such coordinate transformations.

The training data for *SciNet* is generated by sampling θ and $\dot{\theta}$ from the interval $[-0.5, 0.5]$. The network structure for *SciNet* with 2 latent neurons is described in Sect. 9.6. As an input to *SciNet*, we provide the angle and the angular velocity at a starting time and ask *SciNet* to predict the time evolution of these quantities (see Box 6 for the details). The training process is similar as described in Box 4 and we refer to [183] for the details.

Box 6: Time evolution of a nonlinear pendulum [183] (Sect. 10.3)

Problem: Predict the angle and angular velocity of a one-dimensional nonlinear pendulum at different times.

Physical model: Equation of motion: $\ddot{\theta} = -\omega_0^2 \sin(\theta)$ with $\omega_0 = \frac{1}{s}$.
Solution: An analytic solution can be derived in terms of Jacobi elliptic functions [217]. The data set used for training and evaluation of the neural network is created by solving the equation of motion in MATLAB using the *ode45* solver [183].

Observation: $o = [\theta_0, \dot{\theta}_0]$ with $\theta_0, \dot{\theta}_0 \in [-0.5, 0.5]$.

Question: Implicit.

Correct answer: Time series $[a_0, \ldots, a_n] = [(\theta(t_0), \dot{\theta}(t_0)), \ldots, (\theta(t_n), \dot{\theta}(t_n))]$ of $n = 50$ observations, with fixed time steps $\Delta t = t_{i+1} - t_i$ of 0.02 seconds.

Network structure: Essentially, the network structure depicted in Fig. 9.3 with two latent neurons and parametrized update rules of the form

$$u_{\mu,\omega}(r_1, r_2) = e^{\mu \Delta t}([\cos(\omega \Delta t) r_1 - \sin(\omega \Delta t) r_2], \sin(\omega \Delta t) r_1 + \cos(\omega \Delta t) r_2]) ,$$

where μ and ω are given as an output of an additional neural network $\Lambda : \mathbb{R} \mapsto \mathbb{R}^2$ that takes $r_1^2 + r_2^2$ as an input (to enforce circular symmetry in the eigenfunction coordinates). The network sizes can be found in [183].

Training: Similar as described in Box 4. (see [183] for the details).

Key findings:

- *SciNet* finds coordinates that linearize the dynamics of the nonlinear pendulum (see Fig. 10.2).
- The frequency ω decreases continuously in $r_1^2 + r_2^2$ as the energy of the pendulum is increased (see Fig. 10.2).

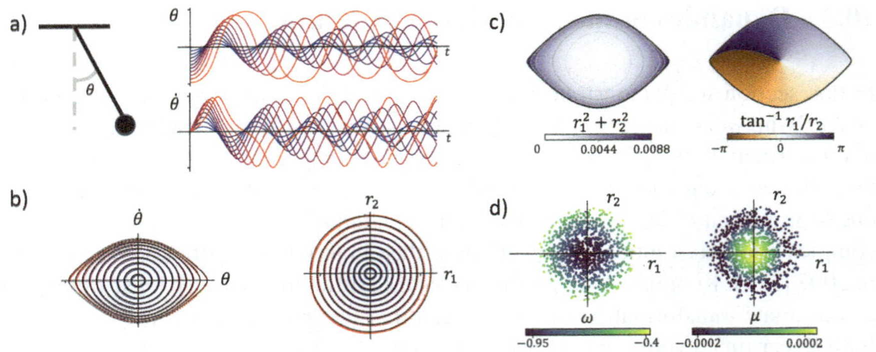

Fig. 10.2 Nonlinear pendulum (Figure reproduced from Lusch et al. *Nature Communications* 2018, [218]). The figure shows the results from [183], where *SciNet* is applied to observation data from a nonlinear pendulum (see Box 6 for the details). *SciNet* learns to accurately predict the time evolution of the nonlinear pendulum (Fig. 10.2a) and finds a coordinate transformation given by an encoder $E : (\theta, \dot{\theta}) \mapsto (r_1, r_2)$ that implements Koopman eigenfunctions (represented in Fig. 10.2c) and maps the angular displacement and velocity to coordinates (r_1, r_2) that are evolved in time Δt according to update rules of the form $u_{\gamma=(\mu,\omega)}(r_1, r_2) = \exp(\mu\Delta t)([\cos(\omega\Delta t)r_1 - \sin(\omega\Delta t)r_2], \sin(\omega\Delta t)r_1 + \cos(\omega\Delta t)r_2])$. Figure 10.2b shows *SciNet*'s prediction for θ and $\dot{\theta}$ and the corresponding values for r_1 and r_2 for ten initial conditions that are evolved in time until the relative prediction error reaches 10%. The dependence of the parameters μ and ω determining the update rule on r_1 and r_2 is shown in Fig. 10.2d. Note that the (absolute value of the) frequency ω decreases if the amplitude of the pendulum increases and that the damping factor μ stays approximately constant

The results are presented in Fig. 10.2 and taken from [183]. *SciNet* is able to predict the time evolution of the nonlinear pendulum with high accuracy. Furthermore, it finds a coordinate transformation from $(\theta, \dot{\theta})$ to (r_1, r_2), such that the representation (r_1, r_2) evolves according to linear dynamics. More concretely, the encoder implements Koopman eigenfunctions with eigenvalues that are allowed to continuously depend on $r_1^2 + r_2^2$. The variation in the eigenvalues allows to handle the continuous spectrum of the nonlinear pendulum in an elegant and interpretable way. As shown in Fig. 10.2, *SciNet* correctly identifies the dependence of the frequency ω on the energy of the pendulum.

10.4 Conservation of Angular Momentum

One of the most important concepts in physics is that of conservation laws, such as conservation of energy and angular momentum. While their relation to symmetries makes them interesting to physicists in their own right, conservation laws are also of practical importance. If two systems interact in a complex way, we can use conservation laws to predict the behaviour of one system from the behaviour of the other, without studying the details of their interaction. For certain types of questions, conserved quantities therefore act as a compressed representation of joint properties of several systems.

We consider the scattering experiment discussed in [22], shown in Fig. 10.3 and described in Box 7, where two point-like particles collide. Given the initial angular momentum of the two particles and the final trajectory of one of them, a physicist can predict the trajectory of the other using conservation of total angular momentum.

To see whether *SciNet* makes use of angular momentum conservation in the same way as a physicist would do, we train it with (simulated) experimental data as described in Box 7 with one latent neuron, and add Gaussian noise to show that the encoding and decoding are robust. Indeed, *SciNet* does exactly what a physicist would do and stores the total angular momentum in the latent representation (Fig. 10.3b). This example shows that *SciNet* can recover conservation laws, and suggests that they emerge naturally from compressing data and asking questions about joint properties of several systems.

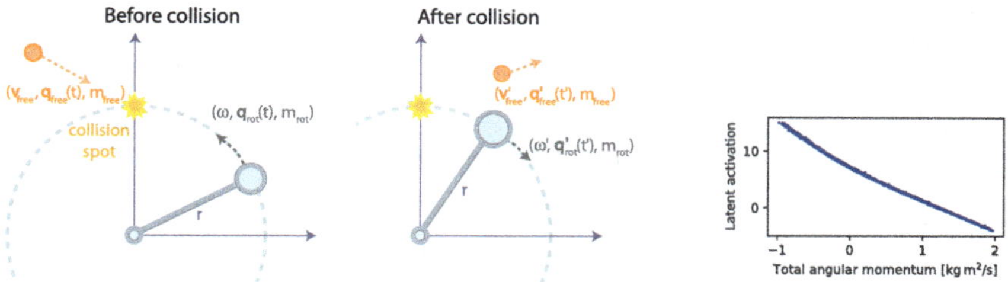

(a) (b)

Fig. 10.3 Collision under conservation of angular momentum (Figure reproduced from Iten and Metger et al., *Physical Review Letters*, 2020 [22]). In a classical mechanics scenario where the total angular momentum is conserved, the neural network learns to store this quantity in the latent representation. **a** Physical setting. A body of mass m_{rot} is fixed on a rod of length r (and of negligible mass) and rotates around the origin with angular velocity ω. A free particle with velocity $\mathbf{v}_{\text{free}}$ and mass m_{free} collides with the rotating body at position $\mathbf{q} = (0, r)$. After the collision, the angular velocity of the rotating particle is ω' and the free particle is deflected with velocity $\mathbf{v}'_{\text{free}}$. **b** Representation learned by *SciNet*. Activation of the latent neuron as a function of the total angular momentum. *SciNet* learns to store the total angular momentum, a conserved quantity of the system

Box 7: Two-body collision with angular momentum conservation [22] (Sect. 10.4)

Problem: Predict the position of a particle fixed on a rod of radius r (rotating about the origin) after a collision at the point $(0, r)$ with a free particle (in two dimensions, see Fig. 10.3a).

Physical model: Given the total angular momentum before the collision and the velocity of the free particle after the collision, the position of the rotating particle at time t'_{pred} (after the collision) can be calculated from angular momentum conservation: $J = m_{\text{rot}} r^2 \omega - r m_{\text{free}} (\mathbf{v}_{\text{free}})_x = m_{\text{rot}} r^2 \omega' - r m_{\text{free}} (\mathbf{v}'_{\text{free}})_x = J'$.

Observation:
Time series of both particles before the collision: $o = [(t_i^{\text{rot}}, \mathbf{q}_{\text{rot}}(t_i^{\text{rot}})), (t_i^{\text{free}}, \mathbf{q}_{\text{free}}(t_i^{\text{free}}))]_{i \in \{1,\ldots,5\}}$, with times t_i^{rot} and t_i^{free} randomly chosen for each training sample. Masses $m_{\text{rot}} = m_{\text{free}} = 1\text{kg}$ and the orbital radius $r = 1\text{m}$ are fixed; initial angular velocity ω, initial velocity $\mathbf{v}_{\text{free}}$, the first component of $\mathbf{q}_{\text{free}}(0)$ and final velocity $\mathbf{v}'_{\text{free}}$ are varied between training samples. Gaussian noise $(\mu = 0, \sigma = 0.01\text{m})$ is added to all position inputs.

Question: Prediction time and position of free particle after collision: $q = \left(t'_{\text{pred}}, [t'_i, \mathbf{q}'_{\text{free}}(t'_i)]_{i \in \{1,\ldots,5\}}\right)$.

Correct answer: Position of rotating particle at time t'_{pred}: $a^\star = \mathbf{q}'_{\text{rot}}(t'_{\text{pred}})$.

Network structure: Network depicted in Fig. 9.1 with one latent neuron and specifications given in Table C.1.

Training: Described in Box 1 with parameters given in Table C.2.

Key findings:
- *SciNet* predicts the position of the rotating particle with root mean square prediction error below 4% (with respect to the radius $r = 1\text{m}$).
- *SciNet* is resistant to noise.
- *SciNet* stores the total angular momentum in the latent neuron.

10.5 Representation of Qubits

Quantum state tomography is an active area of research [219]. Ideally, we look for a *faithful representation* of the state of a quantum system, such as the wave function: a representation that stores all information necessary to predict the probabilities of the outcomes for arbitrary measurements on that system. However, to specify a faithful representation of a quantum system it is not necessary to perform all theoretically possible measurements on the system. If a set of measurements is sufficient to reconstruct the full quantum state, such a set is called *tomographically complete*.

Here we discuss the results from [22, 23] showing that, based only on (simulated) experimental data and without being given any assumptions about quantum theory, *SciNet* recovers a faithful representation of the state of small quantum systems and can make accurate predictions. In particular, this allows us to infer the dimension of the system and distinguish tomographically complete from incomplete measurement sets. Further, *SciNet* can separate parameters in the representation according to locality conditions (using the network structure to find operationally meaningful representations shown in Fig. 9.2) and end up with a similar representation of two qubits as given in [220]. Boxes 8 and 9 summarize the setting and the results.

A pure state on n qubits can be represented by a normalized complex vector $|\psi\rangle \in \mathbb{C}^{2^n}$, where two states $|\psi\rangle$ and $|\psi'\rangle$ are identified if and only if they differ by a global phase factor, i.e., if there exists $\phi \in \mathbb{R}$ such that $|\psi\rangle = e^{i\phi} |\psi'\rangle$. The normalization condition and irrelevance of the global phase factor decrease the number of free parameters of a quantum state by two. Since a complex number has two real parameters, a single-qubit state is described by $2 \times 2^1 - 2 = 2$ real parameters, and a state of two qubits is described by $2 \times 2^2 - 2 = 6$ real parameters.

Box 8: Representation of pure one- and two-qubit states [22] (Sect. 10.5.1)

Problem: Predict the measurement probabilities for any binary projective measurement $|\omega\rangle\langle\omega| \in \mathbb{C}^{2^n}$ on a pure n-qubit state $|\psi\rangle \in \mathbb{C}^{2^n}$ for $n = 1, 2$.
Physical model: The probability to measure 0 on the state $|\psi\rangle \in \mathbb{C}^{2^n}$ performing the measurement $|\omega\rangle\langle\omega| \in \mathbb{C}^{2^n} \times \mathbb{C}^{2^n}$ is given by $p(\omega, \psi) = |\langle\omega\rangle\,\psi|^2$.
Observation: Operational parameterization of a state $|\psi\rangle$: $o = [p(\alpha_i, \psi)]_{i\in\{1,\dots,n_1\}}$ for a fixed set of random binary projective measurements $\mathcal{M} := \{|\alpha_1\rangle\langle\alpha_1|, \dots, |\alpha_{n_1}\rangle\langle\alpha_{n_1}|\}$ ($n_1 = 10$ for one qubit, $n_1 = 30$ for two qubits).
Question: Operational parameterization (see Remark 10.1) of a measurement $|\omega\rangle\langle\omega|$: $q = [p(\beta_i, \omega)]_{i\in\{1,\dots,n_2\}}$ by its outcome probabilities on a fixed set of random states $\mathcal{S} := \{|\beta_1\rangle, \dots, |\beta_{n_2}\rangle\}$ ($n_2 = 10$ for one qubit, $n_2 = 30$ for two qubits).
Correct answer: $a^\star(\omega, \psi) = p(\omega, \psi) = |\langle\omega\rangle\,\psi|^2$.
Network structure: Network depicted in Fig. 9.1 with varying numbers of latent neurons and specifications given in Table C.1.
Training: Described in Box 1 with parameters given in Table C.2.
Key findings:

- *SciNet* can be used to determine the minimal number of parameters necessary to describe the state $|\psi\rangle$ (see Fig. 10.4) without being provided with any prior knowledge about quantum physics.
- *SciNet* distinguishes tomographically complete and incomplete sets of measurements (see Fig. 10.4).

More generally, one may also consider mixed quantum states, which can be considered as a (classical) probabilistic mixture of pure quantum states. Formally, we describe mixed quantum states on n qubits by positive semi-definite matrixes $\rho \in \mathbb{C}^{2^n} \times \mathbb{C}^{2^n}$ with unit trace. Since there exists an eigendecomposition with real eigenvalues for any positive semi-definite matrix ρ, we can write $\rho = \sum_{i=1}^{2^n} p_i |\psi_i\rangle\langle\psi_i|$.[3] Since ρ has unit trace, we have $\sum_i p_i = 1$ and we can consider ρ as a probabilistic mixture of the pure states $|\psi_i\rangle$. A pure state $|\psi\rangle$ can then be considered as a special case of a density matrix $\rho = |\psi\rangle\langle\psi|$ with a single eigenvalue that is equal to one. For further details and a more in depth introduction into the formalism for quantum states we refer to [72].

[3] The notation $|\cdot\rangle$ and $\langle\cdot|$ is common in quantum mechanics and is called the "bra-ket" notation. The reader may consider the ket-vectors $|\cdot\rangle$ as a row vector, where a bra-vector $\langle\cdot|$ is considered as a column vector. Further, the entries from $\langle\psi|$ are the conjugate transposed entries of $|\psi\rangle$.

Here, we consider binary projective measurements on n qubits. These measurements are operators of the form $|\omega\rangle\langle\omega|$, which are fully described by a vector $|\omega\rangle \in \mathbb{C}^{2^n}$, with measurement outcomes labeled by 0 for the projection on $|\omega\rangle$ and 1 otherwise. The probability to get outcome 0 when measuring $|\omega\rangle\langle\omega|$ on a quantum system in a pure state $|\psi\rangle$ is then given by $p(\omega, \psi) = |\langle\omega\rangle\,\psi|^2$, where $\langle\cdot\rangle\,\cdot$ denotes the standard scalar product on $\mathbb{C}^{2^n}$. More generally, the probability to get outcome 0 when measuring ω on a quantum system in a mixed state ρ is given by $p(\omega, \rho) = \mathrm{tr}\,|\omega\rangle\langle\omega|\rho$. Note that this is consistent with the probability stated for pure states setting $\rho = |\psi\rangle\langle\psi|$.

To generate the training data for *SciNet*, we assume that an agent B has access to a preparation experiment consisting of two devices. The first device creates (many copies of) a quantum system in a state ρ, which depends on the setting of the dials and buttons of the device. The second device can perform any binary projective measurements in the set $\mathcal{M} := \left\{|\alpha_1\rangle\langle\alpha_1|, \ldots, |\alpha_{n_1}\rangle\langle\alpha_{n_1}|\right\}$, which we would like to use to determine the state of the quantum system. Agent B performs all measurements in $\mathcal{M}$ several times on the same quantum state ρ to estimate the probabilities $p(\alpha_i, \rho)$ of measuring 0 for the i-th measurement. These probabilities form the observation given to *SciNet*.

Based on a representation for the collected data by agent B, agents C_j are asked to predict the measurement outcomes of measurements $|\omega_j\rangle\langle\omega_j|$ chosen from a set $\mathcal{Q}^j$. To parameterize the measurement $|\omega_j\rangle\langle\omega_j|$ we choose sets of states

$$\mathcal{S}^j := \left\{\left|\beta_1^j\right\rangle, \ldots, \left|\beta_{n_2}^j\right\rangle\right\}.$$

The probabilities $p(\omega_j, \beta_1^j), \ldots, p(\omega_j, \beta_{n_2}^j)$ are provided to the agent C_j as the question input. The agent C_j then has to predict the probability $p(\omega_j, \rho)$ for measuring the outcome 0 on the state ρ when performing the measurement $|\omega_j\rangle\langle\omega_j|$.

Remark 10.1 We could parameterize the set of possible measurements by any parameterization that is natural for the experimental setup, for example the settings of the dials and buttons on an experimental apparatus. Such a natural parameterization is assumed to fully specify the measurement, in the sense that the same settings on the experimental apparatus will always result in the same measurement being performed. Because $\mathcal{S}^j$ only represents our choice for parameterizing the measurement setup, it is natural to assume that $\mathcal{S}^j$ contains enough states to fully characterize all the measurements in $\mathcal{Q}^j$.

10.5.1 Minimal Representations for Pure Quantum States

As described in [22], we use *SciNet* in the quantum setting given above to find the required number of real parameters to describe pure quantum states on one and two qubits. Further, we then demonstrate how to use *SciNet* to determine if a set

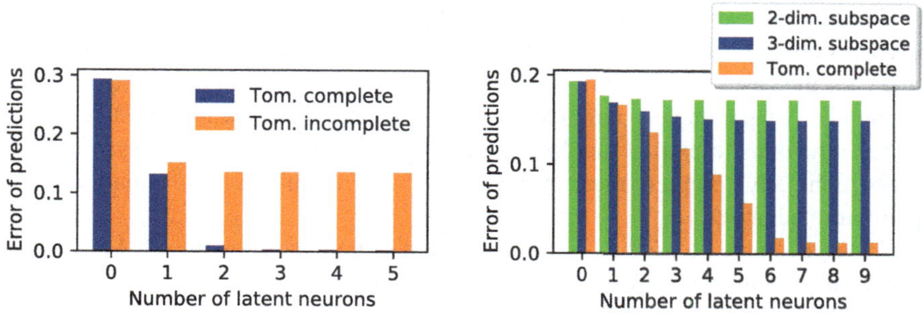

(a) One qubit. (b) Two qubits.

Fig. 10.4 Quantum tomography (Figure taken from Iten and Metger et al., *Physical Review Letters*, 2020 [22]). *SciNet* is given tomographic data for one or two qubits and an operational description of a measurement as a question input and has to predict the probabilities of outcomes for this measurement. We train *SciNet* with both tomographically complete and incomplete sets of measurements, and find that, given tomographically complete data, *SciNet* can be used to find the minimal number of parameters needed to describe a quantum state (two parameters for one qubit and six parameters for two qubits). Tomographically incomplete data can be recognized, since *SciNet* cannot achieve perfect prediction accuracy in this case, and the prediction accuracy can serve as an estimate for the amount of information provided by the tomographically incomplete set. The plots show the root mean square error of *SciNet's* measurement predictions for test data as a function of the number of latent neurons

of measurements is tomographically complete or not. More concretely, we use the network structure for *SciNet* shown in Fig. 9.1 and the training process described in Box 1 to find the minimal number of required parameters to represent quantum states. If the prediction accuracy stays low for any number of latent neurons, we can conclude that the provided observation data does not contain sufficient information, which means in the considered case that the set of measurements $\mathcal{M}$ is tomographically incomplete. How to operationally separate the parameters in the representation is then considered in Sect. 10.5.2.

We consider an agent B performing measurements on pure quantum states $|\psi\rangle$ and one agent $C := C_1$ who is asked to predict the measurement probability $p(\omega, \psi)$ of measuring zero when performing the measurement $|\omega\rangle\langle\omega|$ on $|\psi\rangle$. We train *SciNet* with different pairs $(|\omega\rangle\langle\omega|, |\psi\rangle)$ for one and two qubits, keeping the measurement sets $\mathcal{M}$ and $\mathcal{S} := \mathcal{S}^1$ fixed for each choice of the qubit number. We choose $n_1 = n_2 = 10$ for the single-qubit case and $n_1 = n_2 = 30$ for the two-qubit case. The setting is summarized in Box 8.

The results are shown in Fig. 10.4. Varying the number of latent neurons, we can observe how the quality of the predictions improves as we allow for more parameters in the representation of $|\psi\rangle$. To minimize statistical fluctuations due to the randomized initialization of the network, each network specification is trained three times and the run with the lowest mean square prediction error on the test data is used.

For the cases where $\mathcal{M}$ is tomographically complete, the plots in Fig. 10.4 show a drop in prediction error when the number of latent neurons is increased up to two or six for the cases of one and two qubits, respectively.[4] This is in accordance with the number of parameters required to describe a one- or a two-qubit state. Thus, *SciNet* allows us to extract the dimension of the underlying quantum system from tomographically complete measurement data, without any prior information about quantum mechanics.

SciNet can also be used to determine whether the measurement set $\mathcal{M}$ is tomographically complete or not. To generate tomographically incomplete data, we choose the measurements in $\mathcal{M}$ randomly from a subset of all binary projective measurements. Specifically, the quantum states corresponding to measurements in $\mathcal{M}$ are restricted to random *real* linear superpositions of k orthogonal states, i.e., to a (real) k-dimensional subspace. For a single qubit, we use a two-dimensional subspace; for two quibts, we consider both two- and three-dimensional subspaces.

Given tomographically incomplete data about a state $|\psi\rangle$, it is not possible for *SciNet* to predict the outcome of the final measurement perfectly regardless of the number of latent neurons, in contrast to the tomographically complete case (see Fig. 10.4). Hence, we can deduce from *SciNet's* output that $\mathcal{M}$ is an incomplete set of measurements. Furthermore, this analysis provides a qualitative measure for the amount of information provided by the tomographically incomplete measurements: in the two-qubit case, increasing the subspace dimension from two to three leads to higher prediction accuracy and the required number of latent neurons increases.

Remark 10.2 (Efficient representation of states of many body systems) Quantum many body systems are challenging to describe because of the exponential complexity of the many-body wave function encoding the non-trivial correlations between the particles. In [222–236] neural networks are used to efficiently represent states of quantum many body systems, to find their ground states and even to simulate their unitary time evolution. In particular, in [223], variational autoencoders are used to approximate the distribution of the measurement outcomes of a specific quantum state for a fixed measurement basis and the size of the neural network can provide an estimate for the complexity of the state. In contrast, the approach presented in this section does not focus on an efficient representation of a given quantum state and it is not specifically designed for learning representations of quantum systems. Nevertheless, *SciNet* can be used to produce representations of arbitrary states of simple quantum systems without retraining.

---

[4] In the case of a single qubit, there is an additional small improvement in going from two to three latent neurons: this is a technical issue caused by the fact that any two-parameter representation of a single qubit, for example the Bloch sphere representation, includes a *cyclic parameter*, which cannot be exactly represented by a continuous encoder (see Appendix D). The same likely applies in the case of two qubits, going from 6 to 7 latent neurons. This restriction also makes it difficult to interpret the details of the learned representation. Recent work [221] tackles this problem by using theory of manifolds and splitting manifolds into parts for which there is continuous encoding. Although this work is technically very useful, it is not discussed in detail here because it does not contribute a lot to the discussion at the conceptual level.

10.5.2 Local Representation of Two-Qubit States

In the last section, we investigated the number of parameters required to describe (pure) quantum states. Let us now follow the approach in [23, 237] to separate the parameters for a representation of a (mixed) two-qubit state ρ in an operationally meaningful way, in the sense as described in Sect. 8.4.2.2. More concretely, we use the network structure for *SciNet* shown in Fig. 9.2 and the training process described in Box 3.

Box 9: Local representation of two-qubit states [23] (Sect. 10.5.2)

Problem: Three agents C_1, C_2 and C_3 have to predict the measurement probabilities for binary projective measurements on two qubits. Thereby, agent C_1 and C_2 are asked questions about measurement output probabilities on the first and second qubit, and gent C_3 is asked to predict joint measurement output probabilities on both qubits.

Physical model: The probability to measure 0 on the state ρ performing the measurement $|\omega\rangle\langle\omega|$ is given by $p(\omega, \rho) = \mathrm{tr}\, |\omega\rangle\langle\omega|\, \rho$.

Observation: Operational parameterization of a state ρ: $o = [p(\alpha_i, \rho)]_{i\in\{1,\dots,n_1\}}$ for a fixed set of random binary projective measurements $\mathcal{M} := \{|\alpha_1\rangle\langle\alpha_1|, \dots, |\alpha_{75}\rangle\langle\alpha_{75}|\}$

.

Questions: Operational parameterization (see Remark 10.1) of measurements $|\omega_j\rangle\langle\omega_j|$: $q^j = \left[p(\beta_i^j, \omega_j)\right]_{i\in\{1,\dots,75\}}$ by their outcome probabilities for fixed sets of random states

$$S^j := \left\{\left|\beta_1^j\right\rangle, \dots, \left|\beta_{75}^j\right\rangle\right\},$$

where $\{\left|\beta_i^j\right\rangle\}_{i\in\{1,\dots,75\}}$ are chosen to be of the form $|\beta_i^1\rangle = \left|\tilde{\beta}_i^1\right\rangle \otimes |0\rangle$, $|\beta_i^2\rangle = |0\rangle \otimes \left|\tilde{\beta}_i^2\right\rangle$ for random one-qubit states $\left|\tilde{\beta}_i^1\right\rangle$ and $\left|\tilde{\beta}_i^2\right\rangle$, and where $|\beta_i^3\rangle$ are random two-qubit states. The measurements are chosen to be of the following form: $|\omega_1\rangle\langle\omega_1| = |\psi_1\rangle\langle\psi_1| \otimes \mathrm{id}$, $|\omega_2\rangle\langle\omega_2| = \mathrm{id} \otimes |\psi_2\rangle\langle\psi_2|$ and $|\omega_3\rangle\langle\omega_3| = |\psi_3\rangle\langle\psi_3|$, where $|\psi_1\rangle$ and $|\psi_2\rangle$ are one-qubit states, and $|\psi_3\rangle$ is a two-qubit state.

Correct answers: The decoder D_j should output $a^\star(\omega_j, \rho) = p(\omega_j, \rho) = \mathrm{tr}\left|\omega_j\right\rangle\left\langle\omega_j\right| \rho$.

Network structure: Network depicted in Fig. 9.2 with 15 latent neurons[a] and three decoders and network sizes given in Table C.1.

Training: Described in Box 3 with parameters given in Table C.2.

Key finding: *SciNet* finds a representation for two-qubit states that separates parameters storing local degrees of freedom of each of the qubits. The found representation is similar to the one presented in [220].

---

[a] For simplicity, we have choosen the minimal number of required latent neurons here. This number could be found by successively increasing the number of latent neurons until *SciNet* achieves hight prediction accuracy.

For the preparation experiments, an agent B fixes 75 randomly chosen binary measurements $\mathcal{M} := \{|\alpha_1\rangle\langle\alpha_1|, \ldots, |\alpha_{75}\rangle\langle\alpha_{75}|\}$ that are performed on two-qubit states ρ that are varied during the training. Three agents C_1, C_2 and C_3 are now required to answer different questions about the two-qubit system:

- Agent C_1 and C_2 are asked questions about measurement output probabilities on the first and second qubit, respectively.
- Agent C_3 is asked to predict joint measurement output probabilities on both qubits.

The question inputs consisting of a binary measurement $|\omega_j\rangle\langle\omega_j|$ (on one or two qubits, respectively) are parametrised by 75 randomly chosen states $\mathcal{S}^j := \left\{\left|\beta_1^j\right\rangle, \ldots, \left|\beta_{75}^j\right\rangle\right\}$. The details about the setting are summarized in Box 9.

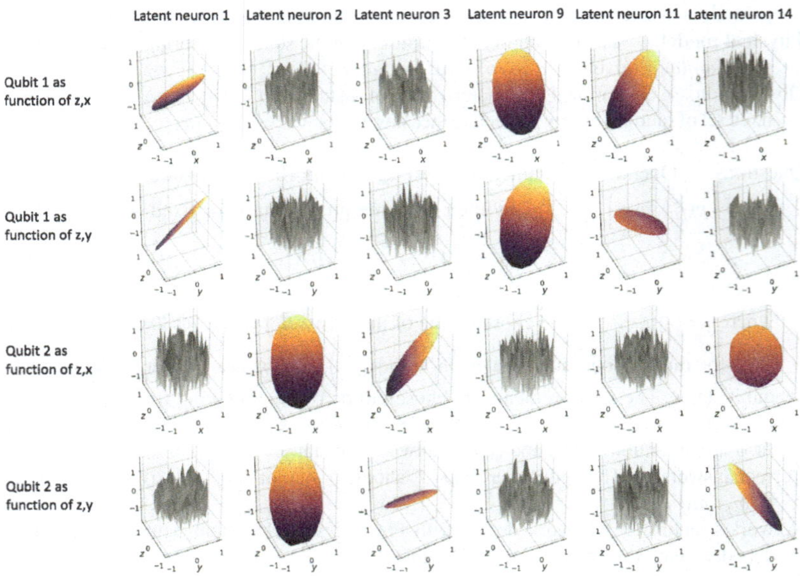

Fig. 10.5 Results for the local representation of two-qubit states (Figure taken from Nautrup, Metger and Iten et al., 2020 [23]). We consider a quantum-mechanical system of two qubits. An encoder maps tomographic data of a two-qubit state to a representation of the state. Three agents C_1, C_2 and C_3 are asked questions about the measurement output probabilities on the two-qubit system, where a question is given as the parameterisation of a measurement. Agents C_1 and C_2 are asked to predict measurement outcome probabilities on the first and second qubit, respectively. The third agent C_3 is tasked to predict measurement probabilities for arbitrary measurements on the full two-qubit system. In total, 15 latent neurons are required to answer the questions of all agents C_1, C_2 and C_3. Agent C_3 requires access to all parameters, while agents C_1 and C_2 need only access to two disjoint sets of three parameters, encoded in latent neurons 1,9,11 and 2,3,14 respectively. The plots show the activation values for these latent neurons in response to changes in the local degrees of freedom of each qubit, with the bottom axes of the plots denoting the components of the reduced one-qubit state $\rho = 1/2(\mathrm{id} + x\,\sigma_x + y\,\sigma_y + z\,\sigma_z)$ on either qubit 1 or 2

We find that three latent neurons are used for each of the local qubit representations as required by agents C_1 and C_2. These local representations store combinations of the x-,y- and z-component of the Bloch sphere representation $\rho = 1/2(\mathrm{id} + x\sigma_x + y\sigma_y + z\sigma_z)$ of a singe qubit (see Fig. 10.5), where σ_x, σ_y, σ_z denote the Pauli matrices. In general, a two-qubit mixed state ρ is described by 15 parameters, since a Hermitian 4×4 matrix is described by 16 parameters, and one parameter is determined by the others due to the unit trace condition. Indeed, we find that the agent who has to predict the outcomes of the joint measurements accesses all 15 latent neurons, including the ones storing the two local representations. These numbers correspond to the numbers found in the analytical approach in [220].

10.6 Charged Particles

In this section, we investigate a toy example from classical mechanics that is considered in [23, 237] to demonstrate that the criterium for a operationally meaningful separation of parameters (Sect. 8.4.2.2) leads to a discovery of the same parameters that we know from physical text books. In contract to the example discussed in Sect. 10.2, the methods used in this section are not biased by the distribution over the training data. We use the network structure for *SciNet* shown in Fig. 9.2 and the training process described in Box 3. Further, we demonstrate how to use several encoding agents instead of one by means of the same example.

We consider the setup shown in Fig. 10.6 consisting of two particles $p_1 = (m_1, q_1)$ and $p_2 = (m_2, q_2)$ with masses m_1, m_2 and charges q_1, q_2. Experimenters (agents B_i) can perform the following reference experiments to gain information about the properties of the particles:

1. The experimenter can elastically collide each of the particles p_i, initially at rest, with an uncharged reference mass m_{ref} moving at a fixed reference velocity v_{ref}. Then, the agent observes a time series of positions of the particle p_i after the collision.
2. We place a particle p_i at the origin of a coordinate system at rest, and place a reference particle $p_{\mathrm{ref}} = (m_{\mathrm{ref}}, q_{\mathrm{ref}})$ with fixed mass and charge at a fixed distance d_0 to the origin. Both particles are free to move. We observe a time series of positions of the particle p_i as it moves due to the Coulomb interaction between itself and the reference particle.

We choose natural questions asked to different decoding agents C_j (see Fig. 8.3). In the physical setting about which we ask the questions, the initial positions of the particles p_i and the positions of two target holes h_i are fixed.

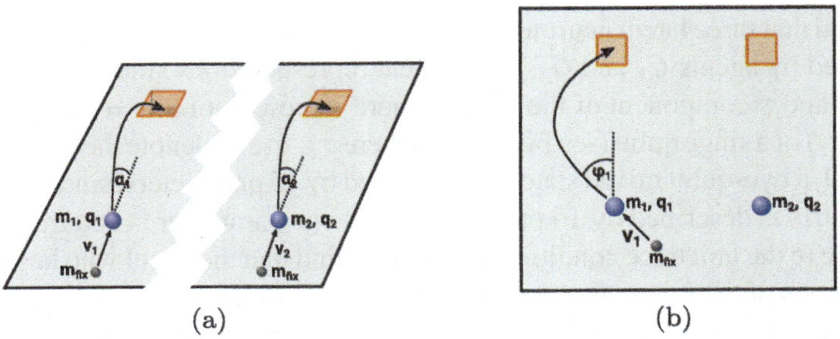

Fig. 10.6 Toy example with charged masses (Figure taken from Nautrup, Metger and Iten et al., 2020 [23, 237]). **a** There are two separated decoding agents C_1 and C_2, so there is no interaction between their individual experimental setups. Each agent is required to *shoot* a particle p_i (of mass m_i and charge q_i) into a hole h_i in the presence of a fixed gravitational field. They do this by elastically colliding a projectile of fixed mass m_{fix} with the particle p_i. The distance to the hole is fixed. Challenging agents choose the velocity v_i with which the projectile is shot, and provide this information to the decoding agents as a question input. The correct answer is the angle α_i out of the plane of the ground, such that if the projectile is fired with the given velocity at this angle, the particle p_i lands directly in the hole. **b** Now we consider two decoding agents C_3 and C_4 in a setting where the two particles are subject to the Coulomb interaction between each other. The agents C_3 and C_4 are again required to shoot a projectile at the particles p_1 and p_2, respectively. The charged particle will move in the Coulomb field of the other agent's particle. Similar to the situation in a), the agents are given velocities v_3 or v_4 as a questions and they have to predict the angles φ_1 or φ_2 in the plane of the ground, respectively, such that if the mass is fired at this angle, it will "roll" on the frictionless ground into the hole while the position of the other charge stays fixed. (The experiment is then repeated with the roles of the agents reversed, i.e., the agent that first fired his mass now fixes it at its starting position, and vice versa)

- Each decoding agent C_1 and C_2 are given projectiles with a fixed mass m_{fix} and a challenging agent chooses a velocity v_j with which the projectile will hit the particle p_j. The decoding agent C_j (described by a decoder D_j) is then asked to predict the correct angle α_j in the yz-plane with which the agent shoots his projectile against the particle p_j, such that the particle p_j lands directly, i.e., without bouncing, in its target hole h_j (under the influence of a constant homogeneous gravitational field).
- Similarly, agents C_3 and C_4 are also given projectiles, whose chosen velocities v_3 and v_4 are given as a question input to decoders D_3 and D_4. The goal of the decoders is to predict the angle φ_1 and φ_2 in the xy-plane so that the particle p_1 and p_2 will fly into its target hole, under the influence of the Coulomb field of the other particle, which stays fixed.

In both cases, we restrict the possible velocities of the projectiles such that there exists a (unique) angle that makes the particle land in the hole. Further, the prediction loss is given by the squared difference between the angle chosen by the agent and the *correct* angle that would have landed the mass directly into the hole; this correct angle can be determined by experiments on the system.

Box 10: Charged particles [23] (Sect. 10.6)

Problem: There are two physical challenges shown in Fig. 10.6 involving two charged particles $p_i = (m_i, q_i)$ with masses m_i and charges q_i. In challenge a), two decoding agents C_1 and C_2 have to predict the shooting-angles α_1 and α_2 in yz-plane for a given schooting-velocity, such that the particle p_1 or p_2, respectively, land in a target hole under the influence of a gravitational field. Similarly, in challenge b), two decoding agents C_3 and C_4 have to predict the shooting-angles φ_1 and φ_2 in the xy-plane for a given shooting-velocity, such that the particles p_1 or p_2, respectively, lands in a target hole.

Physical model: a) A particle with mass m_i moves in a constant gravitational field with strength g in negative z-direction. b) A charged particle p_i moves according to the action of the Coulomb force of the other particle. For a distance r between the particles, the Coulomb force is proportional to $q_1 q_2 / r^2$.

Observations: See the experiments 1 and 2, described at the start of Sect. 10.6. In the case of two encoding agents E_i, each of them has only access to the experiments related to particle p_i.

Questions: In a) and b) the questions are given by the velocity with which the projectile will hit the particle, i.e., $q^j = v_j$ for $j \in \{1, 2, 3, 4\}$.

Correct answers: a) Assuming that the target is placed at $x = d_0$, we find that $\alpha_i = \frac{1}{2} \arcsin \frac{d_0 g}{v_i^2}$. b) The answers in this setting are discussed in Appendix E.

Network structure: **Single encoder:** Network depicted in Fig. 9.2 with one encoder, 3 latent neurons[a] and four decoders. **Multiple encoders:** Network depicted in Fig. 9.2 with two encoders, 4 latent neurons[a] and four decoders. For both cases, the network sizes are given in Table C.1.

Training: Described in Box 3 with parameters given in Table C.2.

Key finding: The key findings for a single and for multiple encoders are:

- **Single encoder:** *SciNet* stores the masses m_1 and m_2 in separate latent neurons (Fig. 10.7). The third latent neuron stores the product of the charges $q_1 q_2$, necessary to quantify the strength of the Coulomb interaction. Crucially, the separation is not biased by the distribution over the training data.
- **Multiple encoders:** There is no way for the encoding agents to directly encode the product of the charges $q_1 \cdot q_2$. Instead, the representation produced by each encoding agent stores q_i individually.

[a] For simplicity, we have choosen the minimal number of required latent neurons here. This number could be found by successively increasing the number of latent neurons until *SciNet* achieves hight prediction accuracy.

10.6.1 Single Encoder

Let us first consider a single encoding agent B, and four decoding agents described by decoders D_1, D_2, D_3, D_4. The setting and the results are summarized in Box 10.

We use the training process described in Box 3 to train *SciNet*. To analyse the learnt representation, we plot the activation of the latent neurons for different examples

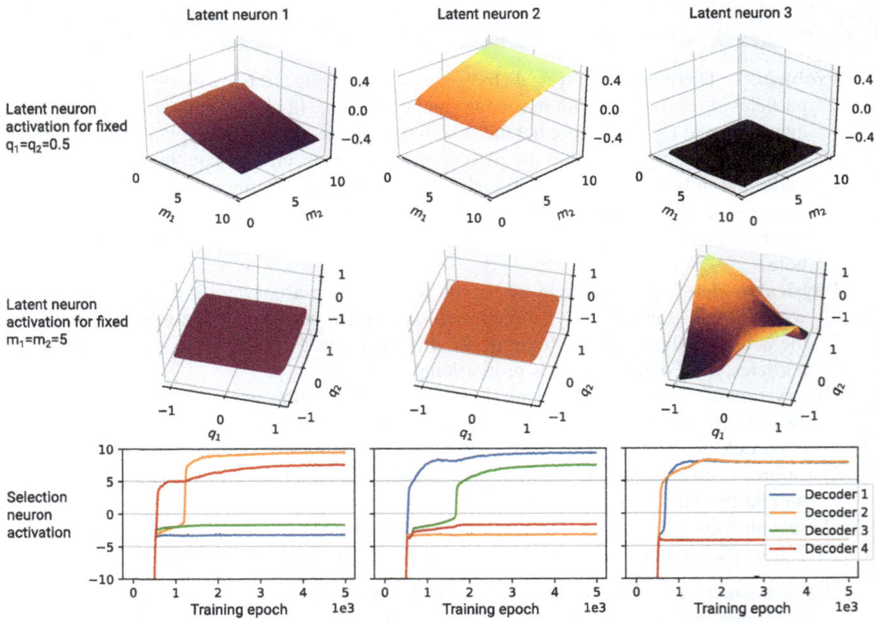

Fig. 10.7 Results for the example with charged masses (single encoder) (Figure taken from Nautrup, Metger and Iten et al., 2020 [23]). The used network (Fig. 9.2) has 3 latent neurons and each column of plots corresponds to one latent neuron. For the first row we generated input data with fixed charges $q_1 = q_2 = 0.5$ (ignoring the units) and variable masses m_1, m_2 in order to plot the activation of latent neurons as a function of the masses. We observe that latent neuron 1 and 2 store the masses m_1, m_2 respectively while latent neuron 3 remains constant. In the second row, we plot the neurons' activation in response to q_1, q_2 with fixed masses $m_1, m_2 = 5$. Here, the third latent neuron approximately stores $q_1 \cdot q_2$, which is the relevant quantity for the Coulomb interaction while the other neurons are independent of the charges.The third row shows which decoder receives information from the respective latent neuron. The y-axis corresponds to $\log(\sigma_i^j)$ for the i-th latent neuron and decoder j. Hence, the y-axis quantifies how much information of the latent neuron is transmitted by the 4 filters to the associated decoder as a function of the training epoch. Positive values mean that the filter does not transmit any information. Decoders 1 and 2 perform non-interacting experiments with particles (m_1, q_1) and (m_2, q_2), respectively. Decoders 3 and 4 perform the corresponding interaction experiments. As expected, we observe that the information about m_1 (latent neuron 1) is received by decoders 1 and 3 and the information about m_2 (latent neuron 2) is used by decoders 2 and 4. Since decoders 3 and 4 answer questions about interaction experiments, the product of charges (latent neuron 3) is received only by them (the green line of decoder 3 in the last plot is hidden below the red one)

with different (known) values of the masses m_1, m_2 and the charges q_1, q_2 against those known values. This corresponds to comparing the learnt representation to a hypothesised representation that we might already have. The plots are shown in Fig. 10.7. The first and second latent neurons are linear in m_1 and m_2, respectively, and independent of the charges; the third latent neuron has an activation that resembles the function $q_1 \cdot q_2$ and is independent of the masses. This means that the first and third latent neurons store the masses individually, as would be expected since the setup in Fig. 10.6a only requires individual masses and no charges. The third neuron roughly stores the product of the charges, i.e., the quantity relevant for the strength of the Coulomb interaction between the charges. This is used by the agents dealing with the setup in Fig. 10.6b, where the particle's trajectory depends on the Coulomb interaction with the other particle.

10.6.2 Multiple Encoders

One can easily generalize the above example to one where one uses several encoding agents. Instead of having a single agent B, we use two agents B_1 and B_2, where each of the agents has only access to one of the charged particles p_1 and p_2, i.e., agent B_i only observes the results of the experiment 10.6 associated with particle p_i. The two encoding agents then have to separately construct a representation of their observations. Then, the two representations are concatenated and treated like in the single-encoder setup; that is, for each decoder, a filter is applied to the concatenated

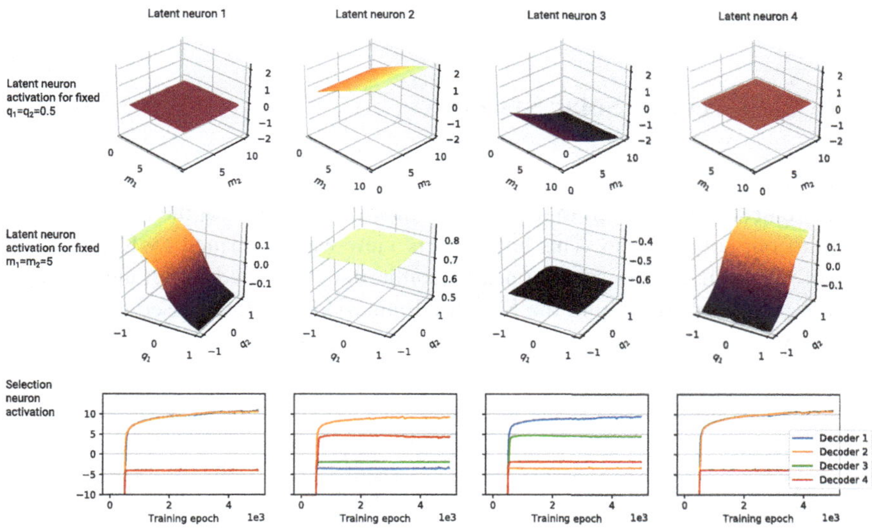

Fig. 10.8 Results for the example with charged masses (two encoders) (Figure taken from Nautrup, Metger and Iten et al., 2020 [23]). The used network (Fig. 9.2) has 4 latent neurons and each column of plots corresponds to one latent neuron. For an explanation of how these plots are generated, see the caption of Fig. 10.7. We observe that latent neurons 2 and 3 store the masses m_1 and m_2, respectively, while latent neurons 1 and 4 are independent of the mass. Latent neurons 1 and 4 store (a monotonic function of) the charges q_1 and q_2, respectively, and are indepependent of m_1 and m_2. The third row shows that the charges q_1 and q_2 are only transmitted to decoders 3 and 4, which are asked to make predictions about interaction experiments (the blue line of decoder 1 and the green line of decoder 3 are hidden under the orange and red lines, respectively, in both of these plots). The mass m_1, stored in the latent neuron 2, is transmitted to decoders 1 and 3, which are the two decoders that make predictions about particle p_1. Analogously, m_2 is transmitted to decoders 2 and 4, which make predictions about particle p_2

representation and the filtered representation is used as input for the decoder. The resulting network structure has similarity to the one in [238] used for summarizing text documents.

The results for this case are shown in Fig. 10.8 and summarized in Box 10. Comparing this result with the single-encoder case, we observe that here, the charges q_1 and q_2 are stored individually in the latent representation, whereas the single encoder stored the product $q_1 \cdot q_2$. This is because, even though the decoders still only require the product $q_1 \cdot q_2$, no single encoder has sufficient information to output this product: the inputs of encoders E_1 and E_2 only contain information about the individual charges q_1 and q_2, respectively, but not their product. Hence, the additional structure imposed by splitting the input among two encoders yields a representation with more structure, i.e., with the two charges stored separately. However the concatenated representation consists of four instead of three latent neurons as in the singe-encoder case, it is still minimal according to Definition 8.8. Indeed, one can see that the chosen structure with two encoders does not allow for a representation with only three latent neurons: If such a representation would exist, one encoder could only output one parameter (and the other encoder outputs two parameters), which is not enough to store the information about the mass and the charge.

10.7 Heliocentric Solar System

In the 16th century, Copernicus used observations of the positions of different planets on the night sky (Fig. 10.9a) to hypothesize that the Sun, and not the Earth, is in the centre of our solar system. This heliocentric view was confirmed by Kepler at the start of the 17th century based on astronomic data collected by Brahe, showing that the planets move around the Sun in elliptical orbits. Here, based on the work in [22], we show that *SciNet* similarly uses heliocentric angles (more exactly, the

Fig. 10.9 Heliocentric model of the solar system. *SciNet* is given the angles of the Sun and Mars as seen from Earth at an initial time t_0 and has to predict these angles for later times. **a** Physical setting. The angles θ_S and θ_M of the Sun and Mars are observed from Earth (relative to the fixed star background). The areas A_E and A_M are swept out by a line segment joining Earth or Mars and the Sun, respectively. These areas are proportional to the mean anomalies $M_E \propto A_E$ and $M_M \propto A_M$ of Earth and Mars, respectively (see (10.1)). **b** Representation learned by *SciNet*. The activations $r_{1,2}$ of the two latent neurons at time t_0 (see Fig. 9.3) are plotted as a function of the mean anomalies M_E and M_M. The plots show that the network stores and evolves parameters that are linear combinations of these heliocentric angles.

mean anomalies) when forced to find a representation for which the time evolution of the variables takes a very simple form, a typical requirement for time-dependent variables in physics.

The observations given to *SciNet* (with the network structure shown in Fig. 9.3 with two latent neurons) are angles $\theta_M(t_0)$ of Mars and $\theta_S(t_0)$ of the Sun as seen from Earth at a starting time t_0 (which is varied during training). The time evolution network $u : \mathbb{R}^2 \mapsto \mathbb{R}^2$ is restricted to addition of a constant (the value of which is learned during training). Hence, we can smoothly parametrize the set of update rules by two parameters $\gamma = (\gamma_1, \gamma_2)$ as $u_\gamma(r_1, r_2) = (r_1 + \gamma_1, r_2 + \gamma_2)$. At each time step i, *SciNet* is asked to predict the angles as seen from Earth at the time $t_i = t_0 + i\Delta t$ based only on its representation at step i. The time step Δt is fixed and is chosen to be 25 d in our implementation. Because the question $\tilde{q}$ is the same at each time step, we do not need to feed it to the decoder explicitly.

Restricting to linear time evolution (by only allowing to add constants γ_i in the update rule) may look like a limitation that does only apply to very particular physical systems. However, such a restriction is actually equivalent to the assumption that the energy of the considered system is conserved. In such a case, there is a variable transformation such that the new canonical coordinates (called *action angles*) evolve linearly in time (see for example [239]).

We simulate the elliptical orbits of Mars and Earth around the sun (note that circular orbits were used in the original work [22]) and train *SciNet* with randomly chosen sequences of observations for around 50 steps of 25 d. Figure 10.9b shows the learned representation and confirms that *SciNet* indeed stores *heliocentric angles*. Indeed, *SciNet* stores a linear combination[5] of the mean anomalies $M_E \in [0, 2\pi)$ and $M_M \in [0, 2\pi)$ of Earth and Mars, i.e., 2π times the fractions of the elliptical orbits' periods that have elapsed since the planets passed their periapsis, the points in the orbits of the planets at which they are nearest to the Sun. The mean anomalies are chosen to lie in the interval $[0, 2\pi)$ because they can be interpreted geometrically as the angles that would be observed from the Sun in the case where Earth and Mars would follow circular orbits (with the same period) around the Sun. An alternative geometrical relation is that the areas A_E and A_M swept out by a line segment joining Earth or Mars and the Sun are proportional to the mean anomalies of Earth and Mars, respectively (see Fig. 10.9). More concretely, we have

$$A_E = \frac{M_E}{2\pi} A_E^{\text{tot}} \text{ and } A_M = \frac{M_M}{2\pi} A_M^{\text{tot}}, \tag{10.1}$$

where A_E^{tot} and A_M^{tot} denote the full area enclosed by the orbit of Earth and Mars, respectively. Hence, an alternative (and physically equivalent) interpretation is that

---

[5] If desired, one could disentangle the mean anomalies of Earth and Mars by using the methods described in Sect. 9.3.2. Indeed, we could introduce two decoding agents. The first should predict the angle θ_S of the Sun, and the second the angle of the Mars θ_M (as seen from Earth). To solve the tasks perfectly, the first agent requires the information about the mean anomaly of Earth and the second agent requires both, the mean anomaly of Earth and the Mars. Hence, the corresponding operationally meaningful representation would naturally separate the two mean anomalies.

SciNet stores linear combination of the areas swept out by the two planets in its latent neurons. The fact that the swept out areas grow linearly in time is stated in Kepler's second law of planet motion. We stress that the training data only contains angles observed from Earth, but *SciNet* nonetheless switches to a heliocentric representation. The details are given in Box 11.

Box 11: Heliocentric model of the solar system (Sect. 10.7)

Problem: Predict the angles $\theta_M(t)$ and $\theta_S(t)$ of Mars and the Sun as seen from Earth, given initial states $\theta_M(t_0)$ and $\theta_S(t_0)$.

Physical model: Earth and Mars orbit the Sun on elliptical orbits according to Kepler's laws of planet motion. The orbit of Earth is approximately a circle (eccentricity: 0.017) and the orbital velocity is approximately constant. The elliptical orbit of Mars has an eccentricity of 0.093 and the orbital velocity varies up to ±10% with respect to the mean orbital velocity.

Observation: Randomly chosen initial angles of Mars and the Sun as seen from Earth: $o = (\theta_M(t_0), \theta_S(t_0))$.

Question: Implicit.

Correct answer: Time series $[a_0, \ldots, a_n] = [(\theta_M(t_0), \theta_S(t_0)), \ldots, (\theta_M(t_n), \theta_S(t_n))]$ of $n = 20$ (later in training: $n = 50$) observations, with fixed time steps $\Delta t = t_{i+1} - t_i$ of 25 days.

Network structure: Network depicted in Fig. 9.3 with two latent neurons and allowing for time updates of the form $u_\gamma(r_1, r_2) = (r_1 + \gamma_1, r_2 + \gamma_2)$. The network sizes are given in Table C.1.

Training: As described in Box 4 with parameters given in Table C.2.

Key findings:

- *SciNet* predicts the angles of Mars and the Sun with a root mean square error below 0.5% (with respect to 2π).
- *SciNet* stores the mean anomalies M_E and M_M of the Earth and Mars with respect to the Sun in the two latent neurons (see Fig. 10.9b).

10.8 Several Particles Connected by Springs

In this section, we consider a physical systems consisting of many interacting objects as an example to demonstrate how physical prior knowledge about the object-structure can help to interpret *SciNet*'s representation (Sect. 8.6).

More concretely, we consider $k \in \{5, 10\}$ particles in a 2-dimensional box with no external forces, apart from the elastic collisions with the walls of the box [184]. Some particles are connected by a string with a fixed spring constant k, and hence, a force $F_{i,j} = k \left\| p_i - p_j \right\|_2$ acts between two particles i and j with positions $p_i \in \mathbb{R}^2$ and $p_j \in \mathbb{R}^2$ that are connected by such a spring. The details and the sample strategy are described in Box 12. The observation input to *SciNet* (with the structure given in Fig. 9.5 using GNNs) is given as time sequences of the 2-dimensional positions and velocities of the particles. The one-hot embeddings $h_{i,j} \in \mathbb{R}^2$ of the edges of the GNN used for the encoder can encode two different types of interactions. The

GNN is then evolved for $L = 2$ steps and representations $r_{i,j}$ for the interaction types between any objects i and j are sampled.

The decoder again consists of a GNN and takes the representation r as an input together with a question q that is described by an initial state of the positions and velocities of the particles at time t' as well as a number $M \in \mathbb{N}$ describing the number of time steps that the state of the particles should be evolved by for a time difference Δt. The decoder then has to predict the future states of the particles at a time $t' + M\Delta t$.

SciNet learns to predict the paths of the particles and the type of interactions between the particles with high accuracy [184]. It finds that there are two types of interactions between two particles, namely "no interaction" and "connected by a spring". This information can be extracted from the one-hot encodings that *SciNet* stores in its representation r.

Box 12: Several particles connected by springs [184] (Sect. 10.8)

Problem: Predict the time evolution of the positions and velocities of $k \in \{5, 10\}$ particles in a 2-dimensional box.

Physical model: The force on particle i at position p_i is given by the sum of the spring-forces $F_{i,j} = k \left\| p_i - p_j \right\|_2$ of the particles j at position p_j that are connected by a spring with particle i. Further, the particles collide elastically with the walls of the box. The trajectories of the particles are then simulated by solving Newton's equation with leapfrog integration [184].

Observation: Time sequences of position and velocities of all particles starting from a randomly chosen initial state: Randomly chosen initial positions p_i, where both components of p_i are samples from a Gaussian distribution $\mathcal{N}(0, 0.5)$ with mean equal to 0 and standard derivation of 0.5. The velocities of the particle are sampled uniformly from the set of 2-dimensional vectors with norm 0.5. A spring is placed between two particles i and $j \neq i$ with probability 50%.

Question: $q = [x_1(t'), \ldots, x_k(t'), M]$ with $k \in \{5, 10\}$, $M \in \mathbb{N}$ and states $x_i(t') = [p_i(t'), q_i(t')]$, where $p_i(t')$ and $q_i(t')$ are the position and velocity of the particle i at time t', respectively.

Correct answer: State $a = [x_1(t' + M\Delta t), \ldots, x_k(t' + M\Delta t)]$ for some time difference Δt and where $x_i(t)$ is the state of particle i given by its position and velocity at time t.

Network structure: Network depicted in Fig. 9.5 with two evolution steps of the GNN implementing the encoder and two-dimensional edge embeddings $h^L_{i,j} \in \mathbb{R}^2$. The details are given in [184].

Training: As described in Sect. 9.7.2. Further details are given in [184].

Key findings: *SciNet* recovers with high accuracy how the particles interact with each other by classifying their interactions as "no interaction" or "connected by a spring".

Chapter 11
Future Research Directions and Further Reading

In the following, we point out two directions for future research on representation learning for physics. The first direction considers the automatization of searching for strategies to collect relevant observation data. The second points out that it can be challenging to interpret *SciNet*'s representation for data where we do not have a hypothesized representation. We suggest that searching for mathematical expressions for the encoding and decoding mappings may help to tackle this challenge. In addition, such an approach could increase *SciNet*'s ability to generalize to observations that are substantially different than the ones seen during training.

11.1 Finding Measurement Strategies and Representations

As mentioned in the introduction (Chap. 1), deciding how to collect observations given some experimental devices can be challenging. In our examples considered in Chap. 10, observing a time series of a particle or performing random measurements on quantum systems provides sufficient information about the physical systems to response to the asked questions. However, in general, more advanced observation strategies might be required. For example, to determine a position of a particle in a box, one may first use a measurement device that scans the whole box and returns the position of the particle with low accuracy. Then, one may use a second, more precise measurement device that can only scan a small part of the box. This device might target the approximate position of the particle found by the first measurement device to determine the position of the particle with high accuracy. Hence, the optimal setting of the second measurement device depends on the output of the first one. Choosing a random part of the box that is scanned by the second measurement device would mostly lead to not finding the particle and hence, the accuracy for the position measurement would stay low.

To find such advanced measurement strategies, one may train a reinforcement learning agent who gets a reward if he predicts the position of the particle in the box with high accuracy. However, to train such an agent in a supervised way, we must

R. Iten, *Artificial Intelligence for Scientific Discoveries*,
https://doi.org/10.1007/978-3-031-27019-2_11

know what variables we are interested in (in the example above, it is the position of the particle). In the setting considered in this part of the book, we only know that the observed information should be sufficient to response to the questions that we ask *SciNet*. Hence, following this spirit, the measurement strategy must be learned in parallel with how to respond to the questions with the collected information. A simple example of a question for the example given above would be to ask *SciNet* to predict the electrical field at a fixed point in the box. Assuming the particle is charged, the electrical field might be a function $f(x)$ of the position of the particle. In [240], it is demonstrated on the basis of some toy examples, that current machine learning tools can be adapted to learn the measurement strategy and the function f in parallel. It would be interesting to further investigate this research direction and to see if one can learn observation strategies and representations in parallel for more complex systems. State representation learning, discussed in Remark 11.1, is a direction in machine learning which is closely related to this goal.

Remark 11.1 (State representation learning) In contrast to representation learning, where relevant features are extracted from a (fixed) data set, in state representation learning one considers interactive settings [241]. Thereby, a reinforcement learning agent can interact with an environment and influence the state of the system by its actions. The goal of state representation learning is then to construct a low dimensional representation of the system that evolves in time accordingly to the chosen actions of the agent. In particular, such a setting provides new possible priors for disentangling the representation because the agent can experiment with different actions and observe their effects [242]. For instance, in [52], the representation is disentangled by an *independence prior*, which encourages the storage of independently controllable features of the environment in separate parameters (see also [243] for a similar approach). As demonstrated in [53, 243], such requirements can lead to natural representations in some scenarios such as creating an abstract representation of a labyrinth for navigational tasks. However, in contrast to an operational meaningful representation that depends on the agents' aims (Sect. 8.4.2.2), such representations depend on which actions (and policies) the agent can perform. In physics, these actions depend on the experimental devices and may be considered less fundamental than the agents' aims. Closer to the approach of disentangling a representation with respect to different questions (Sect. 8.4.2.2), in [244] several reinforcement learning agents with different goals share a common representation which is optimzed during training. Interestingly, they find that learning auxiliary tasks can improve the performance of the agent in learning of the overall objective. A further step was taken in [245], where the auxiliary tasks do not have to be specified by humans, but are learned during the training process. In a similar spirit, it would be interesting to learn the relevant questions for *SciNet* instead of choosing them by hand in future work.

11.2 Interpretability and Generalization of *SciNet*

A lot of effort in machine learning is currently invested in improving the interpretability of artificial neural networks (see e.g. [246] for a recent review). For example, inceptionism techniques [247, 248] can be used to gain some insight into what a neural network has learned. Such techniques were used in [249] to extract some relevant properties from thermodynamic systems. Building prior knowledge about the training data into the network structure often improves interpretability. Usually, the interpretability is related to the generalization power of the neural network, and an improvement in interpretability also increases the generalization power. For example, building in information about the object structure of physical systems, as described in Sect. 8.6 and Sect. 9.7, allows the network to generalize to systems with different numbers and configurations of objects and relations. At the same time, the additional structure in the network helps to extract some information about what the network has learned. Indeed, it is demonstrated in Sect. 10.8, how one can extract information about the interaction types between particles from *SciNet*'s representation. On the other hand, building some mathematical prior knowledge into *SciNet* (see Sect. 9.6 for an example) can help to interpret the found representation as demonstrated for the nonlinear pendulum in [183] (and discussed in Sect. 10.3).

For the long term goal of applying the methods discussed in this book for the foundations of physics, one must be very careful building in prior knowledge into *SciNet*'s network structure. In particular, by assuming an object structure of the observation data as considered in Sect. 8.6, it could be that we separate objects in a way that is not "natural" and that there is a more elegant description of the system with a different partitioning of the system into subsystems. Hence, in this part of the book, we focussed on extracting physical concepts from observation data with minimal prior knowledge about physics and mathematics. Such an approach has the advantage that we do not restrict the set of models that can be learned by the neural network. On the other hand, this is a tradeoff with the generalization power of *SciNet* as well as with its interpretability. To interpret the compressed representation found by *SciNet* without using mathematical or physical prior knowledge, we have to plot it against a hypothesized representation. The challenge for future work is hence to find other methods that use a minimal amount of prior knowledge, and still allow to extract some conceptual information from the machine learning system without any known hypothesis.

One possibility to improve the interpretability as well as the generalization power of *SciNet* would for example be to include a bias that is very common in physics: given the relevant variables and parameters of a physical system, we assume that there is a "simple" mathematical expression that predicts the (future) physical properties of the system (at least for small time steps). Hence, more concretely, we may assume that the decoders of *SciNet* can be described by simple mathematical expressions. A step in this direction was recently done in [250]. The authors in [250] use Equation Learner (EQL) networks, similar to the work in [251, 252], which use activation functions that represent simple mathematical operations such as multiplication or taking the sine or cosine. Further, to find "simple" mathematical expressions, they enforce sparsity for the weights in the network. The more weights are set to zero, the shorter the mathematical expression represented by the EQL network becomes. In [250], EQL networks are then applied to different problems, including a dynamical system analysis of a simple kinematic problem and a harmonic oscillator. They use a similar network structure as the one used in Sect. 9.4 to find representations with simple update rules (Fig. 9.3) with a trivial decoder and where the update rules are parametrized by an EQL network. In both examples, the EQL network is able to find simple mathematical expressions for the time evolution of the physical systems. It would be interesting to replace the decoders in *SciNet* with EQL networks and try to extract simple mathematical expressions from them. Such expressions correspond to applications of physical laws in our setting and hence may help to discover the underlying laws themselves. Further, finding mathematical expressions for the encoder could also help to interpret the representation found by *SciNet*. However, for physical examples, the mapping from observation data to the representation is typically more complex than the one from the representation to the predictions. Hence, it could be challenging to find such expressions with EQL networks. Alternatively, one may use feed-forward neural networks for the encoder and try to extract the equation describing its mapping from the trained network using methods from symbolic regression [24] (see Sect. 6.2.1). A step in this direction was considered in [129] (see Remark 6.1 for the method introduced in [129]). Recently, the additional structure of graph neural networks was exploited to extract equations from different parts of the network [205, 253]. The authors of [205] apply their techniques to a cosmology example and discover a new analytic formula which can predict the concentration of dark matter from the mass distribution of nearby cosmic structures. In [253] a graph neural network is trained to simulate the dynamics of our solar system's Sun, planets, and large moons from thirty years of trajectory data. Then, symbolic regression is used to extract an analytical expression for the force law implicitly learned by the graph neural network, which turns out to be equivalent to Newton's law of gravitation.

Part IV
Future Prospects

In the last part of this book, we discuss future prospects for the automatization of a physicist's discovery process as well as for using machine learning for the foundation of physics. As it is extremely difficult to make time-related predictions about the future development of AI, we will instead focus on pointing out challenging problems that must be solved to fully automatize the physical discovery process. Furthermore, we point out recent steps towards this long-term goal and particularly promising methods that must be further improved in the future to be applicable to real-world problems.

Chapter 12
Future Prospects

In this book, we considered the automatization of several steps in a physicist's discovery process. However, we are still far away from building an AI scientist that is able to replace human scientists. In fact, developing an AI physicist is probably almost as difficult as developing general artificial intelligence, which, according to most AI researchers, will still be decades away [254, 255]. On the one hand, there are many open questions in automatizing each single step of a physicist's discovery process, and on the other hand, combining all steps is also a highly nontrivial challenge. In the following sections, we discuss recent progress in automatizing several steps of a physicist's discovery process together and future challenges in this direction (Sect. 12.1), how learning procedures instead of simple functions may help to build general artificial intelligence (Sect. 12.2) and last but not least, how AI may help to solve fundamental problems in physics in the future.

12.1 AI Physicist

The ultimate long-term goal would be to build AI physicists, i.e., systems that act and learn autonomously from a physical environment. There is no "exact" definition of an AI physicist and finally, one could also end up building AI systems that are specialized for certain domains and can communicate with each other to share information. This would be similar to human physicists, who are also experts in certain fields, but could not perform all the tasks that other physicists can do. For example, my performance in building up an experiment with devices in a laboratory would probably be close to the performance of an agent trying out random actions. However, at the current time we are so far away from creating an AI physicist who could fully replace *any* human physicist that we may ignore the fact that the goal is not completely specified. For concreteness, one may think of the following goal: building an agent that can interact with the real world and for example finds the law of gravitation. To test if the agent really found the law of gravitation, one could let it predict the movement of bodies that move according to the gravitational force in a completely new situation that the

R. Iten, *Artificial Intelligence for Scientific Discoveries*,
https://doi.org/10.1007/978-3-031-27019-2_12

agent has never seen before. Furthermore, the agent should be able to memorize the found insights, and combine them with future insights to unify different theories. In the following, we first describe some recent progress in building AI physicists, and then discuss some essential challenges that are left for future work on building AI physicists.

The first machine in history able to discover new scientific knowledge independently of its human creators was the robot "Adam" [256]. Adam can autonomously perform the scientific cycles of setting up an hypothesis, physically run experiments using a laboratory robot to test the hypothesis and finally, interpret the results. If Adam falsifies the hypothesis, he repeats the cycle. A more recent step towards an artificial intelligence physicist was taken in [4] by also considering several steps of a physicist's discovery process together. In contrast to most other work, where a machine learning system is used to fit all the given data at once, in [4] the environment is separated into several subdomains using unsupervised learning. The subdomains are chosen in a way, such that a physical law is assigned with each subdomain and performs well in predicting the future evolution of the physical system in this domain. This is a typical strategy of physicists; choosing and isolating a physical system that is then investigated further. As a toy example, we may consider two regions in the xy-plane, where we have a harmonic potential or a homogeneous gravitational field in region one or two, respectively. Hence, the equations of motion for a particle of mass m are given by $\ddot{x} = -k/m(x - x_0)$ and $\ddot{y} = -k/m(y - y_0)$ with spring constant k and equilibrium position (x_0, y_0) in the first region, and by $\ddot{x} = g_x$ and $\ddot{y} = g_y$ for a gravitational field $g = (g_x, g_y)$ in the second region. The machine learning method suggested in [4] is then able to recognize the two domains and uses two neural networks to predict the time evolution of the particle in each region separately. To improve generalization power and interpretability, symbolic expressions are extracted for the two neural networks and added to a *theory hub* [4]. In the example above, the following expressions might be found for the region with a homogeneous gravitational field $\ddot{x} = 2$ and $\ddot{y} = 3\ \mathrm{m/s^2}$, where we assume that $g_x = 2$ and $g_y = 3\ \mathrm{m/s^2}$ for concreteness. The theory is kept in the memory of the AI physicist and is used to investigate new environments. Furthermore, some basic unification of theories is considered in [4], where parametrized *master theories* are found. In the example considered above, this unification algorithm would recognize that the two formulas $\ddot{x} = 2$ and $\ddot{y} = 3\ \mathrm{m/s^2}$ are of the same form and can be considered as instances of the master theory $\ddot{u} = g_u\,\mathrm{m/s^2}$, for any coordinate u and a parameter g_u.

The approach to build an AI physicist described in [4] goes beyond finding a model for one given physical system and also points towards several challenging problems that have to be solved to build a general AI physicist and which we discuss in the following. First, it becomes clear that one of the main difficulties for a physicist is to decompose the environment into subsystems in such a way that a "simple" description of the subsystems is possible. Furthermore, one may want to choose subsystems whose investigation likely leads to the discovery of new physical concepts. In fact, a modern physicist may not want to consider a standard pendulum, but preferably colliding particles with high energy, because such scenarios are less well understood and can possible lead to new physical insights. Thinking about all the information a human obtains through the sensory organs every second, recognizing subsystems and choosing and isolating them for experiments that may lead to new physical insights is an extremely hard task. A first step towards this goal might be to find compact representations of parts of the environment, since this simplifies investigations. Furthermore, finding representations that are not sufficient to predict the future evolution of a subsystem, points towards incomplete understanding of these systems. Steps towards extracting representations directly from video data were done in [34, 192]. For example, in [34] the position of an object is extracted from raw visual data. Such methods may turn out useful for building future AI physicists.

Another difficulty in building a fully fledged AI physicist which is often ignored, but considered in [4], lies in building on existing theories and unifying observed laws from different systems. Fully automatizing the unification of theories and searching for underlying physical principles might be extremely hard and can be equally difficult to achieve as general AI. The ultimate goal of unification is to improve the generalization power of a physicist. Hence, one may estimate the generalization performance of an AI physicist by testing its prediction power in a completely new environment not seen during training, but with the same underlying laws as the training environment. An active direction of research in machine learning called *transfer learning* considers reusing learned concepts in an environment to a new environment (see [257] for a comprehensive survey). Progress in this area may also lead to progress in generalizing physical theories that work well in certain environments to a theory that applies to all considered environments.

12.2 Learning Procedures Instead of Simple Functions

Currently, neural networks are the most important tool for AI. In this book, we mostly used feedforward neural networks for fitting functions. However, a general AI agent cannot be built using feedforward neural networks, since general AI should be able to perform any algorithm that could be run on a Turing machine. For example, consider an AI physicist for the task of calculating the electrical field at a given point in space, given an arbitrary number n of charged particles creating the field. For any fixed number n, a feedforward neural network could be constructed that takes the positions and charges of the n particles as an input and calculates the electrical field at the given point. However, if the number of particles n is a priori unknown, given any feedforward neural network with m input neurons, we can always choose $n > m/4$ particles (note that for each particle we have four inputs, the three-dimensional position and the charge) making the AI physicist fail in predicting the electrical field, since it can not handle this input size. In addition, even if the input size could be handled by the feedforward neural network, it fully ignores the fact that the field created by each particle can be calculated in the same way. Applying the same rule to sequential inputs is naturally handled by recurrent neural networks (as for example used in Sect. 9.4). It was even shown that recurrent neural networks are computationally universal (also called *Turing-complete*) *if* properly wired, i.e., they can be used to simulate any Turing machine and hence any procedure [258]. However, if something is possible in principle, this does not necessarily mean that it will work in practice. In [259], the capacity of recurrent neural networks is enriched to simplify the solution of algorithmic tasks by adding a large addressable memory. This is analog to Turing's enrichment of finite-state machines by an infinite memory tape, and the authors of [259] call this structure a *Neural Turing Machine* (NTM). In contract to a (standard) Turing machine, an NTM is not programmed by a human but does learn programs from data by its own. By ensuring that the outputs depend smoothly on the parameters describing the program, one can use stochastic gradient descent to efficiently train an NTM. Preliminary results show that an NTM can infer simple algorithms such as copying and sorting [259], but it remains as a challenging task for future work to learn programs for more complex tasks from data.

As mentioned above, apart from the additional computational power of NTMs compared to feedforward neural networks, they may learn much simpler descriptions of a procedure. In the example where one wants to calculate the electrical field created by n particles, an NTM would essentially need to learn that it can sequentially consider the particles, calculate the distance to the position of interest, estimate the field according to Coulomb's law and adding up the estimated field strengths on the go. Increasing the number of particles would not make this program any more complex, but would just make it run longer. This is a much more elegant approach than fitting a function on the full input at once, and hence, learning Coulomb's law for each pair of particles separately. To motivate an AI agent to learn simple algorithms instead of complex once, we need some measure of complexity. One option is to use the *Kolmogorov complexity*, which is defined as the minimal description length of a program that produces the desired outputs for given inputs. Note that the minimal description length depends on the available commands that the given Turing machine can execute, and hence, the choice of these commands introduces some bias. Unfortunately, there is no way to efficiently calculate the Kolmogorov complexity for given input-output pairs. Nonetheless, we could motivate an NTM to choose the resulting program that has the smallest description length it can produce. Although this complexity measure is coming from computational theory, it might be the fundamental reason why physicists work and think as they do.[1] Furthermore, searching for programs with short description length also improves the interpretability of NTMs. I would expect that approaches in direction of building trainable Turing machines and searching for simple programs will be essential for the long term goal of building general AI, however, the methods used to implement such approaches may change a lot in the next few decades.

[1] A similar complexity measure was used in [4] to find the simplest formulas describing experimental data.

12.3 AI for Foundations of Physics

Let us finally come back to one of the main motivations behind this book, namely to use AI to help solving fundamental problems in modern physics. The content of this book is focussed on methods that use a minimal amount of prior knowledge about the physical system at hand. For example, extracting representations of quantum systems in Sect. 10.5, we ensure that the agent performing the measurements acts operationally, i.e., does not make use of any prior knowledge about quantum mechanics. Moreover, the network structure, *SciNet*, used to extract the representation is not specifically adapted to quantum systems and is demonstrated to work as well for classical systems such as the damped pendulum (Sect. 10.2). Furthermore, we found that searching for operationally meaningful representations, one finds the representations of the systems commonly used in physical textbooks. This suggests that the required properties of operationally meaningful representations (Sect. 8.4) are also required by human physicists, however, physicists may actually not be aware of this fundamental reason why they are working with parameters like the mass or the charge of a particle.

Therefore, following the spirit of this approach in future work may lead to a better understanding of "hidden" assumptions made by physicists. If, in the far future, one could build an AI agent that discovers the formalism of quantum mechanics by interacting with quantum systems and only assuming a few "natural" requirements on the working of the AI agent (such as searching for operationally meaningful representations), one could stepwise remove the assumed principles and retrain the AI agent. This could lead to the discovery of alternative formulations of quantum mechanics, and hence could ultimately help to overcome the conceptual problems we are currently facing with quantum theory and which may be caused by a "hidden" assumption that we should better give up.

Appendix A
Interpretation of the Number of Latent Variables

Abstract In Sect. 8.4.1 we defined minimal representations and required that they should contain a minimal number parameters. In this appendix [taken from (Iten et al. in Phys Rev Lett 124:010508, 2020 [22])] we use techniques from differential geometry to state a formal relation between the number of parameters in a minimal representation and the degrees of freedom in the considered physical data. This relation gives the minimal number of latent neurons required for *SciNet* (consisting of only one encoder and probably several decoders) to make accurate prediction the expected physical meaning, namely that it corresponds to the degrees of freedom in the physical data that are required to reply to all questions that we may ask *SciNet*.

We describe the given data with triples $(\mathcal{O}, \mathcal{Q}, a^\star)$, where $\mathcal{O}$ and $\mathcal{Q}$ are the sets containing the observation data and the questions respectively, and the function $a^\star : (o, q) \mapsto a$ sends an observation $o \in \mathcal{O}$ and a question $q \in \mathcal{Q}$ to the correct reply $a \in \mathcal{A}$. Note that the case where *SciNet* consists of several decoders can be cast into this setting by stacking all the questions and answers together and by considering the resulting tuples as one question or answer, respectively. The following results only take one encoder into account.

Intuitively, we say that the triple $(\mathcal{O}, \mathcal{Q}, a^\star)$ has dimension at least n if there exist questions in $\mathcal{Q}$ that are able to capture n degrees of freedom from the observation data $\mathcal{O}$. Smoothness of this "oracle" is a natural requirement, in the sense that we expect the dependence of the answers on the input to be robust under small perturbations. The formal definition follows.

Definition A.1 (*Dimension of a data set*) Consider a data set described by the triple $(\mathcal{O}, \mathcal{Q}, a^\star)$, where $a^\star : \mathcal{O} \times \mathcal{Q} \to \mathcal{A}$, and all sets are real, $\mathcal{O} \subseteq \mathbb{R}^r$, $\mathcal{Q} \subseteq \mathbb{R}^s$, $\mathcal{A} \subseteq \mathbb{R}^t$. We say that this triple has dimension at least n if there exists an n-dimensional submanifold $\mathcal{O}_n \subseteq \mathcal{O}$ and questions $q_1, \ldots, q_k \in \mathcal{Q}$ and a function.

R. Iten, *Artificial Intelligence for Scientific Discoveries*,
https://doi.org/10.1007/978-3-031-27019-2

$$\begin{aligned} f: \quad \mathcal{O}_n &\to \mathcal{A}^k := \overbrace{A \times A \times \cdots \times A}^{k} \\ o &\mapsto [a^\star(o, q_1), \ldots, a^\star(o, q_k)] \end{aligned}$$

such that $f : \mathcal{O}_n \to f(\mathcal{O}_n)$ is a diffeomorphism.

Proposition A.1 (Minimal representation for *SciNet*) *A (sufficient) latent representation for data described by a triple*$(\mathcal{O} \subset \mathbb{R}^r, \mathcal{Q} \subset \mathbb{R}^s, a^\star : \mathcal{O} \times \mathcal{Q} \to \mathcal{A} \subset \mathbb{R}^t)$ *of dimension at least*n *requires at least* n *latent variables.*

Proof By assumption, there is an n-dimensional submanifold $\mathcal{O}_n \subset \mathcal{O}$ and k questions $q_1, \ldots, q_k$ such that $f : \mathcal{O}_n \to \mathcal{I}_n := f(\mathcal{O}_n)$ is a diffeomorphism. We prove the statement by contradiction: assume that there exists a (sufficient) representation described by an encoder $E : \mathcal{O} \to \mathcal{R}_m \subset \mathbb{R}^m$ with $m < n$ latent variables. By sufficiency of the representation, there exists a smooth decoder $D : \mathcal{R}_m \times \mathcal{Q} \to \mathcal{A}$ such that $D(E(o), q) = a^\star(o, q)$ for all observations $o \in \mathcal{O}$ and questions $q \in \mathcal{Q}$. We define the smooth map

$$\begin{aligned} \tilde{D} : \mathcal{R}_m &\to \mathcal{A}^k \\ r &\mapsto [D(r, q_1), \ldots, D(r, q_k)], \end{aligned}$$

and denote the pre-image of $\mathcal{I}_n$ by $\tilde{\mathcal{R}}_m := \tilde{D}^{-1}(\mathcal{I}_n)$.

By sufficiency of the representation, the restriction of the map $\tilde{D}$ to $\tilde{\mathcal{R}}_m$ denoted by $\tilde{D}|_{\tilde{\mathcal{R}}_m} : \tilde{\mathcal{R}}_m \to \mathcal{I}_n$ is a smooth and surjective map. However, by Sard's theorem (see for example [260]), the image $\tilde{D}(\tilde{\mathcal{R}}_m)$ is of measure zero in $\mathcal{I}_n$, since the dimension of the domain $\tilde{\mathcal{R}}_m \subset \mathbb{R}^m$ is at most m, which is smaller than the dimension n of the image $\mathcal{I}_n$. This contradicts the surjectivity of $\tilde{D}|_{\tilde{\mathcal{R}}_m}$ and finishes the proof.

We can consider an autoencoder as a special case of *SciNet*, where we ask always the same question and expect the network to reproduce the observation input. Hence, an autoencoder can be described by a triple $(\mathcal{O}, \mathcal{Q} = \{0\}, a^\star : (o, 0) \mapsto o)$. As a corollary of Proposition A.1, we show that in the case of an autoencoder, the required number of latent variables corresponds to the "relevant" number of degrees of freedom that describe the observation input. The relevant degrees of freedom, which are called *(hidden) generative factors* in this context in representation learning (see for example [49]), may be described by the dimension of the domain of a smooth nondegenerate data generating function H, defined as follows.

Definition A.2 We say that a smooth function $H : \mathcal{G} \subset \mathbb{R}^d \to \mathbb{R}^r$ is nondegenerate if there exists an open subset $\mathcal{N}_d \subset \mathcal{G}$ such that the restriction $H|_{\mathcal{N}_d} : \mathcal{N}_d \to H(\mathcal{N}_d)$ of H on $\mathcal{N}_d$ is a diffeomorphism.

One may think of H as sending a small dimensional representation of the data onto a manifold in a high dimensional space of observations.

Corollary A.1 *(***Minimal representation for an autoencoder***) Let $H : \mathcal{G} \subset \mathbb{R}^d \to \mathcal{O} \subset \mathbb{R}^r$ be a smooth, nondegenerate and surjective (data generating) function, and let us assume that $\mathcal{G}$ is bounded. Then the minimal representation for data described by a triple $(\mathcal{O}, \mathcal{Q} = \{0\}, a^\star : (o, 0) \mapsto o)$ contains d latent variables.*

Proof First, we show the existence of a (sufficient) representation with d latent variables. We define the encoder mapping (and hence the representation) by $E : o \mapsto \text{argmin}[H^{-1}(\{o\})] \in \mathcal{G}$, where the minimum takes into account only the first vector entry.[1] We set the decoder equal to the smooth map H. By noting that $D(E(o), 0) = o$ for all $o \in \mathcal{O}$, this shows that d latent variables are sufficient.

Let us now show that there cannot exist a representation with less than d variables. By definition of a nondegenerate function H, there exists an open subset $\mathcal{N}_d \subset \mathcal{G}$ in $\mathbb{R}^d$ such that $H|_{\mathcal{N}_d} : \mathcal{N}_d \to H(\mathcal{N}_d)$ is a diffeomorphism. We define the function $f : o \in H(\mathcal{N}_d) \mapsto a^\star(o, 0) \in \mathcal{I}$, where $\mathcal{I} = H(\mathcal{N}_d)$. Since f is the identity map and hence a diffeomorphism, the data described by the triple $(\mathcal{O}, \mathcal{Q} = \{0\}, a^\star : (o, 0) \mapsto o)$ has dimension at least d. By Proposition A.1, we conclude that at least d latent variables are required.

[1] Note that any element in $H^{-1}(\{o\})$ could be chosen.

Appendix B
Variational Autoencoders

Abstract In this appendix [taken from (Iten et al. in Phys Rev Lett 124:010508, 2020 [22])], we describe variational autoencoders (VAEs) (Higgins et al. in ICLR, 2017 [49], Kingma and Welling in 2013 [193]), a tool from machine learning that can be used to find compact representations of data sets. In particular, we consider beta-VAE (Higgins et al. in ICLR, 2017 [49], an extension of VAEs that is meant to disentangle the parameters in the latent representation. We use beta-VAEs in this thesis as a tool to find statistically independent representations (Sect. 8.4.2.1).

The implementation of *SciNet* uses a modified version of so-called variational autoencoders (VAEs) [49, 193]. The standard VAE architecture does not include the question input used by *SciNet* and tries to reconstruct the input from the representation instead of answering a question. VAEs are one particular architecture used in the field of representation learning [27]. Here, we give a short overview over the goals of representation learning and the details of VAEs.

Representation learning.

The goal in representation learning is to map a high-dimensional input vector x to a lower-dimensional representation $z = (z_1, z_2, \ldots, z_d)$, commonly called the *latent vector*.[2] The representation z should still contain all the relevant information about x. In the case of an autoencoder, z is used to reconstruct the input x. This is motivated by the idea that the better the (low-dimensional) representation is, the better the original data can be recovered from it. Specifically, an autoencoder uses a neural network (*encoder*) to map the input x to a small number of latent neurons z. Then, another neural network (*decoder*) is used to reconstruct an estimate of the input, that is $z \mapsto \tilde{x}$. During training, the encoder and decoder are optimized to maximize the reconstruction accuracy and reach $\tilde{x} \approx x$.

[2] The variables x and z correspond to the observation o and the representation r used in the main text.

R. Iten, *Artificial Intelligence for Scientific Discoveries*,
https://doi.org/10.1007/978-3-031-27019-2

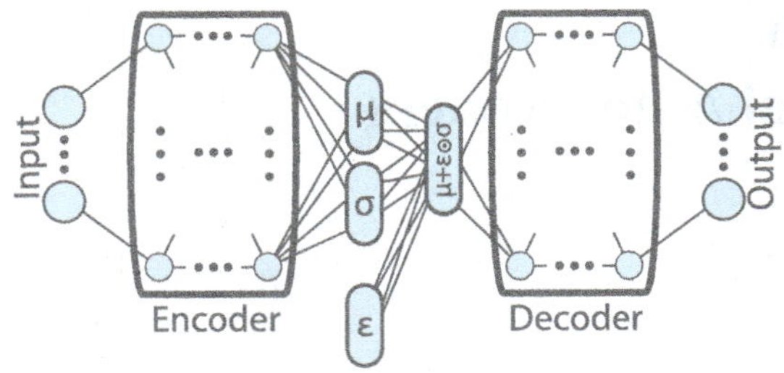

Fig. B.1 Network structure for a variational autoencoder [22]. The encoder and decoder are described by conditional probability distributions $p(z|x)$ and $p(x|z)$ respectively. The output distribution of the encoder are the parameters μ_i and $\log(\sigma_i)$ for independent Gaussian distributions $z_i \sim \mathcal{N}(\mu_i, \sigma_i)$ of the latent variables. The reparameterization trick is used to sample from the latent distribution

Probabilistic encoder and decoder.

Instead of considering deterministic maps $x \mapsto z$ and $z \mapsto \tilde{x}$, we generalize to conditional probability distributions $p(z|x)$ for the encoder and $p(\tilde{x}|z)$ for the decoder. This is motivated by the Bayesian view that the most informative statement the encoder can output a description of a probability distribution over all latent vectors, instead of outputting a single estimate. The same reasoning holds for the decoder. We use the notation $z \sim p(z)$ to indicate that z is picked at random according to the distribution p.

We cannot treat the general case analytically, so we make restricting assumptions to simplify the setting. First we assume that the input can be perfectly compressed and reconstructed by an encoder and decoder which are both neural networks, that is we assume that the ideal distributions $p(z|x)$ and $p(\tilde{x}|z)$ that reach $x = \tilde{x}$ are members of parametric families $\{p_\phi(z|x)\}_\phi$ and $\{p_\theta(\tilde{x}|z)\}_\theta$, respectively. We further assume that it is possible to achieve this with a latent representation where each neuron is independent of the others, $p_\phi(z|x) = \prod_i p_\phi(z_i|x)$. If these distributions turn out hard to find for a given dimension d of the latent representation, we can try to increase the number of neurons of the representation to disentangle them. Finally, we make one more simplifying assumption, which is justified *a posteriori* by good results: that we can reach a good approximation of $p(z|x)$ by using only independent normal distributions for each latent neuron, $p_\phi(z_i|x) = \mathcal{N}(\mu_i, \sigma_i)$, where μ_i is the mean and σ_i the variance. We can think of the encoder as mapping x to the vectors $\mu = (\mu_1, \ldots, \mu_d)$ and $\sigma = (\sigma_1, \ldots, \sigma_d)$.

The optimal settings for ϕ and θ are then learned as follows, see Fig. B.1:

1. The encoder with parameters (weights and biases) ϕ maps an input x to $p_\phi(z|x) = \mathcal{N}[(\mu_1, \ldots, \mu_d), (\sigma_1, \ldots, \sigma_d)]$.
2. A latent vector z is sampled from $p_\phi(z|x)$.
3. The decoder with parameters (weights and biases) θ maps the latent vector z to $p_\theta(\tilde{x}|z)$.
4. The parameters ϕ and θ are updated to maximize the likelihood of the original input x under the decoder distribution $p_\theta(\tilde{x}|z)$.

Reparameterization trick.

The operation that samples a latent vector z from $p_\phi(z|x)$ is not differentiable with respect to the parameters ϕ and θ of the network. However, differentiability is necessary to train the network using stochastic gradient descent. This issue is solved by the reparameterization trick introduced in [193]: if $p_\phi(z_i|x)$ is a Gaussian with mean μ_i and standard deviation σ_i, we can replace the sampling operation using an auxiliary random number $\epsilon_i \sim \mathcal{N}(0, 1)$. Then, a sample of the latent variable $z_i \sim \mathcal{N}(\mu_i, \sigma_i)$ can be generated by $z_i = \mu_i + \sigma_i \epsilon_i$. Sampling ϵ_i does not interfere with the gradient descent because ϵ_i is independent of the trainable parameters ϕ and θ. Alternatively, one can view this way of sampling as injecting noise into the latent layer [261].

β-VAE cost function.

A computationally tractable cost function for optimizing the parameters ϕ and θ was derived in [193]. This cost function was extended in [49] to encourage independency of the latent variables $z_1, \ldots, z_d$ (or to encourage "disentangled" representations, in the language of representation learning). The cost function in [49] is known as the β-VAE cost function,

$$C_\beta(x) = -\left[e_{z\sim p_\phi(z|x)} \log p_\theta(x|z)\right] + \beta\ D_{\mathrm{KL}}\left[p_\phi(z|x)\|h(z)\right],$$

where the distribution $h(z)$ is a prior over the latent variables, typically chosen as the unit Gaussian[3], $\beta \geq 0$ is a constant, and D_{KL} is the Kullback-Leibler (KL) divergence, which is a quasi-distance[4] measure between probability distributions,

$$D_{\mathrm{KL}}\left[p(z)\|q(z)\right] = \sum_z p(z) \log\left(\frac{p(z)}{q(z)}\right).$$

Let us give an intuition for the motivation behind the β-VAE cost function. The first term is a log-likelihood factor, which encourages the network to recover the input data with high accuracy. It asks "for each z , how likely are we to recover the original x after the decoding?" and takes the expectation of the logarithm of this likelihood $p_\theta(x|z)$ (other figures of merit could be used here in an alternative to the logarithm) over z sampled from $p_\phi(z|x)$, in order to simulate the encoding. In practice, this expectation is often estimated with a single sample, which works well enough if the mini-batches are chosen sufficiently large [193].

The second term encourages disentangled representations, and we can motivate it using standard properties of the KL divergence. Our goal is to minimize the amount of correlations between the latent variables z_i: we can do this by minimizing the distance $D_{\mathrm{KL}}\left[p(z)\|\prod_i p(z_i)\right]$ between $p(z)$ and the product of its marginals. For any other distribution with independent z_i, $h(z) = \prod_i h(z_i)$, the KL divergence satisfies

[3] The interpretation of $h(z)$ as a prior is clear only when deriving VAEs as generative networks. For details, see [193].

[4] The KL divergence satisfies all axioms of a metric apart from symmetry.

$$D_{\mathrm{KL}}\left[p(z)\|\prod_i p(z_i)\right] \leq D_{\mathrm{KL}}\left[p(z)\|h(z)\right].$$

The KL divergence is furthermore jointly convex in its arguments, which implies

$$\begin{aligned}D_{\mathrm{KL}}&\left[\sum_x p(x)\; p_\theta(z|x)\|h(z)\right]\\ &\leq \sum_x p(x)\; D_{\mathrm{KL}}\left[p_\theta(z|x)\|h(z)\right].\end{aligned}$$

Combining this with the previous inequality, we obtain

$$\begin{aligned}D_{\mathrm{KL}}&\left[p(z)\;\|\prod_i p(z_i)\right]\\ &\leq e_{x\sim p(x)}\; D_{\mathrm{KL}}\left[p(z|x)\|h(z)\right].\end{aligned}$$

The term on the right hand side corresponds exactly to the second term in the cost function, since in the training we try to minimize $e_{x\sim p(x)}C_\beta(x)$. Choosing a large parameter β also penalizes the size of latent representation z, motivating the network to learn an efficient representation. For an empirical test of the effect of large β see [49], and for another theoretical justification using the information bottleneck approach see [261].

To derive an explicit form of C_β for a simple case, we again assume that $p_\phi(z|x) = \mathcal{N}(\mu, \sigma)$. In addition, we assume that the decoder output $p_\theta(\tilde{x}|z)$ is a multivariate Gaussian with mean $\hat{x}$ and fixed covariance matrix $\hat{\sigma} = \frac{1}{\sqrt{2}}\mathrm{id}$. With these assumptions, the β-VAE cost function can be explicitly written as

$$C_\beta(x) = \|\hat{x} - x\|_2^2 - \frac{\beta}{2}\left(\sum_i \log(\sigma_i^2) - \mu_i^2 - \sigma_i^2\right) + C\,. \tag{B.1}$$

The constant terms C do not contribute to the gradients used for training and can therefore be ignored.

Appendix C
Implementation Details

Abstract In this appendix [which is based on (Iten et al. in Phys Rev Lett 124:010508, 2020 [22])] we provide some details about the implementations used for the examples discussed in Chap. 10. We specify the sizes of the neural networks that were used and provide the most relevant parameters used to train the networks. The full details can be found in the open source code.

The following implementation details refer to Sects. 10.2, 10.4, 10.5, 10.6 and 10.7. For the implementation details for the examples given in Sects. 10.3 and 10.8, we refer to [183, 184], respectively, and their open source code given here:

https://github.com/BethanyL/DeepKoopman and https://github.com/ethanfetaya/nri.

The neural networks used in this work are specified by the input sizes for question and observation inputs, the output sizes, the number of latent neurons, and the sizes of encoders and decoders. The number of neurons in the encoders and decoders are not expected to be important for the results, provided the encoders and decoders are large enough to not constrain the expressivity of the network (and small enough to be efficiently trainable). The training is specified by the number of training examples, the batch size, the number of epochs, the learning rate and the value of the parameter β, which regulates the tradeoff between disentangling and accurate predictions. Note that dependent on the considered example, β corresponds to the parameter in the cost function of the beta-VAE given in (B.1) or the one appearing in the cost function to find operationally meaningful representations given in (9.1), respectively. To test the networks, we use a number of previously unseen test samples. We give the values of these parameters for the examples presented in Chap. 10 in Tables C.1 and C.2.

The source code, all details about the network structure and training process, and pre-trained *SciNets* are available at

https://github.com/eth-nn-physics/nn_physical_concepts and https://github.com/tonymetger/communicating_scinet.

The networks were implemented using the Tensorflow library [212]. For all examples, the training process only takes a few hours on a standard laptop.

R. Iten, *Artificial Intelligence for Scientific Discoveries*,
https://doi.org/10.1007/978-3-031-27019-2

Table C.1 Parameters specifying the network structure for the examples in Chap. 10

Example	Observation input size	Question input size	# Latent neurons	Output size	Encoder	Decoder
Pendulum	50	1	3	1	[500, 100]	[100, 100]
Collision	30	16	1	2	[150, 100]	[100, 150]
One qubit (pure state)	10	10	0–5	1	[100, 100]	[100, 100]
Two qubits (pure state)	30	30	0–9	1	[300, 100]	[100, 100]
Two qubits (mixed state)	75	[75, 75, 75]	20	[1, 1, 1]	[500, 250]	[[500, 100], [500, 100], [500, 100]]
Charged particles (single encoder)	40	[1, 1, 1, 1]	3	[1,1,1,1]	[500, 250]	[[100, 100], [100, 100], [100, 100], [100, 100]]
Charged particles (multiple encoder)	[20, 20]	[1, 1, 1, 1]	[2, 2]	[1, 1, 1, 1]	[[200, 200], [200, 200]]	[[100, 100], [100, 100], [100, 100], [100, 100]]
Solar system	2	0	2	2	[100, 100]	[100, 100]

The notation $[n_1, n_2]$ is used to describe the number of neurons n_1 and n_2 in the first and the second hidden layer of the encoders (or the decoders), respectively. If several instances of an object are used, they are listed in a list, e.g. the entry [[200, 200], [200, 200]] in the column "Encoder" means that two encoders were used, where each of them consists of two hidden layers consisting of 200 neurons

Table C.2 Parameters specifying the training process for the examples in Chap. 10

Example	Batch size	Learning rate	β	# Epochs	# Training samples	# Test samples
Pendulum	512	10^{-3}	10^{-3}	1000	95,000	5000
Collision	500	$(5 \cdot 10^{-4}, 10^{-4})$	0	(100, 50)	490,000	10,000
One qubit (pure state)	512	$(10^{-3}, 10^{-4})$	10^{-4}	(250, 50)	95,000	5000
Two qubits (pure state)	512	$(10^{-3}, 10^{-4})$	10^{-4}	(250, 50)	490,000	10,000
Two qubits (mixed state)	512	$5 \cdot 10^{-4}$	$2 \cdot 10^{-5}$	1000	95,000	5000
Charged particles (single encoder)	512	10^{-3}	10^{-3}	5000	95,000	5000
Charged particles (multiple encoder)	256	10^{-3}	10^{-3}	5000	95,000	5000
Solar system	256–1024	$(0.5–3) \cdot 10^{-3}$	$(0.5–1) \cdot 10^{-2}$	5000	95,000	5000

For training with two phases, the notation (p_1, p_2) refers to the parameters in the first and second phase, respectively. The last example uses five training phases specified in detail in [262]. Some less relevant parameters used for regularization during the training are not listed here, but can be found in [263, 264]

Appendix D
Representations of Cyclic Parameters

Abstract In this appendix [which is taken from (Iten et al. in Phys Rev Lett 124:010508, 2020 [22])] We explain the difficulty of a neural network to learn representations of cyclic parameters, which was alluded to in the context of the qubit example [Sect. 10.5, see (Pitelis et al. in IEEE Conference on Computer Vision and Pattern Recognition, 2013 [265], Korman in 2018 [266]) for a detailed discussion relevant to computer vision]. In general, this problem occurs if the data $\mathcal{O}$ that we would like to represent forms a closed manifold (i.e., a compact manifold without boundary), such as a circle, a sphere or a Klein bottle. In that case, several coordinate charts are required to describe this manifold.

Let us start with an example: We consider data points lying on the unit sphere $\mathcal{O} = \{(x, y, z) : x^2 + y^2 + z^2 = 1\}$, which we would like to encode into a simple representation. The data can be (globally) parameterized with spherical coordinates $\phi \in [0, 2\pi)$ and $\theta \in [0, \pi]$ where $(x, y, z) = f(\theta, \phi) := (\sin\theta\cos\phi, \sin\theta\sin\phi, \cos\theta)$.[5] We would like the encoder to perform the mapping f^{-1}, where we define $f^{-1}((0, 0, 1)) = (0, 0)$ and $f^{-1}((0, 0, -1)) = (\pi, 0)$ for convenience. This mapping is not continuous at points on the sphere with $\phi = 0$ for $\theta \in (0, \pi)$. Therefore, using a neural network as an encoder leads to problems, as neural networks, as introduced here, can only implement continuous functions. In practice, the network is forced to approximate the discontinuity in the encoder by a very steep continuous function, which leads to a high error for points close to the discontinuity.

In the qubit example, the same problem appears. To parameterize a qubit state ψ with two parameters, the Bloch sphere with parameters $\theta \in [0, \pi]$ and $\phi \in [0, 2\pi)$ is used: the state ψ can be written as $\psi(\theta, \phi) = (\cos(\theta/2), e^{i\phi}\sin(\theta/2))$ (see for example [72] for more details). Ideally, the encoder would perform the map $E : o(\psi(\theta, \phi)) := \left(|\langle\alpha_1|\psi(\theta, \phi)\rangle|^2, \ldots, |\langle\alpha_{N_1}|\psi(\theta, \phi)\rangle|^2\right) \mapsto (\theta, \phi)$ for some fixed binary projective measurements $\alpha_i \in \mathbb{C}^2$. However, such an encoder is not con-

[5] The function f is not a chart, since it is not injective and its domain is not open.

R. Iten, *Artificial Intelligence for Scientific Discoveries*,
https://doi.org/10.1007/978-3-031-27019-2

tinuous. Indeed, assuming that the encoder is continuous, leads to the following contradiction:

$$\begin{aligned}(\theta, 0) &= E(o(\psi(\theta, \phi = 0))) \\ &= E(o(\lim_{\phi \to 2\pi} \psi(\theta, \phi))) \\ &= \lim_{\phi \to 2\pi} E(o(\psi(\theta, \phi))) \\ &= \lim_{\phi \to 2\pi} (\theta, \phi) = (\theta, 2\pi),\end{aligned}$$

where we have used the periodicity of ϕ in the second equality and the fact that the Bloch sphere representation and the scalar product (and hence $o(\psi(\theta, \phi))$) as well as the encoder (by assumption) are continuous in ϕ in the third equality.

Appendix E
Classical Mechanics Derivation for Charged Masse

Abstract In Sect. 10.6 we consider a physical setting with two charged particles, where an agent chooses the angle of a projectile that hits one of the particles. The aim of the agent is to choose the angle such that the particle lands in a hole, which is at a fixed position. In this appendix [which is taken from (Nautrup et al. in Mach Learn Sci Technol 3(4):045025, 2022 [23], Nautrup et al. in Mach Learn Sci Technol, 3:045025 2022 [237])], we derive the analytical solution for this setting, which is then used for supervised training of *SciNet*'s predictions.

In this section, we provide the analytic solution to the charged masses example in Sect. 10.6 that we use to evaluate the cost function for training the neural networks. This is a fairly direct application of the generic Kepler problem, but we include the derivation for the sake of completeness. We use the notation of [267].

The setup we consider is shown in Fig. E.1. Our goal is to derive a function $v_0(\phi)$ that, for fixed q, Q, d_0 and given ϕ, outputs an initial velocity for the left mass such

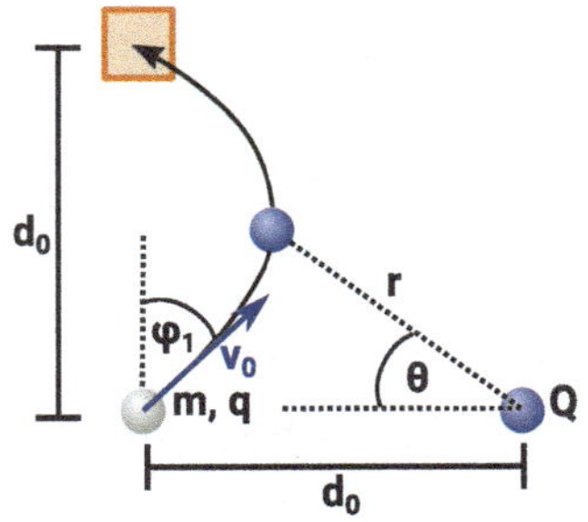

Fig. E.1 Setup and variable names for a charged mass being shot into a holes [23, 237]. A charged particle with mass m and charge q moves in the electrostatic field generated by another charge Q at a fixed position. The initial conditions are given by the velocity v_0 and the angle ϕ. We want to determine the value for ϕ that will result in the particle landing in the target hole, given a velocity v_0

R. Iten, *Artificial Intelligence for Scientific Discoveries*,
https://doi.org/10.1007/978-3-031-27019-2

that the mass will reach the hole. Introducing the inverse radial coordinate $u = \frac{1}{r}$, the orbit $r(\theta)$ of the left mass obeys the following differential equation (see for example [267, Sect. 4.3]):

$$\frac{d^2u}{d\theta^2} + u = \frac{k}{l^2}, \tag{E.1}$$

with the constant

$$k = \frac{-qQ}{4\pi\epsilon_0 m} \tag{E.2}$$

and the mass-normalised angular momentum

$$l = r^2\frac{d\theta}{dt}. \tag{E.3}$$

This is a conserved quantity and we can determine it from the initial condition of the problem

$$l = d_0 v_0 \cos\phi. \tag{E.4}$$

The general solution to Eq. (E.1) is given by

$$u = A\cos(\theta - \theta_0) + \frac{k}{l^2}, \tag{E.5}$$

where A and θ_0 are constants to be determined from the initial conditions. The initial conditions are

$$r(\theta = 0) = \frac{1}{A\cos(\theta_0) + \frac{k}{l^2}} = d_0, \tag{E.6}$$

$$\left.\frac{dr}{d\theta}\right|_{\theta=0} = \frac{-A\sin\theta_0}{\left(A\cos\theta_0 + \frac{k}{l^2}\right)^2}\frac{v_0\cos\phi}{d_0} = v_0\sin\phi. \tag{E.7}$$

Combining these yields

$$A\cos\theta_0 = \frac{1}{d_0} - \frac{k}{l^2}, \tag{E.8}$$

$$A\sin\theta_0 = -\frac{1}{d_0}\tan\phi. \tag{E.9}$$

The condition that the mass reaches the hole is expressed in terms of $r(\theta)$ as follows:

$$r\left(\theta = \frac{\pi}{4}\right) = \frac{1}{A\cos(\frac{\pi}{4} - \theta_0) + \frac{k}{l^2}} = \sqrt{2}d_0. \tag{E.10}$$

Using $\cos(\pi/4 - \theta_0) = \cos(\theta_0)/\sqrt{2} + \sin(\theta_0)/\sqrt{2}$ and the definition of l as well as Eqs. (E.8) and (E.9), we can solve this for v_0:

$$v_0^2 = \frac{(\sqrt{2} - 1)k}{d_0} \frac{1}{\cos\phi \sin\phi} . \tag{E.11}$$

Restricting ϕ to a suitably small interval, this function is injective and has a well-defined inverse $\phi(v_0)$. The neural network has to compute this inverse from operational input data. To generate valid question-answer pairs, we evaluate $v_0(\phi)$ on a large number of randomly chosen ϕ (inside the interval where the function is injective).

References

1. Schrittwieser, J., Antonoglou, I., Hubert, T., Simonyan, K., Sifre, L., Schmitt, S., Guez, A., Lockhart, E., Hassabis, D., Graepel, T., Lillicrap, T., & Silver, D. (2020). *Nature, 588*(7839), 604.
2. Segler, M. H. S., Preuss, M., & Waller, M. P. (2018). *Nature, 555*(7698), 604.
3. Dalgaard, M., Motzoi, F., Sørensen, J. J., & Sherson, J. (2020). *NPJ Quantum Information, 6*(1), 6. https://doi.org/10.1038/s41534-019-0241-0
4. Wu, T., & Tegmark, M. (2018). http://arxiv.org/abs/1810.10525
5. Boyle, R. (1772). *The works of the honourable Robert Boyle* (2nd ed, Vol. 6).
6. Wigner, E. P. (1995). The collected works of Eugene Paul Wigner. In J. Mehra (Ed.), *Philosophical reflections and syntheses* (pp. 247–260). Springer. https://doi.org/10.1007/978-3-642-78374-6_20
7. Deutsch, D., & Penrose, R. (1985). Proceedings of the Royal Society of London. *A. Mathematical and Physical Sciences, 400*(1818), 97.
8. Brukner, Č. (2017). The frontiers collection. In R. Bertlmann & A. Zeilinger (Eds.), *Quantum [Un]speakables II: Half a Century of Bell's Theorem* (pp. 95–117). Springer International Publishing. https://doi.org/10.1007/978-3-319-38987-5_5
9. Frauchiger, D., & Renner, R. (2018). *Nature Communications, 9*(1), 3711.
10. Proietti, M., Pickston, A., Graffitti, F., Barrow, P., Kundys, D., Branciard, C., Ringbauer, M., & Fedrizzi, A. (2019). *Science Advances, 5*(9). https://advances.sciencemag.org/content/5/9/eaaw9832
11. Bong, K. W., Utreras-Alarcón, A., Ghafari, F., Liang, Y. C., Tischler, N., Cavalcanti, E. G., Pryde, G. J., & Wiseman, H. M. (2020). *Nature Physics*. https://www.nature.com/articles/s41567-020-0990-x. Publisher: Nature Publishing Group.
12. Hawking, S. W. (1976). *Physics Review D, 14*, 2460.
13. Preskill, J. (1993). *International Symposium on Black Holes*. Membranes: Wormholes, and Superstrings.
14. Dunjko, V., & Briegel, H. J. (2017). http://arxiv.org/abs/1709.02779
15. Roscher, R., Bohn, B., Duarte, M. F., & Garcke, J. (2019). http://arxiv.org/abs/1905.08883
16. Alhousseini, I., Chemissany, W., Kleit, F., & Nasrallah, A. (2019). http://arxiv.org/abs/1905.01023
17. Carleo, G., Cirac, I., Cranmer, K., Daudet, L., Schuld, M., Tishby, N., Vogt-Maranto, L., & Zdeborová, L. (2019). http://arxiv.org/abs/1903.10563
18. Crutchfield, J. P., & McNamara, B. S. (1987). *Complex Systems, 1*(3), 417.

R. Iten, *Artificial Intelligence for Scientific Discoveries*,
https://doi.org/10.1007/978-3-031-27019-2

19. Schmidt, M., & Lipson, H. (2009). *Science, 324*, 81. https://doi.org/10.1126/science.1165893
20. Schmidt, M., Vallabhajosyula, R. R., Jenkins, J. W., Hood, J. E., Soni, A. S., Wikswo, J. P., & Lipson, H. (2011). *Physical Biology, 8*(5), 055011. https://doi.org/10.1088/1478-3975/8/5/055011
21. Koza, J. R. (1994). *Statistics and Computing, 4*(2), 87.
22. Iten, R., Metger, T., Wilming, H., del Rio, L., & Renner, R. (2020). *Physics Review Letter 124*, 010508. https://doi.org/10.1103/PhysRevLett.124.010508
23. Nautrup, H. P., Metger, T., Iten, R., Jerbi, S., Trenkwalder, L. M., Wilming, H., Briegel, H. J., & Renner, R. (2022). *Machine Learning: Science and Technology, 3*(4), 045025.
24. Koza, J. R. (1994). *Statistics and Computing, 4*(2), 87.
25. Forrest, S. (1993). *Science, 261*(5123), 872.
26. Krizhevsky, A., Sutskever, I., & Hinton, G. E. (2012). In F. Pereira, C. J. C. Burges, L. Bottou, & K. Q. Weinberger (Eds.), *Advances in Neural Information Processing Systems 25*, (pp. 1097–1105). Curran Associates, Inc.
27. Bengio, Y., Courville, A., & Vincent, P. (2012). *IEEE Transactions on Pattern Analysis and Machine Intelligence 35*. https://arxiv.org/abs/1206.5538?context=cs
28. Geman, S., Bienenstock, E., & Doursat, R. (1992). *Neural Computation*. https://doi.org/10.1162/neco.1992.4.1.1
29. Bates, C., Yildirim, I., Tenenbaum, J. B., & Battaglia, P. W. (2015). *Proceedings of the 37th Annual Conference of the Cognitive Science Society* (pp. 172–77). Pasadena, CA.
30. Wu, J., Yildirim, I., Lim, J. J., Freeman, B., & Tenenbaum, J. (2015). In C. Cortes, N. D. Lawrence, D. D. Lee, M. Sugiyama, & R. Garnett (Eds.), *Advances in Neural Information Processing Systems 28* (pp. 127–135). Curran Associates, Inc.
31. Bramley, N. R., Gerstenberg, T., Tenenbaum, J. B., & Gureckis, T. M. (2018). *Cognitive Psychology, 105*, 9.
32. Rempe, D., Sridhar, S., Wang, H., & Guibas, L. J. (2019). http://arxiv.org/abs/1901.00466
33. Kissner, M., & Mayer, H. (2019). http://arxiv.org/abs/1905.09891
34. Ehrhardt, S., Monszpart, A., Mitra, N. & Vedaldi, A. (2018). http://arxiv.org/abs/1805.05086
35. Ye, T., Wang, X., Davidson, J., & Gupta, A. (2018). http://arxiv.org/abs/1808.10002
36. Lerer, A., Gross, S., & Fergus, R. (2016). http://arxiv.org/abs/1603.01312
37. Nielsen, M. A. (2018). *Neural Networks and Deep Learning*. http://neuralnetworksanddeeplearning.com/
38. Tishby, N., & Zaslavsky, N. (2015). *IEEE Information Theory Workshop (ITW)* 1–5. https://doi.org/10.1109/ITW.2015.7133169
39. Nair, V., & Hinton, G. (2010). *Proceedings of the 27th International Conference on Machine Learning, Haifa* (pp. 807–814).
40. Glorot, X., Bordes, A. & Bengio, Y. (2010). *International Conference on Artificial Intelligence and Statistics, 15*.
41. Clevert, D. A., Unterthiner, T. & Hochreiter, S. (2015). http://arxiv.org/abs/1511.07289
42. Lu, Z., Pu, H., Wang, F., Hu, Z., & Wang, L. (2017) In I. Guyon, U. V. Luxburg, S. Bengio, H. Wallach, R. Fergus, S. Vishwanathan, & R. Garnett (Eds.), *Advances in neural information processing systems* (Vol. 30). Curran Associates, Inc. https://proceedings.neurips.cc/paper/2017/file/32cbf687880eb1674a07bf717761dd3a-Paper.pdf
43. Cybenko, G. (1989). Mathematics of control. *Signals and Systems, 2*(4), 303.
44. Hornik, K., Stinchcombe, M., & White, H. (1989). *Neural Networks, 2*(5), 359.
45. Sonoda, S., & Murata, N. (2017). *Applied and Computational Harmonic Analysis 43*(2), 233. https://doi.org/10.1016/j.acha.2015.12.005. http://arxiv.org/abs/1505.03654. ArXiv: 1505.03654
46. Pascanu, R., Montufar, G., & Bengio, Y. (2014). arXiv:1312.6098. http://arxiv.org/abs/1312.6098
47. Lee, H., Grosse, R., Ranganath, R., & Ng, A. Y. (2009). *Proceedings of the 26th Annual International Conference on Machine Learning - ICML '09* (pp. 1–8). Montreal, Quebec, Canada: ACM Press. https://doi.org/10.1145/1553374.1553453. http://portal.acm.org/citation.cfm?doid=1553374.1553453

48. Zeiler, M .D., & Fergus, R. (2014). Lecture notes in computer science. In D. Fleet, T. Pajdla, B. Schiele, & T. Tuytelaars (Eds.), *Computer Vision—ECCV 2014* (pp. 818–833). Cham: Springer International Publishing. https://doi.org/10.1007/978-3-319-10590-1_53
49. Higgins, I., Matthey, L., Pal, A., Burgess, C., Glorot, X., Botvinick, M., Mohamed, S., & Lerchner, A. (2017). *ICLR*. https://openreview.net/references/pdf?id=Sy2fzU9gl
50. Chen, T. Q., Li, X., Grosse, R. B., & Duvenaud, D. (2018). http://arxiv.org/abs/1802.04942
51. Kim, H. & Mnih, A. (2018). https://arxiv.org/abs/1802.05983
52. Thomas, V., Bengio, E., Fedus, W., Pondard, J., Beaudoin, P., Larochelle, H., Pineau, J., Precup, D., & Bengio, Y., (2018). http://arxiv.org/abs/1802.09484
53. François-Lavet, V., Bengio, Y., Precup, D., & Pineau, J. (2019). *The Thirty-Third AAAI Conference on Artificial Intelligence, AAAI 2019* (pp. 3582–3589). https://doi.org/10.1609/aaai.v33i01.33013582
54. Bengio, Y. (2017). http://arxiv.org/abs/1709.08568
55. Krenn, M., Malik, M., Fickler, R., Lapkiewicz, R., & Zeilinger, A. (2016). *Physics Review Letter, 116*, 090405.
56. Melnikov, A. A., Nautrup, H. P., Krenn, M., Dunjko, V., Tiersch, M., Zeilinger, A., & Briegel, H. J. (2018). *Proceedings of the National Academy of Sciences, 115*(6), 1221.
57. Krenn, M., Erhard, M., & Zeilinger, A. (2020). *Nature Reviews Physics*, *2*(11), 649–661. https://doi.org/10.1038/s42254-020-0230-4
58. Gao, X., Krenn, M., Kysela, J., & Zeilinger, A. (2019). *Physical Review A, 99*(2), 023825. https://doi.org/10.1103/PhysRevA.99.023825
59. Malik, M., Erhard, M., Huber, M., Krenn, M., Fickler, R., & Zeilinger, A. (2016). *Nature Photonics, 10*(4), 248. https://doi.org/10.1038/nphoton.2016.12
60. Schlederer, F., Krenn, M., Fickler, R., Malik, M., & Zeilinger, A. (2016). *New Journal of Physics, 18*(4), 043019. https://doi.org/10.1088/1367-2630/18/4/043019
61. Wang, F., Erhard, M., Babazadeh, A., Malik, M., Krenn, M., & Zeilinger, A. (2017). *Optica, 4*(12), 1462. https://doi.org/10.1364/OPTICA.4.001462
62. Babazadeh, A., Erhard, M., Wang, F., Malik, M., Nouroozi, R., Krenn, M., & Zeilinger, A. (2017). *Physical Reviews, 119*(18), 180510. https://doi.org/10.1103/PhysRevLett.119.180510
63. Erhard, M., Malik, M., Krenn, M., & Zeilinger, A. (2018). *Nature Photonics, 12*(12), 759. https://doi.org/10.1038/s41566-018-0257-6
64. Kysela, J., Erhard, M., Hochrainer, A., Krenn, M., & Zeilinger, A. (2020). *Proceedings of the National Academy of Sciences, 117*(42), 26118.
65. Krenn, M., Hochrainer, A., Lahiri, M., & Zeilinger, A. (2017). *Physical Reviews, 118*(8), 080401. https://doi.org/10.1103/PhysRevLett.118.080401
66. Krenn, M., Gu, X., & Zeilinger, A. (2017). *Physics Review Letter, 119*, 240403.
67. Gao, X., Erhard, M., Zeilinger, A., & Krenn, M. (2020). *Physical Reviews, 125*(5), 050501. https://doi.org/10.1103/PhysRevLett.125.050501
68. Krenn, M., Kottmann, J., Tischler, N., & Aspuru-Guzik, A. (2020). arXiv:2005.06443 [physics, physics:quant-ph] http://arxiv.org/abs/2005.06443. ArXiv: 2005.06443
69. Friederich, P., Krenn, M., Tamblyn, I. & Aspuru-Guzik, A. (2020). Scientific intuition inspired by machine learning generated hypotheses.
70. Briegel, H. J., & Cuevas, G. D. l. (2012). *Scientific Reports, 2*, 400. https://doi.org/10.1038/srep00400, https://www.nature.com/articles/srep00400
71. Bell, J. S. (1964). *Physics Physique Fizika, 1*, 195.
72. Nielsen, M. A., & Chuang, I. L. (2010). *Quantum Computation and Quantum Information: 10th Anniversary Edition*. Cambridge University Press. https://doi.org/10.1017/CBO9780511976667
73. Sachdev, S. (2000). *Quantum Phase Transitions*. Cambridge University Press. https://doi.org/10.1017/CBO9780511622540
74. Huber, M., & de Vicente, J. I. (2013). *Physical Reviews 110*(3), 030501. https://doi.org/10.1103/PhysRevLett.110.030501. http://arxiv.org/abs/1210.6876. ArXiv: 1210.6876

75. Horodecki, R., Horodecki, P., Horodecki, M., & Horodecki, K. (2009). *Review Model Physics, 81*, 865.
76. Amico, L., Fazio, R., Osterloh, A., & Vedral, V. (2008). *Review Model Physics 80*, 517.
77. Pan, J. W., Chen, Z. B., Lu, C. Y., Weinfurter, H., Zeilinger, A., & Żukowski, M. (2012). *Review Model Physics, 84*, 777.
78. Hein, M., Eisert, J., & Briegel, H. J. (2004). *Physics Review A, 69*, 062311.
79. Scarani, V., & Gisin, N. (2001). *Physics Review Letter, 87*, 117901.
80. Scarani, V., Bechmann-Pasquinucci, H., Cerf, N. J., Dušek, M., Lütkenhaus, N., & Peev, M. (2009). *Review Model Physics, 81*, 1301.
81. Mautner, J., Makmal, A., Manzano, D., Tiersch, M., & Briegel, H. J. (2015). *New Generation Computing, 33*(1), 69.
82. Melnikov, A. A., Makmal, A., & Briegel, H. J. (2014). arXiv:1405.5459 [cs]. http://arxiv.org/abs/1405.5459. ArXiv: 1405.5459
83. Melnikov, A. A., Makmal, A., Dunjko, V., & Briegel, H. J. (2017). *Scientific Reports, 7*(1), 14430.
84. Melnikov, A. A., Makmal, A., & Briegel, H. J. (2018). *IEEE Access, 6*, 64639. https://doi.org/10.1109/ACCESS.2018.2876494
85. Hangl, S., Ugur, E., Szedmak, S., & Piater, J. (2016). *2016 IEEE/RSJ International Conference on Intelligent Robots and Systems (IROS)* (pp. 2799–2804). https://doi.org/10.1109/IROS.2016.7759434
86. Hangl, S., Dunjko, V., Briegel, H. J., & Piater, J. (2020). *Frontiers in Robotics and AI, 7*, 42. https://doi.org/10.3389/frobt.2020.00042, https://www.frontiersin.org/article/10.3389/frobt.2020.00042
87. Nautrup, H. P., Delfosse, N., Dunjko, V., Briegel, H. J. & Friis, N. (2019). *Quantum, 3*, 215. https://doi.org/10.22331/q-2019-12-16-215
88. Tiersch, M., Ganahl, E. J., & Briegel, H. J. (2015). *Scientific Reports, 5*(1), 12874.
89. Wallnöfer, J., Melnikov, A. A., Dür, W., & Briegel, H. J. (2020). *PRX Quantum, 1*, 010301.
90. Melnikov, A. A., Sekatski, P., & Sangouard, N. (2020). *Physics Review Letter 125*, 160401.
91. Ried, K., Müller, T., & Briegel, H. J. (2019). *PLOS ONE, 14*(2), e0212044.
92. López-Incera, A., Ried, K., Müller, T., & Briegel, H. (2020). *PloS one, 15*(12), e0243628.
93. López-Incera, A., Nouvian, M., Ried, K., Müller, T., & Briegel, H. J. (2020). Collective defense of honeybee colonies: Experimental results and theoretical modeling.
94. Paparo, G. D., Dunjko, V., Makmal, A., Martin-Delgado, M. A., & Briegel, H. J. (2014). *Physics Review X, 4*, 031002.
95. Dunjko, V., Friis, N., & Briegel, H. J. (2015). *New Journal of Physics, 17*(2), 023006.
96. Friis, N., Melnikov, A. A., Kirchmair, G., & Briegel, H. J. (2015). *Scientific Reports, 5*(1), 18036.
97. Dunjko, V., Taylor, J. M., & Briegel, H. J. (2016). *Physics Review Letter 117*, 130501. https://link.aps.org/doi/10.1103/PhysRevLett.117.130501
98. Dunjko, V., Taylor, J. M., & Briegel, H. J. (2017). *2017 IEEE International Conference on Systems, Man, and Cybernetics (SMC)* (pp. 282–287). https://doi.org/10.1109/SMC.2017.8122616
99. Sriarunothai, T., Wolk, S., Giri, G. S., Friis, N., Dunjko, V., Briegel, H. J., & Wunderlich, C. (2018). *Quantum Science and Technology, 4*(1), 015014.
100. Jerbi, S., Trenkwalder, L. M., Poulsen Nautrup, H., Briegel, H. J., & Dunjko, V. (2021). *PRX Quantum, 2*, 010328. https://doi.org/10.1103/PRXQuantum.2.010328, https://link.aps.org/doi/10.1103/PRXQuantum.2.010328
101. Briegel, H. J., & Müller, T. (2015). *Minds and Machines, 25*(3), 261.
102. Briegel, H. J. (2012). *Scientific Reports, 2*(1), 522.
103. Müller, T., & Briegel, H. J. (2018). *Dialectica, 72*(2), 219.
104. Tulving, E. (1972). *Organization of Memory* (pp. xiii, 423–xiii, 423). Academic Press.
105. Ingvar, D. H. (1985). *Human Neurobiology, 4*(3), 127.
106. Makmal, A., Melnikov, A. A., Dunjko, V., & Briegel, H. J. (2016). *IEEE Access, 4*, 2110. https://doi.org/10.1109/ACCESS.2016.2556579

107. Klyshko, D. N. (1988). *Physics-Uspekhi 31*(1), 74. https://ufn.ru/en/articles/1988/1/f/
108. Leach, J., Padgett, M. J., Barnett, S. M., Franke-Arnold, S., & Courtial, J. (2002). *Physics Review Letter, 88*, 257901.
109. Schütt, K. T., Chmiela, S., von Lilienfeld, O. A., Tkatchenko, A., Tsuda, K. & Müller, K. R. *Machine Learning Meets Quantum Physics*. https://link.springer.com/book/10.1007/978-3-030-40245-7#toc
110. Decelle, A., Martin-Mayor, V., & Seoane, B. (2019). http://arxiv.org/abs/1904.07637
111. Gilmer, J., Schoenholz, S. S., Riley, P. F., Vinyals, O., & Dahl, G. E. (2017). arXiv:1704.01212, http://arxiv.org/abs/1704.01212
112. De Cao, N., & Kipf, T. (2018). arXiv:1805.11973, http://arxiv.org/abs/1805.11973
113. Krenn, M., Häse, F., Nigam, A., Friederich, P., & Aspuru-Guzik, A. (2020). *Machine Learning: Science and Technology, 1*(4), 045024.
114. Sanchez-Lengeling, B., & Aspuru-Guzik, A. (2018). *Science, 361*(6400), 360. https://doi.org/10.1126/science.aat2663. https://science.sciencemag.org/content/361/6400/360. Publisher: American Association for the Advancement of Science Section: Review.
115. Gromski, P. S., Henson, A. B., Granda, J. M., & Cronin, L. (2019). *Nature Reviews Chemistry, 3*(2), 119. https://doi.org/10.1038/s41570-018-0066-y. https://www.nature.com/articles/s41570-018-0066-y. Number: 2 Publisher: Nature Publishing Group.
116. Bongard, J., & Lipson, H. (2007). *Proceedings of the National Academy of Sciences, 104*(24), 9943. https://doi.org/10.1073/pnas.0609476104, https://www.pnas.org/content/104/24/9943
117. Daniels, B. C., & Nemenman, I. (2015). *Nature Communications, 6*, 8133. https://doi.org/10.1038/ncomms9133
118. Brunton, S. L., Proctor, J. L., & Kutz, J. N. (2016). *Proceedings of the National Academy of Sciences, 113*(15), 3932.
119. Reinbold, P. A. K., Kageorge, L. M., Schatz, M. F., & Grigoriev, R. O. (2021). *Nature Communications, 12*(1), 3219.
120. Lu, P. Y., Ariño, J. B., & Soljačić, M. (2022). *Communications Physics, 5*(1), 1.
121. Cubitt, T. S., Eisert, J., & Wolf, M. M. (2012). *Physical Reviews, 108*(12), 120503. https://doi.org/10.1103/PhysRevLett.108.120503, https://link.aps.org/doi/10.1103/PhysRevLett.108.120503
122. Raissi, M., Perdikaris, P., & Karniadakis, G. E. (2019). *Journal of Computational Physics 378*, 686. https://doi.org/10.1016/j.jcp.2018.10.045, http://www.sciencedirect.com/science/article/pii/S0021999118307125
123. Raissi, M. (2018). http://arxiv.org/abs/1801.06637
124. Raissi, M., & Karniadakis, G. E. (2018). *Journal of Computational Physics, 357*, 125. https://doi.org/10.1016/j.jcp.2017.11.039. http://www.sciencedirect.com/science/article/pii/S0021999117309014
125. Zhang, D., Guo, L., & Karniadakis, G. E. (2019). http://arxiv.org/abs/1905.01205
126. Goeßmann, A., Götte, M., Roth, I., Sweke, R., Kutyniok, G., & Eisert, J. (2020). Tensor network approaches for learning non-linear dynamical laws.
127. Forrest, S. (1993). *Science 261*(5123). https://doi.org/10.1126/science.8346439, https://science.sciencemag.org/content/261/5123/872
128. Udrescu, S. M., & Tegmark, M. (2020). *Science Advances, 6*(16). https://doi.org/10.1126/sciadv.aay2631, https://advances.sciencemag.org/content/6/16/eaay2631
129. Neuweiler, P. (2019). Semester thesis, ETH.
130. Anderson, P. W. (1972). *Science, 177*(4047), 393.
131. Noether, E., König Gesellsch, N. D., & Göttingen, D. W. Z. (1918). *Math-Physics Klasse, 235*
132. Hanc, J., Tuleja, S., & Hancova, M. (2004). *American Journal of Physics, 72*(4), 428. https://doi.org/10.1119/1.1591764
133. Guimerà, R., Reichardt, I., Aguilar-Mogas, A., Massucci, F.A., Miranda, M., Pallarès, J., & Sales-Pardo, M., *Science Advances, 6*(5), 6971. https://doi.org/10.1126/sciadv.aav6971, https://www.science.org/doi/abs/10.1126/sciadv.aav6971

134. Patrignani, C. (2016). *Chinese Physics C, 40*(10), 100001. https://doi.org/10.1088/1674-1137/40/10/100001. https://iopscience.iop.org/article/10.1088/1674-1137/40/10/100001/meta. Publisher: IOP Publishing.
135. Aaltonen, T., et al. (2008). *Physics Review D, 78*, 012002.
136. Aaltonen, T., et al. (2009). *Physical Review D, 79*(1), 011101.
137. Meyer, A. (2010). *AIP Conference Proceedings, 1200*, 293.
138. Choudalakis, G. (2011). *PHYSTAT 2011, Workshop on Statistical Issues Related to Discovery Claims in Search Experiments and Unfolding*. CERN.
139. Aaboud, M., Aad, G., Abbott, B., Abdinov, O., Abeloos, B., Abidi, S. H., AbouZeid, O. S., Abraham, N. L., Abramowicz, H., et al. (2019). *The European Physical Journal C, 79*(2). https://doi.org/10.1140/epjc/s10052-019-6540-y
140. Asadi, P., Buckley, M. R., DiFranzo, A., Monteux, A., & Shih, D. (2017). *Journal of High Energy Physics, 2017*(11), 194. https://doi.org/10.1007/JHEP11(2017)194
141. D'Agnolo, R. T., & Wulzer, A. (2019). *Physical Review D, 99*(1), 015014.
142. De Simone, A., & Jacques, T. (2019). *The European Physical Journal C, 79*(4), 289.
143. Sugiyama, M., Suzuki, T., & Kanamori, T. (2012). *Density Ratio Estimation in Machine Learning*. Cambridge University Press. https://doi.org/10.1017/CBO9781139035613
144. Poggio, T., Mhaskar, H., Rosasco, L., Miranda, B., & Liao, Q. (2017). *International Journal of Automation and Computing, 14*, 1. https://doi.org/10.1007/s11633-017-1054-2
145. Bach, F. (2017). *Journal Machine Learning Resources, 18*(1), 629.
146. Montanelli, H., & Du, Q. (2017). https://arxiv.org/abs/1712.08688
147. Schilling, M. F. (1986). *Journal of the American Statistical Association 81*(395), 799. https://doi.org/10.1080/01621459.1986.10478337, https://www.tandfonline.com/doi/abs/10.1080/01621459.1986.10478337
148. Henze, N. (1988). *The Annals of Statistics, 16*(2), 772.
149. Wang, Q., Kulkarni, S. R., & Verdu, S. (2005). *IEEE Transactions on Information Theory, 51*(9), 3064. https://doi.org/10.1109/TIT.2005.853314
150. Wang, Q., Kulkarni, S. R., & Verdu, S. (2006). *2006 IEEE International Symposium on Information Theory* (pp. 242–246). https://doi.org/10.1109/ISIT.2006.261842
151. Dasu, T., Krishnan, S., Venkatasubramanian, S., & Yi, K. (2006). *Proceeding Symposium on the Interface of Statistics, Computing Science, and Applications*.
152. Pérez-cruz, F. (2008). *Proceedings of IEEE International Symposium on Information Theory* (pp. 1666–1670).
153. Kremer, J., Gieseke, F., Steenstrup Pedersen, K., & Igel, C. (2015). *Astronomy and Computing, 12*, 67. https://doi.org/10.1016/j.ascom.2015.06.005
154. Qing Wang, S. R. Kulkarni, & Verdu, S. (2005). *IEEE Transactions on Information Theory, 51*(9), 3064. https://doi.org/10.1109/TIT.2005.853314
155. Wang, Q., Kulkarni, S. R., & Verdu, S. (2006). *2006 IEEE International Symposium on Information Theory* (pp. 242–246). https://doi.org/10.1109/ISIT.2006.261842
156. Pérez-cruz, F. (2008). *Proceedings of IEEE International Symposium on Information Theory* (pp. 1666–1670).
157. Manly, B. F. J. (2007). *International Statistical Review, 75*(2), 269. https://doi.org/10.1111/j.1751-5823.2007.00015_21.x, https://onlinelibrary.wiley.com/doi/abs/10.1111/j.1751-5823.2007.00015_21.x._eprint: https://onlinelibrary.wiley.com/doi/pdf/10.1111/j.1751-5823.2007.00015_21.x
158. Vaart, A. W. V. D. (1998). Cambridge series in statistical and probabilistic mathematics. In*Asymptotic Statistics*. Cambridge University Press. https://doi.org/10.1017/CBO9780511802256
159. Bell, J. S. (1964). *Physics Physique Fizika, 1*(3), 195. https://doi.org/10.1103/PhysicsPhysiqueFizika.1.195, https://link.aps.org/doi/10.1103/PhysicsPhysiqueFizika.1.195. Publisher: American Physical Society
160. Tishby, N., Pereira, F. C. & Bialek W. (2000). http://arxiv.org/abs/physics/0004057
161. Shwartz-Ziv, R., & Tishby, N. (2017). http://arxiv.org/abs/1703.00810
162. Bahdanau, D., Cho, K., & Bengio, Y. (2015). https://arxiv.org/abs/1409.0473

163. Firat, O., Cho, K., Sankaran, B., Yarman Vural, F. T., & Bengio, Y. (2017). *Computer Speech Languge*, *45*(C), 236–252. https://doi.org/10.1016/j.csl.2016.10.006, https://doi.org/10.1016/j.csl.2016.10.006
164. Celikyilmaz, A., Bosselut, A., He, X. & Choi, Y. (2018). *Proceedings of the 2018 Conference of the North American Chapter of the Association for Computational Linguistics: Human Language Technologies, Volume 1 (Long Papers)* (pp. 1662–1675). Association for Computational Linguistics, New Orleans, Louisiana. https://doi.org/10.18653/v1/N18-1150. https://www.aclweb.org/anthology/N18-1150
165. Koopman, B. O. (1931). *Proceedings of the National Academy of Sciences, 17*(5), 315.
166. Koopman, B. O., & Neumann, J. V. (1932) *Proceedings of the National Academy of Sciences, 18*(3), 255. https://doi.org/10.1073/pnas.18.3.255, https://www.pnas.org/content/18/3/255. Publisher: National Academy of Sciences Section: Mathematics
167. Yeung, E., Kundu, S., & Hodas, N. (2017). http://arxiv.org/abs/1708.06850
168. Wehmeyer, C., & Noé, F. (2018). *The Journal of Chemical Physics, 148*(24), 241703.
169. Mardt, A., Pasquali, L., Wu, H., & Noé, F. (2018). *Nature Communications, 9*(1), 5.
170. Takeishi, N., Kawahara, Y., & Yairi, T. (2017) Learning Koopman invariant subspaces for dynamic mode decomposition. http://arxiv.org/abs/1710.04340
171. Otto, S. E., & Rowley, C. W. (2017). http://arxiv.org/abs/1712.01378
172. Li, Q., Dietrich, F., Bollt, E. M., & Kevrekidis, I. G. (2017) *Chaos: An Interdisciplinary Journal of Nonlinear Science, 27*(10). https://doi.org/10.1063/1.4993854, http://arxiv.org/abs/1707.00225
173. Champion, K., Lusch, B., Kutz, J. N., & Brunton, S. L. (2019). *Proceedings of the National Academy of Sciences, 116*(45), 22445.
174. Mezić, I., & Banaszuk, A. (2004). *Physica D: Nonlinear Phenomena, 197*(1–2), 101. https://doi.org/10.1016/j.physd.2004.06.015, https://linkinghub.elsevier.com/retrieve/pii/S0167278904002507
175. Mezić, I. (2005). *Nonlinear Dynamics, 41*(1), 309.
176. Mezic, I. (2019). arXiv:1702.07597. http://arxiv.org/abs/1702.07597
177. Budišić, M., Mohr, R., & Mezić, I. (2012). *Chaos: An Interdisciplinary Journal of Nonlinear Science, 22*(4), 047510. https://doi.org/10.1063/1.4772195, https://aip.scitation.org/doi/10.1063/1.4772195. Publisher: American Institute of Physics.
178. Mezić, I. (2013). *Annual Review of Fluid Mechanics, 45*(1), 357. https://doi.org/10.1146/annurev-fluid-011212-140652, http://www.annualreviews.org/doi/10.1146/annurev-fluid-011212-140652
179. Schmid, P. J. (2010). *Journal of Fluid Mechanics, 656*, 5. https://doi.org/10.1017/S0022112010001217, https://www.cambridge.org/core/journals/journal-of-fluid-mechanics/article/dynamic-mode-decomposition-of-numerical-and-experimental-data/AA4C763B525515AD4521A6CC5E10DBD4. Publisher: Cambridge University Press.
180. Rowley, C. W., Mezić, I., Bagheri, S., Schlatter, P., & Henningson, D. S. (2009). *Journal of Fluid Mechanics, 641*, 115. https://doi.org/10.1017/S0022112009992059, https://www.cambridge.org/core/product/identifier/S0022112009992059/type/journal_article
181. Brunton, S. L., Proctor, J. L., Tu, J. H., & Kutz, J. N. (2015). *Journal of Computational Dynamics, 2*(2), 165. https://doi.org/10.3934/jcd.2015002, https://www.aimsciences.org/article/doi/10.3934/jcd.2015002. Company: Journal of Computational Dynamics Distributor: Journal of Computational Dynamics Institution: Journal of Computational Dynamics Label: Journal of Computational Dynamics Publisher: American Institute of Mathematical Sciences.
182. Arbabi, H. (2018).
183. Lusch, B., Kutz, J. N., & Brunton, S. L. (2018). *Nature Communications, 9*, 4950.
184. Kipf, T., Fetaya, E., Wang, K. C., Welling, M., & Zemel, R., (PMLR, 2018). *Proceedings of Machine Learning Research* (Vol. 80, pp. 2688–2697). http://proceedings.mlr.press/v80/kipf18a.html
185. Battaglia, P., Pascanu, R., Lai, M., Rezende, D. J., & Kavukcuoglu, K. (2016). *Proceedings of the 30th International Conference on Neural Information Processing Systems* (pp. 4509–4517). http://dl.acm.org/citation.cfm?id=3157382.3157601

186. Zheng, D., Luo, V., Wu, J., & Tenenbaum, J. B. (2018). http://arxiv.org/abs/1807.09244
187. Chang, M. B., Ullman, T., Torralba, A., & Tenenbaum, J. B. (2016). http://arxiv.org/abs/1612.00341
188. Raposo, D., Santoro, A., Barrett, D., Pascanu, R., Lillicrap, T., & Battaglia, P. (2017). http://arxiv.org/abs/1702.05068
189. Hinton, G. E., & Salakhutdinov, R. R. (2006). *Science, 313*(5786), 504. https://doi.org/10.1126/science.1127647
190. Eslami, S. M. A., Rezende, D. J., Besse, F., Viola, F., Morcos, A. S., Garnelo, M., Ruderman, A., Rusu, A. A., Danihelka, I., Gregor, K., Reichert, D. P., Buesing, L., Weber, T., Vinyals, O., Rosenbaum, D., Rabinowitz, N., King, H., Hillier, C., Botvinick, M., D. Wierstra, K. Kavukcuoglu, & Hassabis, D. (2018). *Science, 360*(6394), 1204.
191. Hornik, K. (1991). *Neural Networks, 4*(2), 251.
192. Chen, B., Huang, K., Raghupathi, S., Chandratreya, I., Du, Q., & Lipson, H. (2022). *Nature Computational Science, 2*(7), 433.
193. Kingma, D. P., & Welling, M. (2013). https://arxiv.org/abs/1312.6114
194. Elman, J. L. (1990). *Cognitive Science, 14*(2), 179.
195. Liu, S., Yang, N., Li, M., & Zhou, M. (2014). *Proceedings of the 52nd Annual Meeting of the Association for Computational Linguistics (Volume 1: Long Papers)* (pp. 1491–1500). Baltimore, Maryland: Association for Computational Linguistics. https://doi.org/10.3115/v1/P14-1140. https://www.aclweb.org/anthology/P14-1140
196. Auli, M., Galley, M., Quirk, C., & Zweig, G. (2013). *Proceedings of the 2013 Conference on Empirical Methods in Natural Language Processing* (pp. 1044–1054). Seattle, Washington, USA: Association for Computational Linguistics. https://www.aclweb.org/anthology/D13-1106
197. Sutskever, I., Vinyals, O., & Le, Q. V. (2014). In Z. Ghahramani, M. Welling, C. Cortes, N. D. Lawrence, & K. Q. Weinberger *Advances in Neural Information Processing Systems 27* (pp. 3104–3112). Curran Associates, Inc. http://papers.nips.cc/paper/5346-sequence-to-sequence-learning-with-neural-networks.pdf
198. Scarselli, F., Gori, M., Tsoi, A. C., Hagenbuchner, M., & Monfardini, G. (2009). *IEEE Transactions on Neural Networks, 20*(1), 61.
199. Li, Y., Tarlow, D., Brockschmidt, M., & Zemel, R. (2017). arXiv:1511.05493. http://arxiv.org/abs/1511.05493
200. Sanchez-Gonzalez, A., Heess, N., Springenberg, J .T., Merel, J., Riedmiller, M., Hadsell, R., & Battaglia, P. (2018). arXiv:1806.01242. http://arxiv.org/abs/1806.01242
201. Li, Y., Wu, J., Tedrake, R., Tenenbaum, J. B., & Torralba, A. (2019). *ICLR.*
202. Mrowca, D., Zhuang, C., Wang, E., Haber, N., Fei-Fei, L., Tenenbaum, J. B., & Yamins, D. L. K. (2018). arXiv:1806.08047. http://arxiv.org/abs/1806.08047
203. Bapst, V., Keck, T., Grabska-Barwińska, A., Donner, C., Cubuk, E. D., Schoenholz, S. S., Obika, A., Nelson, A. W. R., Back, T., Hassabis, D., & Kohli, P. (2020). *Nature Physics, 16*(4), 448.
204. Sanchez-Gonzalez, A., Godwin, J., Pfaff, T., Ying, R., Leskovec, J., & Battaglia, P. W. (2020). Learning to simulate complex physics with graph networks. https://arxiv.org/abs/2002.09405
205. Cranmer, M., Sanchez-Gonzalez, A., Battaglia, P., Xu, R., Cranmer, K., Spergel, D., & Ho, S. (2020) arXiv:2006.11287, http://arxiv.org/abs/2006.11287
206. Hamilton, W. L., Ying, R., & Leskovec, J. (2018). arXiv:1706.02216, http://arxiv.org/abs/1706.02216
207. Kipf, T. N., & Welling, M. (2017) arXiv:1609.02907. http://arxiv.org/abs/1609.02907
208. Zaheer, M., Kottur, S., Ravanbakhsh, S., Poczos, B., Salakhutdinov, R. R., & Smola, A. J. (2017). In I. Guyon, U. V. Luxburg, S. Bengio, H. Wallach, R. Fergus, S. Vishwanathan, & R. Garnett (Eds.), *Advances in Neural Information Processing Systems 30* (pp. 3391–3401). Curran Associates, Inc. http://papers.nips.cc/paper/6931-deep-sets.pdf
209. Herzig, R., Raboh, M., Chechik, G., Berant, J., & Globerson, A. (2018). arXiv:1802.05451, http://arxiv.org/abs/1802.05451

210. Liu, Z., & Zhou, J. (2020). Synthesis lectures on artificial intelligence and machine. *Learning, 14*(2), 1.
211. Maddison, C. J., Mnih, A., & Teh, Y. W. (2017). arXiv:1611.00712, http://arxiv.org/abs/1611.00712
212. Abadi, M., Agarwal, A., Barham, P., Brevdo, E., Chen, Z., Citro, C., Corrado, G. S., Davis, A., Dean, J., Devin, M., Ghemawat, S., Goodfellow, I., Harp, A., Irving, G., Isard, M., Jia, Y., Jozefowicz, R., Kaiser, L., Kudlur, M., Levenberg, J., Mané, D., Monga, R., Moore, S., Murray, D., Olah, C., Schuster, M., Shlens, J., Steiner, B., Sutskever, I., Talwar, K., Tucker, P., Vanhoucke, V., Vasudevan, V., Viégas, F., Vinyals, O., Warden, P., Wattenberg, M., Wicke, M., Yu, Y., & Zheng, X. (2015) TensorFlow: Large-scale machine learning on heterogeneous systems. https://www.tensorflow.org/
213. Ross, B. C. (2014). *PLOS ONE, 9*(2), e87357.
214. Kraskov, A., Stoegbauer, H., & Grassberger, P. (2004). *Physical Review E, 69*(6), 066138.
215. Kozachenko, L. F., & Leonenko, N. N. (1987). *Problems Inform. Transmission, 23*, 95.
216. Pedregosa, F., Varoquaux, G., Gramfort, A., Michel, V., Thirion, B., Grisel, O., Blondel, M., Prettenhofer, P., Weiss, R., Dubourg, V., Vanderplas, J., Passos, A., Cournapeau, D., Brucher, M., Perrot, M., & Duchesnay, E. (2011). *Journal of Machine Learning Research, 12*, 2825.
217. Beléndez, A., Pascual, C., Méndez, D. I., Beléndez, T., & Neipp, C. (2007). *Revista Brasileira de Ensino de Física, 29*(4), 645.
218. This figure is composed of figures in "B. Lusch, J. N. Kutz, and S. L. Brunton, Deep learning for universal linear embeddings of nonlinear dynamics, Nat. Commun. 9, 4950, https://doi.org/10.1038/s41467-018-07210-0, 2018". The work is published under the terms of the Creative Commons Attribution 4.0 license (https://creativecommons.org/licenses/by/4.0/). The labels in the image are modified to fit the context of this book.
219. Paris, M., & Reháček, J. (eds.), (2004). Lecture notes in physics. In *Quantum state estimation*. Berlin, Heidelberg: Springer. https://doi.org/10.1007/b98673. http://link.springer.com/10.1007/b98673
220. Gamel, O. (2016). *Physical Review A, 93*(6). https://doi.org/10.1103/physreva.93.062320, http://dx.doi.org/10.1103/PhysRevA.93.062320
221. Floryan, D., & Graham, M. D. (2022). *Nature Machine Intelligence, 4*(12), 1113.
222. Carleo, G., & Troyer, M. (2017). *Science, 355*(6325), 602.
223. Rocchetto, A., Grant, E., Strelchuk, S., Carleo, G., & Severini, S. (2018). *NPJ Quantum Information, 4*(1), 28. https://doi.org/10.1038/s41534-018-0077-z, https://www.nature.com/articles/s41534-018-0077-z
224. Glasser, I., Pancotti, N., August, M., Rodriguez, I. D., & Cirac, J. I. (2018). *Physical Review X, 8*(1), 011006.
225. Carleo, G., Nomura, Y., & Imada, M. (2018). *Nature Communications, 9*(1), 5322.
226. Cai, Z., & Liu, J. (2018). *Physical Review B 97*(3). https://doi.org/10.1103/PhysRevB.97.035116, http://arxiv.org/abs/1704.05148
227. Huang, Y., & Moore, J. E., (2017). http://arxiv.org/abs/1701.06246
228. Deng, D. L., Li, X., & Sarma, S. D. (2017). *Physical Review B, 96*(19). https://doi.org/10.1103/PhysRevB.96.195145. http://arxiv.org/abs/1609.09060
229. Schmitt, M., & Heyl, M. (2018). *SciPost Physics, 4*(2). https://doi.org/10.21468/SciPostPhys.4.2.013, http://arxiv.org/abs/1707.06656
230. Torlai, G., Mazzola, G., Carrasquilla, J., Troyer, M., Melko, R., & Carleo, G. (2018). *Nature Physics, 14*(5), 447.
231. Nomura, Y., Darmawan, A. S., Yamaji, Y., & Imada, M. (2017). *Physical Review B, 96*(20). https://doi.org/10.1103/PhysRevB.96.205152. http://arxiv.org/abs/1709.06475
232. Deng, D. L., Li, X., & Sarma, S. D. (2017). *Physical Review X, 7*(2). https://doi.org/10.1103/PhysRevX.7.021021. http://arxiv.org/abs/1701.04844
233. Gao, X., & Duan, L. M. (2017). *Nature Communications, 8*(1).https://doi.org/10.1038/s41467-017-00705-2, http://arxiv.org/abs/1701.05039
234. Carrasquilla, J., Torlai, G., Melko, R. G., & Aolita, L. (2019). *Nature Machine Intelligence, 1*(3), 155.

235. Beach, M. J. S., De Vlugt, I., Golubeva, A., Huembeli, P., Kulchytskyy, B., Luo, X., Melko, R.G., Merali, E., & Torlai, G. (2019). *SciPost Physics, 7*(1), 009. https://doi.org/10.21468/SciPostPhys.7.1.009, http://arxiv.org/abs/1812.09329
236. Torlai, G., & Melko, R. G. (2019). http://arxiv.org/abs/1905.04312
237. We use material from "Hendrik Poulsen Nautrup, Tony Metger, Raban Iten, Sofiene Jerbi, Lea M. Trenkwalder, Henrik Wilming, Hans J. Briegel, and Renato Renner, Operationally meaningful representations of physical systems in neural networks, Mach. Learn.: Sci. Technol. 3 045025, https://doi.org/10.1088/2632-2153/ac9ae8, 2022". The work is published under the terms of the Creative Commons Attribution 4.0 license (https://creativecommons.org/licenses/by/4.0/). The material was modified to fit the context of this book (but no essential changes were made to the content).
238. Celikyilmaz, A., Bosselut, A., He, X., & Choi, Y. (2018). arXiv:1803.10357. http://arxiv.org/abs/1803.10357
239. Goldstein, H., Poole, C., & Safko, J. (2002). *Classical mechanics* (3rd edn). Addison-Wesley.
240. González, E. (2020). Master thesis, ETH.
241. Lesort, T., Díaz-Rodríguez, N., Goudou, J. F., & Filliat, D. (2018). *Neural Networks, 108*, 379.
242. Bengio, E., Thomas, V., Pineau, J., Precup, D., & Bengio, Y. (2017). http://arxiv.org/abs/1703.07718
243. Jonschkowski, R., & Brock, O. (2015). *Autonomous Robots, 39*(3), 407.
244. Jaderberg, M., Mnih, V., Czarnecki, W. M., Schaul, T., Leibo, J. Z., Silver, D., & Kavukcuoglu, K. (2017). *5th International Conference on Learning Representations, ICLR 2017, Toulon, France*, April 24–26, 2017, Conference Track Proceedings. https://openreview.net/forum?id=SJ6yPD5xg
245. Veeriah, V., Hessel, M., Xu, Z., Lewis, R., Rajendran, J., Oh, J., van Hasselt, H., Silver, D., & Singh, S. (2019). arXiv:1909.04607, http://arxiv.org/abs/1909.04607
246. Fan, F., Xiong, J., & Wang, G. (2020). arXiv:2001.02522, http://arxiv.org/abs/2001.02522
247. Inceptionism: Going Deeper into Neural Networks. http://ai.googleblog.com/2015/06/inceptionism-going-deeper-into-neural.html
248. Schindler, F., Regnault, N., & Neupert, T. (2017). *Physical Review B, 95*(24), 245134.
249. Seif, A., Hafezi, M., & Jarzynski, C. (2020). *Nature Physics,* 1–9 (2020). https://doi.org/10.1038/s41567-020-1018-2. https://www.nature.com/articles/s41567-020-1018-2. Publisher: Nature Publishing Group.
250. Kim, S., Lu, P. Y., Mukherjee, S., Gilbert, M., Jing, L., Čeperić, V., & Soljačić, M. (2019). https://arxiv.org/abs/1912.04825
251. Martius, G., & Lampert, C. H. (2016). arXiv:1610.02995, http://arxiv.org/abs/1610.02995
252. Sahoo, S. S., Lampert, C. H., & Martius, G. (2018). arXiv:1806.07259, http://arxiv.org/abs/1806.07259
253. Lemos, P., Jeffrey, N., Cranmer, M., Ho, S., & Battaglia, P. (2022). Rediscovering orbital mechanics with machine learning. https://doi.org/10.48550/ARXIV.2202.02306. https://arxiv.org/abs/2202.02306
254. Müller, V. C. (2016). *Fundamental issues of artificial intelligence*. Springer.
255. Grace, K., Salvatier, J., Dafoe, A., Zhang, B., & Evans, O. (2018). *Journal of Artificial Intelligence Research, 62*, 729. https://doi.org/10.1613/jair.1.11222, https://www.jair.org/index.php/jair/article/view/11222
256. King, R. D., Whelan, K. E., Jones, F. M., Reiser, P. G. K., Bryant, C. H., Muggleton, S. H., Kell, D. B., & Oliver, S. G. (2004). *Nature, 427*(6971), 247.
257. Zhuang, F., Qi, Z., Duan, K., Xi, D., Zhu, Y., Zhu, H., Xiong, H., & He, Q. (2021). *Proceedings of the IEEE, 109*(1), 43. https://doi.org/10.1109/JPROC.2020.3004555
258. Siegelmann, H., & Sontag, E. (1995). *Journal of Computer and System Sciences, 50*(1), 132. https://doi.org/10.1006/jcss.1995.1013, https://www.sciencedirect.com/science/article/pii/S0022000085710136
259. Graves, A., Wayne, G., & Danihelka, I. (2014). *Neural turing machines*.

260. Lee, J. (2012). Graduate texts in mathematics . In *Introduction to smooth manifolds* (2nd edn). Springer-Verlag. https://doi.org/10.1007/978-1-4419-9982-5, www.springer.com/de/book/9781441999818
261. Achille, A., & Soatto, S. (2018). *IEEE Transactions on Pattern Analysis and Machine Intelligence, 8828*, 1. https://doi.org/10.1109/TPAMI.2017.2784440
262. https://github.com/eth-nn-physics/nn_physical_concepts/blob/copernicus/analysis/copernicus_analysis_elliptic.ipynb
263. https://github.com/eth-nn-physics/nn_physical_concepts
264. https://github.com/tonymetger/communicating_scinet
265. Pitelis, N., Russell, C., & Agapito, L. (2013). *IEEE Conference on Computer Vision and Pattern Recognition,* (pp. 1642–1649). https://doi.org/10.1109/CVPR.2013.215
266. Korman, E. O. (2018). http://arxiv.org/abs/1803.00156
267. Tong, D. http://www.damtp.cam.ac.uk/user/tong/relativity.html

Climate Displacement in Mexico

Armelle Gouritin

Climate Displacement in Mexico

Towards Vulnerable Population Protection

FLACSO
MÉXICO

Springer

Armelle Gouritin
Mexican National Center for Science
and Technology
Mexico City, Mexico

ISBN 978-3-031-10334-6 ISBN 978-3-031-10335-3 (eBook)
https://doi.org/10.1007/978-3-031-10335-3

The edition is not for sale in Mexico. Customers from Mexico please order the print book from: FLACSO, México. ISBN of the FLACSO edition: 978-607-8817-02-3
Original edition in Spanish: Migrantes climáticos en México ©Facultad Latinoamericana de Ciencias Sociales, Sede México, 2021

This Springer imprint is published by the registered company Springer Nature Switzerland AG
The registered company address is: Gewerbestrasse 11, 6330 Cham, Switzerland

Foreword

Migrations and climate change have been connected throughout the history of our species. Many studies report on the impact of environmental changes and the role of natural disasters in human distribution globally, across human history. Indeed, this book makes reference to such a phenomenon, thousands of years ago.

The world's current complexity makes the attribution of migrations to single factors, with the exception of, perhaps war problematic. However, it is a fact that the climate change we are currently experiencing, of which, more severe impacts are beginning to be acknowledged will greatly modify human settlement patterns all over the world.

The Intergovernmental Panel on Climate Change has recognized that human migration could be one of the biggest impacts of climate change. It foresees, with a high level of confidence that "[d]isplacement and involuntary migration are often temporary. With increasing climate risks, displacement is more likely to involve permanent migration."[1]

As we know, global warming is the most obvious consequence of the anthropogenic change of climate systems. However, its impacts on natural and human systems are more profound still. Climate change directly and indirectly impacts fundamental human relationships and the fulfillment of a society's basic needs. Starting from the territory where the people live, to access to quality water, energy use, food security, ecosystems, and health as well as the economy, among other things. This global phenomenon influences decisions as varied as energy supply, how cities and communities organize mobility or deal with and mitigate heat waves, face up to extreme hydrometeorological events, or even how they articulate social organization and work. It boosts the search for sustainable development models that

[1] Intergovernmental Panel on Climate Change (IPCC) (2014). *Climate Change 2014: Synthesis Report. Cambio climático 2014: Informe de síntesis. Contribution of Working Groups I, II, and III to the Fifth Assessment Report of the Intergovernmental Panel on Climate Change.* [Core Writing Team, R.K. Pachauri and L.A. Meyer (eds.)]. Geneva, Switzerland: IPCC, p. 71. https://archive.ipcc.ch/pdf/assessment-report/ar5/syr/AR5_SYR_FINAL_All_Topics.pdf.

reduce the detrimental impacts of climate change while current and future populations acquire the adaptive capacities to identify and adopt decisions in order to face climate change's local and global impact.

In today's world, it is difficult to attribute migration exclusively to climate change. The legislation at an international level does not protect those who flee because of climate change. It is therefore complicated to quantify climate migration. Furthermore, as the book's authors comment, the incentives and drivers behind migration are multifaceted.

With this in mind, it is important to note that the text coordinated by Armelle Gouritin breaks down the climate-related impulses behind population movement, and more specifically for vulnerable populations (women, Indigenous Peoples, and those with fewer economic resources). Adding to the poverty, insecurity, and the lack of opportunities discussed, the authors also present documented and analyzed data from a human rights perspective. This illustrates the cases of those forced to migrate, the people displaced, or emigrating to other countries in cases that can be clearly linked to climate-related anomalies.

The book fills a gap in the knowledge that empirically results from the simple logical implication drawing a link between climate and migration. However, the scientific verification of said link is fairly complex. Armelle Gouritin and the researcher who wrote the book have made great efforts to progress in the highly complex answers to such a premise and to the other questions which guide the development of the research. From their study, it is possible to appreciate that migrants displaced due to climate-related reasons are forced to leave behind their homes, families, property, and work because of the exacerbating impacts of climate change on environmental deterioration. This is very much a quantified and accepted phenomenon.

On the other hand, internally displaced people or migrants, spurred on by environmental factors, leave behind community life, experiences, and memories. It weakens cultural inherence and although such losses are unquantifiable in economic terms they are invaluable when it comes to the cultural richness and ancestral knowledge which could be lost with climate-induced displacement or migration.

The question is very much whether or not Mexican legislation and public policies address internal climate displacement. The reflections and potential answers on how to approach and tackle those public policies are, indeed, central to this book.

The book proposes several elements which impact on the formation of public policy. These include identifying and distinguishing the public problem that generates cost, the extensive use of resources, and inconvenience to public authorities. Potential human rights violations for displaced people and communities, most especially for women, Indigenous Peoples, and those lacking financial recourse. Alongside the description of the State's means and mechanisms in a broad and holistic vision of its ability to deal with this global phenomenon that so very directly has a devastating effect on local communities.

The reader will reflect on whether or not the options the legislation and public policies provide are, in fact, enough, or if it is instead necessary to provide more and better instruments and mechanisms to address the problem. The reader will also analyze how different social and natural sciences interact to outline the scenarios that would enable a better decision-making process.

While some questions addressed in the book seek to be answered in its development, it is also clear that climate migration and displacement are not and should not be considered a form, type, or kind of adaptation. It is important to be straightforward on this point: if anyone is forced to leave their home, family, belongings, and culture behind because of climate anomalies, we are not talking about adaptation. In any case, it would be a negative adaptation and accrue many losses and much damage from an economic point of view.

The work coordinated by Armelle Gouritin succeeds in relaunching the debate on this problem in Mexico, in spite of the limitations surrounding access to sources of information and the scarce nature of literature on this subject at a national level, as well as similar problems when carrying out research in other countries. This work has been instrumental in rendering visible the phenomenon in the National Special Climate Change Program together with the recently updated Mexican Nationally Determined Contributions delivered to the United Nations Framework Convention on Climate Change on December 30, 2020.

This study is not the finishing line when it comes to such investigations but the starting one, and is an invitation to further study the phenomenon analyzed. It deserves all of our attention, in order to identify and promote public policies aimed at enhancing the population's well-being while aiming to prevent further climate-related breakdown and address such problem with the best tools available.

Lastly, we recognize the great value and role that the National Autonomous University of Mexico's PINCC played in fostering and enabling the research and development of studies that address contemporary and complex problems related to climate change. We are extremely impressed by this program's leadership and the importance of linking such research to the improvement of our country's public policies. Congratulations to the authors and the PINCC.

Mexico City, Mexico

Dr. María Amparo Martínez Arroyo
General Director of the National Institute on Ecology and Climate Change (2013–2021)
(Dirección General del Instituto Nacional de Ecología y Cambio Climático—INECC)

Dr. Marco Antonio Heredia Fragoso
National Institute on Ecology and Climate Change (2014–2020)
(Instituto Nacional de Ecología y Cambio Climático—INECC)
General Director of Climate Change Policies, Environment and Natural Resources Ministry (2020–2021)
(Director General de Políticas para el Cambio Climático, Secretaría del Medio Ambiente y Recursos Naturales—SEMARNAT)

Acknowledgements

This book presents the results of an investigation carried out in 2019. The National Autonomous University of Mexico (UNAM)'s Climate Change Research Program (PINCC) funded the research coordinated by Armelle Gouritin. We wish to very much thank the PINCC for its support.

We also thank the reviewers of the Spanish and English versions of the book. Their comments have definitely helped us to improve the final manuscript. Similarly, Gisela González Guerra and Julio Rubén Roldán Estrada (FLACSO México, Publications)'s continuous and kind support has been essential for both the Spanish and English versions of the book.

We also thank Dr. María Amparo Martínez Arroyo and Dr. Marco Antonio Heredia Fragoso who kindly agreed to write the book's foreword.

General Introduction

In his book *The Myth of Deterritorialization: From the 'Territories End' to Multiterritoriality*, Haesbaert (2011) relies on the territory's etymology and argues that it refers to "earth domination and terror" or a "place from where people are expelled or where they are warned not to enter" (p. 37).[2] Etymology is particularly relevant to forced internal climate migrations in Mexico where it has everything to do with terror and expulsion. Unlike disasters caused by floods and hurricanes, terror can sometimes be gradual. For example, it happens with the gradual and silent loss of biodiversity amplified by climate change. This gradual terror rarely appears on the front pages of the newspapers or on social media. Likewise, grabbing lands where social relationships were forged and world views established also generates terror.

Climate migration is not new. However, its breadth characterizes the current situation. Researchers have been addressing the phenomenon for some 30 years (Martin and Warner, 2012), but the research of academics and international institutions has been accelerated the last two decades. Most of the "first-generation" studies were devoted to empirically establishing, using case studies, that climate change drives migratory movements. There were still many doubts that climate change (or environmental deterioration) was a valid dimension to understanding migratory movements, i.e., a public problem. The "second-generation" studies generally focused on refining the knowledge gained in the first-generation studies. They investigated the specificities of climate-related mobility and generated nuanced and fresh knowledge to better understand the phenomenon (Hunter, 2018). For example, the climate dimension alone cannot explain mobilities. Instead a set of social, economic, political, and cultural elements generate mobilities in a climate change context. The research reported in this manuscript belongs to the second generation or the beginning of a third generation which aims to further understand climate change as a driver of migration and proposes public policy and legal elements. In this book, we do not discuss the adoption of a, more or less, binding international legal framework, and we do not dedicate ourselves to generating empirical data to quantify internal climate migration in Mexico (the World Bank did it in 2018, we report this data at the beginning of the

[2] Unless otherwise mentioned, all the translations are those of the author.

first chapter). We instead aim to critically analyze the Mexican legal framework as well as public policies and formulate proposals to face their shortcomings.

In this context, this multidisciplinary research (law and political science) is laced with numerous tensions (Piguet, Pécoud and De Guchteneire, 2011). The sheer diversity of disciplines, approaches, or dogmas can go some way to explain them. The multiplicity of studies in recent years meets with growing attention in Mexico on the nature of the forced internal migration phenomenon (or, at least, greater visibility of the issue). "The future is already here" in the words of Ana María Aragonés (2016). Increased attention and visibility came about when this project started. We hope this book will participate in this growing attention focusing on a particular driver for forced internal migration: climate change.

The research's main objectives are to answer the following questions: Do the Mexican legal framework and public policies address forced internal climate mobility? If not, what could be the elements of a legal framework and related public policies to address the phenomenon?

We organize our multidisciplinary research into three stages. In the first stage (Chaps. 1 and 2), we report on empirical data that acknowledges the breadth of the phenomenon in Mexico. According to a 2018 World Bank report, by 2050 (and in the most pessimistic scenario), climate change will induce up to 3.1 million forced internal migrants in Mexico. This empirical data is complemented by a taxonomy which allows us to disaggregate the phenomenon and answer the research questions. We develop a typology of six forced internal climate migration scenarios based on academic literature: (1) sudden-onset disasters, (2) slow-onset phenomena, (3) "sinking" islands and coasts, (4) conflicts or violence caused by the scarcity of natural resources, (5) areas designated unfit for human settlements, and (6) "development projects" related to renewable energies. We also identify three particularly vulnerable populations that receive special attention throughout this work: Indigenous Peoples, women, and low-income people.

We then propose and justify definitions, three categories of forced internal migrants (emergency forced climate migrants, forced climate migrants, trapped forced climate migrants), and criteria for those.

The six scenarios, identification of three particularly vulnerable populations, definitions of three migrant categories and related criteria make it possible to further develop the analytical grid (the territory and migrants' protection requirements). We base those requirements on a differentiated rights-based approach (towards Indigenous Peoples, women, and those with limited economic resources) and environmental rights.

We then proceed with the research's second stage and direct our attention to the Mexican political-legal framework (Chap. 3 and 4) to determine if it addresses the phenomenon. When reviewing the Mexican federal and states legal framework (Chiapas and Guerrero), we find that the Mexican legislation scarcely mentions the phenomenon and consequently puts at risk the protection of internal climate migrants. We confront this shortcoming with federal public policies to determine whether they address the phenomenon despite the legal shortcomings. We find this is not the case. However, they do provide some relevant elements (e.g., in terms of pre-existing social vulnerability). For the political-legal framework to address the phenomenon, it would be necessary to "connect the dots" and also apply these elements to internal forced climate mobility.

We subsequently turn our attention to the public policies adopted at the state and municipal levels to identify whether or not they address the phenomenon despite the limitations at federal level. We choose to analyze the policies of Tabasco state as well as two municipal Tabasco case studies: the 2007 floods in the Centro Municipality (a sudden-onset disaster) applying a differentiated approach to women, and sea-level rise in the Centla municipality alongside a differentiated approach to Indigenous Peoples. We find, once again, that public policies are inadequate and do not consider the protection of migrants *de iure* in the legal framework and *de facto* in the implementation of public policies related to the 2007 flood.

With the results of this second stage which answered our first question in the negative, we propose elements for reflection on the adoption and content of a political-legal framework that would address the phenomenon in the third stage of research. The starting point is a critical analysis of the tools potentially adopted in Mexico when addressing the phenomenon (Chap. 5). The critical approach based on the territory and the protection requirements enables us to highlight the risks and limitations of the tools in place in Mexico. We focus on four measures: (1) recognition of the phenomenon, (2) forced relocation, (3) preventive adaptation measures (*in situ* adaptation organized around the concept of resilience), and (4) reactive adaptation measures (*ex situ* adaptation: mobility considered a success). We subsequently reflect on the potential of the Socio-Ecological Systems approach to guide public policies aimed at addressing the phenomenon.

February 2022 Armelle Gouritin

References

Aragonés, A. M. (2016). *Consecuencias del cambio climático sobre la migración*. Ciudad de México: Universidad Nacional Autónoma de México.

Haesbaert, R., (2011). *El mito de la desterritorialización: del "fin de los territorios" a la multiterritorialidad*. Ciudad de México: Siglo XXI.

Hunter, L. M. (2018). Where do we go from here? Reflections on the future of environmental migration and displacement research. In R. McLeman & F. Gemenne, *Routledge Handbook of Environmental Displacement and Migration* (pp. 430–435). London: Routledge/GSE Research.

Martin, S., & Warner, K. (2012). Climate change, migration, and development. In I. Omelaniuk, *Global perspectives on migration and development* (pp. 153–172). Dordrecht: Springer/OIM.

Piguet, E., Pécoud, A., & De Guchtlneire, P. (2011). Migration and climate change: An overview. *Refugee Survey Quarterly*, *30*(3), 1–23.

Contents

About the Authors

Armelle Gouritin (Ph.D., Investigadora por México CONACyT, Facultad Latinoamericana de Ciencias Sociales-FLACSO—Sede México) holds a Ph.D. in international law (Free University of Brussels, Belgium). She is a National Investigator, level 1 (*Sistema Nacional de Investigadores*—CONACyT). Her main lines of research are environmental rights, socio-environmental conflicts, the tension between anthropocentrism and ecocentrism, climate migration, and environmental liability. She has published extensively along those lines. This book presents the results of the research project she coordinated on internal climate migrants in Mexico. e-mail: armelle.gouritin@flacso.edu.mx

José Antonio León Mella holds a Bachelor's degree in biology (Universidad Veracruzana generación 1999–2002). He has been working both in the private sector (petroleum industry) and in the public sector for the last 14 years. In 2020, he graduated from the Energy and Environmental management and Public Policies Master (*Maestría en Política y Gestión Energética y Medioambiental*, Facultad Latinoamericana de Ciencias Sociales-FLACSO—Mexico). He currently works as Projects Evaluation Director for the National Agency on Petroleum Security, Energy and Environment (*Agencia Nacional de Seguridad, Energía y Ambiente del sector hidrocarburos*—ASEA) regarding environmental impacts. e-mail: jose.leon@asea.gob.mx

Chapter 1
Examining the Phenomenon: The Link Between Migration and Climate Change

Armelle Gouritin

This chapter aims to provide different elements to approaching forced internal climate mobilities. We expose the extent of the phenomenon at a global as well as a Mexican level (1), propose a six-scenario typology (2), point out the sectors of the population that are particularly vulnerable (3), and provide a definition of forced internal climate migrants as well as criteria (4).

1.1 Internal Forced Climate Migration in Mexico

The vast majority of the academic literature and reporting from international organizations converge in recognizing a climate-triggered migratory phenomenon (Hicke et al. 2022; Mycoo et al. 2022; Bednar-Friedl et al. 2022; Castellanos et al. 2022; Begum et al. 2022; Lawrence et al. 2022; Caretta et al. 2022; Ali et al. 2022; New et al. 2022; O'Neill et al. 2022; Schipper et al. 2022; Shaw et al. 2022; Cissé et al. 2022; Birkmann et al. 2022; Trisos et al. 2022; Shaw et al. 2022).[1] Any attempt to quantify and formulate projections thereof are subject to much criticism (Bettini and Andersson 2014; Gemenne 2011b). The qualitative research presented within this book, alongside the projections and quantifications presented, are, of course, subject to this same criticism. They nevertheless provide an idea of the scale of the phenomenon which is, in itself, essential for defining a public problem and developing resultant public policies (Gemenne 2011b).

[1] An ongoing debate presents two opposing viewpoints: the "alarmists" and the "skeptics." Alarmists warn of waves of climate migrants, while skeptics reject the existence of climate-driven migrants and estimates and projections that emphasize a large-scale phenomenon. Generally speaking, environmental academics are the "alarmists," and migration specialists are the "skeptics." On this debate, see Dun and Gemenne (2008); Clark and Bettini (2017), pp. 36–37; Durand-Delacre et al. (2020); and Piguet (2013), pp. 154–156. Academics recently advocate to go beyond this debate. For example, Issova et al. (2020) propose analyzing environmental migration through the 2016 Sustainable Development Goals' lens.

A. Gouritin, *Climate Displacement in Mexico*,
https://doi.org/10.1007/978-3-031-10335-3_1

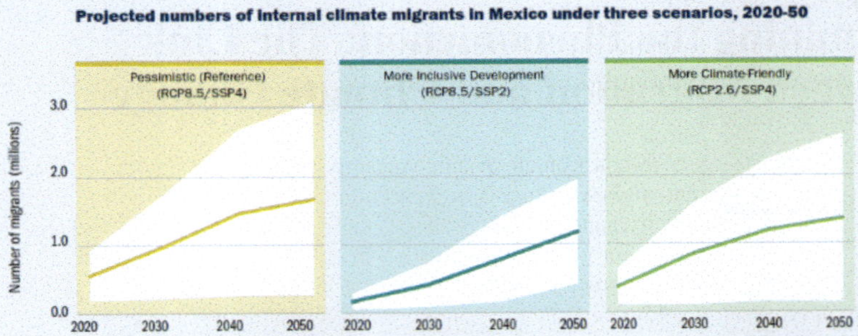

Climate migrants as a percentage of the total population

Year	2020	2030	2040	2050	2020	2030	2040	2050	2020	2030	2040	2050
%	0.45	0.45	1.08	1.23	0.14	0.30	0.55	0.80	0.31	0.64	0.88	1.01

Note: Dark lines represent the average of all model runs for each scenario. White unshaded areas represent the highest and lowest model runs for that scenario. The large intervals partly reflect the fact that climate and sectoral models were selected to represent the widest possible range of outcomes.

Fig. 1.1 Projected numbers of internal climate migrants in Mexico under three scenarios, 2020–50. *Note* From Rigaud et al. (2018, p. 162)

Numerous reports across organizations and academic papers provide quantitative data.[2] We report elements of the 2018 World Bank study "Groundswell: Preparing for internal climate migration" (Rigaud et al. 2018) for several reasons. It is recent, focuses on internal climate migration (not international migration),[3] adopts definitions similar to the definitions used in this project,[4] and relies on past experience to develop its methodological framework. The report points out that by 2050, according to the most pessimistic scenario, up to 3.1 million people will be internal migrants for climate reasons in Mexico (see Fig. 1.1). Climate migrants could, then, represent 11% of internal migrants in this scenario (p. 126).

Visual representations of the data allow a more precise understanding of the extent of the phenomenon in Mexico. Identifying places of origin and destination is particularly useful for this project. For example, in Map 1.1, we can clearly see the Valley of Mexico will be a destination territory.

---

[2] The reported data differs, since the methodology to quantify and qualify climate mobility differs. Piguet (2021) reports on methodologies and their respective use in academic literature. He mentions six "research methods on the basis of chronology and data typification" (p. 3): "spatial analysis, multilevel, survey, historical analogues, hot spots, and qualitative."

[3] Quantitatively estimating internal displacement is particularly difficult. See Gemenne (2011b), p. S46.

[4] Rigaud et al. (2018) provide the definitions at p. vii.

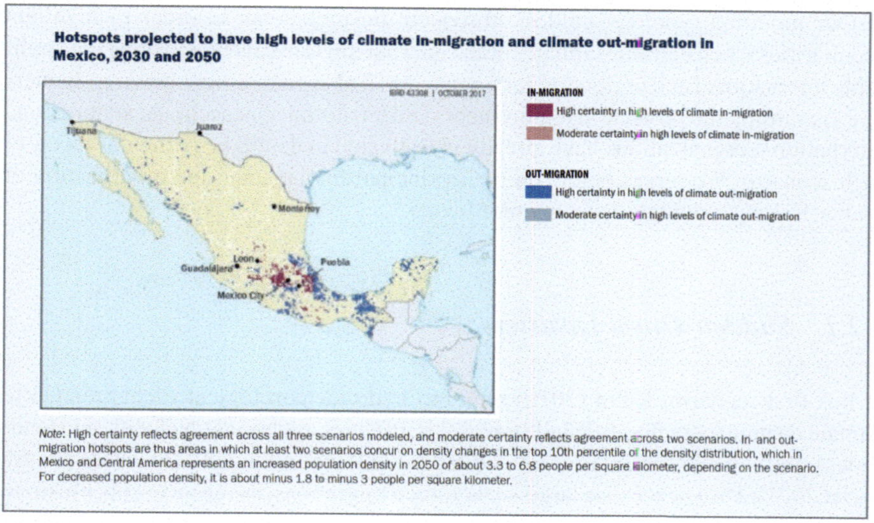

Map 1.1 Projection of places of origin ("out-migration") and places of destination ("in-migration"), Mexico, 2030 and 2050. *Note* From Rigaud et al. (2018, p. 166)

1.2 Typology: Six Scenarios that Generate Climate Internal Forced Migration

The information available in the reports and academic literature can be systematized to clarify climate migration triggers. The systematization does not attempt to reflect the reality of the phenomenon in its entirety, nor does it remain static.[5] It is a framework which allows the investigation of the phenomenon in a more orderly and analytical way.[6]

Several authors and institutions offer typologies of climate mobility triggers (see, for example, the typology of the Inter-American Commission on Human Rights, CIDH 2013, p. 40).[7] Pr. Walter Kälin (2010) proposes a typology according to five scenarios: sudden-onset disasters, slow-onset phenomena, "sinking" island states, areas designated unfit for human settlement, and conflicts or violence. We make use of it as it focuses on the scenarios themselves and does not confuse them with mobility specificities (temporary versus permanent, internal versus international, voluntary versus forced) or definition elements, criteria, and typology. We, in fact, modify it to reflect the specificities of the Mexican situation and go on to add another scenario: the mobility induced by development projects related to renewable energy.

[5] Also, the scenarios can also add up (on environmental degradation and disasters as cumulative factors, see e.g. Dasgupta et al. 2022).

[6] In addition, the identification of the scenarios aligns with McLeman et al. (2021) who argue climate mobility should be situated within the "climate risk" concept instead of the "human security" and "socioeconomic processes" concepts "that prevail at the IPCC level."

[7] For an overview of these typologies, see Ivanov and Bekyashev (2016), pp. 39–43.

Our modified typology guides subsequent developments in order to highlight peculiarities when defining climate forced internal migrants and identifying the applicable international and regional legal framework. It also guides the following chapters (e.g., regarding the protection requirements and territorial approach). Academics and international organizations have already investigated in detail the particular facets of each scenario. We report below on their principal findings and the modification of Walter Kälin's typology, focusing on Mexico.[8]

1.2.1 *Sudden-Onset Disasters*

In this first scenario, Kälin (2010) refers to hydrometeorological events related to climate change (e.g., floods, wind storms—hurricanes, typhoons, cyclones, and landslides). We can add forests fires (Hicke et al. 2022), and heat and cold waves (Ibarra Sarlat 2021). Disasters cause large-scale forced migrations, as recent events illustrate (Hurricane Katrina in the USA, Tsunami in Asia, etc.). The theoretical and empirical works by academics widely demonstrate the link between these extreme events related to climate change and climate mobilities (Simatele and Simatele 2015). The complex adaptation to such events stands out (on floods, see among others Wilby and Keenan 2012). They also report on low elevation coastal area settlements in a precarious situation (McGranahan et al. 2007). Academics and international organizations also emphasize these disasters can generate conflicts (Ide et al. 2021).

In the context of Mexico, a considerable proportion of the disasters have a link to climate change: "According to the database of the Natural Disaster Fund, 91% of the resources authorized by the declaration of disaster, from 1999 to 2017, are weather-related (...) Between 1999 and 2017, for every geological disaster, there is an occurrence of 13 climate-related disasters, and their cost has been shown to be ten times higher" (SEMARNAT and INECC 2018, pp. 465–466). Extreme hydrometeorological events are particularly prevalent in Mexico:

> The great extension of its coasts and its location between mid and tropical latitudes favors the intertropical convergence zone's influence during the hurricane season. Hydrometeorological phenomena such as tropical cyclones constantly impact the country. They impact more than 60% of the national territory.
>
> Tropical cyclone seasons are generally from June to November in the Atlantic Ocean, the Caribbean Sea, and the Mexican Gulf, and from May to November in the Pacific Ocean. Between 1970 and 2017, 269 tropical cyclones impacted the coasts of Mexico. These phenomena are expected to be more intense with climate change and have greater impacts on the population, infrastructure, and natural resources. (SEMARNAT and INECC 2018, p. 34)

[8] The particular case of Mexico should be nuanced. The Mexican territory is so heterogeneous that some of the effects of climate change can be identified in some areas and not in others. Climate change can also generate opposing effects (e.g., a decrease in energy consumption). For a nuanced panorama, see SEMARNAT and INECC (2018).

The National Risk Atlas (CENAPRED 2020) confirms this exposure: 92.57% of Emergency Declarations and natural disasters (updated data—July 2020) relate to hydrometeorological events.

1.2.2 Slow-Onset Phenomena

Kälin (2010) mentions environmental degradation and more specifically slow-onset phenomena ("increased salinization of groundwater and soil, the long-term effects of recurrent flooding, the thawing of permafrost, as well as droughts and desertification or other forms of reduced water availability," p. 85). Academics and international, regional and national organizations also tell of soil deterioration, desertification, deterioration of water quality and quantity, droughts, genetic diversity loss, and the deterioration of ecosystems and the benefits they provide. An essential aspect of which is the accumulation of slow-onset phenomena (e.g., eutrophication of a river and deforestation occurs at the same site).[9] Changes in protected and potentially dangerous species for the geographical distribution of human settlements is another slow-onset phenomenon thus far not sufficiently studied. For example, Reid (2020) mentions the phenomenon of polar bears which approach the Inuit settlements in Canada.[10] Reid also focuses on the tension between an anthropocentric or an ecocentric approach. This kind of gradual change can trigger the relocation scenario to guarantee the survival of protected species (we elaborate on this scenario below).

Academics and international institutions extensively establish the link between slow-onset phenomena and climate mobilities (both theoretically and empirically, Renaud et al. 2011; Reuveny 2005; Kazcan and Orgill-Meyer 2019; O'Neill et al. 2022).[11] They have studied and criticized deforestation and desertification in particular (see e.g. Carr 2009).[12] The inability to access drinking water, the disappearance of fish, or desertification are apparent reasons which trigger mobility of a forced nature. Similarly, fieldworks have unveiled a more subtle effect (Renaud et al. 2011, p. E21): progressive (and more or less rapid) environmental degradation has economic consequences related to the access to resources for their exploitation.[13]

[9] Although the accumulation of slow-onset phenomena is a crucial aspect in studying the environmental and (or) climate impulses of migratory movements, it appears to be studied very scarcely or not even mentioned. See, e.g., Reuveny (2005). The author mentions this cumulative dimension at p. 4.

[10] For another example, see Trisos et al. (2022).

[11] However, the slow-onset scenario receives less attention than the rapid-onset scenario (Serraglio et al. 2021).

[12] For a critique, see Bettini and Andersson (2014, pp. 168–169): the authors point out that some of the studies lack validity because they do not sufficiently consider the distinctions between nearby phenomena (long-term trends—"secular rainfall decrease," sporadic events—droughts, and cyclical fluctuations—"endemic rainfall variety"). On the difficulty to formulate models and predictions regarding the desertification process, see Lioubimtseva (2004).

[13] We explore this topic in detail in Sect. 1.4.2.

In addition, reports and academic papers establish that climate change amplifies environmental degradation (Renaud et al. 2011, p. E8). The latter subsequently enhances vulnerability to other climate mobility scenarios. For example, the degradation to the ecosystems of coastal areas caused by increasing demographic pressure makes them more vulnerable to sudden-onset disasters (McGranahan et al. 2007).[14]

We can distinguish four particularly relevant aspects in the context of Mexico. Firstly, periods of drought highly impact Mexico:

> Drought is considered a temporary climatic condition in which the precipitation level is significantly lower than average, which can cause severe hydrological imbalances that negatively affect ecological and productive systems.
>
> In Mexico, drought generally begins in November and ends when the rainy season begins, between May and June. In recent decades, there have been severe periods of drought: between 2000 and 2003, in 2006, between 2007 and 2008, in 2009 and between 2010 and 2012. In May 2011, droughts affected more than 90% of the country's territory. (...)
>
> Between 2010 and 2015, 45% of the territory suffered at least two years of droughts (not necessarily consecutive), mainly in the northern half of the country and almost in the entire Yucatan Peninsula. (SEMARNAT and INECC 2018, p. 35)

Droughts have several adverse effects. Among which are the role they play in the multiplication of forest fires and the acceleration of soil erosion (SEMARNAT and INECC 2018, p. 449; IPCC 2019, pp. 14–17). The resultant consequences can be devastating for agriculture and livestock alike, both important to the Mexican economy: "[i]n between 2012 and 2016, agriculture represented 61.4% of agricultural GDP, with a growth of 17.2% during the same period" (SEMARNAT and INECC 2018, p. 59). It is reported that "One of the most documented (*climate change impacts on agriculture*) is increasingly frequent droughts" (SEMARNAT and INECC 2018, p. 341). The importance of the agricultural sector in Mexico points to an amplified risk in terms of food security and hunger: "Climate change can severely affect food security and, by the end of this century, could put between 5 and 170 million at risk of hunger of additional people, depending on the socio-economic development scheme that is assumed" (SEMARNAT and INECC 2018, p. 348).

Secondly, the conservation of ecosystems and biodiversity is more than deficient in Mexico: all ecosystems in Mexico have loss percentages, and some loss percentages are "very high such as mangroves (46.6%), cloud forests (40.9%) or rain forests (40.5%)" (SEMARNAT and INECC 2018, p. 38). In that context, sea-level rise further participates in their degradation (SEMARNAT and INECC 2018, p. 473). Furthermore, sea warming has consequences for the ecosystems, including in Mexico (SEMARNAT and INECC 2018, p. 313) and an increase in sea acidity and corrosiveness caused by CO2 has also been reported. It also has detrimental effects on marine ecosystems (SEMARNAT and INECC 2018, pp. 314–315). There are though, some species which do not suffer but instead benefit from sea warming and their population increases (e.g., sardines) (SEMARNAT and INECC 2018, p. 315).

---

[14] Simultaneously, the risks in terms of phenomena slow-onset and sudden-onset disasters can be linked to sea-level rise.

Thirdly, desertification is very problematic in Mexico: "The arid and semi-arid zones of Mexico represent approximately 60% of the country's continental territory (…) In general, all ecoregions present trends of increasing temperatures (both maximum and minimum), as well as high levels of extreme climatic events incidence" (SEMARNAT and INECC 2018, p. 471). Finally, deforestation is also a large-scale phenomenon in Mexico (FAO 2020, p. 139).

Within this scenario, we address climate internal forced mobilities including environmental mobilities insofar as it is possible to demonstrate that climate change amplifies environmental degradations which drive environmental mobility.[15]

1.2.3 *"Sinking" Islands and Coasts*

Walter Kälin's third scenario (2010) states that climatic mobilities are "a consequence of rising sea levels." Because of the low-lying topology of certain islands, "such areas may become uninhabitable" (p. 85). Sea-level rise is increasingly documented. Many international organizations studies and academic papers report and provide projections on the extent of the phenomenon (Kulp and Strauss 2019). McGranahan et al. (2007) signal that

> The estimates of global-mean sea-level rise in the Special Report on Emissions Scenarios of the Intergovernmental Panel on Climate Change (IPCC) range from 22 to 34 cm, between 1990 and the 2080s. Far faster sea-level rise (more than a meter per century) could result from accelerated melting of the Greenland ice sheet or the collapse of the West Antartic ice-sheet, although this is not considered likely during the twenty-first century. It has been estimated that, in the absence of any other changes, a sea-level rise of 38 cm would increase five-fold the number of people flooded by storm surges. (p. 20)

We modify this scenario in the context of the book: it applies to islands within the Mexican territory and not island states (since the focus is on internal mobilities) and Mexican islands.

Departing from Walter Kälin, we also include coastal areas. Academics signal that people living in coastal areas are particularly exposed to forced climate mobilities. McGranahan et al. (2007) mention worldwide settlements in low elevation coastal areas are particularly vulnerable to the risks generated by sea-levels rise, especially considering the population density of these areas is high, and continuing to rise (in the same sense, see Glavovic et al. 2022). They report human settlements at risk from rising seas represent globally 2% of earth territory, but 10% of the world population lives in those areas, which is 13% of the urban population. Hugo (2011) points out the link between sea-level rise and flooding in valley areas with rivers and deltas, and McGranahan et al. (2007) specify environmental degradation in these areas due to urbanization which amplifies the effect of sea-level rise. In Mexico, low elevation coastal areas (and the impacts they face) are significant: they represent about 93 000 km^2 (5% of the territory), where more than 24 million people lived in 2015

---

[15] Contra, Mayer (2012) argues p. 115 that environmental migrations encompass climate migrations.

(SEMARNAT and INECC 2018, p. 349). Regarding Mexican islands, "Between 1% and 3.8% of the national island surface would be submerged in the event of scenarios with a sea-level increase of 1 m and 5 m, respectively" (SEMARNAT and INECC 2018, pp. 472–473).

The Mexican Secretariat of Environment and Natural Resources (SEMARNAT) and National Institute of Ecology and Climate Change (INECC) recognize the magnitude of sea-level rise impacts in Mexico:

> A consequence associated with the increase in the temperature of the seas and the polar ice caps melting due to global warming is (*sea-level rise*) and the exacerbation of the erosion processes on the sandy coasts, giving rise to the degradation of important coastal ecosystems such as mangroves and wetlands, dunes and beaches, as well as damage to coastal infrastructure, homes, roads, docks, ports, bridges, etc., the effect being much greater both in low-lying coastal areas and on islands (…)
>
> In another study on the evaluation of climate change impacts for Mexican islands and their area of influence, it was estimated that in the face of a hypothetical scenario of a five meters sea-level rise, the country runs the risk of losing 4.3% of its exclusive economic zone (...).
>
> The areas that could face the most significant impact in the event of a static sea-level rise with scenarios of one and two meters of increase are: the coastal plains of the Mexican Gulf, the Pacific, and the Yucatan Peninsula. (SEMARNAT and INECC 2018, pp. 313 and 463)

The degradation of coastal ecosystems (partly caused by climate change) is particularly worrisome since they play a crucial role in mitigating various effects of climate change (among others, sea-level rise and extreme hydrometeorological events) which

> include storm flood protection, flood control, erosion prevention along shorelines, groundwater recharge, freshwater supply, nutrient recycling, and capture and storage of blue carbon in its sediments, living biomass (aerial and underground), and necromass in magnitudes greater than terrestrial ecosystems. (…) [C]oastal blue carbon ecosystems conservation and restoration also contribute to climate change adaptation of the communities that inhabit them. They are natural protective barriers that retain the erosion winds and tides generate and reduce extreme natural phenomena impacts, therefore increasing coastal communities' resilience. They also provide protection, breeding, and spawning areas for commercial fish species that support coastal populations' food security. (SEMARNAT and INECC 2018, pp. 316–317)

1.2.4 Areas Designated Unsuitable for Human Settlement

According to Kälin (2010), the scenario refers to cases where "[g]overnments may designate *areas as high-risk zones* too dangerous for human habitation on account of environmental dangers" (pp. 85–86, emphasis in the text). Which occurred in Sri Lanka after the 2004 Indian Ocean tsunami (Renaud et al. 2011), China (Tan

2017),[16] Canada (Thompson et al. 2014),[17] and Alaska (Manrique et al. 2018).[18] It also happened in New Zealand, with debates surrounding the possibility of a "managed retreat" (Hayward 2008, pp. 52–53). The same debates occurred after the Asian Tsunami of 2004 (Kolmannskog and Trebbi 2010). In the Maldives, the government exploited climate change to justify mobility and resettlement policies (it tried to gather on ten islands a dispersed population spread over 200 islands). These policies were already on the political agenda but were then triggered by economic rationality (to reduce costs). They were not triggered by the impact of climate change (Kothari 2014). These scenarios generate much criticism and debate, as they also do in the case of Mexico. We analyze them in detail in Chap. 5. In the book we focus on forced relocations.

1.2.5 Conflicts or Violence Caused by the Scarcity of Natural Resources

Kälin (2010) mentions "*unrest seriously disturbing public order, violence or even armed conflict* may be triggered, at least partially, by a decrease in essential resources due to climate change (such as water, arable land or grazing grounds)" (p. 86, emphasis in the text). In case studies, academics identify climate change as a factor known to amplify or generate socio-environmental conflicts (Tebas and García 2016; O'Neill et al. 2022; Hicke et al. 2022; Castellanos et al. 2022; Dodman et al. 2022; Bezner Kerr et al. 2022; Caretta et al. 2022; Cissé et al. 2022; Mycoo et al. 2022; Shaw et al. 2022; IPCC 2022; Castellanos et al. 2022; Trisos et al. 2022; Birkmann et al. 2022; New et al. 2022).[19] However, it is difficult to systematically identify the causal relationship (Koubi et al. 2012).[20] The difficulty is also present with climate change issues where it is complicated to identify mono-causal relationships. We see it in more detail below when we define internal climate forced mobility.

[16] In this article, the author demonstrates the risks posed in terms of unstable resettlements. By the same author and on the extent of the relocations decided by the government (7.7 million persons relocated) see Tan (2018).

[17] This is a particular case: in the face of floods that threatened a large part of the Manitoba Province territory, public authorities decided to divert the waters to a lake and permanently flood some communities and Indigenous Peoples' lands to safeguard urban and agricultural areas.

[18] On designated areas not suitable for human settlement with an international migration dimension, see McAdam (2015) and Derani and Ribeiro Vieira (2016).

[19] In the same sense, see Freeman (2017); SEMARNAT and INECC (2018). Aremu and Abraham (2018) mention that human mobility driven by climate change is one of the factors that generate violent conflicts between pastoralists and farmers in Nigeria.

[20] See more specifically pp. 115–116, where the authors present a literature review and report on the opposition between pessimists ("Neo-Malthusianists") and optimists ("Cornucopians"). For a pessimistic reading from the perspective of the European Union, see Werz and Hoffman (2016). Abel et al. (2019) establish the link but note that it is limited to specific periods and contexts.

The fifth scenario is particularly relevant to Mexico. There are already numerous socio-environmental conflicts (FLACSO and De Montfort University 2019). A historical approach further informs the link between the scarcity of natural resources and conflicts in Mexico:

> [D]ry spells have been present or have preceded important social movements in Mexico. Like the period 100 years before—in 1810, during the war of independence—the severe drought of 1910 is considered by some historians as a crucial factor that contributed to the Mexican revolution. Recent studies show that Mexico went into a new drought period from 1994 onwards: 2009 saw the worst drought for the last 70 years—comparable with the drought of the 1050s—and combined with an economic crisis and food scarcity, this creates an explosive socio-political context for the future. (Cohen et al. 2013, p. 56)

1.2.6 "Development Projects" Related to Renewable Energies

This scenario does not appear in the typology of scenarios established by Kälin (2010). We add it within the framework of analysis applied in the book. It refers to cases where development projects to generate energy from renewable sources, in turn, induce forced mobility at an internal level (e.g., the construction of a dam). At first sight, it can appear similar to the fourth scenario (areas designated unfit for human settlement). However, it is distinct: it does not concern the environmental risks *stricto* sensu, but, instead, the risks generated by human activity in the construction and operation of the said "development projects."[21]

In the book, we exclusively consider projects linked to climate change mitigation. More concretely, we focus on development projects to produce "clean energy" within the Mexican mitigation policy (hydroelectric dams, for example).[22] There is currently no national monitoring system for people displaced by dams. Monitoring the displacements would have to consider the people and communities forcibly displaced by the flooding induced by the dam and the displacement induced by the project as a whole. For example, people below the dam's curtain could be forced to move because they would no longer have access to suitable land for agriculture

---

[21] Academia pays a growing attention to the so far under-researched link between renewable energy and displacement. For such study, see e.g. Kramarz et al. (2021) who identify "three broad types of displacement" (p. 4): displacement by (i) dispossession, (ii) degradation, and (iii) through commodity-dependent development. Also, other types of activities officially aimed at mitigating climate change that induce forced displacement (e.g., the dispossession of land to develop agrofuels or reforestation). Like "development projects," they involve forcefully displacing the people located on the lands attributed to these activities. For a critique and the projects' legitimation in the name of climate change mitigation, see Schade (2012). Domínguez (2014) distinguishes five development projects: "forced internal displacement whose main cause is the acquisition and occupation of land reserves for the construction of physical infrastructure (... energy projects, etc.) or urban (...); environmental infrastructure (... wind farms, etc.); social and economic projects" (p. 5). In the book, we focus on physical and environmental infrastructure.

[22] For another example of the extensive conception that encompasses "development projects" see Duyck and Gouritin (2018).

Table 1.1 Forcefully displaced and affected persons by three hydroelectrical dams

Name of the hydroelectrical dam	State	Number of forcefully displaced persons	Number of affected persons
La Parota	Guerrero	25 000	75 000
Las Cruzes	Nayarit	54	6 137
El Zapotillo	Jalisco	1 500	15 000

Note Data from Jurado del Tribuna de los Pueblos (2012, pp. 18, 21 and 29)

or livestock activities—slow-onset scenario, or because of the conflicts or violence induced by the scarcity of natural resources.

In 2012 and "only considering five of the megaprojects (hydroelectric dams) (...) more than 58 thousand people were dispossessed of their lands," and the five projects would affect "another 162 thousand" people in Mexico (Rámirez 2012). Cohen et al. (2013) report on conflicts related to water scarcity in Mexico where agriculture continues to use 78% of the available water. We present a table below with three cases to illustrate the scale of the impact of the hydroelectric dam projects and the forceful displacement they have generated (Table 1.1).

In the subsequent section, we further complete our analysis identifying the most vulnerable people and communities we particularly focus on in the book.

1.3 Particularly Vulnerable Populations

Vulnerability is a critical element of all research dealing with climate mobilities and it is a constant theme within this book. When we mention the concept of vulnerability, we refer to vulnerability according to social and economic reference factors. If we evoke another reference factor (e.g. the environmental one), it is specified.

Climate migrants do not react "mechanically" to the climate mobility scenarios. Social (Piguet et al. 2011; Rubio Díaz-Leal 2017), gender, and demographic (Kam et al. 2021) factors are also involved (Dodman et al. 2022; Cissé et al. 2022; Birkmann et al. 2022; Schipper et al. 2022; Castellanos et al. 2022; O'Neill et al. 2022).[23] Accordingly, vulnerability is key to understanding and addressing climate mobility. The vulnerability of specific sectors of the population is an essential aspect in the probability of them being forced to migrate for climate reasons (Renaud et al. 2011).[24]

[23] In the same sense, see Hauer et al. (2020).

[24] This vulnerability has another effect in terms of protecting the people and communities forced to migrate. For more on methodological issues concerning climate change vulnerability identification, see Birkenholtz (2012, and particularly p. 296 for a methodology overview). Ferris (2011) mentions p. 10 the specific risks forced migrants face in development projects contexts: referring to Michael Cernea, she mentions "landlessness, joblesness, homelessness, marginalization, food insecurity, increased morbidity and mortality, loss of access to common property, and social disintegration." The National Institute of Ecology and Climate Change (INECC 2018) points out the LGCC (2020) and other climate instruments do not use the same notion of vulnerability (pp. 15–16, 218).

Kälin (2008a, no page) puts it very clearly: "The most vulnerable among them (*the displaced*) are particularly at risk. The consequences of natural disasters exacerbate pre-existing inequalities and patterns of discrimination, further marginalize the poor, single women, the elderly or people with disabilities or living with HIV/AIDS and chronic illnesses, and affect the rights of minorities or indigenous peoples."

In this book, we select three sectors of the population academics and international organizations identify as being particularly vulnerable. They are populations with scarce economic resources, Indigenous Peoples and women.[25] They are not to be portrayed as victims,[26] but rather the holders of rights according to a differentiated approach.

1.3.1 People with Limited Economic Resources

Climate change impacts countries with lower economic development more severely than countries with higher economic development (Cournil 2010). The same applies within countries: environmental degradation and climate change impact disproportionally populations with scarce resources (Rojas Hernández 2016; SEMARNAT and INECC 2018; SEGOB 2019).[27] Accordingly, the people and communities with scarce economic resources more intensively perceive the impacts of climate change which trigger forced internal mobilities (McGranahan et al. 2007). The limited ability of communities to adapt to climate change impacts on their dependence on the environment for sustenance, and in turn explains their greater sensitivity to climate migration triggers (Reuveny and Moore 2009; Gray and Bilsborrow 2013).

In Mexico, the National Development Plan (*Plan Nacional de Desarrollo*, PND) (SEGOB 2019) recognizes that climate change impacts on the most vulnerable populations and regions (p. 38) increasing and deepening poverty and inequality in Mexico

[25] In contrast to Indigenous Peoples and people with limited economic resources, academics do not systematically identify children as particularly vulnerable. See e.g., Bose and Lunstrum (2014). However, a growing body of literature addresses children's specific vulnerability. See e.g., Szaboova (2020), Uddin et al. (2021), and Selby et al. (2021). In the same sense, the specific situation of people who live with disabilities are not is only scarcely addressed currently (for the time being). On people who live with disabilities and their specific vulnerabilities to climate change impacts (and the necessity to include them and not to "[leave them] behind in activism efforts"), see Novack and Frias (2022).

[26] For arguments against victimization, Clark and Bettini (2017) emphasize the role women play in informal support networks for climate disaster victims. Regarding Indigenous Peoples, Kelman (2018) develops "counter-narratives" to the predominant passive vision. In the same sense, see Bronen and Cochran (2021).

[27] Academics have already established the link between poverty and climate mobility: they have refined the notion of poverty, addressed inequality studies in a more general way, and developed increasingly detailed analytical grids (see e.g. Tan et al. 2016). On the process of introducing social inequalities in the analysis of climate and environmental mobility, see Ackerly (2016).

(p. 185). It establishes the objective of a response to and the coordination of climate-induced emergencies and disasters (p. 77). It recognizes the lack of economic, political, and social development alternatives which trigger forced migration. It also refers to the relationship between the impacts of climate change and migration and finds that environmental degradation contributes to migration (p. 104). Rubio Díaz-Leal (2017) identifies the particular vulnerability of "marginalized regions and communities" in Mexico. Unfortunately, a very high proportion of the population lives in poverty in Mexico, despite reported decreasing trends: "Although the population in a situation of poverty remains high—43.6% of the total population in 2016—in recent years there has been said to be a decrease. The figure reported was 46.1% in 2010." (SEMARNAT and INECC 2018, p. 46).

The need to consider this sector of the population applies to both rural and urban populations (Torres and Casey 2017). The particularly vulnerable situation of populations living in Mexican cities is well established: "the projections of climate change costs for the inaction scenario indicate they will not be distributed evenly between countries or inside them. In general, the highest costs occur in urban-type areas" (SEMARNAT and INECC 2018, p. 339). As regards the rural areas, as already pointed out, climate change has a disastrous impact on Mexican agriculture. Academics solidly document the vulnerable situation of farmers. Among other factors, they signal that

> [r]ural Mexican households are particularly vulnerable to climate change. Although not entirely dependent on agricultural production, this livelihood activity contributes to up to two-thirds of rural Mexican household income. Because of its small, often non-commercial scale, much of Mexico's rural agricultural sector lacks even the most basic technological buffers to climate change. For example, in 2001 only about one quarter permanently cropped land was irrigated. Due to the lack of technological buffers, climate and weather events have led to major economic losses within the agricultural sector. Adding to this, the liberalization of the Mexican economy has increased household sensitivity to changes in climate and market conditions. (Nawrotzki et al. 2015, p. 463)

Two specific aspects further justify focusing on this sector of the population. Climate change will induce an increase in energy consumption in several areas of the country. It will be problematic for the populations that will not be able to afford access to energy (i.e. energy poverty) (SEMARNAT and INECC 2018).[28] In addition to this, limited economic resources is also one of the factors which explains why some people or communities do not migrate despite climate triggers that compel them to do so. Communities who cannot move for economic reasons ("trapped migrants") are one of the categories of internal migrants addressed in the book.

1.3.2 Indigenous Peoples

The Sixth Mexican National Communication and Second Biennial Update Report to the United Nations Framework Convention on Climate Change (SEMARNAT

[28] In contrast to this, some areas of the country will see energy consumption decrease, and forcibly displaced persons might take into account this element when choosing their destination territory.

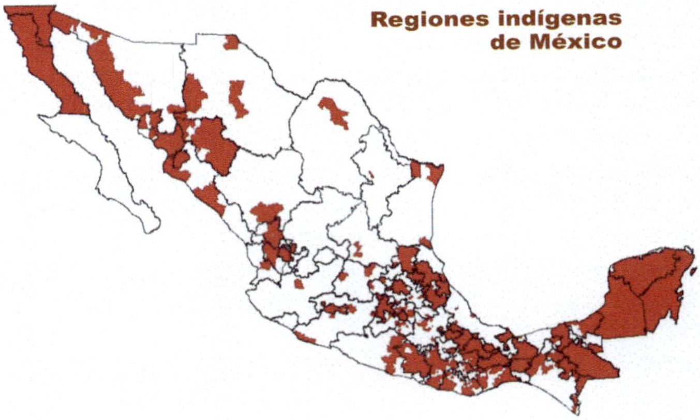

Map 1.2 Mexican indigenous regions, 2005. *Note* From Serrano Carreto (2006, p. 1)

and INECC 2018) does not mention the vulnerable situation of Indigenous Peoples. We regret this omission on several grounds. Firstly, Indigenous Peoples represent a substantial part of the nation's population: in 2015, "26 million (*people*) (21.5%) considered themselves indigenous according to their culture, history and traditions" (SEMARNAT and INECC 2018, p. 43). Secondly, they live in a situation of ex-ante vulnerability amplified by climate change (Gouritin 2018).[29] Finally, their unique relationship with their lands and territories has significant negative impact for them, such as the loss of livelihood and negative effects on physical and mental health (Torres and Casey 2017). Indigenous Peoples' ex-ante vulnerability is amplified by climate change and their unique relationship to their lands, which is not, of course, specific to Mexico: academics also signal similar situations in the United States and Canada (Kieval 2020; Merrill 2013; Manrique et al. 2018; Reid 2020).

Few studies illustrate the impacts climate change have on Indigenous Peoples and the resultant impacts in terms of mobility. An example is Klooster (2018): "They (*villagers*) also note that climate change is affecting maize agriculture, and all this could create powerful incentives to migrate in the future. 'The climate is out of control. Rains come early, or they come late'" (p. 119). In spite of the scarcity of available studies, four maps and a graph below show the vulnerability of Indigenous Peoples in the context of the climate mobility scenarios in Mexico. We present elements of understanding in four maps: we contrast the presence of Indigenous Peoples (Map 1.2) with the incidence of forest fires (Map 1.3), deforestation (Fig. 1.2), biodiversity loss (Map 1.4), and the conflicts in which Indigenous Peoples are involved (Map 1.5).

Indigenous Peoples are present in almost all Mexican states, mainly in the Southeast, South, Northwest of the country (in Yucatán, Quintana Roo, Puebla, Veracruz, and Hidalgo States—Map 1.2). Many Indigenous Peoples settle in coastal areas and

[29] Klooster (2018) argues (pp. 109–110) the Santa Cruz community's problematic access to health explains why many community members left. See also Rubio Díaz-Leal (2017, p. 145).

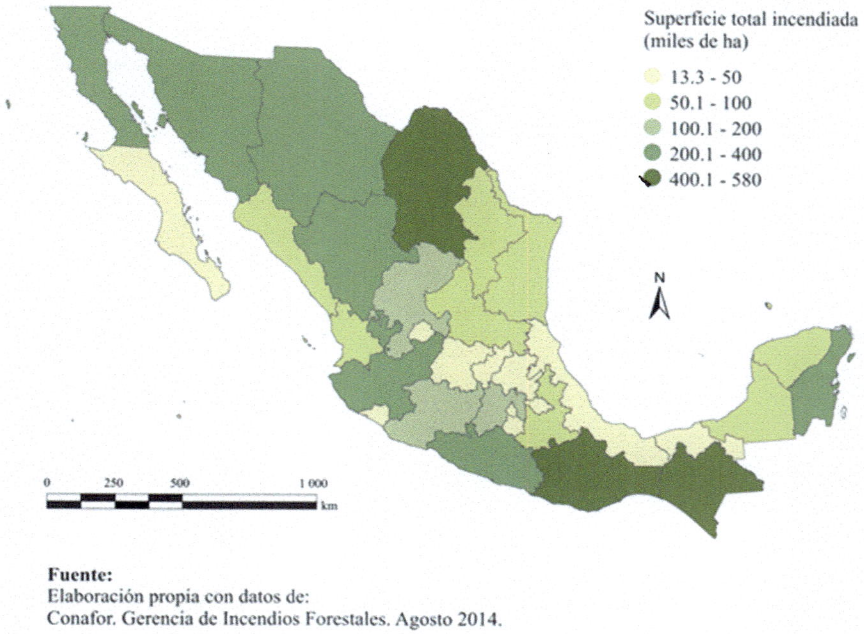

Map 1.3 Burnt forest surface per federate entity, 1998–2013. *Note* From SEMARNAT (n.d.)

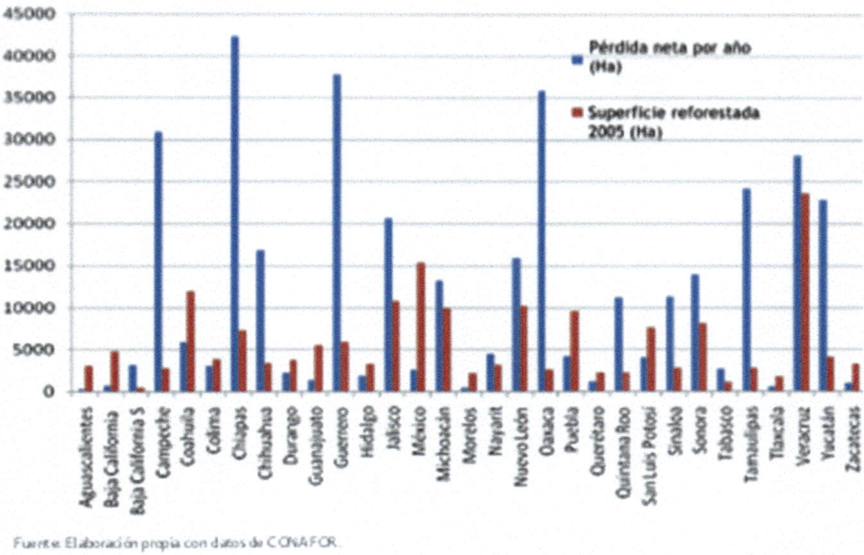

Fig. 1.2 Reforestation and forest resource net loss per Mexican federate entity. *Note* From Céspedes-Flores and Moreno-Sánchez (2010, p. 12)

Map 1.4 Natural capital sustainability rating. *Note* From Mora (2019, p. 5)

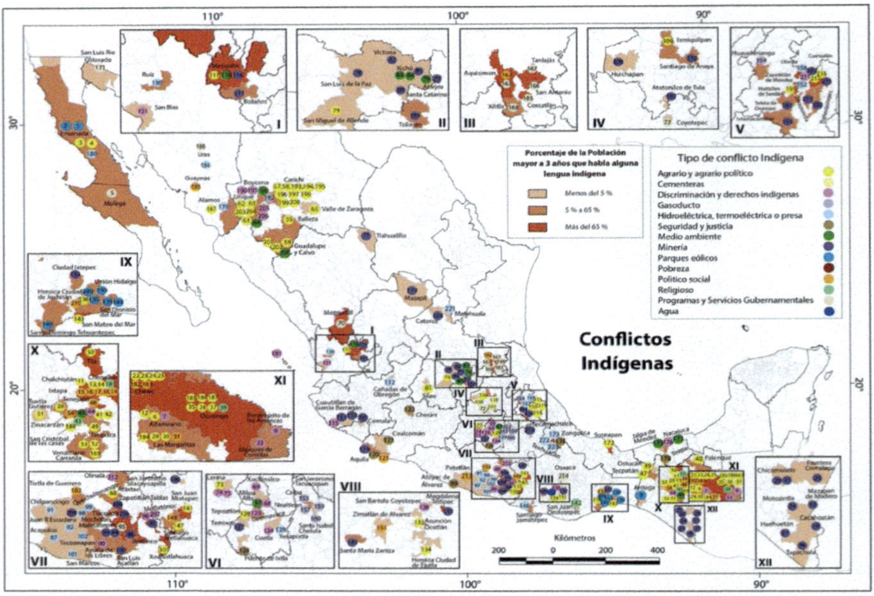

Map 1.5 Indigenous conflicts. *Note* From CDPI (n.d.)

are particularly exposed to sea-level rise, soil erosion, and extreme hydrometeorological events. Also, areas affected by forest fire and forest resources loss severely affect States with a strong indigenous presence (Yucatán, Chiapas, Campeche, Oaxaca, Guerrero, Quintana Roo, Chihuahua—Map 1.3, Fig. 1.2). This is highly problematic. We have seen that climate change amplifies forest cover loss, which triggers other aspects identified above: rising conflicts to access natural resources (forest resources) and negative impacts for biodiversity, ecosystems, and the services they provide (Rubio Díaz-Leal 2017).

In terms of ecological degradation, several areas where Indigenous Peoples settle have suffered significant degradation (among them, the Southeast of the country and the States of Yucatán, Puebla, Quintana Roo, Veracruz, and Hidalgo—Map 1.4). Accordingly, the slow-onset scenario applies particularly acutely to Indigenous Peoples and, again, this set of circumstances triggers other scenarios (e.g., conflicts over access to natural resources).

To further illustrate the above, Map 1.5, showing the conflicts in which Indigenous Peoples are involved is particularly illustrative of the vulnerability of Indigenous Peoples in the context of the mobility scenarios. It indicates a high number of conflicts related to water, agriculture, and "clean energy" projects (wind farms, hydroelectric, thermoelectric).

1.3.3 Women

From the outset, identifying women as a vulnerable segment in forced internal climate migrations is open to criticism. We recognize the relevance and validity of the criticisms towards the "monolithic" identification of women as a category used for analytical purposes:

> [A]ll the case studies highlight women-specific vulnerabilities and use gender as a synonym for women. This essentialises women as a fictitious monolithic category, ignoring questions of scale, context-specificity, generational processes, power relations, belonging and membership, which are necessary for gender analysis. (…) Scholars should be conscious to avoid simplistic descriptions of women in heteronormative, universal and static terms. (Gioli and Milan 2018, pp. 138, 143)

In this manuscript, we do not intend to address the "power structures that reproduce" "real gendered vulnerabilities" (Gioli and Milan 2018, p. 138). However, several elements justify this choice, as the situation of women in Mexico is worrying. Among other issues is the fact that most of the people in poverty in Mexico are women (Clark and Bettini 2017). A gender approach to climate change is also undertaken in Mexico (SEMARNAT and INECC 2018) and globally (Ogra and Badola 2014; Onwutuebe 2019).[30] Gioli and Milan (2018) provide a comprehensive list of the challenges women face. Alongside other challenges, climate change negatively impacts

[30] Also, a growing body of academics apply a gender approach to forced climate mobility. See e.g. Folkebrant (2021) and Bleeker et al. (2021). For example, Ayeb-Karlsson (2020) signals how

women's health dramatically (SEMARNAT and INECC 2018; Oswald Spring 2008) and empirical data signals women's particular vulnerability in terms of forced climate mobilities (Schipper et al. 2022; Shaw et al. 2022; O'Neill et al. 2022; Mycoo et al. 2022; Cissé et al. 2022; Castellanos et al. 2022). Academics report exacerbated health and hygiene challenges, violence, and sexual abuse. Women are also affected by human trafficking (Clark and Bettini 2017).

1.4 Definition and Criteria

Some academics express doubts when it comes to defining climate migrants. Kothari (2014) signals the usȩ of the dominant narrative to capture and encapsulate climate migrants in conceptual boxes and the risks associated therein (in the same sense, see Venkataraman 2020; Nash 2020).[31] Mayrhofer (2020) identifies the same risks at a United Nations level, and Strunk (2020) identifies security narrative mobilization in academic works. Dun and Gemenne (2008) express it very clearly:

> The need for a definition is a crucial step in the conceptualisation of environmental migration, and the development of policy responses to address these flows. However, two main factors driving the need for a definition could hinder its development.
>
> Firstly, many scholars would like to establish environmental migration as a specific field within migration studies. There is a tendency to fence o this area and consider it apart from classical migration theories, as if environmental migration were of another kind. More would be gained by trying to integrate environmental factors into existing migration studies.
>
> Secondly, there is a widespread appetite for numbers and forecasts amongst journalists and policymakers. In order to make their research policy-relevant, many feel compelled to provide some estimation of the number of those who are or may become 'environmentally displaced.' These numbers, obviously, need to rely on a clear definition of who is an environmental migrant. Larger definitions draw bigger numbers; there is a tendency to enlarge the definition so as to encompass as many people as possible. However, defining environmental migration too widely would be damaging for those in need of the most protection. (pp. 10–11)

However, it is necessary to establish a definition and criteria to generate valid scientific knowledge. Providing definitions is also crucial for the design, implementation, and evaluation of public policies.[32] We distinguish between the forced internal migration (Sect. 1.4.1) and climate dimensions (Sect. 1.4.2). Regarding climate migration, we distinguish between emergency migrants, forced migrants,

---

gendered subjectivities explain women immobility in the sudden-onset disaster scenario in a case-study in Bangladesh.

[31] In the same sense, Fruh (2021) proposes to attend climate mobilities through the lens of reparations.

[32] In the same vain, see Bettini and Andersson (2014): "The lack of clear and common definitions may open up for incoherent strategies and ineffective policies, or inaction" (p. 169), and Rubio Díaz-Leal (2017). The authors are very critical of definition and criteria elaboration. For Mexico specifically, see Bobes (2017). See also Vlassopoulos (2013), who identifies the historical evolution of tensions related to the definition of climate migration and identifies the inherent premises in terms of public policies, problem construction, political relevance, and solutions identification.

and trapped migrants. We complement these definitions with criteria to assess the forced nature of mobility (Sect. 1.4.3).

1.4.1 Internal Migration

A priori, defining an internal migrant is relatively easy: they are people and communities who migrate without crossing international borders. The United Nations Guiding Principles on Internal Displacement provide the following definition:

> For the purposes of these Principles, internally displaced persons are persons or groups of persons who have been forced or obliged to flee or to leave their homes or places of habitual residence, in particular as a result of or in order to avoid the effects of armed conflict, situations of generalized violence, violations of human rights or natural or human-made disasters, and who have not crossed an internationally recognized state border. (UN 1998, Introduction: scope and purpose, para. 2)

Despite the apparent simplicity of defining internal migrants, academics state difficulties when deciding upon categories. They refer to the criteria applied to internal migrants: How far does a person have to be displaced in order to qualify as an internal migrant? When does a person cease to be an internal migrant?[33] When returning to the place of origin, or when having settled in the destination area? These questions will be answered in the next chapter when elaborating on the territorial grid applied as a unit of analysis and the associated protection requirements.

1.4.2 Climate Migration

Academics provide typologies that distinguish between forced environmental and climate migration on the one hand and voluntary environmental and climate migration on the other. Renaud et al. (2007) differentiate between "environmental refugees," "forced environmental migrants," and "migrants motivated by environmental reasons" (in our case, it would be "climate refugees," "forced climate migrants," and "climate-motivated migrants"). This distinction is useful as it allows us to set aside the categories that do not fit within the scope of our research: the last category does not apply (it would be voluntary migration), and the first two categories would fit a priori within the scope of our research. However, we do not apply the climate refugee concept for several reasons despite it still being in fairly common use in academic literature.

---

[33] Mooney (2003) points out the 1998 United Nations Guiding Principles on Internally Displaced Persons establish three scenarios for internal displacement termination (emphasis omitted): "(i) return to their home areas or place of habitual residence; (ii) (re)settlement in the localities where they go to once displaced; (iii) resettlement in another part of the country." See also Cohen (2003). On protracted internally displaced persons, see Mundt and Ferris (2008).

We focus on internal migration, and the notion of refugee has an international dimension: it applies to migrants who cross borders. In that regard, Barutciski (1998) elaborates on the risks in the widespread application of the notion of refugee in the context of internal displacement. Deciding whether or not to apply the definition of Refugee to international climate migrations is beyond the scope of our research. We simply mention that the majority of academics and international organizations reject its application to climate migration (Black 2001; Castles 2002; Hulme et al. 2008; Kibreab 1997; McGregor 1993; Renaud et al. 2011).[34]

Of the above, the "forced climate migrants" category is still included within our project. We disaggregate it into three categories: emergency forced migrants, forced migrants, and people who do not migrate despite meeting the definitions and criteria that we explore subsequently (the "trapped forced migrants").

Emergency Forced Climate Migrants

Some authors who used the term "refugee" no longer use it and instead mention "emergency environmental migrants" (Renaud et al. 2011). It refers to those migrants who move temporarily and predominantly for environmental reasons (Renaud et al. 2011). Piguet et al. (2011) report a temporal distinction inspired by the United Nations: temporary displacement (less than three months), short-term migration (from three months to one year), and long-term migrations (more than a year). According to academics they represent a substantial part of environmental and climate mobility (Raleigh et al. 2008). This reflects the notion of forced migration, and is thusly included in our project as a category. We nevertheless modify the concept, adding the possible cases where the climate dimension is not the main reason to migrate, but is the only one which prompted the migration. Such a case is unlikely in reality, but for a complete analysis it is pertinent to specify it.

Forced Climate Migrants

The first distinction between emergency climate migration and forced climate migration refers to the temporal dimension of the phenomenon. Forced climate migration refers to permanent or temporary migration, whereas the concept of emergency climate migration applies to temporary migration (Raleigh et al. 2008). The second distinction between emergency climate migration and forced climate migration is the weighting between the climate factor as an impetus to migrate and other factors. Contrary to emergency climate migration, forced climate migration encompasses other factors that drive migration. These factors can be as preponderant as the climate factor or yet more preponderant (Raleigh et al. 2008; Cissé et al. 2022; Mycoo et al. 2022).

Academics report that together, the environmental and climate dimension is only one of the dimensions that drive migration (Piguet 2013; Renaud et al. 2011). Even

---

[34] Contra, see Bettini and Andersson (2014) and Bose and Lunstrum (2014). Some authors also propose to adapt or redefine the concept of refugee to apply it to environmental and/or climatic migrants. See Renaud et al. (2007); Bates (2002); Renaud et al. (2011). Against this proposal, see Bettini and Andersson (2014).

though environmental degradation or climate change impacts can be demonstrated with certainty, this dimension is only one of a number of factors which triggers migration. The other factors can be economic, social or cultural. They have variable importance compared to the climate factor in the impulse to migrate on a case-by-case basis. There is a debate surrounding the distinction between climate factors and "purely" environmental factors (Mayer 2012).[35] Within the framework of this project, we consider that environmental migration is covered by the phenomenon of climate migration, as long as a link between environmental degradation and climate change can be demonstrated.

Academics formulate methodological proposals to identify the environmental or climate triggers against the other triggers (for an overview, see e.g. Kniveton et al. 2008). They rely on the socio-ecological systems or the notion of ecosystem services and conduct interviews (Renaud et al. 2011). They also attempt to apply a "probability-based risk-management framework" (Okeowo 2013) and put forward a "sensitivity" analysis of non-climatic factors ("integrated assessment"), "agent-based modeling," and "scenario building" (Black et al. 2011).

We argue that the combination of distinct triggers cannot rule out climate mobility. As we saw at the beginning of this chapter, the projections point to an amplification of the environmental or climate triggers in migration patterns. Also, ignoring this dimension could have extreme consequences: it would deprive the affected people of protection. This protection is a requirement (Renaud et al. 2011) (we cover the protection requirements in detail in the next chapter). Consequently, in the book, the protection requirements and the *pro-homine* principle[36] justify an extensive definition of the forced climate migration concept and category when the climate factor can be identified, although it is among other factors.

Trapped Climate Migrants

Trapped migrants are "involuntarily immobile" (Lubkemann 2008), "displaced in place" (Lubkemann 2010). "Trapped populations are those people who not only aspire but also need to move for their own protection but who nevertheless lack the ability" (Black and Collyer 2014, p. 54).[37] The notion is subject to criticism: Ayeb-Karlsson et al. (2018) identify risks when using it: there is a "potential for direct or inadvertent policy abuse/misuse of the concept in its current form," "[i]f definitions of what constitutes a 'trapped' population are applied with too broad a brush, the rights of affected people could be threatened and existing inequalities and vulnerabilities further extended by placing the burden of adaptation on already fragile individuals" (p. 560). Niemi (2020) emphasizes the distinction between trapped migrants and voluntary immobility:

---

[35] Another debate environmental migrants in the context of development projects (e.g., the construction of a dam that forcibly causes the displacement of populations) and whether they are environmental or "development projects" migrants. See Renaud et al. (2011, p. E16).

[36] On the principle's ground, see among others the Political Constitution of the United Mexican States (CPEUM 2020, article 1), the American Convention on Human Rights (OAS 1969, article 29), and Morales Sánchez (2019, pp. 106–112).

[37] See also Islam and Shamsuddoha (2017), who identify this phenomenon in their case studies.

> voluntary immobility refers to a person choosing, on a voluntary basis, to stay in place in spite of climate change impacts. A person who is voluntarily immobile is able to choose if s/he moves or not—thus s/he has agency, s/he is able to refuse migration. This is a crucial aspect when voluntary immobility is compared to so-called trapped populations. Trapped populations consist of people who aspire to move and who need to move for their own protection but who nevertheless are not able to do so due to lacking, for example, financial or social means to migrate. (p. 14)

Trapped migrants fit into the preceding categories and meet the corresponding criteria but cannot migrate for several reasons: "social vulnerability may be a consequence of displacement, as displaced people lose their social, economic, cultural and linguistic, networks. Yet the danger is that attention goes to people who have been displaced, but not to the more discrete victims of environmental change who are not able to move" (Mayer 2012, p. 117).[38] Economic vulnerability can be an obstacle to mobility and environmental degradation can generate or accentuate this: "poor environmental conditions can reduce migration by limiting access to capital" (Gray 2010, p. 693).[39] The situation of women is particularly critical in this context (Gray 2010). Place attachment also explains why some people are trapped (Blondin 2021).

Further to this, François Gemenne outlines the phenomenon and the implications a 4 °C global heating induced temperature rise would have (2011a):

> [N]ot everyone moves when confronted with environmental changes. Another consequence of a temperature rise of 4 °C+ might be, paradoxically and in some cases, a decrease in the number of people on the move. Numerous studies show that migration flows tend to decrease when environmental crises peak. This is especially true in the case of droughts, as people tend to allocate their income primarily to meet their household's basic needs rather than to moving. People will move only if they have the resources that allow them to do so: this includes financial resources—moving is a costly process—but also access to social networks facilitating mobility. Furthermore, empirical evidence shows that the most vulnerable are often unable to move when faced with an environmental crisis. For example, prior to Hurricane Katrina, about 60 000 people were unable to leave the city of New Orleans: evacuation required money for food, gas and lodging, and many poor families were unable to afford the expense. Furthermore, the hurricane struck at the end of the month: many of the poorest residents were awaiting pay cheques, leaving even fewer resources available for their evacuation.
>
> If vulnerability and poverty increase in some regions, as has been the case in recent decades, one might expect that the number of people who would find themselves unable to move in

---

[38] In the same vein, see Black et al. (2011). Black and Collyer (2014) present other reasons. Similarly, Adams and Kay (2019) expose psychological factors' role in explaining immobility. The importance of these factors is relative in our research since the subjective dimension is taken into account to qualify the forced nature of internal migration. However, it is not the only dimension applied (we also use an objective criterion). Rubio Díaz-Leal (2017) mentions elderly nursing home residents who were not displaced in the United States of America in the context of storm Sandy in 2012 as the government had decided not to evacuate them. Zickgraf (2018) signals other causes and the lack of social networks that extend outside the place of origin, demographic variables, and the combination of all these factors.

[39] Similarly, see Gray and Bilsborrow (2013). In the same sense, but focusing on international migration from rural areas of Mexico to the United States, see Hunter et al. (2013).

the event of an environmental crisis would also be on the rise. An increasing number of people might thus find themselves *forced to stay*. (p. 188, emphasis in the text)[40]

Considering climate change amplifies environmental degradation, the trapped migrants' phenomenon is relevant in this book. In terms of protection requirements of forced climate migrants, it concerns the prevailing situation in the place of origin. By definition, trapped migrants do not go through places of transit or reach a destination territory.[41] When approaching the trapped migrants' phenomenon in the book, we are aware of the resultant risks to public policies. For example, the phenomenon may legitimize relocation plans designed in an authoritarian manner (Clark and Bettini 2017).[42]

Academics and international reports acknowledge the trapped migrants' phenomenon (Lubkemann 2008; Black and Collyer 2014; Kothari 2014; Mycoo et al. 2022; Cissé et al. 2022; Castellanos et al. 2022; Shaw et al. 2022; Cissé et al. 2022; Dodman et al. 2022; New et al. 2022; O'Neill et al. 2022; Hicke et al. 2022). The Inter-American Court of Human Rights (IACtHR) mentions trapped migrants in the *Indigenous Communities Members of the Lhaka Honhat Association (Our Land) vs. Argentina* case (2020, paragraph 285). It refers to Indigenous Peoples trapped in a context which corresponds with the slow-onset scenario (environmental degradation), and Simon (2018) identified the phenomenon within Mexico. However, it has not been studied in detail yet and no quantification is currently available despite the obvious relevance of further research on the phenomenon:

> The justification for a concern with the immobile is that particularly vulnerable populations will be trapped. Yet the potentially extreme vulnerability of the involuntarily immobile justifies greater attention to this group anyway. It also justifies some attempt to extrapolate existing information to gain some understanding of how those who are trapped might respond to progressively more severe crises or shocks and how these responses could be supported. (…) A consistent focus on movement "renders the involuntarily immobilised invisible". (…) The combination of multiple constraints on opportunities for mobility is likely to compound the impact of enforced immobility. This is just as true for the combined effects of environmental disasters and restrictive migration policy in Bangladesh as it is for conflict-related violence and drought in Mozambique: the greatest burden falls on those who are least able to cope. (Black and Collyer 2014, pp. 52, 54)

Similarly, its acknowledgement as a public problem is recent (Black and Collyer 2014) and public policies are yet to address it.

Several aspects are essential for research and reflection on public policy elements to address the phenomenon. The climate and environmental dimensions require

---

[40] On Hurricane Katrina and the trapped people, Fussell (2018) points out p. 282 the case of the migrants trapped at the evacuation site since they did not have access to a home to return to or the necessary financial resources.

[41] Contra and on the case of migrants trapped in transit, see Black and Collyer (2014). In this research, we do not cover this case: we prefer to have a restrictive conception of the trapped migrant phenomenon and address this aspect within the framework of the challenges for transit places. It has the advantage of being more explicit and not adding complexity to our already complex analytical grid.

[42] See also Adams (2016) and Black and Collyer (2014, p. 55).

taking into account the impact of climate change and the specificities of environmental degradation. The natural resources affected also have their own reality which needs to be taken into account. For example, a significant time can elapse between an event and the materialization of the damage or between the resultant damage and its identification or acknowledgment, the necessary time an ecosystem needs to recover, or the inability of an ecosystem to recover (irreversibility) (Gouritin 2016, pp. 68–69).

Also, setting categories and criteria is not enough: public authorities cannot limit themselves to carrying out desk office work. The phenomenon is as diverse as the people and communities which migrate. Fieldwork (e.g., through interviews) is essential in legislative frameworks and public policy design and implementation. It also allows for the leaving behind of the tendency to create or feed "environmental myths" (in our case, "climate myths"), taking into account "site-specificity" and "scale-sensitive aspects" (Bettini and Andersson 2014). It requires identifying the authority responsible or authorities to carry it out and allocate sufficient human resources in order to do so thoroughly.

Additionally we identify a tension between an objective approach (office work) or a subjective one (the perspective of the people affected themselves) in applying the criteria to determine the forced nature of migration. We subsequently elaborate on those.

1.4.3 Criteria to Assess the Forced Character

In contrast to the wealth of academic literature on climate mobility, academics have not yet dealt in detail with the development of criteria to deal with climate migration of a forced nature. Ottonelli and Torresi (2013) propose four criteria: coercion, insufficiency, no "exit options" (or they are too expensive), and incomplete information (or not adequate: it is not trusted, understandable, relevant, or up-to-date). Reuveny (2005) addresses the issue but does not provide criteria to concretely operationalize the concept: "migration is truly involuntary only if migrants are expelled against their will" (p. 3). Furthermore, Becker and Ferrara (2019) indicate the distinction between voluntary or forced mobility is complicated to establish, and a binary approach is not adequate. They instead portray the forced or voluntary nature along a spectrum that goes from forced to voluntary, with a whole area of nuance: "there is a spectrum between perfectly voluntary migration and forced migration and every context deserves careful attention to where on that spectrum a migration episode is located" (p. 14). Graeme Hugo (2008) similarly argues the difference between forced and voluntary is not binary, but can be conceptualized as a "continuum" that goes from totally forced to totally voluntary (p. 16). In the same sense and contrary to what is stated in this section, McAdam (2012) rejects the choice criterion to distinguish between voluntary or forced mobility (p. 6). Eventually, Assaduzzaman et al. (2020) propose the distinction according to the capability approach developed by Amartya Sen.

To delimit the scope of this research, we propose two operational criteria: coercion and the absence of an alternative. We recognize the limits of this categorization, which apply in a similar way to the other categorizations which we capture in the conceptual grid within this research. On these limits, see for example Erdal and Oeppen (2018): in a context of "migration management," "[b]ureaucratic distinctions and labels have discursive powers have discursive power, and the labelling of individuals as forced or voluntary impacts their migration experience" (p. 982), and "a migrant might be someone who is doing migration, who is being a migrant or someone who is being described as a migrant" (p. 983, emphasis omitted), which influences migrants' treatment.

In this instance, we adopt both a subjective[43] and an objective approach to the coercion criterion.[44] The absence of alternatives is the operative criterion of the volition absence. Some authors provide a grid with four scenarios of unacceptable alternatives (Renaud et al. 2011)[45]:

(1) in cases of disaster, the impacted areas socially, economically and physically recover slowly and ineffectively; in cases of slow-onset phenomenon; or concerning sea-level rise,
(2) an alternative livelihood is possible in the impacted area, but it takes a long time to come to fruition;
(3) there is no alternative livelihood in the impacted area; and/or
(4) the impacted area no longer exists.

These distinctions are particularly useful as each scenario carries specific protection requirements. For example, the human rights at risk in the context of a hurricane or desertification differ regarding emergency and access to shelter or food.

In applying these criteria, a tension arises between subjectivity and objectivity, similar to the criteria outlined above for identifying emergency climate migrants, forced climate migrants, and trapped climate migrants. This is to be expected as it touches upon the conceptualization and application of migration law (for example, Moussalli 1992 elaborates on the difficulty when defining and providing criteria

[43] In the same sense, Kälin (2008b) proposes three criteria ("permissibility", "possibility", and "reasonableness"), and Domínguez (2014, p. 16) refers to the absence of agency capacity. Contra, see Mayer (2018).

[44] Another distinction refers to agency as opposed to structure. This distinction is commonly used in academic literature on environmental policy. On environmental studies, see for example, Coburn (2016), McLaughlin and Dietz (2008) who focus on vulnerability, Kuhmonen (2020) who focuses on resilience. On agency and structure in the context of migration and displacement, see e.g. Squire (2017) who expresses doubts on the dichotomy in the context of unauthorized migration, Bakewell (2010), Randell (2016), and Triandafyllidou (2017). The distinction is extremely useful to distinguish (in the context of the book) climate migrants' perceptions ("agency") from the mobility scenarios and socio-economic factors that trigger mobility ("structure"). However, in the book we prefer to distinguish subjectivity and objectivity, as those terms are less anchored in political science and, thereby, facilitate interdisciplinarity. For example, the distinction more easily resonates with legal studies.

[45] On the acceptable alternative concept see Erdal and Oeppen (2018, pp. 986–987).

[paras. 37–38, 40–41]). Ottonelli and Torresi (2013) argue the definition of forced or voluntary lags behind in all migration studies.

It is also true of the reception of the subjective dimension contained within research on climate migration and the definition and criteria therein. Despite the importance of climate change perceptions in climate mobility (DeLongueville et al. 2020), empirical studies remain scarce (Ibid.). As a result, "[s]tudies on environmental migrations thus traditionally use conventional climate data" (Ibid., p. 125).

As Laurie Parsons puts it (2018), "movement is not just movement" (p. 2). In the same sense, Ottonelli and Torresi (2013) advocate the need to define and consider the forced dimension, which would enable the taking into account of agency as regards migrants. Consequently, it becomes necessary to focus on how migrants perceive the environment. There are very few studies on this point (see e.g. Koubi et al. 2016).

The coercion element can only be understood, applied objectively and verified by studies and scientific reports that physically address movement, carried out by a secretariat or international organizations, for example. Alternatively, it could emphasize the experiences of migrants and the perceptions thereof by combining an objective and a subjective dimension (the people felt obliged to migrate because of their experience and perception).

The perception of coercion grounded in the personal relationship a person has with his/her origin territory and the changes he/she identifies in it are no less valid than objective elements. Acknowledging the subjective dimension avoids the risk of perceiving migrants as passive victims (Rathod 2020; Mascia 2020) and adopt a binary or deterministic approach,[46] which remains the dominant approach to climate and environmental migrations, but is criticized by certain academics in what, we feel, is a very relevant way:

> As a result of this imbalance (*between "Western scientific" and local, subjective conceptions*), 'unreflexive approaches that barely consider the social contexts and conditions are still dominant in the debate'. (…) Denigration and alienation are as much a result of the changing climate as economic adaptation, but whilst a scattering of scholars have considered the emotional and behavioural dimensions of climate change, any efforts to explore these areas have invariably lacked sensitivity to both mobility and the agency it entails. Dominated as climate science has been by Eurocentric and 'masculinist discourses,' the transition from viewing environmental change as an 'abstract statistical index' towards a recognition that its importance is highly subjective has been a slow one. Indeed, in many cases, causal reasoning has predominated to the extent that 'one wonders whether current approaches to climate change adaptation represent a new form of environmental determinism.' (…) The arrival of a perspective whose priority is the subjective dimensions of movement is both welcome and necessary; a reaction to 'the dominance of economic and policy analyses of migration, which tend to downplay emotional factors or overlook them altogether.' (Parsons 2018, p. 5)

The doctrine on climate mobilities tries to overcome these criticisms by articulating the link between structural and emotional dimensions of the phenomenon (Parsons 2018). With this approach, culture and the subjective dimension are no

---

[46] See Piguet (2013) for another criticism on taking into account the environmental (or climate) dimension according to a deterministic and mono-causal approach, and Ogra and Badola (2014, p. 1516) for a compilation of the subjective dimension associated with climate change.

longer an "intermediate variable," but play a central role in the analysis of climate change driven movement, along with more objective dimensions. Laurie Parsons (2018) very clearly exposes that "even the most intimate emotional landscapes are rooted in physical realities, which in turn reflect the nuance of lived experience" (p. 13). She explains, in detail, climate mobilities based on the approaches of translocality, mobilities, and migration.[47] Hence the mention in the first chapter of this element of "climate mobility" as opposed to "climate migrations". This choice is illustrative of the tension between migratory studies and environmental studies that generally have the objective of understanding the link between environmental and/or climate changes and mobilities, while the objective of the research in this book is to address the phenomenon from the protection and territory dimensions (i.e., the phenomenon's prevention and consequences). More specifically, we adopt Jacques Levy's definition of mobility reported in Haesbaert (2011, p. 196): "social relationship linked to change of place, that is, as the set of modalities by which members of a society address the possibility that they themselves or others successively occupy various places." We do not limit ourselves to a merely "objective" or "metaphorical" notion.

This tension implicitly infiltrates another debate, applying a group approach versus an individual one (Ivanov and Bekyashev 2016, p. 33). In the book, we adopt a group approach to identifying specific sectors of forced migrants' particular vulnerabilities. However, we do not adopt it in the definition and criteria because, as justified in the preceding lines, we also adopt a subjective approach when applying the coercion criterion. Contra, Mayer (2012) justifies his preference for the group approach. He argues that the individual approach presents two problems: it refers to the environmental refugee concept and would deny the migratory phenomenon's distinct triggers.

1.5 Conclusion

In this chapter, we have established the extent of the phenomenon in Mexico. This allows us to understand the resultant challenges at present and in the medium and long-term. We provide a six-scenario typology to analyze the phenomenon in the subsequent chapters: sudden-onset disasters, slow-onset phenomena, "sinking" islands and coasts, areas designated unsuitable for settlements, conflicts or violence caused by the scarcity of natural resources, and "development projects" linked to renewable energies. We have reported on the relevance of the scenarios in Mexico and distinguished three sectors of the population that are particularly vulnerable: people with scarce economic resources, Indigenous Peoples, and women.

Intense debates surround the very concept of climate mobility. We have taken them into account and proposed the definitions and criteria applied in the book. Two

[47] In the same sense, see Wrathall et al. (2014).

tensions arise. Firstly, there are debates regarding the weighting of the climate dimension as compared to other dimensions that trigger forced mobility. Secondly, some tension occurs between objectivity and subjectivity when identifying the categories and applying the criteria. We propose the concepts of emergency climate migrants (the climate dimension mostly drives migration that is generally temporary), forced climate migrants (climate change and other factors drive migration, migration is temporary or permanent), and trapped climate migrants (they are unable to carry out the migratory phenomenon among others for economic reasons).

Once we have established the definitions and criteria to apply as well as the scope of our research, we can refine the theoretical framework and provide an analytical grid in the next chapter (protection requirements and territorial axis).

References

Abel GJ, Brottrager M, Cuaresma JC, Muttarak R (2019) Climate, conflict and forced migration. Glob Environ Chang 54:239–249

Ackerly BA (2016) Hidden in plain sight: social inequalities in the context of environmental change. In: McLeman R, Schade J, Faist T (eds) Environmental migration and social inequality. Springer, New York, pp 131–149

Adams H (2016) Why populations persist: mobility, place attachment and climate change. Popul Environ 37(4):429–448

Adams H, Kay S (2019) Migration as a human affair: integrating individual stress thresholds into quantitative models of climate migration. Environ Sci Policy 93:129–138

Ali E, Cramer W, Carnicer J, Georgopoulou E, Hilmi NJM, Le Cozannet G, Lionello P (2022) Cross-chapter paper 4: mediterranean region. In: Pörtner H-O, Roberts DC, Tignor M, Poloczanska ES, Mintenbeck K, Alegría A, Craig M, Langsdorf S, Löschke S, Möller V, Okem A, Rama B (eds) Climate change 2022: impacts, adaptation, and vulnerability. Contribution of working group II to the sixth assessment report of the intergovernmental panel on climate change. Cambridge University Press (in press). https://www.ipcc.ch/report/ar6/wg2/

Ara Begum R, Lempert R, Ali E, Benjaminsen TA, Bernauer T, Cramer W, Cui X, Mach K, Nagy G, Stenseth NC, Sukumar R, Wester P (2022) Point of departure and key concepts. In: Pörtner H-O, Roberts DC, Tignor M, Poloczanska ES, Mintenbeck K, Alegría A, Craig M, Langsdorf S, Löschke S, Möller V, Okem A, Rama B (eds) Climate change 2022: impacts, adaptation, and vulnerability. Contribution of working group II to the sixth assessment report of the intergovernmental panel on climate change. Cambridge University Press (in press). https://www.ipcc.ch/report/ar6/wg2/

Aremu T, Abraham P (2018) Herdsmen on the move: the burdens of climate change and environmental migration in Nigeria. In: Leal Filho W (ed) Handbook of climate change resilience. Springer, New York, pp 1–11

Assaduzzaman M, Filatova T, Coenen F, Lovett J (2020) Freedom of choice to migrate: adaptation to climate change in Bangladesh. Int J Sustain Dev World Ecol 27:652–661

Ayeb-Karlsson S (2020) When the disaster strikes: gendered (im)mobility in Bangladesh. Clim Risk Manag 29:100237. https://doi.org/10.1016/j.crm.2020.100237

Ayeb-Karlsson S, Smith CD, Kniveton D (2018) A discursive review of the textual use of 'trapped' in environmental migration studies: the conceptual birth and troubled teenage years of trapped populations. Ambio 47(5):557–573

Bakewell O (2010) Some reflections on structure and agency in migration theory. J Ethn Migr Stud 36(10):1689–1708. https://doi.org/10.1080/1369183X.2010.489382

Barutciski M (1998) Tensions between the refugee concept and the IDP debate. Forced Migr Rev 3:11–14

Bates DC (2002) Environmental refugees? Classifying human migrations caused by environmental change. Popul Environ 23(5):465–477

Becker SO, Ferrara A (2019) Consequences of forced migration: a survey of recent findings. Labour Econ

Bednar-Friedl B, Biesbroek R, Schmidt DN, Alexander P, Børsheim KY, Carnicer J, Georgopoulou E, Haasnoot M, Le Cozannet G, Lionello P, Lipka O, Möllmann C, Muccione V, Mustonen T, Piepenburg D, Whitmarsh L (2022) Europe. In: Pörtner H-O, Roberts DC, Tignor M, Poloczanska ES, Mintenbeck K, Alegría A, Craig M, Langsdorf S, Löschke S, Möller V, Okem A, Rama B (eds) Climate change 2022: impacts, adaptation, and vulnerability. Contribution of working group II to the sixth assessment report of the intergovernmental panel on climate change. Cambridge University Press (in press). https://www.ipcc.ch/report/ar6/wg2/

Bettini G, Andersson E (2014) Sand waves and human tides: exploring environmental myths on desertification and climate-induced migration. J Environ Dev 23(1):160–185

Bezner Kerr R, Hasegawa T, Lasco R, Bhatt I, Deryng D, Farrell A, Gurney-Smith H, Ju H, Lluch-Cota S, Meza F, Nelson G, Neufeldt H, Thornton P (2022) Food, fibre, and other ecosystem products. In: Pörtner H-O, Roberts DC, Tignor M, Poloczanska ES, Mintenbeck K, Alegría A, Craig M, Langsdorf S, Löschke S, Möller V, Okem A, Rama B (eds) Climate change 2022: impacts, adaptation, and vulnerability. Contribution of working group II to the sixth assessment report of the intergovernmental panel on climate change. Cambridge University Press (in press). https://www.ipcc.ch/report/ar6/wg2/

Birkenholtz T (2012) Network political ecology: method and theory in climate change vulnerability and adaptation research. Prog Hum Geogr 36(3):295–315

Birkmann J, Liwenga E, Pandey R, Boyd E, Djalante R, Gemenne F, Leal Filho W, Pinho PF, Stringer L, Wrathall D (2022) Poverty, livelihoods and sustainable development. In: Pörtner H-O, Roberts DC, Tignor M, Poloczanska ES, Mintenbeck K, Alegría A, Craig M, Langsdorf S, Löschke S, Möller V, Okem A, Rama B (eds) Climate change 2022: impacts, adaptation, and vulnerability. Contribution of working group II to the sixth assessment report of the intergovernmental panel on climate change. Cambridge University Press (in press). https://www.ipcc.ch/report/ar6/wg2/

Black R (2001) Environmental refugees: myth or reality? UNHCR

Black R, Collyer M (2014) Poblaciones "atrapadas" en épocas de crisis. Rev Migr Forzadas 45

Black R, Kniveton D, Schmidt-Verkerk K (2011) Migration and climate change: towards an integrated assessment of sensitivity. Environ Plan A 43(2):431–450

Bleeker A et al. (2021) Advancing gender equality in environmental migration and disaster displacement in the Caribbean, Studies and Perspectives series-ECLAC Subregional Headquarters for the Caribbean, No. 98 (LC/TS.2020/188-LC/CAR/TS.2020/8), Santiago, Economic Commission for Latin America and the Caribbean (ECLAC)

Blondin S (2021) Staying despite disaster risks: place attachment, voluntary immobility and adaptation in Tajikistan's Pamir Mountains. Geoforum 126:290–301. https://doi.org/10.1016/j.geoforum.2021.08.009

Bobes VC (2017) Política migratoria en México: legislación, imaginarios y actores. FLACSO, México

Bose P, Lunstrum E (2014) Introduction environmentally induced displacement and forced migration. Refug: Canada's J Refug 29(2):5–10

Bronen R, Cochran P (2021) Decolonize climate adaptation research. Science 372(6548):1245. https://doi.org/10.1126/science.abi9127

Caretta MA, Mukherji A, Arfanuzzaman M, Betts RA, Gelfan A, Hirabayashi Y, Lissner TK, Liu J, Lopez Gunn E, Morgan R, Mwanga S, Supratid S (2022) Water. In: Pörtner H-O, Roberts DC, Tignor M, Poloczanska ES, Mintenbeck K, Alegría A, Craig M, Langsdorf S, Löschke S, Möller V, Okem A, Rama B (eds) Climate change 2022: impacts, adaptation, and vulnerability. Contribution of working group II to the sixth assessment report of the intergovernmental panel on climate change. Cambridge University Press (in press). https://www.ipcc.ch/report/ar6/wg2/

Carr D (2009) Population and deforestation: why rural migration matters. Prog Hum Geogr 33(3):355–378

Castellanos E, Lemos MF, Astigarraga L, Chacón N, Cuvi N, Huggel C, Miranda L, Moncassim Vale M, Ometto JP, Peri PL, Postigo JC, Ramajo L, Roco L, Rusticucci M (2022) Central and South America. In: Pörtner H-O, Roberts DC, Tignor M, Poloczanska ES, Mintenbeck K, Alegría A, Craig M, Langsdorf S, Löschke S, Möller V, Okem A, Rama B (eds) Climate change 2022: impacts, adaptation, and vulnerability. Contribution of working group II to the sixth assessment report of the intergovernmental panel on climate change. Cambridge University Press (in press). https://www.ipcc.ch/report/ar6/wg2/

Castles S (2002) Environmental change and forced migration: making sense of the debate. UNHCR

Centro Nacional de Prevención de Desastres (CENAPRED) (2020) Visualización de Datos. Sistema de Consulta de Declaratorias 2000–2020. Atlas Nacional de riesgos de la República Mexicana. http://www.atlasnacionalderiesgos.gob.mx/archivo/visualizacion-datos.html. Accessed 30 Nov 2021

Céspedes-Flores SE, Moreno-Sánchez E (2010) Estimación del valor de la pérdida de recurso forestal y su relación con la reforestación en las entidades federativas de México. Investig Ambient 2(2):5–13

Cissé G, McLeman R, Adams H, Aldunce P, Bowen K, Campbell-Lendrum D, Clayton S, Ebi KL, Hess J, Huang C, Liu Q, McGregor G, Semenza J, Tirado MC (2022) Health, wellbeing, and the changing structure of communities. In: Pörtner H-O, Roberts DC, Tignor M, Poloczanska ES, Mintenbeck K, Alegría A, Craig M, Langsdorf S, Löschke S, Möller V, Okem A, Rama B (eds) Climate change 2022: impacts, adaptation, and vulnerability. Contribution of working group II to the sixth assessment report of the intergovernmental panel on climate change. Cambridge University Press (in press). https://www.ipcc.ch/report/ar6/wg2/

Clark N, Bettini G (2017) 'Floods' of migrants, flows of care: between climate displacement and global care chains. Sociol Rev 65(suppl 2):36–54

Coburn CE (2016) What's policy got to do with it? How the structure-agency debate can illuminate policy implementation. Am J Educ 122(3):465–475. https://doi.org/10.1086/685847

Cohen R (2003) Sovereignty as responsibility: the guiding principles on internal displacement. In: Unpublished presentation held at Jadavpur University, 5. https://www.brookings.edu/wp-content/uploads/2016/06/20031205.pdf. Accessed 30 Nov 2021

Cohen IS, Spring ÚO, Padilla GD, Paredes JC, Inzunza Ibarra MA, López RL, Díaz JV (2013) Forced migration, climate change, mitigation and adaptive policies in Mexico: some functional relationships. Int Migr 51(4):53–72

Comisión para el Diálogo con los Pueblos Indígenas (CDPI) (n.d.) Conflictos Indígenas. https://agua.org.mx/wp-content/uploads/2018/01/Noticias-355-disputas-indigenas.jpg. Accessed 30 Nov 2021

Constitución Política de los Estados Unidos Mexicanos (CPEUM) (2020) Last reform Diario Oficial de la Nación, 8 May de 2020. http://www.diputados.gob.mx/LeyesBiblio/pdf_mov/Constitucion_Politica.pdf. Accessed 30 Nov 2021

Cournil C (2010) Les défis du droit international pour protéger les "réfugiés climatiques": réflexions sur les pistes actuellement proposées. In: Cournil C, Colard-Fabregoule C (eds) Changements climatiques et défis du droit. Bruylant, Brussels, pp 345–372

Dasgupta S, Wheeler D, Bandyopadhyay S, Ghosh S, Roy U (2022) Coastal dilemma: climate change, public assistance and population displacement. World Dev 150:105707. https://doi.org/10.1016/j.worlddev.2021.105707

DeLongueville F, Ozer P, Gemenne F, Henry S, Mertz O, Nielsen J (2020) Comparing climate change perceptions and meteorological data in rural West Africa to improve the understanding of household decisions to migrate. Clim Chang 160(1):123–141. https://doi.org/10.1007/s10584-020-02704-7

Derani C, Ribeiro Vieira L (2016) Displaced by development: analysis on the implications of the principle of permanent sovereignty over natural resources. Veredas Do Direito 13:89

Dodman D, Hayward B, Pelling M, Castan Broto V, Chow W, Chu E, Dawson R, Khirfan L, McPhearson T, Prakash A, Zheng Y, Ziervogel G (2022) Cities, settlements and key infrastructure. In: Pörtner H-O, Roberts DC, Tignor M, Poloczanska ES, Mintenbeck K, Alegría A, Craig M, Langsdorf S, Löschke S, Möller V, Okem A, Rama B (eds) Climate change 2022: impacts, adaptation, and vulnerability. Contribution of working group II to the sixth assessment report of the intergovernmental panel on climate change. Cambridge University Press (in press). https://www.ipcc.ch/report/ar6/wg2/

Domínguez JC (2014) Desplazamiento forzado por proyectos de desarrollo: retos para la cooperación internacional en América Latina. Instituto de Investigaciones Dr. José María Luis Mora

Dun OV, Gemenne F (2008) Defining 'environmental migration.' Forced Migr Rev 31:10–11

Durand-Delacre D, Farbotko C, Fröhlich C, Boas I (2020) Climate migration: what the research shows is very different from the alarmist headlines. The Conversation Trust, London. https://theconversation.com/climate-migration-what-the-research-shows-is-very-different-from-the-alarmist-headlines-146905. Accessed 30 Nov 2021

Duyck S, Gouritin A (2018) The road from Paris to sustainable development. Integrating human rights and gender equality into EU external policies. https://eu.boell.org/sites/default/files/visu15june.pdf. Accessed 30 Nov 2021

Erdal MB, Oeppen C (2018) Forced to leave? The discursive and analytical significance of describing migration as forced and voluntary. J Ethn Migr Stud 44(6):981–998

Facultad Latino Americana de Ciencias Sociales Sede México (FLACSO), De Montfort University (2019) Proyecto Conversando con Goliat. Resultados Principales Base de Datos Hemerográfic, Boletín a presa. https://conversingwithgoli.wixsite.com/misitio/publicaciones-y-resultados. Accessed 30 Nov 2021

Ferris E (2011) Climate change and internal displacement: a contribution to the discussion. In: Brookings-Bern Project on Internal Displacement February 2011. Prepared for UNHCR Roundtable Bellagio, 22–26 Feb 2011. https://www.brookings.edu/wp-content/uploads/2016/06/0228_cc_displacement_ferris.pdf. Accessed 30 Nov 2021

Folkebrant W (2021) Environmental security–for everyone? An ecofeminist and intersectional case study of climate displacement in Tuvalu. https://lup.lub.lu.se/student-papers/search/publication/9045174

Food and Agriculture Organization of the United Nations (FAO) (2020) Global forest resources assessment 2020: main report. Rome. http://www.fao.org/3/ca9825en/CA9825EN.pdf. Accessed 30 Nov 2021

Freeman L (2017) Environmental change, migration, and conflict in Africa: a critical examination of the interconnections. J Environ Dev 26(4):351–374

Fruh K (2021) Climate change driven displacement and justice: the role of reparations. Essays Philos 22(1/2):102–121. https://doi.org/10.5840/eip20211292

Fussell E (2018) Population displacements and migration patterns in response to Hurricane Katrina. In: McLeman R, Gemenne F (eds) Routledge handbook of environmental displacement and migration. Routledge/GSE Research, London, pp 277–288

Gemenne F (2011a) Climate-induced population displacements in a 4 C+ world. Philos Trans R Soc a: Math, Phys Eng Sci 369(1934):182–195

Gemenne F (2011b) Why the numbers don't add up: a review of estimates and predictions of people displaced by environmental changes. Glob Environ Chang 21:S41–S49

Gioli G, Milan A (2018) Gender, migration and (global) environmental change. In: McLeman R, Gemenne F (eds) Routledge handbook of environmental displacement and migration. Routledge/GSE Research, London, pp 135–159

Glavovic B, Dawson R, Chow W, Garschagen M, Haasnoot M, Singh C, Thomas A (2022) Cross-chapter paper 2: cities and settlements by the sea. In: Pörtner H-O, Roberts DC, Tignor M, Poloczanska ES, Mintenbeck K, Alegría A, Craig M, Langsdorf S, Löschke S, Möller V, Okem A, Rama B (eds) Climate change 2022: impacts, adaptation, and vulnerability. Contribution of working group II to the sixth assessment report of the intergovernmental panel on climate change. Cambridge University Press (in press). https://www.ipcc.ch/report/ar6/wg2/

Gouritin A (2016) EU environmental law, international environmental law, and human rights law: confirmation, complement and conflicts. The case of environmental responsibility. International environmental law series. Brill, Leiden

Gouritin A (2018) Human rights violations in the context of socio-environmental conflicts: wind farms in San Dionisio del Mar, Oaxaca (Mexico). In: Duyck S, Gouritin A (eds) The road from Paris to sustainable development. Integrating human rights and gender equality into EU external policies. https://eu.boell.org/sites/default/files/visu15june.pdf. Accessed 30 Nov 2021

Gray CL (2010) Gender, natural capital, and migration in the southern Ecuadorian Andes. Environ Plan A 42(3):678–696

Gray C, Bilsborrow R (2013) Environmental influences on human migration in rural Ecuador. Demography 50(4):1217–1241

Haesbaert R (2011) El mito de la desterritorialización: del "fin de los territorios" a la multiterritorialidad. Siglo XXI, Ciudad de México

Hauer ME, Fussell E, Mueller V, Burkette M, Call M, Abel K, McLeman R, Wrathall D (2020) Sea-level rise and human migration. Nat Rev Earth Environ 1:28–39. https://doi.org/10.1038/s43017-019-0002-9

Hayward B (2008) 'Nowhere far from the sea': political challenges of coastal adaptation to climate change in New Zealand. Polit Sci 60(1):47–59

Hicke JA, Lucatello S, Mortsch LD, Dawson J, Domínguez Aguilar M, Enquist CAF, Gilmore EA, Gutzler DS, Harper S, Holsman K, Jewett EB, Kohler TA, Miller K (2022) North America. In: Pörtner H-O, Roberts DC, Tignor M, Poloczanska ES, Mintenbeck K, Alegría A, Craig M, Langsdorf S, Löschke S, Möller V, Okem A, Rama B (eds) Climate change 2022: impacts, adaptation, and vulnerability. Contribution of working group II to the sixth assessment report of the intergovernmental panel on climate change. Cambridge University Press (in press). https://www.ipcc.ch/report/ar6/wg2/

Hugo G (2008) Migration, development and environment. International Organization for Migration, Geneva. https://publications.iom.int/books/mrs-no-35-migration-development-and-environment. Accessed 30 Nov 2021

Hugo G (2011) Future demographic change and its interactions with migration and climate change. Glob Environ Chang 21:S21–S33

Hulme M, Biermann F, Boas I (2008) Climate refugees: cause for a new agreement? Environment 50(6):50

Hunter LM, Murray S, Riosmena F (2013) Rainfall patterns and US migration from rural Mexico. Int Migr Rev 47(4):874–909

Ibarra Sarlat R (2021) Desplazados Climáticos. Evolución de su Reconocimiento y Protección Jurídica. UNAM, Mexico

Ide T, Kristensen A, Bartusevičius H (2021) First comes the river, then comes the conflict? A qualitative comparative analysis of flood-related political unrest. J Peace Res 58(1):83–97

Instituto Nacional de Ecología y Cambio climático (INECC) (2018) Evaluación Estratégica del Avance Subnacional de la Política Nacional de Cambio Climático. https://www.gob.mx/cms/uploads/attachment/file/418239/EVALUACION_ESTRATEGICA_AVANCE_SUBNACIONAL_PNCC.pdf. Accessed 30 Nov 2021

Inter-American Commission on Human Rights (IAComHR) (2013) Derechos Humanos de los migrantes y otras personas en el contexto de la movilidad humana en México, OEA/Ser.L/V/II. Doc 48/13. http://www.oas.org/es/cidh/migrantes/docs/pdf/Informe-Migrantes-Mexico-2013.pdf. Accessed 30 Nov 2021

Inter-American Court of Human Rights (IACtHR) (2020) Caso Comunidades Indígenas Miembros de la Asociación Lhaka Honhat (Nuestra Tierra) vs. Argentina. Fondo, Reparaciones y Costas. Sentencia de 6 de febrero de 2020. Serie C No. 400

Intergovernmental Panel on Climate Change (IPCC) (2019) Summary for policymakers. In: Shukla PR, Skea J, Calvo Buendia E, Masson-Delmotte V, Pörtner H-O, Roberts DC, Zhai P, Slade R,

Connors S, van Diemen R, Ferrat M, Haughey E, Luz S, Neogi S, Pathak M, Petzold J, Portugal Pereira J, Vyas P, Huntley E, Kissick K, Belkacemi M, Malley J (eds) Climate change and land: an IPCC special report on climate change, desertification, land degradation, sustainable land management, food security, and greenhouse gas fluxes in terrestrial ecosystems (in press)

Intergovernmental Panel on Climate Change (IPCC) (2022) Summary for policymakers. In: Pörtner H-O, Roberts DC, Tignor M, Poloczanska ES, Mintenbeck K, Alegría A, Craig M, Langsdorf S, Löschke S, Möller V, Okem A, Rama B (eds) Climate change 2022: impacts, adaptation, and vulnerability. Contribution of working group II to the sixth assessment report of the intergovernmental panel on climate change. Cambridge University Press (in press). https://www.ipcc.ch/report/ar6/wg2/

Islam MR, Shamsuddoha M (2017) Socioeconomic consequences of climate induced human displacement and migration in Bangladesh. Int Sociol 32(3):277–298

Issova L, Kulbayeva A, Gubaidullina M, Idrysheva Z, Kaipzhan Z (2020) Environmental migration through the prism of United Nations sustainable development goals. In: E3S web of conferences, vol 159. EDP Sciences, p 01011

Ivanov DV, Bekyashev DK (2016) Environmental migration in international law. Cambridge Scholars Publishing, Newcastle

Jurado del Tribunal de los Pueblos (2012) Presas, Derechos de los Pueblos e Impunidad. Dictamen final Mesa de devastación ambiental y derechos de los pueblos. Capítulo Mexicano del tribunal Permanente de los Pueblos. http://rosalux.org.mx/docs/Dictamen_Pre-audiencia_TPP.pdf. Accessed 30 Nov 2021

Kälin W (2008a) The climate change–displacement nexus. In: ECOSOC panel on disaster risk reduction and preparedness: addressing the humanitarian consequences of natural disasters, New York, 15–17 July 2008

Kälin W (2008b) Displacement caused by the effects of climate change–who will be affected and what are the gaps in the normative framework for their protection. Brookings Institution

Kälin W (2010) Conceptualising climate-induced displacement. In: Climate change and displacement: multidisciplinary perspectives, pp 81–103

Kaczan D, Orgill-Meyer J (2019) The impact of climate change on migration: a synthesis of recent empirical insights. Clim Chang 158(3–4):281–300. https://doi.org/10.1007/s10584-019-02560-0

Kam PM, Aznar-Siguan G, Schewe J, Milano L, Ginnetti J, Willner S, McCaughey JW, Bresch DN (2021) Global warming and population change both heighten future risk of human displacement due to river floods. Environ Res Lett 16(4):044026. https://doi.org/10.1088/1748-9326/abd26c

Kelman I (2018) Islandness within climate change narratives of small island developing states (SIDS). Isl Stud J 13(1):149–166

Kibreab G (1997) Environmental causes and impact of refugee movements: a critique of the current debate. Disasters 21(1):20–38

Kieval M (2020) On thin ice: exploring solutions for climate-induced displacement in the face of disappearing permafrost. Arctic Yearbook 2020

Klooster D (2018) More space and more constraint: migration and environment in Santa Cruz Tepetotutla. In: Robson JP, Klooster D, Hernández-Díaz J (eds) Communities surviving migration. Routledge, New York, pp 107–121

Kniveton D, Schmidt-Verkerk K, Smith C, Black R (2008) Climate change and migration: improving methodologies to estimate flows. IOM

Kolmannskog V, Trebbi L (2010) Climate change, natural disasters and displacement: a multi-track approach to filling the protection gaps. Int Rev Red Cross 92(879):713–730

Kothari U (2014) Political discourses of climate change and migration: resettlement policies in the Maldives. Geogr J 180(2):130–140

Koubi V, Bernauer T, Kalbhenn A, Spilker G (2012) Climate variability, economic growth, and civil conflict. J Peace Res 49(1):113–127

Koubi V, Spilker G, Schaffer L, Böhmelt T (2016) The role of environmental perceptions in migration decision-making: evidence from both migrants and non-migrants in five developing countries. Popul Environ 38(2):134–163

Kramarz T, Park S, Johnson C (2021) Governing the dark side of renewable energy: a typology of global displacements. Energy Res Soc Sci 74:101902. https://doi.org/10.1016/j.erss.2020.101902
Kuhmonen I (2020) The resilience of Finnish farms: exploring the interplay between agency and structure. J Rural Stud 80:360–371. https://doi.org/10.1016/j.jrurstud.2020.10.012
Kulp SA, Strauss BH (2019) New elevation data triple estimates of global vulnerability to sea-level rise and coastal flooding. Nat Commun 10(1):1–12
Lawrence J, Mackey B, Chiew F, Costello MJ, Hall N, Hennessy K, Nidumolu UB, Pecl G, Rickards L, Tapper N, Woodward A, Wreford A (2022) Australasia. In: Pörtner H-O, Roberts DC, Tignor M, Poloczanska ES, Mintenbeck K, Alegría A, Craig M, Langsdorf S, Löschke S, Möller V, Okem A, Rama B (eds) Climate change 2022: impacts, adaptation, and vulnerability. Contribution of working group II to the sixth assessment report of the intergovernmental panel on climate change. Cambridge University Press (in press). https://www.ipcc.ch/report/ar6/wg2/
Ley General de Cambio Climático (LGCC) (2020) Last reform Diario Oficial de la Federación, 6 Nov 2020. http://www.diputados.gob.mx/LeyesBiblio/ref/lgcc.htm. Accessed 30 Nov 2021
Lioubimtseva E (2004) Climate change in arid environments: revisiting the past to understand the future. Prog Phys Geogr 28(4):502–530
Lubkemann SC (2008) Involuntary immobility: on a theoretical invisibility in forced migration studies. J Refug Stud 21(4):454–475
Lubkemann SC (2010) Culture in chaos: an anthropology of the social condition in war. University of Chicago Press, Chicago
Manrique DR, Corral S, Pereira ÂG (2018) Climate-related displacements of coastal communities in the Arctic: engaging traditional knowledge in adaptation strategies and policies. Environ Sci Policy 85:90–100
Mascia R (2020) Complications of the climate change narrative within the lives of climate refugees. Consilience 22:31–38
Mayer B (2012) Sustainable development law on environmental migration: the story of an obelisk, a bag of marbles, and a tapestry. Environ Law Rev 14(2):111–133
Mayer B (2018) Definitions and concepts. In: McLeman R, Gemenne F (eds) Routledge handbook of environmental displacement and migration. Routledge/GSE Research, London, pp 323–328
Mayrhofer M (2020) Victims, security threats or agents? Framing climate change-related mobility in international human rights documents. Int J Law, Lang Discourse 8(2):9–33
McAdam J (2012) Climate change, forced migration, and international law. Oxford University Press, Oxford
McAdam J (2015) Relocation and resettlement from colonisation to climate change: the perennial solution to "danger zones." Lond Rev Int Law 3(1):93–130
McGranahan G, Balk D, Anderson B (2007) The rising tide: assessing the risks of climate change and human settlements in low elevation coastal zones. Environ Urban 19(1):17–37
McGregor J (1993) Refugees and the environment. In: Black R, Robinson V (eds) Geography and refugees: patterns and processes of change. Belhaven Press, Londres, Nueva York, pp 157–170
McLaughlin P, Dietz T (2008) Structure, agency and environment: toward an integrated perspective on vulnerability. Glob Environ Chang 18(1):99–111. https://doi.org/10.1016/j.gloenvcha.2007.05.003
McLeman R, Wrathall D, Gilmore E, Thornton P, Adams H, Gemenne F (2021) Conceptual framing to link climate risk assessments and climate-migration scholarship. Clim Chang 165(1–2). https://doi.org/10.1007/s10584-021-03056-6
Merrill J (2013) Climate change and its effect on indigenous peoples of the southwest. Am Indian Law Rev 38:225–259
Mooney E (2003) Bringing the end into sight for internally displaced persons. Forced Migr Rev 17:4–7
Mora F (2019) The use of ecological integrity indicators within the natural capital index framework: the ecological and economic value of the remnant natural capital of México. J Nat Conserv 47:77–92

Morales Sánchez J (2019) Migración irregular y Derechos Humanos. Tirant Lo Blanch, Ciudad de México

Moussalli M (1992) Handbook on procedures and criteria for determining refugee status under the 1951 convention and the 1967 protocol relating to the status of refugees. Office of the United Nations High Commissioner for Refugees, Geneva

Mundt A, Ferris E (2008) Durable solutions for IDPs in protracted situations: three case studies. The Brookings Institution

Mycoo M, Wairiu M, Campbell D, Duvat V, Golbuu Y, Maharaj S, Nalau J, Nunn P, Pinnegar J, Warrick O (2022) Small islands. In: Pörtner H-O, Roberts DC, Tignor M, Poloczanska ES, Mintenbeck K, Alegría A, Craig M, Langsdorf S, Löschke S, Möller V, Okem A, Rama B (eds) Climate change 2022: impacts, adaptation, and vulnerability. Contribution of working group II to the sixth assessment report of the intergovernmental panel on climate change. Cambridge University Press (in press). https://www.ipcc.ch/report/ar6/wg2/

Nash SL (2020) Disrupting the knowledge-power politics of human mobility in the context of climate change. Questioning established categories. In: Brzoska M, Scheffran J (eds) Climate change, security risks, and violent conflicts. Hamburg University Press, Hamburg, pp 195–207

Nawrotzki RJ, Riosmena F, Hunter LM, Runfola DM (2015) Amplification or suppression: social networks and the climate change—migration association in rural Mexico. Glob Environ Chang 35:463–474

New M, Reckien D, Viner D, Adler C, Cheong S-M, Conde C, Constable A, Coughlan de Perez E, Lammel A, Mechler R, Orlove B, Solecki W (2022) Decision making options for managing risk. In: Pörtner H-O, Roberts DC, Tignor M, Poloczanska ES, Mintenbeck K, Alegría A, Craig M, Langsdorf S, Löschke S, Möller V, Okem A, Rama B (eds) Climate change 2022: impacts, adaptation, and vulnerability. Contribution of working group II to the sixth assessment report of the intergovernmental panel on climate change. Cambridge University Press (in press). https://www.ipcc.ch/report/ar6/wg2/

Niemi S (2020) Mobility and immobility in climate migration in the Pacific Islands. Siirtolaisuus-Migr 46(3):13–16

Novack V, Frias D (2022) Climate change, environmental activism, and disability. Stanf Soc Innov Rev 20(1):(Sp)7–(Sp)8

Ogra MV, Badola R (2014) Gender and climate change in the Indian Hindu-Kush Himalayas: global threats, local vulnerabilities. Earth Syst Dyn Discuss 5(2):1491–1520

Okeowo D (2013) Examining the link: climate change. Environ Law Rev 15(4):273–289

O'Neill B, van Aalst M, Zaiton Ibrahim Z, Berrang Ford L, Bhadwal S, Buhaug H, Diaz D, Frieler K, Garschagen M, Magnan A, Midgley G, Mirzabaev A, Thomas A, Warren R (2022) Key risks across sectors and regions. In: Pörtner H-O, Roberts DC, Tignor M, Poloczanska ES, Mintenbeck K, Alegría A, Craig M, Langsdorf S, Löschke S, Möller V, Okem A, Rama B (eds) Climate change 2022: impacts, adaptation, and vulnerability. Contribution of working group II to the sixth assessment report of the intergovernmental panel on climate change. Cambridge University Press (in press). https://www.ipcc.ch/report/ar6/wg2/

Onwutuebe CJ (2019) Patriarchy and women vulnerability to adverse climate change in Nigeria. SAGE Open 9(1):1–7

Organizacion of American States (OAS) (1969) Convención Americana sobre Derechos Humanos (Pacto de San José), adoptada el 22 de noviembre de 1969. https://www.refworld.org.es/docid/57f767ff14.html. Accessed 30 Nov 2021

Oswald Spring Ú (2008) Gender and disasters: human, gender and environmental security: a HUGE challenge. Studies of the University: Research, Counsel, Education No. 8/2008, United Nations University, Institute for Environment and Human Security, Bonn, Germany

Ottonelli V, Torresi T (2013) When is migration voluntary? Int Migr Rev 47(4):783–813

Parsons L (2018) Structuring the emotional landscape of climate change migration: towards climate mobilities in geography. Prog Hum Geogr 43:670–690

Piguet E (2013) From "primitive migration" to "climate refugees": the curious fate of the natural environment in migration studies. Ann Assoc Am Geogr 103(1):148–162

Piguet E (2021) Linking climate change, environmental degradation, and migration: an update after 10 years. Wiley Interdiscip Rev: Clim Chang 13:e746. https://doi.org/10.1002/wcc.746
Piguet E, Pécoud A, De Guchteneire P (2011) Migration and climate change: an overview. Refug Surv Q 30(3):1–23
Raleigh C, Jordan L, Salehyan I (2008) Assessing the impact of climate change on migration and conflict. In: Paper commissioned by the World Bank Group for the social dimensions of climate change workshop, Washington, DC
Rámirez E (2012) Presas e hidroeléctricas: negocio y despojo. Contralínea. https://www.contralinea.com.mx/archivo-revista/2012/11/08/presas-hidroelectricas-negocio-despojo/. Accessed 30 Nov 2021
Randell H (2016) Structure and agency in development-induced forced migration: the case of Brazil's Belo Monte Dam. Popul Environ 37(3):265–287. https://doi.org/10.1007/s11111-015-0245-4
Rathod J (2020) Legal protections for environmental migrants: expanding possibilities and redefining success. CLAS Working Paper Series No. 29, Center for Latin American and Latino Studies, American University, Washington
Reid J (2020) Constructing human versus non-human climate migration in the Anthropocene: the case of migrating polar bears in Nunavut, Canada. Anthropocenes–Hum, Inhum, Posthum 1(1)
Renaud FG, Bogardi JJ, Dun O, Warner K (2007) Control, adapt or flee: how to face environmental migration? UNU-EHS
Renaud FG, Dun O, Warner K, Bogardi J (2011) A decision framework for environmentally induced migration. Int Migr 49:e5–e29
Reuveny R (2005) Environmental change, migration and conflict: theoretical analysis and empirical explorations. In: Human security and climate change: an international workshop, 21–23 June 2005, vol 2
Reuveny R, Moore WH (2009) Does environmental degradation influence migration? Emigration to developed countries in the late 1980s and 1990s. Soc Sci Q 90(3):461–479
Rigaud KK, de Sherbinin A, Jones B, Bergmann J, Clement V, Ober K, Schewe J, Adamo S, McCusker B, Heuser S, Midgley A (2018) Groundswell: preparing for internal climate migration. World Bank, Washington, DC
Rojas Hernández J (2016) Society, environment, vulnerability, and climate change in Latin America: challenges of the twenty-first century. Lat Am Perspect 43(4):29–42
Rubio Díaz-Leal L (2017) Desplazamiento ambiental. Experiencia global, realidad mexicana. Comisión Mexicana de Defensa y Promoción de los Derechos Humanos, Heinrich Böll Stiftung México, Centroamérica y el Caribe. http://cmdpdh.org/project/desplazamiento-ambiental-experiencia-global-realidad-mexicana/. Accessed 30 Nov 2021
Schade J (2012) Les migrants des politiques climatiques: nouveaux défis face aux déplacements générés par le changement climatique. Cult Confl 88:85–110
Schipper ELF, Revi A, Preston BL, Carr ER, Eriksen SH, Fernandez-Carril LR, Glavovic B, Hilmi NJM, Ley D, Mukerji R, Muylaert de Araujo MS, Perez R, Rose SK, Singh PK (2022) Climate resilient development pathways. In: Pörtner H-O, Roberts DC, Tignor M, Poloczanska ES, Mintenbeck K, Alegría A, Craig M, Langsdorf S, Löschke S, Möller V, Okem A, Rama B (eds) Climate change 2022: impacts, adaptation, and vulnerability. Contribution of working group II to the sixth assessment report of the intergovernmental panel on climate change. Cambridge University Press (in press). https://www.ipcc.ch/report/ar6/wg2/
Secretaría de Gobernación (SEGOB) (2019) Plan Nacional de Desarrollo 2019–2024 (PND). Diario Oficial de la Federación, 12 julio de 2019. https://www.dof.gob.mx/nota_detalle.php?codigo=5565599&fecha=12/07/2019. Accessed 30 Nov 2021
Secretaría de Medio Ambiente y Recursos Naturales (SEMARNAT) (n.d.) El medio ambiente en México, 2013–2014. https://apps1.semarnat.gob.mx:8443/dgeia/informe_resumen14/00_mensajes/00_intro.html. Accessed 30 Nov 2021
Secretaría de Medio Ambiente y Recursos Naturales (SEMARNAT), Instituto Nacional de Ecología y Cambio Climático (INECC) (2018) Sexta Comunicación Nacional y Segundo Informe Bienal

de Actualización ante la Convención Marco de las Naciones Unidas sobre el Cambio Climático. https://cambioclimatico.gob.mx/sexta-comunicacion/introduccion.php. Accessed 30 Nov 2021

Selby S, Jiwanji A, Clarey T, Cabot-Venton C (2021) Walking into the eye of the storm: how the climate crisis is driving child migration and displacement. Save the Children, November 22. https://apo.org.au/sites/default/files/resource-files/2021-10/apo-nid314906.pdf. Accessed 30 Nov 2021

Serraglio DA, Aleksandrova M, Schraven B (2021) Human mobility in the context of LFDD in Latin America and the Caribbean: a review of national climate and desertification policies. Curr Opin Environ Sustain 50:197–207. https://doi.org/10.1016/j.cosust.2021.04.005

Serrano Carreto E (2006) Regiones indígenas de México. Comisión Nacional para el Desarrollo de los Pueblos Indígenas, Programa de las Naciones Unidas para el Desarrollo. https://www.gob.mx/cms/uploads/attachment/file/35735/cdi-regiones-indigenas-mexico.pdf. Accessed 30 Nov 2021

Shaw R, Luo Y, Cheong TS, Abdul Halim S, Chaturvedi S, Hashizume M, Insarov GE, Ishikawa Y, Jafari M, Kitoh A, Pulhin J, Singh C, Vasant K, Zhang Z (2022) Asia. In: Pörtner H-O, Roberts DC, Tignor M, Poloczanska ES, Mintenbeck K, Alegría A, Craig M, Langsdorf S, Löschke S, Möller V, Okem A, Rama B (eds) Climate change 2022: impacts, adaptation, and vulnerability. Contribution of working group II to the sixth assessment report of the intergovernmental panel on climate change. Cambridge University Press (in press). https://www.ipcc.ch/report/ar6/wg2/

Simatele D, Simatele M (2015) Migration as an adaptive strategy to climate variability: a study of the Tonga-speaking people of Southern Zambia. Disasters 39(4):762–781

Simon DH (2018) Environmental migration in Mexico. In: McLeman R, Gemenne F (eds) Routledge handbook of environmental displacement and migration. Routledge/GSE Research, London, pp 257–269

Squire V (2017) Unauthorised migration beyond structure/agency? Acts, interventions, effects. Politics 37(3):254–272. https://doi.org/10.1177/0263395716679674

Strunk JD (2020) Climate change, migration, and European security. Ph.D. thesis, Naval Postgraduate School, Monterey, CA

Szaboova L (2020) Concepts, contexts and categorisations of climate mobility. In: Climate mobility and children: a virtual symposium, 3–4 Nov 2020. UNICEF, IOM. https://www.unicef.org/globalinsight/media/1806/file/Concepts,%20Contexts%20and%20Categorizations%20of%20Climate%20Mobility.pdf. Accessed 30 Nov 2021

Tan Y (2017) Resettlement and climate impact: addressing migration intention of resettled people in west China. Aust Geogr 48(1):97–119

Tan YL (2018) Environmental stressors and population mobility in China. In: McLeman R, Gemenne F (eds) Routledge handbook of environmental displacement and migration. Routledge/GSE Research, London, pp 238–256

Tan Y, Liu X, Hugo G (2016) Exploring the relationship between social inequality and environmentally-induced migration: evidence from urban household surveys in Shanghai and Nanjing of China. In: McLeman R, Schade J, Faist T (eds) Environmental migration and social inequality. Springer, New York, pp 73–90

Tebas JM, García MDMH (2016) La incidencia del cambio climático en las migraciones y en la seguridad. Tiempo Paz 120:44–55

Thompson S, Ballard M, Martin D (2014) Lake St. Martin First Nation community members' experiences of induced displacement: "we're like refugees." Refug: Canada's J Refug 29(2):75–86

Torres JM, Casey JA (2017) The centrality of social ties to climate migration and mental health. BMC Public Health 17(1):600

Triandafyllidou A (2017) Beyond irregular migration governance: zooming in on migrants' agency. Eur J Migr Law 19(1):1–11. https://doi.org/10.1163/15718166-12342112

Trisos CH, Adelekan IO, Totin E, Ayanlade A, Efitre J, Gemeda A, Kalaba K, Lennard C, Masao C, Mgaya Y, Ngaruiya G, Olago D, Simpson NP, Zakieldeen S (2022) Africa. In: Pörtner H-O, Roberts DC, Tignor M, Poloczanska ES, Mintenbeck K, Alegría A, Craig M, Langsdorf S, Löschke S, Möller V, Okem A, Rama B (eds) Climate change 2022: impacts, adaptation, and vulnerability. Contribution of working group II to the sixth assessment report of the intergovernmental panel on climate change. Cambridge University Press (in press). https://www.ipcc.ch/report/ar6/wg2/

Uddin R, Philipsborn R, Smith D, Mutic A, Thompson LM (2021) A global child health perspective on climate change, migration and human rights. Curr Probl Pediatr Adolesc Health Care. https://doi.org/10.1016/j.cppeds.2021.101029

United Nations (UN) (1998) Principios Rectores De Los Desplazamientos Internos, UN doc.E/CN.4/1998/53/Add.2, 11 de febrero de 1998, Oficina de Coordinación de Asuntos Humanitarios de las Naciones Unidas. https://www.right-to-education.org/sites/right-to-education.org/files/resource-attachments/OCHA_Principios_Rectores_%20Desplazamientos_Internos_2007_ES.pdf. Accessed 30 Nov 2021

Venkataraman N (2020) The absent victims: an ecolinguistic study of environmental refugees in The New York Times. Jam It (j Am Stud Italy) 3:109–141

Vlassopoulos CA (2013) Defining environmental migration in the climate change era: problem, consequence or solution? In: Faist T, Schade J (eds) Disentangling migration and climate change. Springer, Dordrecht, pp 145–163

Werz M, Hoffman M (2016) Europe's twenty-first century challenge: climate change, migration and security. Eur View 15(1):145–154

Wilby RL, Keenan R (2012) Adapting to flood risk under climate change. Prog Phys Geogr 36(3):348–378

Wrathall DJ, Bury J, Carey M, Mark B, McKenzie J, Young K, Baraer M, French A, Rampini C (2014) Migration amidst climate rigidity traps: resource politics and social–ecological possibilism in Honduras and Peru. Ann Assoc Am Geogr 104(2):292–304

Zickgraf C (2018) Immobility. In: McLeman R, Gemenne F (eds) Routledge handbook of environmental displacement and migration. Routledge/GSE Research, London, pp 71–84

Chapter 2
Analytical Grid: Territory and Human Rights

Armelle Gouritin

Taking into account the scenarios and categories identified in the previous chapter, we go on to elaborate on the research grid of analysis in the current one. We elaborate on two pillars: the territory (1) and the protection requirements regarding forced internal climate migrants (2). From the outset, we must acknowledge the chapter's limitations: we justify the protection requirements based on formal validity (i.e. sources of the identified rights, within positive law), but the sociological component would allow for the consolidating of human rights contributions to address the phenomenon (e.g. through fieldwork).

2.1 The Territory: Elements of Analysis and Challenges

2.1.1 Notion of Territory

We do not apply a restrictive notion of territory encapsulated in an administrative or geographical concept. The academic literature criticizes this as being dangerous and obsolete. For example, Surrallés and Hierro (2004) criticize the risk of dispossession this restrictive conception entails. Most academics favor a more expansive concept which includes all of the territorial dimensions: environmental, symbolic, economic, social, organizing, cultural, and historical (Valle 2012). Haesbaert (2011) proposes four dimensions ("basic aspects," p. 35):

- the political "referring to space-power relations in general,"
- the cultural or "symbolic-cultural": "the territory is seen, above all, as the product of the symbolic appropriation/valuation of a group concerning its lived space,"
- the economic: "special dimension of economic relations, the territory as a source of resources or incorporated into the conflict between social classes, and in the capital-labor relationship", and

A. Gouritin, *Climate Displacement in Mexico*,
https://doi.org/10.1007/978-3-031-10335-3_2

- the "natural" related to the "'natural' behavior of men in relation to their physical environment."

As to the territory's "relational approach" (Haesbaert 2011, p. 68), academics specify the territory is both the setting for social relations and the result of a social construction (Gómez and Mahecha 1998). In the same sense, Avcı and Fernández-Salvador (2016) point out the contributions of the territory in their broad sense:

> It (*the concept of territory*) provides an analytical lens to study the interactions between material practices, socio-economic structures, and cultural-political institutions. It enables us to interrogate what are the power relations among the actors involved in struggles over territory, and the material impact of these relations on territory. (p. 913, references omitted)

In the book we apply the extensive notion of territory and territoriality that "can be activated and deactivated" (Haesbaert 2011, p. 68). It is crucial for the research presented in the book as it indicates a strategy and implies flexibility and mobility. Flexibility refers to the non-static character mentioned in the previous chapter (e.g., from one mobility scenario to another one).

We subsequently justify the territory as a unit of analysis (Sect. 2.1.2), present the challenges in terms of forced internal climate migration for origin and destination territories (Sect. 2.1.3) and more precisely identify the challenges for cities and Indigenous Peoples (Sect. 2.1.4). The territorial element fulfills several essential functions within the research framework: it informs the protection requirements developed in the following section and the Socio-Ecological Systems' approach to guide public policy measures (Chap. 6).

2.1.2 The Territory as a Unit of Analysis

Territory when portrayed in all of its dimensions makes it possible to delineate the phenomenon studied. We left some questions open in the previous chapter on the definition of internal migrants and now move to answer them.

The territory is an integral part in the definition of the scope of the present research: it covers the protection and prevention requirements from origin to destination and transit territories. It does not seek to distinguish between whether the person or community has traveled 1 km or 1,000 km. If they are found to be in a different territory (be it in its cultural dimension), the person (or community) qualifies as an internal migrant. The protection and prevention requirements and the territorial dimension also make it possible to define the mobility's temporal scope. Protection (and to a lesser extent prevention) issues trigger the internal migrant qualification. What matters is that a person not in his/her "territory of origin" faces the risk of having his/her rights violated. It is not of importance if a month or a year has elapsed since the persons forcibly left their territory of origin.[1]

[1] In the same vein (regarding the termination of displacement), see Kälin (2002). The literature proposes other criteria. E.g., Mooney (2003) proposes three criteria: "cause-based criteria," "solutions-based criteria," and "needs-based criteria.".

In addition, the territorial element helps us to better understand climate mobility. The analyses that start based on the territory generally focus on pointing out the conflicts related to climate migration (see e.g. Reuveny 2005). However, it is only one aspect of the phenomenon. Additionally, the academic literature generally focuses on adopting public policies and a legal framework at an international level. Having said this, several authors provide strong arguments that justify a national or sub-national approach (Bettini and Andersson 2014). This is likely to be because quantitative studies dedicated to the national or local level are very robust (Gemenne 2011, p. S47; Birkenholtz 2012). However, relatively little research is devoted to understanding the national picture (Gemenne and Magnan 2010). In this context, a territorial dimension makes it possible to consider the specificities of the sites where forced internal climate migration occurs, that is, to direct attention to the national and sub-national levels. The academic literature, for example, points out that the territory is an essential element for low elevation coasts that are particularly vulnerable to rising sea levels (McGranahan et al. 2007). This is particularly relevant in Mexico due to the extensive nature of these densely populated areas.

The territorial dimension also takes into account that the phenomenon is not static: climate change and its effects are, indeed, neither static nor linear (Blitz 2014), the six climate mobility scenarios themselves are also not static (e.g., a territory can be impacted by the slow-onset scenario that later develops the generalized violence scenario due to the scarcity of natural resources), and climate migrants (and trapped migrants)' vulnerabilities are not shown to be static either. Similarly, migrants can become immobile, or alternatively, immobile people can migrate (Zickgraf 2018). And climate change as a driver of migration depends on the economic and institutional conditions in the affected territory (Ambrosetti and Petrillo 2016), and those conditions can change over time. In this regard, Rogério Haesbaert's concept of deterritorialization (2011) is instrumental: it is broad, dynamic, and highlights cases where deterritorialization can occur without mobility (e.g. in the case of populations impacted due to the installation of a hydroelectric dam that affects the territory, or in the case of trapped migrants) and, conversely, territorialization may occur "on the move" (2011, p. 195).

The territorial dimension also considers the cultural, social, and historical context of climate migration in the six climate mobility scenarios. For example, it enables reflection on all of the facets of the notion of livelihood and the assets it comprises of (human assets, natural assets, physical assets, and financial assets).[2] As for the subjective dimension of the coercion criterion, the territory in all of its dimensions and not only those which are geographical or administrative allow for the formulating of bottom-up rather than top-down proposals.

Finally, socio-environmental conflicts related to access to land is particularly acute in Mexico, and environmental degradation (ecological vulnerability) is a crucial element in the occurrence of the six scenarios identified within this research (e.g., soil erosion, deforestation or mangrove destruction that amplifies the damage caused by extreme hydrometeorological events).

---

[2] The absence of assets is a mobility trigger. On livelihood assets, see Thompson et al. (2014).

In the following sections, we cover the challenges pointed out in the literature and reports that address climate and environmental migration. We focus on the challenges that will arise for the territory considering the protection requirements: the objective is that territories do not become (or cease to be) "places of injustice" (Skillington 2015, p. 291). We identify challenges from the perspective of territories of origin and destination (Sect. 2.1.2) and specific challenges for cities and Indigenous Peoples (Sect. 2.1.3).

2.1.3 Challenges for Origin and Destination Territories

Firstly, the academic literature and available data, signal challenges for territories of origin and differentiate between the challenges generated by permanent or temporary migration. In this section, we provide an overview of those challenges. The literature comes from migration studies (rather than environmental studies) and pessimistic social theories regarding the phenomenon of migration. The more optimistic literature points to positive impacts (e.g. remittances and a greater numbers of women participating in matters of governance matters because many men migrated or because certain federal laws imposed it) (Hernández-Díaz 2018).

Academics point out changes to the traditional institutions that regulate "village life and agriculture," as well as affinities and support networks (De Haas 2010) and family structures (Hernández-Díaz and Robson 2018). In terms of protection, they report the effect on mental health (Torres and Casey 2017) and the breakdown of social ties if some community members stay or if the communities are dispersed (Thompson et al. 2014). Another challenge is infrastructural deficiencies (Thompson et al. 2014), including for trapped migrants (Becker and Ferrara 2019). Becker and Ferrara (2019) point out two cases: (i) people with more economic resources who have the means to face the difficult conditions that generate the migratory impulse on one hand, and (ii) people with limited economic resources who cannot move for that reason. In such cases, the challenges in terms of prevention are particularly acute.

The place of origin can also be the place of return, and the literature points to specific challenges in this regard. In terms of prevention, one can mention the adaptation of agriculture to new climate conditions (e.g., changes in rain patterns). Academics have extensively researched this aspect (Fischer et al. 2005; Howden et al. 2007; Kalra et al. 2007; Nelson et al. 2009; Piao et al. 2010; Anglés Hernández and Maldonado Madin 2020). Kalra et al. (2007) emphasize the need to work with biophysical data and socio-economic aspects. This resonates with the section on forced migration regarding the tension between objectivity and subjectivity. Access to land and "durable responses" are other challenges regarding return (Rubio Díaz-Leal 2017).[3]

[3] It is interesting to note that land scarcity is both an element that drives climate migration (see e.g. Radel et al. 2018), and a challenge for return. On access to land and (not specifically climate) migration see Obeng-Odoom (2017): the author advocates the focus should not lie on the countries

Finally, some challenges are specific to rural places of origin. The literature indicates profound changes in rural areas in Latin America and the Global South, and how climate mobility impacts them. Academics indicate a decrease in traditional agriculture, the diversification of activities outside the farm, and demographic change (the rural population tends to age, decrease, and families tend to be smaller) (Robson and Klooster 2018a). The changes generate a "new rurality" impacted by the mobilities leaving rural territories: they exacerbate demographic changes, capital flows, and model new relationships between urban and rural areas. The new rurality approach emerged in the 1990s and focused on rural development in Latin America (Kay 2009).[4] The new rurality approach has been applied to Mexico (Cadena Iñiguez et al. 2013; Barkin 2006).[5]

These impacts generate social and economic reorganization (Robson and Klooster 2018a). Alterations in the management of common lands are particularly significant:

> [M]igration removes the hearts and minds and backs and hands of the commoners, presumably weakening the institutions and organizational structures expected to manage rural resource commons. In this way, the temporary or permanent movement of people away from their places of origin can present a dilemma for the sustainability of local land management systems, which typically rest on collective decision-making, and the labour of those who access these resources. (Robson and Klooster 2018a, p. 6)

In Mexico, the above can apply to *ejidos* (a common land regime) and Indigenous Peoples' lands.

Despite the significance of these changes, the effects on the institutions and structures of the communities for self-governance, cultural reproduction and the use of territories and management remain poorly studied (Robson and Klooster 2018a).[6] It is an important loophole in the knowledge, be it for its essential role when protecting biodiversity and ecosystems. The new rurality phenomenon and its negative impact on the prevention of climate migration scenarios can be seen in Mexico. Slow-onset environmental degradation illustrates this:

> Mexico's main migrant-sending regions remain burdened by reduced village populations. (…) More than one-third of Mexico's land base and approximately two-thirds of its forests are found on community lands. (…) They shelter globally-important biodiversity and carbon, regionally-important ecosystem services, and resources vital for local livelihoods. (Robson and Klooster 2018a, p. 7)

---

of origin's shortcomings or the "calculation" of migrants concerning their interest ("push–pull framework"), but on access to land and guaranteeing access to land (following Henry George's approach). This approach applies to territories of origin and territories of destination.

[4] On the approach's territorial dimension and the identification of new rural territories see Pisani and Franceschetti (2011).

[5] For a critical appraisal of the new rurality approach, see Ramírez-Miranda (2014) and Kay (2015).

[6] For recent research on this aspect, see Yates et al. (2021). On weighting climate mobility's respective benefits and costs for "source, destination and transit" territories, see Gemenne and Bloquer (2017) who pinpoint research voids and propose a detailed methodology "to advance knowledge on the respective benefits and costs.".

In addition, current data point out the potentially damaging effects on biodiversity induced by land-use changes (previously cultivated lands become forests, the disappearance of home gardens, and disappearance of cultivated native species).[7] In other words, the decline in agricultural activities does not always have a positive impact on biodiversity (Robson and Klooster 2018b).

Secondly, from the perspective of recipient territories, academics have reflected a lot on the conflicts that may arise (or be amplified) by climate and environmental migrations (Reuveny and Moore 2009; Koubi et al. 2021; Castells-Quintana et al. 2021; Cissé et al. 2022).[8] Koubi et al. (2021) found internal environmental migrants in Kenya who had experienced extreme hidrometeorological events in their (rural) territories of origin were more prone to participate in social movements in cities. *Contra*, Petrova (2021) carried out several case studies in Bangladesh and found mobilities linked with risk of disaster do not enhance the frequency of protests in destination territories.

The doctrine identifies four conflict scenarios between people who migrate for environmental reasons (in our case, it would be people who migrate for climate reasons) and people who live in the destination territory (Reuveny 2005).[9] Those scenarios can overlap. Others criticize the link between migration and conflict, pointing out methodological shortcomings in case studies and "environmental security researchers misunderstanding both typical migration patterns in the developing world and the conditions that create conflict" (Raleigh et al. 2008, pp. 34–35). Raleigh et al. also identify scenarios which can generate tension and conflict (2008, pp. 36–27). In this research, we apply Reuveny's analytical framework (2005), much as it has been applied in the reports and literature subsequent to Raleigh et al. (2008) (among others regarding case studies).

The arrival of migrants can represent a burden "promoting native-migrant competition for economic and natural resources" (Reuveny 2005, p. 4). It is particularly relevant if the destination territory already suffers from economic and natural-resource shortages. In these cases, when pressures are high, "some people may try to protect resources by force, especially if institutions regarding property rights are underdeveloped" (Reuveny 2005, p. 5).

This is particularly relevant in Mexico: the Mexican Environmental and Natural Resources Ministry (*Secretaría de Medio Ambiente y Recursis Naturales*—SEMARNAT) and National Ecology and Climate Change Institute (*Instituto Nacional de Ecología y Cambio Climático*—INECC) point out the very high pressures cities suffer (2018, pp. 477–482). More particularly, the World Bank (Rigaud et al. 2018) identifies, as destination territories, some that already suffer from a

[7] *A contrario*, Robson and Klooster (2018a) also mention a positive aspect: the pressure on these territories decreases.

[8] In that respect, Stojanov et al. (2021) outline development cooperation actors to "safe and orderly mobility and the decent reception and long-term inclusion of migrants and displaced persons under conditions of more severe climate hazards, and in the context of rising nationalism and xenophobia, poses significant challenges" (p. 209) and identify "critical" areas "for further exploration" (p. 224).

[9] One of these cases does not apply within the research framework since it concerns international migration and the exploitation of the migratory phenomenon by the States involved.

shortage of natural resources and very high pressure on them (state of Mexico, East of Puebla, Oaxaca, Baja Northern California).

Turning to migrants' "ethnic" origins, "when environmental migrants and residents belong to different ethnic groups, the arrival of many newcomers over a relatively short period of time may upset a precarious ethnic balance" (Reuveny 2005, p. 5). In the same sense, Becker and Ferrara (2019) report studies dedicated to identifying whether a link can be established between the arrival of forced migrants and an increase in votes for racist parties. They identify contradictory cases. Also, Reuveny (2005) argues pre-existing ethnic conflicts are an aggravating factor. This scenario can have racist overtones. It is necessary to tread carefully with these issues. As several authors have pointed out, there is a racist dimension that can arise from the link between mobilities and climate change. As researchers, we must constantly question our work to avoid such racist dimensions, whether consciously or unconsciously. Baldwin (2013, 2016) elaborates on the inherent racist dimension of the link between climate change and mobility. To be very clear, within the present research framework, we are keen to mention that Indigenous Peoples indeed do face racism and discrimination in Mexico (CNDH 2018).

Reuveny (2005) also mentions the exploitation of pre-existing tensions ("fault lines") (p. 5) and provides examples thereof: herders and farmers may compete for land, and migrants and residents may compete for jobs. One example is particularly relevant: migrants from rural areas can be brought to compete with residents of urban areas: "Political entrepreneurs in urban areas may exploit rural migrants' frustration and poverty and mobilize them to challenge the State. Threatened, the State may respond with force" (p. 5). In that respect, the World Bank's projections (Rigaud et al. 2018) identify, in Mexico, a flow of internal climate migrations from rural to urban areas. In the next section, we report on the challenges cities face in more detail, having already mentioned the risk that accommodation prices may increase and generate difficulties when receiving forced migrants (Becker and Ferrara 2019).[10]

Indigenous Peoples' situation is peculiar regarding fault-lines. Academics signal the specific challenges they face, starting with their right to self-determination: "land accessibility, environmental and cultural differences with the host community, governance and funding" (Pascoe 2015, p. 72). Interestingly, Sophie Pascoe highlights the problematic access to lands when governed by common property systems (that is, other than private property) in the destination territory, in the context of the Carteret Islands (Papua New Guinea): these lands "cannot be bought or sold" (pp. 78–79).

This is particularly relevant in Mexico as, in terms of the availability of destination lands and their tenure, the *ejido* regime regulates a large portion of the territory. Also, Indigenous Peoples have (or at least should have) ownership over traditional lands and territories. The challenges are also cultural and spiritual: Indigenous Peoples have (for the most part) their way of life, worldview, traditions, and values strongly linked to their places of origin. The challenge then arises of being able to reproduce these characteristics at the destination. Conflicts may arise related to these cultures'

[10] The authors mention p. 6 the case of forced internal migrants in Colombia and cases where the opposite happened: prices for lodging fell.

coexistence with the cultures and spiritualities of the people or communities located in the destination territory (Pascoe 2015).

Another specific case occurs when forced migrants cannot choose their destination when public authorities arbitrarily assign them (Thompson et al. 2014) and deny migrants' participation and consultation. Consequently, such assignation may not be suitable in terms of the adequacy of the designated place in terms of infrastructure, cultural and social suitability, and the potential to make a livelihood. Regarding this last aspect, adverse effects of the arbitrary location of forced migrants in economic terms are well-documented (e.g., the possibility for them to find a job, wages, and working hours). Factors such as language/dialect in the assigned place and if forced migrants speak this language/dialect, and networks and composition of those are essential. Where Indigenous Peoples are forcibly displaced and cannot choose their destination, discrimination goes hand in hand with the problems that arise in terms of their own norms and institutions (due to imposing governance structures and institutional arrangements) (Becker and Ferrara 2019).

2.1.4 *Particular Challenges for Cities and Indigenous Peoples*

Various challenges arise for cities as places of origin, transit, or destination in the context of climate change. There is a wealth of literature on the challenges for informal neighborhood facilities (if not entire cities) in territories not suitable for human settlements and where the most socio-economic vulnerable sectors of the population settle (see e.g. Rana and Ilina 2021). These territories are available precisely because of the inherent danger living their brings or the low suitability they had for human settlements (Hardoy and Pandiella 2009). Such territories are not only prone to disasters. Hardoy and Pandiella (2009) also mention the lack of stable property rights, houses built with unsuitable materials in terms of thermal efficiency, and very high population density.

The sewer system is a good example of infrastructure and service deficiencies and climate change impacts. It can have dramatic consequences regarding flooding, or, often favor diseases.[11] Access to basic healthcare services is crucial. However, academic works and reports that address the link between the three elements of the public problem in question (climate change—forced mobility—health) are scarce.[12] Schütte et al. (2018) propose researching this public problem relying on the discipline of "planetary health."

[11] On the challenges that arise with informal settlements and proposals to "re-evaluate" their potential for cities, see Lombard (2014). On forced evictions and demolition of informal settlements' socio-economic effects see Roberts and Okanya (2020).

[12] For a pioneer work on this topic, see McMichael et al. (2012), and for very recent research see Hunter et al. (2021) (in this article, the authors confirm little research on the subject, and propose a conceptual model for future research) and Burrows et al. (2021).

These challenges take on a dimension of their own when considering climate change driven migration. In the first place, they can amplify the coercion and absence of alternatives that trigger mobility (places of origin). Climate change multiplies environmental risks (more particularly for cities located in coastal areas and with low elevation above sea level), environmental and ecosystems degradation (including for ecosystem services and natural resources), and the difficulty for people living in these areas to have the resources to avoid or limit the effects of climate change. For example, regarding heatwaves, installing air conditioning could potentially reduce the mortality of sectors particularly vulnerable to heatwaves, such as the very young, the elderly, and people with illnesses (Hardoy and Pandiella 2009). Costa et al. (2013) also mention the dangers posed in terms of mitigation, as cities are known to be major greenhouse gases (GHG) emitters.

Secondly, the challenges are amplified for transit and destination territories. This phenomenon reflects one of the central ideas in the work of Graeme Hugo, a pioneering scholar on the link between migration and the environment. Graeme Hugo talks about this phenomenon as follows: "migration is both a consequence and a cause of environmental change" (cited in Gamlen et al. 2018, p. 304). By receiving migrants, the density of cities increases, which intensifies the already very high competition for access to natural resources and the magnitude of any disasters. The literature also mentions social cohesion (Tebas and García 2016) and access to public services (for example, health services). Araya et al. (2019) apply the "place attachment theory" and propose concrete elements to guarantee that the destination population does not perceive an overload for the infrastructures that would result from the arrival of migrants.

In Mexico, starting in the 1950s, three migration phases from rural to urban areas have generated intense pressure on water resources: 77% of the population lives in cities located in the arid northern part of the country or semi-arid areas and in the central valley, while there is 31% of the total available water in the country in these areas (Cohen et al. 2013, p. 58).

Within this context, the situation in Mexico City and the Valley of Mexico is particularly worrying. The World Bank identified them as a privileged destination for climate forced internal migrations (Rigaud et al. 2018). It is all the more worrying as numerous studies and academic literature identify the challenges already present in Mexico City and the Valley. For example, water scarcity, floods, and water injustice are commonly reported in the capital: "Following a trend in existence since the Porfiriato era, decisions regarding infrastructure provision have been maintained and consolidated in a manner that benefits wealthy zones and contributes to a pattern of unequal spatial access to water. (…) Those areas lacking sewers simply dump their wastewater and pollute sub-soil and aquifers" (Lankao 2010, pp. 171–172).

In this book we particularly focus on Indigenous Peoples and their struggles. They maintain a firm bond with their lands and territories. We noted, in the previous section, some challenges that result from this bond. As to human rights, the inter-American system firmly grounds the obligation to maintain the bond between Indigenous Peoples and their lands and territories. The notion of territory it adopts encompasses its cultural and symbolic dimensions in an historical approach. The Inter-American Court of Human Rights' case-law consistently points out that the territory involves matters of livelihood and encompasses the worldview held by Indigenous Peoples as well as their cultural and spiritual identity (IACtHR 2001, para. 149; 2012a, para. 155; 2005a, para. 131).

Consequently, the obligation States have to maintain the relationship between Indigenous Peoples and their lands and territories is based on the survival of Indigenous Peoples' identity. The case law emanating from the Inter-American Court of Human Rights on their rights to consultation and consent in matters of forced displacement in the context of development projects and demarcation of their lands and territories illustrates as much (among them: IACtHR 2007). In such a context, academics point out mobility effects for members who stayed in the origin territories, more specifically in rural territories (Robson and Klooster 2018a). It is particularly relevant in Mexico, where a significant part of the territory belongs to indigenous communities. The case of Oaxaca case illustrates this: 87.50% of the territory belongs to agrarian communities, and 18.42% belongs to ejidos (Robson and Klooster 2018a, p. 25).

In the section that follows, we identify the scope of the obligations of States based on a broad concept of Indigenous Peoples' lands and territories according to a differentiated approach and environmental rights.

2.2 Introduction to the Protection Requirements

2.2.1 Challenges When Establishing the Protection Requirements

The aim of this research is not to discuss the concept and application of climate justice. We aim to determine the protection requirements that apply to people and communities who forcibly migrate within Mexican borders because of climate change and operate within a paradigm shift: forced climate migrants are "subjects of legal protection" instead of "subjects of legal domination" (Skillington 2015, p. 290).

The vulnerability of migrants in terms of environmental, climate, and economic factors are well documented in Mexico. Indeed none of this is new. It relates to general aspects, not just specific to the dimension of mobility (e.g., drug trafficking or legal frameworks and their shortcomings in terms of implementation) (Garza-Guerra 2015; CNDH 2011) and denounced regarding internal mobilities in Mexico (IAComHR 2013, pp. 35–37).

In other words, we do not argue that internal migrants should enjoy more privileged protection in comparison with other people or communities that are also under

Mexican jurisdiction. Instead, we intend to address the specific risks that result from mobility and the causes that generate forced mobility in a framework of rights protection and enjoyment by the individuals and communities that forcefully migrate. For this reason and based on the principle of non-discrimination and human dignity, the demand for the protection of internal forced climate migrants is not limited to internal migrants of Mexican nationality. It extends to internal migrants with other nationalities, without considering their potential irregular situation in Mexico.[13]

Another challenge for the protection requirements is that they does not cease when the migration trigger ceases, nor when migrants settle in the place of destination, or return to their place of origin. The protection requirements extend *rationae temporis* and *ratione loci*: forced internal migrants "usually face continuing problems, requiring support until they achieve a durable solution to their displacement," and "A durable solution is achieved when internally displaced persons no longer have any specific assistance and protection needs that are linked to their displacement and can enjoy their human rights without discrimination on account of their displacement" (HCHR 2010, p. 1).[14]

The requirements detailed below must also be taken into account and applied differently according to the six scenarios identified, the three specific categories of migrants covered in the project, and the territorial specificities. For example, disasters require immediate relief measures whereas slow-onset phenomena require longer-term structural policies. Finally, the difficulty is identifying whether the specificity of forced climate migration compared to other forms of forced migration can be grounded in the basis, limitation, and context of the protection requirements. Put simply: does the protection of climate forced internal migrants have to be different from the protection of people and communities who migrate, for example, due to violent conflict or the dispossession of their lands? Once the migration process begins, should the protection requirements be any different? We answer the above in the sections and tables below.

In the subsequent sections, we first establish the international and regional legal panorama that applies to internal climate migrants (Sect. 2.2) and explain why we ground the protection requirements in a rights-based approach despite the evident limitations (3). More specifically, we detail this approach and apply a differentiated approach to reflect the situation of Indigenous Peoples and women. We also explain why and how an environmental rights approach should complement the differentiated human rights-based approach.

---

[13] On the principle of non-discrimination applied to irregular international migrants, see Morales Sánchez (2019, pp. 60, 62–78).

[14] The criteria to determine a durable solution is achieved are summarized p. 3, and in more detail pp. 18–29.

2.2.2 *Deficiencies Within the International and Regional Legal Panorama*

We present the legal panorama according to the two dimensions of the phenomenon: *climate* migrations and *internal forced* migration.[15] At a regional level, we limit ourselves to the American region. We do not refer to other regional human rights protection systems (e.g., the African or European ones). Our analysis of the international and regional framework reveals shortcomings regarding the demand for protection, with the notable exception of the 1998 United Nations Guiding Principles on Internal Displacement (hereinafter "the Guiding Principles"—UN 1998b).

No internal or international hard-law instrument addresses climate migration and protects migrants' rights (Skillington 2015).[16] Academics debate on how to address this loophole. A "maximalist" approach advocates for the adoption of international standards and a "minimalist" approach emphasizes the local dimension (understood as a national/regional/intraregional dimension) (Bettini and Andersson 2014, p. 163).[17]

This does not mean, however, that the issue is not on the international agenda, on the contrary, it continues to gain more prominence (McAdam 2014; Tebas and García 2016; Clark and Bettini 2017).[18] Several proposals have been put forward in this regard: the adoption of an *ad-hoc* convention on climate displaced persons, of a Protocol to the UN Convention on Climate Change, of a Protocol to the UN Geneva Convention Relating to the Status of Refugees, of regional instruments (Kälin 2019), or the adoption of a *soft-law* instrument (that is, a non-legally binding instrument) (Cournil 2010; Mayer 2012; Okeowo 2013; Hodgkinson and Young 2013).[19] Academics comment extensively on the adoption of an *ad-hoc* international Convention. Several even proposed the text of a Convention on the International Status of Environmentally Displaced Persons (see e.g. Prieur et al. 2008).

---

[15] Rubio Díaz-Leal points out this parallel perspective and its adverse effects (2017, p. 40).

[16] This absence does not impede the carrying out of the necessary tasks to analyze the non-binding regulations that deal with the phenomenon in order to determine the balance between the approaches of "dominant Global Environmental Management" and its opposite, the "populist" discourse (see Bettini and Andersson, 2014, pp. 162–163); the tension between top-down and bottom-up approaches; or a "critical environmental justice" approach (the author plans to develop the latter in subsequent research).

[17] For an in-depth critical analysis of climate displacement regulation at international and regional levels, see Ibarra Sarlat (2021).

[18] Nor does it mean that climate migrants are not protected by any legal framework. On environmental migration, see Mayer (2012). Additionally, as we will see later, the rights-based approach provides a protection grid.

[19] Also, the "Nansen initiative" was developed since 2012. It was a consultation process and information sharing initiative on international climate migration related to disasters. It brought together various actors and various States led and financed it: "The Nansen Initiative is a bottom-up, State-led consultative process with multi-stakeholder involvement," https://www.nanseninitiative.org/secretariat/. On the postulates and results of the Initiative, see Ruiz and Carvajal (2015, pp. 40–42).

In sharp contrast to the climate migration aspect, the rights of internally displaced people are codified at an international level. The Rome Statute of the International Criminal Court (UN 1998a) ratified by Mexico in 2005 prohibits "forcible transfer of population" (Article 7(1)(d)). Also, the UN 1998 Guiding Principles on Internal Displacement (a *soft-law* instrument) is a key instrument regarding internal forced displacement. The thirty Guiding Principles provide for a series of obligations binding States and rights of internally displaced people. The Principles' legal force does not bind States (as a Convention, for example). It does not mean that they have no legal value (Kälin 2001). As with other international law instruments that do not *a priori* have a binding value, they can slip towards a binding value. For example, the Inter-American Court of Human Rights has already grounded its case-law on the principles (among others, IACtHR 2012b, para. 256).[20] Also, the Principles are meant to reflect current human rights standards.

Cohen and Deng (2008) mention tensions between an approach "focused on the needs of the displaced" and another approach "focused on rights." The Principles reflect the "skillful" union of the two approaches. The tension that arose in the elaboration process is also mentioned: "For the first time, international experts outside a traditional intergovernmental process drafted, revised and completed an international legal document of great significance" (pp. 4–5). The Guiding Principles were acclaimed, and their contributions in terms of protection and guidance for States' action cannot be questioned. However, it does not mean they cannot be criticized. Among others, Sánchez Mojica (2009) criticizes the Principles for not having as their objective "the protection of the well-being of those forced to move, but rather the containment of their exodus within the borders of the States in which they have originated" (p. 11). Other Principles did not obtain as much visibility as the United Nations Guiding Principles and are not widely referred to in international instruments or literature, such as the 2013 Peninsula Principles (Solutions 2016).[21]

The Guiding Principles apply, at least partly, to climate migration. This application stems from the Introduction to the Guiding Principles:

> [I]nternally displaced persons are persons or groups of persons who have been forced or obliged to flee or to leave their homes or places of habitual residence, in particular as a result of or in order to avoid the effects of (...) situations of generalized violence, violations of human rights or natural or human-made disasters, and who have not crossed an internationally recognized State border. (United Nations 1998b, introduction, para. 2)

The Guiding Principles undoubtedly apply to three of the scenarios identified in the first chapter: sudden-onset disasters, "sinking" islands and coasts, and conflicts

---

[20] Another example illustrates the possible transformation of non-binding international instruments towards binding norms: the 2006 Protocol on the Protection and Assistance to Internally Displaced Persons from the Great Lakes Region (Protocol available on February 5, 2019 at https://www.unhcr.org/protection/idps/5ad5a3aa4/international-conference-great-lakes-region-protocol-protection-assistance.html) attributes binding value to the Guiding Principles (see e.g. Article 2).

[21] Other United Nations Principles with a more limited purpose apply (the Housing and Property Restitution for Refugees and Displaced Persons Principles, the "Pinheiro Principles") (UN 2005).

or violence generated by the scarcity of natural resources. However, doubts arise regarding the application of the Principles to the three other scenarios:

(1) slow-onset phenomena (since it is not a natural or human-made catastrophe if the latter is understood as an industrial catastrophe),[22]
(2) areas designated unsuitable for human settlement (as long as sudden-onset disasters do not cause the unsuitability), and
(3) "development projects" linked to renewable energy (except for catastrophes induced by these projects).[23]

With respect to the latter, Principle 6(2)(c) of the Guiding Principles mentions development projects in the context of arbitrary prohibition of displacement. Doubts nevertheless arise regarding the application to projects related to renewable energies since "compelling and overriding public interests" may justify them (Principle 6(2)(c) *in fine*). In the Mexican case, energy-related projects receive this status and, consequently, the prohibition of arbitrary displacement principle does not apply. For example, the Hydrocarbons Law (*de Hidrocarburos* 2016) provides for the possibility of establishing certain easements if the conflict is not resolved despite the mediation process (Articles 96, 101 Fr. V, 106, 108–110).

The limited applicability of the Guiding Principles to the six mobility scenarios justifies why, in this research, we prefer to mention climate mobility and forced internal climate migrations rather than "forced climate displacements" (an expression used by the Guiding Principles). The Guiding Principles do not provide a limited list (Kälin 2008b). Therefore, we cannot rule out their application to all of the scenarios identified in this research. This aspect is instead left open. However, we prefer to avoid confusion and distance ourselves from the Principles. Maintaining a healthy perspective as regards the Guiding Principles can also be justified by the premises or objectives thereof. A wealth of academic literature criticizes the Principles for not pursuing "humanitarian" goals but rather preventing and trying to limit international flows (Vidal López 2007).

2.3 The Differentiated Rights-Based Approach

2.3.1 A Rights-Based Approach

Two distinct approaches are put forward to protect climate migrants: The obligations of States ("obligations approach") or migrants' rights ("rights approach"). Some academics give preference to the obligations approach. The objective is to

[22] In the same sense, see Bradley and Cohen (2013, pp. 211, 219–220) and van der Vliet (2018, pp. 20–21).

[23] The United Nations Basic Principles and Guidelines on Development-based Evictions and Displacement apply to the last scenario (Kothari 2007).

put the center of gravity on the responsibility of States in the context of environmental or climate migration. And, according to these authors, a rights-based approach dilutes this responsibility (Mayer 2012). The critical arguments towards human rights of academics uphold the obligations approach. They do not develop their arguments regarding environmental or climate migrations. However, it does not deprive their criticisms of validity about how to guarantee the protection of climate migrants. The "traditional" criticisms point out the lack of respect for human rights and their programmatic nature. Other criticisms are more advanced and derive from a rigorous critical analysis (López 2007; Estévez and Vázquez 2017). They refer to the State administering or managing suffering and "public necropolitics" (Estévez and Vázquez 2017, pp. 187–189). Ariadna Estévez (2015) exposes the foundations of the "suffering management" as follows:

> To contain the social explosiveness derived from this overwhelming suffering (*individual and collective suffering of overwhelming proportions "that results from" the crisis of human rights*), the Mexican government has introduced other types of public policies that manage suffering with a view to cushioning the conflict and the mobilization that results from massive suffering. These policies integrate what we name here the suffering management device. (p. 8)

These criticisms undoubtedly apply to climate migrations. The tension arises between controlling mobility flows on one hand and the protection of migrants on the other. These detractions also address the neoliberal context and disapprove of material dispossessions (e.g., lands) and human rights abuses, resulting in the objectification of the victim (Celorio 2017). The criticisms against human rights in the neoliberal context resonate a lot within the framework of this research: the root causes of climate change and environmental degradation can be pinpointed in the prevailing neoliberal model (e.g., over-exploitation of natural resources, substances and practices that generate environmental damage motivated by the desire for short-term profit, lands dispossession for neo-extractivism activities, exploitation of the periphery).

Despite these arguments and criticisms, in this book we rely on the rights-based approach. Other authors also support a rights-based approach to address climate migration (Zetter 2009; Schade 2012; Marshall 2016; Cullen 2020; Scott and Salamanca 2020; Ibarra Sarlat 2021). The rights-based approach applied to internal migration is also supported by the Organization of American States (OAS) (OAS 2007, point 7; 2011, point 6; 2014, point 7). At least six elements justify this approach. Firstly, the obligations approach based on the sovereignty principle may be more specific than the rights approach of internal migrants (Cohen 2003; Deng 2003). However, this specificity can also undermine their protection: each situation a migrant finds himself/herself in is unique, and a migrants' rights-based approach can ensure that each of these situations is appreciated flexibly (casuistic approach).

Secondly, and linked to the preceding argument, a rights-based approach guarantees taking into account internal migrants' subjectivity. In other words, this approach ensures that protection does not depend on squares or grids established more or less arbitrarily by the State but on the peculiarities of the situation of each migrant (Kälin 2008a).

Thirdly, human rights' tools can be utilized. For example, a rights-based standpoint can bring into play a differentiated approach (e.g., towards women and Indigenous

Peoples' rights). Other human rights tools are the positive obligations of States, human rights based interpretation (*pro-homine*), the requirement for the effective enjoyment of human rights, the horizontal effect of human rights, and the non-regression principle ("standstill effect") (Gouritin 2016, pp. 8–36). In this sense, we do not think that a rights approach evades the responsibility of States. On the contrary: human rights tools give flesh to the obligations of States. This is particularly true of positive obligations. They require States not only to refrain from unduly interfering with human rights, but also to adopt the administrative and legal measures that guarantee the effective enjoyment of rights. In other words, the States' discretion to address the phenomenon and provide protection to the people and communities affected is limited.

Fourthly, in the previous section, we determined that the protection requirements are set into motion when the person or community is forcibly migrating to a distinct territory in any of its dimensions (the distinct destination territory criterion). The rights-based approach complements this criterion by expanding its application to migrants who move very short distances: if a risk of a human rights violation is identified, the protection requirement apply, including in cases where the destination territory is not distinct from the origin territory. Such cases are not merely hypothetical. For example, in the sudden-onset disaster scenario, the people and communities may settle within 10 m of their destroyed house. In this case, the distinct destination territory criterion would not allow the protection requirement to apply, but the risk of human rights violation criterion would.

Fifthly and regarding the scope of the rights-based approach, it allows not only for the critical analysis of public policies and legal frameworks in force. It can also orientate public policies and the elaboration of legal frameworks, with respect for human rights at its heart.

Finally, the rights-based approach enables the bringing of the protection requirements of climate migrants before human rights bodies, and those can develop, in making their decisions, into autonomous norms (i.e. rights that do not depend on the definitions of States and their interpretation of those rights) and a series of obligations through their decisions. In turn, those decisions enable rights and obligations to be consolidated and disseminated at a national, regional, or international level.

We systematize the protection requirements in the table below which aims to map the protection requirements in the context of forced climate mobility (Table 2.1).

2.3.2 A Differentiated Approach

In the research, we identify three sectors of the population which are most particularly vulnerable: people with limited economic resources, Indigenous Peoples, and women. We accordingly adopt a differentiated approach in our human rights

Table 2.1 Protection requirements systematization. The rights-based approach

Guaranteed rghts	Legal grounds
	UN Guiding Principles (1998) ("GP") American Convention on Human Rights (OAS, Pact of San José, 1969) ("Pact of San José") Additional Protocol to the American Convention on Human Rights in the Area of Economic, Social and Cultural Rights "Protocol of San Salvador" (OAS, Protocol of San Salvador, 1988) ("Prot. San Salvador") UN Declaration of Human Rights (UN, 1948) ("UNDHR") UN International Covenant on Civil and Political Rights (UN, 1966) ("ICPR") UN International Covenant on Economic, Social and Cultural Rights (UN, 1966) ("ICESCR") Mexican Constitution (*Constitución Política de los Estados Unidos Mexicanos*, CPEUM, 2020) ("CPEUM") Rome Status of the International Criminal (UN, 1998a) ("Rome Status")
General principles and rights that apply irrespective of the mobility phase and without a differentiated approach	
Protection Principle (concerning internal displacement)	GP: Pple 3 CPEUM: Art. 11
Prevention Principle (concerning internal displacement)	GP: Pple 5
Equality	GP: Pple1(1) [b] Pact of San José: Art. 24 UNDHR: Art. 1 ICPR: Art. 23(4), Art. 26 ICESCR: Art. 3 CPEUM: Art. 4 (men/women)

(continued)

Table 2.1 (continued)

	No discrimination	GP: Pple 1(1), Pple 4(1) Pact of San José: Art. 1(1), Art. 17(2), Art. 24 Art. 27(1) Prot. San Salvador: Art. 3 UNDHR: Art. 2, Art. 7 ICCPR: Art. 2(1), Art. 20(2), Art. 26 ICESCR: Art 2(2) CPEUM: Art. 1
	Prohibition of arbitrary displacement	GP: Pple 6(2) (it specifies "The prohibition of arbitrary displacement includes displacement: (…) (c) In cases of large-scale development projects, which are not justified by compelling and overriding public interests; (d) In cases of disasters, unless the safety and health of those affected requires their evacuation") Estatuto Roma: Art. 7(1)(d)
	Rights when displacement is decided by public authorities	GP: Pple 7
	Culture	Prot. San Salvador: Art. 14 UNDHR: Art. 27 ICESCR: Art. 15(1)(a), Art. 15(1) CPEUM: Art. 4
	Education	Prot. San Salvador: Art. 13 UNDHR: Art. 26 ICESCR: Art. 13 CPEUM: Art. 3
	Food	Prot. San Salvador: Art. 12 ICESCR: Art. 11, pár. 2 CPEUM: Art. 4
	Healthy environment	Prot. San Salvador: Art. 11 ICESCR: Art. 12(2)(b) CPEUM: Art. 4
	Water	CPEUM: Art. 4 ICESCR: Art. 11, Art. 12 (General Observation No. 15)
	Health	Prot. San Salvador: Art. 10 ICESCR: Art. 12 UNDHR: Art. 25 CPEUM: Art. 4

(continued)

Table 2.1 (continued)

	Social security	Prot. San Salvador: Art. 9 UNDHR: Art. 22 ICESCR: Art. 9
	Participation in public decision-making	Pact of San José: Art. 23(1)(a) UNDHR: Art. 21 ICCPR: Art. 25(a)
	Access to public functions	Pact of San José: Art. 23(1)(c) UNDHR: Art. 21 ICCPR: Art. 25(c) CPEUM: Art. 35 VI
	Vote and being elected	Pact of San José: Art. 23(1)(b) UNDHR: Art. 21 ICCPR: Art. 25(b) CPEUM: Art. 35 (I), (II), (VIII)
	Work	Prot. San Salvador: Art. 6, Art. 7, Art. 8 UNDHR: Art. 23 ICESCR: Art. 6, Art. 7 CPEUM: Art. 5
	Life	GP: Pple 8 Pact of San José: Art. 4 UNDHR: Art. 3 ICCPR: Art. 6(1) CPEUM: Art. 29
	Dignity	GP: Pple 8 Pact of San José: Art. 11 UNDHR: Art. 5 ICCPR: Art. 7 CPEUM: Art. 4, Art. 25
	Freedom	GP: Pple 8 UNDHR: Art. 3 ICCPR: Art. 9(1) Pact of San José: Art. 7 CPEUM: Art. 25
	Security	GP: Pple 8 UNDHR: Art. 3 CPEUM: Art. 16, Art. 29
	Prohibition of slavery	Pact of San José: Art. 6 UNDHR: Art. 4 ICCPR: Art. 8 CPEUM: Art. 1
	Personal freedon	Pact of San José: Art. 7 UNDHR: Art. 9 ICCPR: Art. 9

(continued)

Table 2.1 (continued)

	Conscience and religion freedom	Pact of San José: Art. 12 UNDHR: Art. 18 ICCPR: Art. 18 CPEUM: Art. 24
	Throught and expression freedom	Pact of San José: Art. 13 UNDHR: Art. 19 ICCPR: Art. 19 CPEUM: Art. 6, Art. 7
	Meeting freedom	Pact of San José: Art. 15 UNDHR: Art. 20 ICCPR: Art. 21 CPEUM: Art. 9
	Association fredom	Pact of San José: Art. 16 UNDHR: Art. 20 ICCPR: Art. 22 CPEUM: Art. 9
	Nacionality	Pact of San José: Art. 20 UNDHR: Art. 15 CPEUM: Art. 37
	Personal physical, psychological and moral integrity	Pact of San José: Art. 5 UNDHR: Art. 9 CPEUM: Art. 29
	Judicial protection—Access to courts and tribunals	Pact of San José: Art. 25 UDHR: Art. 8, Art. 10 ICCPR: Art. 2(3), Art. 14 CPEUM: Art. 8, Art. 17
	Family	Prot. San Salvador: Art. 15 Pact of San José: Art. 11(2), Art. 17 UNDHR: Art. 16(3), Art. 12 ICCPR: Art. 17(1), Art. 23 ICESCR: Art. 10 CPEUM: Art. 16
	Property	Pact of San José: Art. 21 UNDHR: Art. 17 CPEUM: Art. 27
	Adecuate level of life	UDHR: Art. 25 ICESCR: Art. 7(a)(ii), Art. 11(1)
	Legal personality	Pact of San José: Art. 3 UNDHR: Art. 6 ICCPR: Art. 16 CPEUM: Art. 4, Art. 29

(continued)

Table 2.1 (continued)

Protection during mobility	General principles and rights that apply according to the mobility phase	
	Life	GP: Pple 10(1)
	Dignity	GP: Pple 11
	Physical, mental or moral integrity	GP: Pple 11 Pact of San José: Art. 5
	Freedom	GP: Pple 12(1)
	Security	GP: Pple 12(1)
	Freedom to circulate	GP: Pple 12(2), Pple 14 UNDHR: Art. 13 ICCPR: Art. 12
	Freedom to chose place of residence	GP: Pple 14, Pple 15 Pact of San José: Art. 22(1), Art. 22(2) [Exceptions: Art. 22(3), Art. 22(4)], Art. 22(5–9) UNDHR: Art. 13, pár. 1: derecho de elegir su residencia Art. 13, pár. 2: derecho de salir de cualquier país y regresar a su país Art. 14: derecho de buscar asilo ICCPR: Art. 12, Art. 13 CPEUM: Art. 9, Art. 11
	To know the fate and whereabouts of missing relatives	GP: Pple 16
	Family life	GP: Pple 17
	Adecuate level of life	GP: Pple 18
	Health	GP: Pple 18, Pple 19
	Recognition of legal personality	GP: Pple 20
	Property	GP: Pple 21
	Freedom of thought, conscience, religion or belief, opinion and expression	GP: Pple 22
	To seek freely opportunities for employment and to participate in economic activities	GP: Pple 22
	To associate freely and participate equally in community affairs	GP: Pple 22
	To vote and to participate in governmental and public affairs, including the right to have access to the means necessary to exercise this right	GP: Pple 22

(continued)

Table 2.1 (continued)

	To communicate in a language they understand	GP: Pple 22 ICCPR: Art. 27
	Education	GP: Pple 23
Humanitarian assistance	To receive humanitarian assistance in accordance with the principles of humanity and impartiality and without discrimination	GP: Pple 3, Pple 24
Return, resettlement and integration	To return to homes or places of habitual residence or to resettle voluntarily	GP: Pple 28(1)
	Reintegration	GP: Pple 28(1)
	Participation	GP: 28(2), 29(1)
	To receive humanitarian assistance	GP: Pple 30
	Prohibition of discrimination	GP: Pple 29(1)
	Acces to public services	GP: Pple 29(1)
	Property	GP: 29(2)

Note 1 As regards women and Indigenous Peoples, we limit ourselves to mentioning the specific requirements provided by instruments dedicated to the differentiated approach (e.g., United Nations Declaration on the Rights of Indigenous Peoples). We only report instruments not dedicated to this approach when the requirements or standards explicitly refer to women and Indigenous Peoples
Note 2 We organize the table following the distinctions of the Guiding Principles between four protection requirements categories: (1) general principles, (2) principles that apply during displacement, (3) humanitarian assistance, and (4) principles that apply regarding return, resettlement, and reintegration. The rights mentioned in the "general principles" partly apply to all mobility stages
Note 3 Five Guiding Principles are not reported in these tables because they are not directly relevant for our purpose (they address children's particular situation and specific aspects regarding the scope of State obligations). They are respectively Guiding Principles 2, 13, and 25–27
[a] Peguero Moreno and Laura (2019, pp. 50–51) apply an analogy reasoning with the jurisprudence of the Inter-American Court of Human Rights and conclude Article 11 of the CPEUM "implicitly" protects "against internal displacement."
[b] Kälin (2008b, p. 11) explains the distinction between Principles 1 and 4: Principle 1 prohibits discrimination between displaced people and the general population (that is, people who were not displaced). Principle 4 prohibits discrimination between displaced persons

systematization.[24] The differentiated approach allows the taking into account of the specificities that flow from the forced character of the migration. The differences between voluntary and forced climate migrants imply specificities regarding the migration process and adaptation to the new place of residence. These specificities have consequences in terms of protection and applicable criteria.

Unlike voluntary migrants, Graeme Hugo (2008, p. 18) points out that forced migrants "are unlikely to be as prepared to adjust to circumstances of their destination as are voluntary migrants," "are likely to maintain a greater commitment to their place of origin and retain a strong desire to return to it regardless of how

[24] The OAS also adopts the differentiated approach. See e.g. OAS (2011, points 2 and 7) and OAS (2014, points 3 and 4).

un-realistic that might be" and do not commit themselves to destination places as much as voluntary migrants do. Also, they "are more likely to be in a state of emotional and physical stress because of the losses of family, friends and property they have suffered and the uncertainty about their future." Similarly, Lagos-Gallego, Gutierrez-Segura, Lagos-Grisales and Rodriguez-Morales (2017) establish a causal link between internal forced displacement and post-traumatic disorder in the Colombian case. Forced migrants are very rarely able to take their property with them, "are less likely to have established linkages with people and institutions at their destination" (Hugo 2008, p. 18), and most voluntary migrants "rely on the traditional kinship system and on ethnic identity to help them during the transition to life in the city. On arrival they often live with a relative or close friend who gives them meals, assistance in the job search, information about urban life and moral support" (Kols and Lewison 1983, M-257). This constitutes a major difference between voluntary and involuntary migrants (Hugo 2008, p. 18).

These differences harden their adaptation to the new place of residence: "involuntary migrants are more likely to suffer difficulty in adjusting to, and establishing themselves at, their destination than do voluntary migrants. Clearly this means that they are deserving of, and should attract, special attention from relevant agencies" (Hugo 2008, p. 18).[25]

Also, the Global Protection Cluster Working Group (2007, p. 9) mentions the following specificities:

- IDPs (*Internally Displaced Persons*) have lost their homes and, as a result, they may be in need of shelter. In some cases, they may be compelled to seek shelter in crowded camps or settlements, which can give rise to various protection risks.
- They have often lost access to their land and other property and are cut off from their normal livelihoods and sources of income. As a result they may suffer poverty, marginalization, exploitation and abuse.
- Access to adequate food, safe water and public services, such as education and health care becomes difficult, often leading to high levels of hunger, malnutrition and disease.
- Family and community-structures often collapse and family members become separated. Unaccompanied and separated children, single-headed households (in particular when headed by women or children), older people and people living with disabilities are often at heightened risk of abuse, including sexual exploitation, child labor or forced recruitment into armed forces or groups.
- Identification documents are often lost, destroyed or confiscated in the course of displacement. As a result IDPs often face difficulties in accessing public services, such as education and health care, limits on freedom of movement and heightened risk of harassment, as well as exploitation or arbitrary arrest and detention.
- In many cases, IDPs are displaced to areas where they face marginalization, discrimination and hostility. They are exposed to landmines or explosive remnants of war, or are targeted for abuse and attack. In addition, tensions in these areas can

[25] In the same sense, see Becker and Ferrara (2019, p. 2), and Adri and Simon (2018).

be exacerbated by, for instance, competition over scarce resources or an increased risk of attacks owing to the presence of IDP settlements.

Adding to these specificities, Becker and Ferrara (2019)[26] point out the effects of economic integration upon arrival at the destination. The authors provide the example of difficult entry into the labor market (p. 11). However, the authors note that it does not happen in all cases. For example, in the case of internally displaced migrants by Hurricane Katrina: "Hurricane Katrina in 2005 had large and persistent effects on the location choice of individuals, and only small and transitory negative effects on income and employment" (p. 12). They also point out the possibility for mobility to generate a socio-economic improvement for migrants. In such cases, they mention young people who may find better opportunities to study at the destination territory (p. 12). The literature also teaches us that forced migrants are more likely to face health problems than voluntary migrants (Tebas and García 2016, p. 49; Cissé et al. 2022). In the particular case of mental health, migrants tend to suffer more from mental health problems than populations that do not migrate (Torres and Casey 2017, p. 5). Thompson et al. (2014) provide a list of the effects of displacement on physical and mental health. As for mental health, the authors mention anxiety, depression, post-traumatic stress disorder, and suicide. They manifest not only in the short term but also in the long term. Similarly, the Inter-American Court of Human Rights notes the particular vulnerability of forced internally displaced people (notably IACtHR 2013, para. 321; IACtHR 2016a, p. 4 [iv]).

2.3.3 *Legal Basis*

International, Regional and National Law

In our systematization in the table below (Table 2.2), we integrate the 1998 UN Guiding Principles (UN 1998b), the 1969 American Convention on Human Rights (Pact of San José, OAS 1969), the 1988 Additional Protocol to the American Convention on Human Rights in the area of Economic, Social and Cultural Rights (OAS 1988), the 1948 United Nations Universal Declaration of Human Rights (UN 1948), the 1966 International Covenant on Civil and Political Rights (UN 1966a); and the 1966 International Covenant on Economic, Social and Cultural Rights (UN 1966b).[27]

[26] The author builds on studies already carried out by others and highlight the specificities of forced migrations not exclusively driven by climate change. In addition, they focus on migratory movements almost exclusively on a large scale. It rules out the phenomenon in its less "dramatic" dimension: forced mobilities of small groups of people (when added together, they have a quantitative dimension no less important than those of large-scale mobilities). Individual mobilities (as opposed to mass mobilities) tend to be invisible. See Peguero Moreno and Laura (2019, p. 20), and Ruiz and Carvajal (2015) on invisibility related to the impacts of natural phenomena.

[27] Mexico is a State Party to these instruments (Pact of San José, Additional Protocol to the American Convention on Human Rights in the Area of Economic, Social and Cultural Rights, and United Nations Pacts of 1966). The Mexican State's obligations and the rights of migrants registered in

Table 2.2 Protection requirements systematization. Differentiated approach—Indigenous Peoples and women

Guaranteed rights	Legal grounds
	UN Guiding Principles (1998a, b) ("GP") American Declaration on the Rights of Indigenous Peoples (OAS 2016) ("ADRIP") UN Declaration on the Rights of Indigenous Peoples (UN 2007) ("DRIP") International Labour Organization Indigenous and Tribal Peoples Convention No. 169 (ILO 1989) ("Conv. ILO") UN Declaration of Human Rights (UN 1948) ("UNDHR") UN International Covenant on Civil and Political Rights (UN 1966a, b) ("ICPR") UN International Covenant on Economic, Social and Cultural Rights (UN 1966a, b) ("ICESCR") American Convention on Human Rights (OAS, Pact of San José, 1969) ("Pact of San José") Inter-American Convention on the Granting of Civil Rights to Women (OAS 1948a) ("Conv. OAS CR Women") Inter-American Convention on the Granting of Political Rights to Women (OAS 1948b) ("Conv. OAS PR Women") Inter-American Convention on the Prevention, Punishment, and Eradication of Violence against Women (OAS 1994) ("Convention of Belem Do Para") UN Convention on the Elimination of All Forms of Discrimination against Women (UN 1979) ("Conv. Against discrimination") UN Convention on the Political Rights of Women (UN 1952) ("Conv. UN PR women") UN Convention on Consent to Marriage, minimum age for marriage and Registration of Marriages (UN 1962) ("Conv. marriage") UN Supplementary Convention on the Abolition of Slavery, the Slave Trade, and Institutions and Practices Similar to Slavery (UN 1956) ("Sup. Conv. slavery") Mexican Constitution (*Constitución Política de los Estados Unidos Mexicanos*, CPEUM 2020) ("CPEUM")

(continued)

Table 2.2 (continued)

Women	
Dignity	GP: Pple 11(2) Pact of San José: Art. 17(3) Conv. ILO: Art. 20(3)(d) UNDHR: Art. 16 ICCPR: Art. 23(3) ICESCR: Art. 10(1) Conv. Against discrimination: Art. 16(1)(b) Conv. marriage: Art. 1 CPEUM: Art. 2(B)(V) Sup. Conv. slavery: Art. 1
Prohibition of physical, sexual and psychological violence	GP: Pple 11(2) Convention of Belem Do Para: Art. 3 Conv. Against discrimination: Art. 6
Participation	GP: Pple 18(3)
Health	GP: Pple 19(2) UNDHR: Art. 25(2) ICESCR: Art. 10, Art. 11, Art. 12(2) Conv. Against discrimination: Art. 12
No discrimination	Convention of Belem Do Para: Art. 6 Conv. Against discrimination: Art. 2, Art. 14
Recognition of legal personality	GP: Pple 20(3) Conv. Against discrimination: Art. 16
Educaction	GP: Pple 23(3)(4) Convention of Belem Do Para: Art. 6(b) Conv. Against discrimination: Art. 10
Rights equally guaranteed between men and women	ADRIP: Art. 32 ICCPR: Art. 3 ICESCR: Art. 7(a)(i) Conv. OAS CR Women: Art. 1 Conv. OAS PR Women: Art. 1 Convention of Belem Do Para: Art. 4, Art. 5 Conv. Against discrimination: Art. 7 (civil and political rights), Art. 8 (nationality), Art. 10 (education), Art. 11 (jobs), Art. 12 (health), Art. 13 (economic and social life), Art. 15 Conv. UN PR women: Art. 1, Art. 2, Art. 3 CPEUM: Art. 4
Indigenous Peoples	
Self-identification	ADRIP: Art. 1(2) Conv. ILO: Art. 1(2) CPEUM: Art. 2

(continued)

Table 2.2 (continued)

Free determination	ADRIP: Art. 3 DRIP: Art. 3 CPEUM: Art. 2, Art. 2(B)(VIII)
Gender equality	ADRIP: Art. 7 Conv. ILO: Art. 20(3)(d) CPEUM: Art. 2(A)(III)
To belong to Indigenous Peoples	ADRIP: Art.8 DRIP: Art. 8
Legal personality	ADRIP: Art. 9
Rejecting asimilation	ADRIP: Art. 10 DRIP: Art. 8(1)
Protection against genocide	ADRIP: Art. 11 DRIP: Art. 7(2)
Guarantees against racism, racial discrimination, xenophobia and other forms of intolerance	ADRIP: Art. 12 DRIP: Art. 2, Art. 8(2)(e) Conv. ILO: Art. 3(1), Art. 24, Art. 31
Identity and cultural integrity	ADRIP: Art. 13 DRIP: Art. 11, Art. 33 Conv. ILO: Art. 5 ICCPR: Art. 27 CPEUM: Art. 2(A)(IV)
Preservar, usar, desarrollar, promover, revitalizar y transmitir a generaciones futuras sistemas de conocimientos, lenguaje y comunicación	ADRIP: Art. 14 DRIP: Art. 12, Art. 13, Art. 16 ICCPR: Art. 27 CPEUM: Art. 2(A)(IV)
Education	ADRIP: Art. 15 DRIP: Art. 14 Conv. ILO: Art. 21–22, Art. 26, Art. 27, Art. 28, Art. 29, Art. 30 CPEUM: Art. 2(B)(II)
Indigenous spirituality	ADRIP: Art. 16, DRIP: Art. 12, Art. 13 ICCPR: Art. 27
Indigenous family	ADRIP: Art. 17 DRIP: Preámbulo
Health	ADRIP: Art. 18 DRIP: Art. 24 Conv. ILO: Art. 25 CPEUM: Art. 2(B)(III)
Healthy environment	ADRIP. Art. 19 DRIP: Art. 29 Conv. ILO: Art. 7(4) CPEUM: Art. 2(A)(V)
Association, meeting, expression, thought	ADRIP: Art. 20 DRIP: Art. 16(2) CPEUM: Art. 2(B)(VI)

(continued)

Table 2.2 (continued)

Autonomy and self-government	ADRIP: Art. 21 DRIP: Art. 4 CPEUM: Art. 2(A)(I)
Indigenous law and jurisdiction	ADRIP: Art. 22 DRIP: Art. 33(2), Art. 34 Conv. ILO: Art. 8 CPEUM: Art. 2(A)(II)(III)(VII)
Indigenous Peoples participation and indigenous legal systems and organization inputs (includes rights to be consulted)	ADRIP: Art. 23 DRIP: Art. 5, Art. 18 Conv. ILO: Art. 5(c), Art. 6(2), Art. 7, Art. 15(2), Art. 17(2), Art. 22(3), Art. 27(3), Art. 28 CPEUM: Art. 2(B)(IX), Art. 2(A)(III)
Lands, territories and resources	GP: Pple 9 ADRIP: Art. 25 DRIP: Art. 8(2)(b), Art. 25, Art. 26, Art. 27 Conv. ILO: Art. 13, Art. 14, Art. 17(1)(3), Art. 18, Art. 15(1), Art. 19 CPEUM: Art. 2(A)(V)(VI)
To stay in voluntary isolation o initial contact	ADRIP: Art. 26
Labor rights	ADRIP: Art. 27 DRIP: Art. 17 Conv. ILO: Art. 20
Protection of cultural patrimony and intellectual property	ADRIP: Art. 28 DRIP: Art. 31
Development	ADRIP: Art. 29 DRIP: Art.20, Art. 23 Conv. ILO: Art. 7 CPEUM: Art. 2(B)
To be consulted and cooperate concerning projects that impact their lands o territories and other resources	ADRIP: Art. 29(4) DRIP: Art. 19, Art. 32(2) Conv. ILO: Art. 6, Art. 7, Art. 15(2)
Restitution o compensation for lands or territories loss	ADRIP: Art. 29(5) DRIP: Art. 28, Art. 32(3)
Peace, security and protection	ADRIP: Art. 30 DRIP: Art. 7(1) Conv. ILO: Art. 3(2), Art. 11, Art. 12 CPEUM: Art. 2(B)(IV), Art. 2(A)(VIII)
Return	Conv. ILO: Art. 16(3)
Adequate resources	ADRIP: Art. 33 DRIP: Art. 21 Conv. ILO: Art. 23 CPEUM: Art. 2(B)(I)(IV)(VII)
Equality (principle)	ADRIP: Art. 36 DRIP: Art. 2
Nationality	DRIP: Art. 6

(continued)

Table 2.2 (continued)

Life	ADRIP: Art. 19(1), Art. 26(2) DRIP: Art. 7(1)
Migrants	CPEUM: Art. 2(B)(VIII)
Prohibition of forced relocation from their lands or territories	DRIP: Art. 8(2)(c) Conv. ILO: Art. 16(1)
Prohibition of relocation without the free, previous and informed consent, an agreement on a fair compensation and, where possible, a return option	DRIP: Art. 10 Conv. ILO: Art. 16(2)(5)

Note 1 We mention the requirements provided by instruments dedicated to the differentiated approach (e.g., United Nations Declaration on the Rights of Indigenous Peoples). We only report instruments not dedicated to this approach when the requirements or standards explicitly refer to women and Indigenous Peoples

Note 2 We do not integrate the rights and guarantees related to criminal aspects since the literature does not mention it as particularly relevant. Also, the norms we report do not have the same legal value. Some are legally binding, and some are not

The differentiated approach is endorsed in the UN Guiding Principles on Internal Displacement (1998b), Principle 4(2):

> Certain internally displaced persons, such as children, especially unaccompanied minors, expectant women, mothers with young children, women heads of household, persons with disabilities and the elderly persons, shall be entitled to protection and assistance required by their condition and to treatment which takes into account their special needs.

From the categories we identified in the first chapter, the paragraph mentions women but there is no reference to Indigenous Peoples. However, they enjoy specific protection, stemming from Principle 9: "States are under a particular obligation to protect against the displacement of Indigenous Peoples, minorities, peasants, pastoralists and other groups with a special dependency on and attachment to their lands."

Other specific legal bases for the protection of Indigenous Peoples are the 2016 American Declaration on the Rights of Indigenous Peoples (OAS 2016), the 2007 United Nations Declaration on the Rights of Indigenous Peoples (UN 2007); and the 1989 International Labour Organization's Indigenous and Tribal Peoples Convention 169 (ILO 1989). Mexico is a State Party to the ILO Convention.

As for women, seven international and regional conventions protect their specific rights: the Inter-American Convention on the Granting of Civil Rights to Women (OAS 1948a), Inter-American Convention on the Granting of Political Rights to Women (OAS 1948b), Inter-American Convention on the Prevention, Punishment,

---

this international legal framework apply and bind the national framework (CPEUM 2020, Article 1). However, Mexico's many shortcomings when complying with them cannot be ignored. On this matter and more specifically regarding the case of international migrant law, see Martínez Quezada, Ramírez and Sletza (2015).

and Eradication of Violence against Women ("Convention of Belem Do Para") (OAS 1994), United Nations Convention on the Elimination of All Forms of Discrimination against Women (UN 1979), United Nations Convention on the Political Rights of Women (UN 1952), United Nations Convention on Consent to Marriage, minimum age for marriage and Registration of Marriages (UN 1962); and the 1956 United Nations Supplementary Convention on the Abolition of Slavery, the Slave Trade, and Institutions and Practices Similar to Slavery (UN 1956). Mexico is a State Party to these instruments.

In the systematization of the protection requirements, we also report on the rights guaranteed by the political constitution of the United Mexican States. At this point, we have not integrated all the laws or programs that could apply to the phenomenon studied. We do so in the next chapter.

Inter-American Court of Human Rights Case-Law

Finally, we complement this study with the Inter-American Court of Human Rights' ("the Court") case-law on forced displacement. This *corpus Juris* is particularly crucial in order to have an overview of the protection requirements. It emerged from forced displacements induced by a series of five causes: establishing a protected natural area, armed conflict, the exploitation of natural resources, "development projects," and Indigenous Peoples putting in claims to recover their lands (occupied by third parties in several cases).[28]

First, the Court highlights some obligations to protect which apply to the general population (that is, they are not limited to Indigenous Peoples) in cases of forced displacement induced by armed conflict or widespread violence. In *Members of the Chichupac Village and neighboring communities of the Municipality of Rabinal v. Guatemala* (IACtHR 2016b), the Court had the opportunity to recall and establish the obligations that apply in the event of forced displacement and the obligations' legal basis. Based on Article 22(1) of the Inter-American Convention (freedom of movement and residence), the Court had already established in its previous case-law that it "also protects the right not to be forcibly displaced within a State Party," and that "the United Nations Guiding Principles on Internal Displacement are particularly relevant to determine its content and scope" (para. 172).[29]

The particularly vulnerable and "defenseless" situation of the internally displaced places them in a "de facto" situation with a "lack of protection" (para. 173).[30] It calls for application of the positive obligations doctrine "to revert the effects of the aforementioned condition of weakness, vulnerability and defenselessness, including with respect to the actions and practices of private third parties" (para. 173).[31] The human rights' horizontal effect doctrine also applies:

---

28 Also within the framework of the OAS, the reader can see IAComHR (2009, paras. 151, 220–267), (2015, note on page number 299), and (2018, pp. 43–59).

29 In the same sense, see IACtHR (2013, para. 219).

30 In the same sense, see IACtHR (2013, para. 315).

31 A difference can be noted with IACtHR (2012b). In the 2012 case, the Court concluded the freedom of movement and residence had been violated due to the mere existence of forced displacement: "this provision (*Article 22[1], right of movement and residence*) also protects the right not

> this Court has indicated that the right of movement and residence may be *de facto* violated if the State has not established the conditions or provided the means that allow it to be exercised, for example, when a person is a victim of threats or harassment and the State does not provide the necessary guarantees so that they can move and reside freely in the territory in question, *even when the threats and harassment come from non-State actors*. (par. 174, emphasis mine)

The positive obligations are to bring about the adoption of measures to actively prevent displacement, and to ensure the right to return "or, where appropriate, adequate compensation" (para. 189) "without the risk of their rights being violated" (para. 175),[32] and reintegration (para. 175). Specifically, the States have the obligation to "provide the necessary conditions for a dignified and safe return to their place of habitual residence or their voluntary resettlement in another part of the country" (para. 175). Applying a differentiated approach and taking into account the particular risks Indigenous Peoples face, the Court establishes the requirement to "adopt specific protection measures, considering Indigenous Peoples' specificities, as well as their customary law, values, uses and customs, to prevent and reverse the effects of said situation" (para. 176).

In this case, the Court concluded that the freedom of movement and right to residence had been violated.[33] It noted the impossible return of the displaced people "due to the well-founded fear of being the object of violations of their rights to life and personal integrity," including "during the armed conflict" (para. 179), the "fear and insecurity of the surviving population due to the current presence of those responsible for the violence" (para. 180), and the fact that third parties occupied their plots after the conflict (para. 181).

The Court also noted that Guatemala had not adopted measures to reverse the effects of displacement: the absence of "effective investigation on the violent events and displacement suffered by the members of the village" (para. 183), as well as the difficulty of expediting the recovery of lost documents (including possession titles) (para. 184). In terms of compensation, the Court noted several deficiencies regarding the participation of displaced people, the impact of displacement, and the absence of return or resettlement measures which would have taken into account that they are Indigenous Peoples. Once again, the Court applies a differentiated approach to these impacts and considers the specific situation of women (para. 197).

Furthermore, in *Indigenous Communities Members of the Lhaka Honhat (Our Land) Association vs. Argentina* (IACtHR 2020), the Court established the obligation of States to consult Indigenous Peoples "whenever it is necessary to transfer indigenous communities (…) from their traditional lands to other places" (footnote no. 161). This requirement applies to the forced relocation scenario.

---

to be displaced forcibly within a State Party to the Convention" (para. 255), and para. 267. In the same sense, see IACtHR (2010a, para. 149).

[32] Return is essential for internal forced climate migrations and more particularly in the disaster scenario since many persons and communities return to their place of origin. It is reported by Raleigh et al. (2008, p. 24), among others.

[33] The Court also found the violation of this right in IACtHR (2005c).

This case law leads to several comments. *Mutatis mutandis,* it would apply to the conflicts or violence caused by the scarcity of natural resources. It can also be applied to other scenarios. The Court found forced displacements to fall within the scope of the application of the right of movement and residence without limiting it to generalized violence. Regarding the categories that we identify in the first chapter, the case-law applies to two of them (emergency forced climate migrant and forced climate migrant) as the criterion the Court applies is the forced nature of the displacement. The Court indirectly addresses trapped migrants when it notes the special situation of populations particularly exposed to human rights violations due to their socio-economic situation (IACtHR 2012b, para. 273). Finally, the Court's case-law reflects the differentiated approach: it signals Indigenous Peoples and the specific case of women and the protection requirements in the context of forced internal migration.

Firstly, the Court identifies rights and obligations that specifically apply to Indigenous Peoples in cases of conflicts related to the establishment of protected natural areas and the exploitation of resources located on their lands and territories. The differentiated approach drives those obligations. The Court grounds its case-law in property law and it provides that:

> (1) the traditional possession of Indigenous Peoples over their lands has effects equivalent to the title of freehold granted by the State; (2) traditional possession grants indigenous people the right to demand official recognition of property and its registration; (3) *the members of the Indigenous Peoples who, for reasons beyond their control, have left or lost possession of their traditional lands maintain the right to property over them, even in the absence of legal title, except when the lands have been legitimately transferred to third parties in good faith* ; (4) the State must delimit, demarcate, and grant collective title to the lands to the members of the indigenous communities; (*5) the members of Indigenous Peoples who have involuntarily lost possession of their lands, and these have been legitimately transferred to third parties in good faith, have the right to recover them or to obtain other lands of equal size and quality*; (6) The State must guarantee the effective property of the Indigenous Peoples and refrain from performing acts that may lead the agents of the State itself, or third parties who act with their acquiescence or tolerance, to affect the existence, value, use or the enjoyment of its territory; (7) the State must guarantee the right of Indigenous Peoples to effectively control and own their territory without any external interference from third parties, and (8) the State must guarantee the right of Indigenous Peoples to control and use their territory and natural resources. Regarding the aforementioned, the Court has held that it is not a matter of a privilege to use the land, which can be stripped by the State or overshadowed by the property rights of third parties, but rather a right of the members of Indigenous Peoples and tribals to obtain the title to their territory in order to guarantee the permanent use and enjoyment of said land. (IACtHR 2018, para. 128, emphasis mine)[34]

[34] In the same sense, see IACtHR (2010b, para. 109), (2015a, para. 173), and (2015c, para. 131).

The rights stipulated in (3) and (5) also apply within the framework of this book: the criterion to identify the three categories of migrants is the forced nature of internal migration. Also, the six scenarios identified in this research apply: they can all induce Indigenous Peoples to leave or lose possession of their lands or territories. A notable aspect in *Xákmok Kásek Indigenous Community v. Paraguay* (IACtHR 2010b) is the State had issued a declaration for a private natural reserve. It was, therefore, not about a de facto situation, but a situation generated by an act of the State. It also resonates with the fourth scenario (the State declares areas to be unfit for human settlement).

In terms of Indigenous Peoples recovering their lands and the possibility of obtaining alternative lands, the Court specified the right is not absolute. In *Kaliña and Lokono peoples v. Suriname* (IACtHR 2015c), the Court refers to its previous case-law to determine that "if, for objective, specific and justified reasons, the State is unable to take measures to return the traditional territory and the communal resources (…) the State may offer alternative lands of the same or greater size and quality or payment of fair compensation, or both, by mutual agreement with the peoples concerned" (para. 158, references omitted).

We also consider that the possibility of the State not returning lands to Indigenous Peoples very clearly applies within this book. If their lands are flooded (due to a sudden-onset disaster, sea-level rise, or a development project), the State would have to offer alternative lands. Some may question the analogy: in the disaster or lands lost due to sea-level rise scenarios those lands are not in the hands of other private people or the State. However, in the opinion of the author, the analogy can be based on the requirement to return the lands or offer alternative lands ("for reasons beyond their control they have left or lost possession"/"they have involuntarily lost possession of their lands") together with the objective to preserve the special bond between Indigenous Peoples and lands or territories.[35]

Finally, a case directly applies to forced internal migration induced by development projects related to the renewable energy scenario. In *Kuna Indigenous Peoples of Madungandí and Emberá de Bayano and their members v. Panama* (IACtHR 2014), a hydroelectric dam project required part of an Indigenous People's lands to be flooded. They had been relocated to new lands after indigenous representatives signed agreements. The Court, recalling its previous case-law, found the obligations of the State regarding Indigenous Peoples' collective property[36] also applies to alternative lands. Accordingly, whenever a forced climate migration scenario implies the State must grant alternative lands, those obligations would apply.

[35] See among others IACtHR (2006, para. 118) and (2005a, para. 131).

[36] Delimitation, demarcation, ownership, protection against the intrusion of third parties.

2.4 An Environmental Rights Approach

2.4.1 Contributions of Environmental Rights to the Protection Requirements

There is a wealth of academic literature on environmental rights. Providing an in-depth analysis of those rights is outside the scope of this chapter.[37] We instead present an overview and focus on the link between the environmental rights and the protection requirements in the context of forced internal climate migration. The debate on how to encapsulate environmental rights remains open. The author gives preference to a distinction that has the advantage of clarity. Three series of rights can be distinguished: the autonomous right to a healthy environment would recognize the environment's intrinsic value autonomously (there is no need for interference with rights guaranteed to men—e.g., property or water),[38] material environmental rights (e.g., water), and procedural environmental rights. The latter is traditionally broken down into three pillars: the right to access environmental information, the right to participate in environmental decision-making, and the right to access justice in environmental matters.

Similar to the criticisms made towards the rights-based approach, environmental rights are also heavily criticized (Francioni 2010; Gouritin 2016). Among the criticism is the fact that environmental protection can conflict with human rights. For example, the United Nations' Reducing Emissions from Deforestation and Forest Degradation in Developing Countries (REDD+) program to improve the carbon stock of forests illustrates an instrument aimed at protecting the environment can have adverse effects on human rights. Various reports and academics have mentioned the violations of the rights of Indigenous Peoples and communities regarding their lands and territories (including violations of the right to consultation) (Gouritin 2016, p. 9).

Another common criticism is the anthropocentric and utilitarian approach to human and environmental rights texts and their resultant interpretation and articulation (Gouritin 2016). Finally, the predominantly individual dimension of human rights is regularly criticized for not fitting with environmental matters that tend to have a collective dimension. For example, air pollution in Mexico City does not affect just one person, but a multitude of people. Also, some environmental effects have a collective dimension per se (e.g., the disappearance of an ecosystem generates damage for humanity in general or the recognition of the rights of future generations). This last criticism is not as valid in the inter-American context as the Court

---

[37] For in-depth analysis, see among others Gouritin (2016); IACtHR (2017) and Cancino et al. (2018).

[38] The debates about the right to a healthy environment are still intense. For an example of categorization distinct from the one proposed by the author see Cancino et al. (2018, p. 83), where the authors apply another concept of "right to a healthy environment" (an anthropocentric one).

solidly established the collective dimension of various Indigenous Peoples' rights (e.g., collective property).

Despite these criticisms, we incorporate environmental rights to complement the rights-based approach and concretize the protection requirements for several reasons.[39] Among them, it brings about the positive development of the potential of the positive effect doctrine. The doctrine has been essential for developing environmental rights.[40] For forced internal climate migration, there is the potential to oblige States to adopt measures that prevent migration and ensure the effective respect of migrants' rights.

Additionally, other tools have essential, practical applications. For example, the horizontal effect doctrine implies that not only the State and public authorities can violate human rights (vertical effect). Private individuals can also precipitate the violation of another private person's rights (horizontal effect): the responsibility of the State can be questioned for the actions of a private person. For example, where a company contaminates a river and hampers an adjacent community from having access to water or fish necessary for its survival, or damages mangroves that further exposes a community settled on the coastal area to extreme hidrometeorological events. Communities would consequently be forced to leave their places of residence.

Another tool is the non-regression principle (or standstill effect/*effet cliquet*): human rights standards cannot be weakened. Protection standards can only get stronger. It is crucial for environmental protection standards related to human rights: they cannot be modified to be weaker. They can only be more extensive. Environmental rights enable the systemization of protection requirements. Grouping the requirements according to the three environmental rights groups allows a more straightforward analysis of the situation in Mexico, and at the same time, a more analytical one. The systematization is not a list of rights that appear disconnected, but a set of "blocks" that allow identifying areas of deficiency or compliance in terms of protection requirements for forced internal climate migrants in Mexico.

In the following lines, we propose a systematization of these rights, provide their legal foundations and make explicit their specific contributions to climate forced internal migration. Since international, regional or Mexican law do not fully endorse the intrinsic value of the environment (Gouritin 2016, pp. 14–17)[41] we do not consider

---

[39] To avoid any misunderstanding, we incorporate the environmental rights approach as an analytical approach. It belongs to and is part of the rights-based approach, it is not an additional approach in itself.

[40] On the doctrine in environmental matters, see Gouritin (2011, pp. 134–164). On the recognition of the application of the positive obligations doctrine and environmental rights at the Mexican level see Seventh Collegiate Court in administrative matters of the First Circuit (*Séptimo Tribuna Colegiado en Materia Administrativa del Primer Circuito*) (2016a), and on the recognition of the horizontal effect and environmental rights at the Mexican level see among others Second Collegiate Court on Administrative Matters of the First Circuit (*Segundo Tribunal Colegiado en Materia Administrativa del Primer Circuito* 2007).

[41] An exception is IACtHR (2017, paras. 62–63), but the Court's jurisprudence itself reflects the utilitarian approach it denounces to date (except for a recent case: IACtHR 2020). Another exception is the 1972 Stockholm Declaration (United Nations Conference on the Human Environment)

the right to a healthy environment. Instead, we focus on procedural and material environmental rights.

2.4.2 *Environmental Rights: Content and Legal Foundations*

Three pillars form the procedural environmental rights: the right to environmental justice, the right to access environmental information, and the right to participate in environmental decision-making. They complement each other, and are interdependent: there cannot be effective participation without access to information, there cannot be adequate access to information without the guarantee of access to a court if public authorities do not respect the right, and there cannot be effective access to justice for the repair of environmental damage if citizens do not have adequate information regarding, for example, contamination.

Procedural environmental rights' legal foundations are very solid in international, regional, and national law. *Ad-hoc* instruments guarantee them (see Table 2.1). Included within them are non-binding (e.g., the 1992 Rio Declaration on Environment and Development—UN 1992) and binding norms (e.g., the Inter-American Court of Human Rights' case law and the 2018 Escazú Agreement—ECLAC, CEPAL 2018).[42]

Access to environmental justice is "the possibility of obtaining an expeditious and complete solution by the judicial authorities to a legal conflict" (Cancino et al. 2018, p. 14) and "the possibility that every person (...) has for a satisfactory solution before the judicial or administrative authorities of a conflict of an environmental nature through a legal resolution that resolves a controversy or that forces the application of the law" (Cancino et al. 2018, p. 148). In Mexico, a comprehensive series of theses (both jurisprudence and isolated) specify the scope of this right along with its content. Several international instruments guarantee the right: the United Nations

---

and its Principle 1. At the national level, the CPEUM (2020) enshrines the right to a healthy environment in its article 4. The article's interpretation demonstrates the confusing situation of this right regarding the tensions mentioned above. It has several labels: "fundamental right to have an adequate environment" (SCJN 2014), "protection of the clean environment" (Eighteenth Collegiate Court in Administrative Matters of the First Circuit [*Décimo Octavo Tribunal Colegiado en Materia Administrativa del Primer Circuito*] 2013). Conversely, these interpretations tend to recognize its collective dimension. On recognizing the collective dimension, see e.g. Seventh Collegiate Court in administrative matters of the First Circuit (*Séptimo Tribunal Colegiado en Materia Administrativa del Primer Circuito* 2016b). However, this progressive position fails to recognize an ecocentric content to the right in question: the anthropocentric conception is predominant (see e.g. First Collegiate Court on Administrative Matters of the First Circuit [*Primer Tribunal Colegiado en Materia Administrativa del Primer Circuito* 2017]). Also, the right is conceived as a relative right (see e.g. First Collegiate Court in Administrative Matters of the First Circuit [*Primer Tribunal Colegiado en Materia Administrativa del Primer Circuito* 2013]). See UN (2020) on the limits of the right to a healthy environment.

[42] Cancino et al. (2018, p. 148) mention others, among them: the Stockholm Declaration, the World Charter for Nature, Agenda 21, the Access Principles (PP10), and the Earth Charter.

Guiding Principles on Internal Displacement (UN 1998b), American Convention on Human Rights (OAS 1969), United Nations Universal Declaration of Human Rights (UN 1948), United Nations International Covenant on Civil and Political Rights (UN 1966a), and the Political Constitution of the United Mexican States (CPEUM 2020).

The right to access environmental information is also firmly established in *ad-hoc* instruments and the Mexican Constitution (CPEUM 2020) which guarantees the right of access to information in article 6 in a generic way (that is to say, it is not the right to specific environmental information). The right has been widely commented on and developed by various international judicial and non-judicial bodies.[43] For its part, the Inter-American Court of Human Rights specified the scope of the right to information is not limited to "supplying the information requested by a particular person. It implies, among others, the obligation to make public management transparent and to provide timely, accessible, and complete information *ex officio*" (Cancino et al. 2018, pp. 140–141). In other words, the doctrine of positive obligations applies in the field of environmental rights. Also, the differentiated approach applies to the right. Inter-American human rights law articulates the right as a "key instrument" for "the most vulnerable subjects" (Cancino et al. 2018, p. 141) and more specifically, in the context of this book, women, Indigenous Peoples and people in situations of extreme poverty.

Academics (Anglés Hernández 2014; Gouritin and Aguilar 2017) and various reports (IAComHR 2009) systematize the right to participate in environmental decision-making. The right to consultation is of particular importance. The inter-American human rights protection system has principally directed its attention to consultation and more particularly the consultation of Indigenous Peoples. International and regional law firmly guarantee Indigenous Peoples' right to culturally appropriate and good faith consultation (ILO 1989, Articles 6, 7, 15 and 17; UN 2007, Articles 11, 19, 27, 28(1) and 32(2); OAS 2016, Articles 21(2), 23, and 29(4); and IACtHR 2001, 2005a, b, 2007, 2008, 2010b, 2012a, 2013, 2014, 2015a, b, c, and 2020). The consultation has to be organized in order to obtain their free, prior, and informed consent each time natural resources are exploited on their lands, the development of infrastructure projects affect their lands and territories, and the adoption of normative instruments alter their life or customs (IAComHR 2009, pp. 108–128; IACtHR 2017, p. 91).

This right is highly contentious. First of all, there are many specific challenges in Mexico: the academic literature and various national and international reports signal deficiencies in the consultation of Indigenous Peoples (Gouritin 2018, pp. 43–59). Secondly, there remains to be seen whether the obligation of consultation grants a veto to Indigenous Peoples:

> The relevant question here is whether Indigenous Peoples should give their formal consent for an exploitation project to take place on their traditional lands and territories. In other words, do Indigenous Peoples have the right to veto the project? Is the State's obligation of

---

[43] On the work of the Inter-American Commission of Human Rights see for example Cancino et al. (2018, pp. 139–140).

> results (to obtain the veto consent of Indigenous Peoples) or of means (to use all the means at their disposal to obtain consent)? (Gouritin and Aguilar 2017, p. 309).

To date, there is still significant confusion regarding this point, and it is hard to affirm that Indigenous Peoples have a generic right of veto grounded in international or regional law (Gouritin and Aguilar 2017, pp. 314–317).[44]

Further to the three procedural rights, the recent Inter-American Court of Human Rights' case-law has developed another requirement: the State must supervise the performance of an environmental and social-impact assessment before granting concessions within the territory of Indigenous Peoples (IACtHR 2020, para. 174). This study "is not equivalent to the exercise of the right to free, prior and informed consultation" (IACtHR 2020, footnote no. 162) but rather, adds to the right to consultation.

Turning to material environmental rights, unlike environmental procedural rights that have an ad hoc legal basis (Aarhus Agreement, Escazú Agreement), they are not guaranteed as such by ad hoc instruments except for the recognition of a right to a healthy environment[45] and certain specific rights (e.g., the right to water). Despite this limitation, the institutions in charge of monitoring the respect for human rights and their effective enjoyment have recognized that environmental degradation can interfere with the enjoyment of a series of material rights. It is particularly the case for the rights of "groups in a situation of vulnerability," "among which are Indigenous Peoples" and "communities that depend, economically or for their survival, (…) on environmental resources" (IACtHR 2020, para. 209). This special vulnerability applies to two of the groups identified in the research: Indigenous Peoples and people with limited economic resources. The "greening" of Human Rights in case-law allowed the identification of material environmental rights. The material rights most commonly mobilized to establish the link between environmental degradation and the enjoyment of human rights are the right to health, private and family life, food, and life (Gouritin 2016, pp. 11–13).

The recognition of an environmental origin in an interference with human rights (e.g., contamination) does not *ipso facto* imply their violation. As is the case for all relative human rights (e.g. rights to health and to private and family life), the interference may not constitute a violation of the human right in question. Departing from absolute rights (e.g. the right to life) which are violated from the moment they are interfered with, the authorities in charge of monitoring the protection of human rights operate a series of verifications to determine whether the interference constitutes a

[44] For cases in which there is a right of veto, see among others IAComHR (2009, p. 128).

[45] When recognized in legal norms, this right is an umbrella for linking the environment and human rights. Its legal effect is at least limited since it confirms human rights "greening" (carried out by national, regional and international case-law). Thus, it does not add in terms of States' rights or obligations to date. In this regard, a recent case of the Inter-American Court of Human Rights may bring about a paradigm shift: in the *Indigenous Communities Members of the Lhaka Honhat Association (Our Land) vs. Argentina* case (IACtHR 2020), the court (based on Advisory Opinion OC 23/17 IACtHR 2017) advances towards the recognition of a right to an autonomous healthy environment with an ecocentric approach.

violation of the right in question (legality, proportionality and requirements, and the pursuing of a legitimate objective in a democratic society).

Finally, environmental and procedural rights are not hermetic: the institutions in charge of determining if material environmental rights have been violated also look to see where procedural requirements have been infringed. Infringement of the latter is an element that can support human rights institutions to conclude the State or public authority have not duly acknowledged the person or community's interests when balanced against public interest (Gouritin 2020; Gouritin and Aguilar 2017).

2.4.3 Environmental Rights' Potential: A Panorama

This section illustrates the potential for environmental rights to complement the rights-based approach. The environmental rights approach addresses both the factors that drive forced internal climate migrations and the demands that arise in the mobility process. We successively address procedural and material environmental rights.

The potential of procedural environmental rights within the framework of the research is extensive. Academics and reports mention the need to guarantee the participation and consultation of forced climate migrants (e.g. HCHR 2010, p. 2: "All relevant actors need to respect the right of internally displaced persons to make an informed and voluntary choice on what durable solution to pursue and to participate in the planning and management of durable solutions"). The case studies carried out to date point to very strong deficiencies and their consequences in terms of climate justice (Thompson et al. 2014, pp. 83–84; IACtHR 2016a, p. 3). They also mention the problematic Mexican context in terms of environmental rights and identify deficiencies that arise both in the legislative and public policy design and in practice (Cancino et al. 2018, pp. 16–17).

Material environmental rights can be brought into play in various ways. Climate change can cause soil, water, and air pollution (e.g., a flood endangers facilities that handle substances dangerous to the environment or human health). Also, climate change can amplify the effects of pollution. For example, where contamination impacts an area in which climate change has already damaged biodiversity. Another illustration of the potential of environmental rights refers to the environmental protection requirement. It is particularly relevant in terms of trapped populations in a highly degraded or dangerous environment.

In another order of things, land planning is crucial for various internal climate migration scenarios: land use (and especially the issue of rural land urbanization), reserves, urban development, cultural heritage protection, mobility, territorial planning, urban settlements regulation,[46] environmental and human health. Those issues are essential for slow-onset environmental degradation, sudden-onset disasters, and

[46] Cancino et al. (2018, pp. 190–197) identify the elements of the right to the city are "a combination of existing human rights understood as an integral unit" (p. 191) and would be: (i) the right to life and sustenance, (ii) the right to an adequate standard of living, (iii) the right to adequate housing, (iv) the right to public services, (v) the right to choose the place of residence, (vi) the right to freedom

conflicts generated by the scarcity of natural resources. The situation of cities is particularly crucial in this context. We previously reported on the available projections that identify cities as the privileged place of destination in Mexico, and the challenges resulting from this pattern (e.g., an even more acute poverty concentration, greater exposure to natural hazards,[47] competition to access natural resources, infrastructures). Against that background, environmental rights have tremendous potential: there is a wealth of case-law that illustrates how environmental rights trigger land-planning and ground States' obligations (e.g., to adopt the measures aimed at regulating a natural protected area, to regulate industrial emissions in order not to contaminate an adjacent neighborhood's air or water, the proper implementation of environmental law—e.g. regarding waste treatment) (Gouritin 2016, Cancino et al. 2018).

Likewise, in terms of risk management, the environmental rights approach and obligation to regulate certain activities and to enforce environmental legislation can apply to a whole series of cases that concern natural (e.g., a flood—sudden-onset disaster scenario) or industrial risks (e.g., spills from a plant that handles hazardous substances in the vicinity of protected or already degraded areas—slow-onset environmental degradation scenario).

Lastly, in a former section of the present chapter we saw that climate change causes the appearance of diseases (by the presence of vectors such as mosquitoes or difficult access to water—slow-onset degradation). In that respect, the right to health in the environmental context can unfold its effects within the framework of this research.

Illustration: The Right to Water and Sanitation

The right to water and sanitation can illustrate the potential mentioned above. This illustration relies on the analytical reading of literature and a theoretical application of the scenarios and categories identified in this research. Its validity would be more robust if checked against fieldwork and more empirical data (to date, they are not available).

At an international level, the right to water and sanitation has been developed in numerous Resolutions of the United Nations General Assembly, the United Nations High Commissioner for Human Rights, and is further detailed in the Special Rapporteur on the Human Right to Safe Drinking Water and Sanitation's numerous reports (for a list see Cancino et al. 2018, pp. 198–205). Also, the UN Committee on Economic, Social and Cultural Rights' General Comment No. 15 on the right to

---

of movement, (vii) the right not to suffer any form of discrimination, (viii) the right to privacy, (ix) the right to environmental hygiene, (x) the right to health, and (xi) the special rights of women and children (pp. 191–192).

[47] Cancino et al. (2018): "Mexico, due to its geographical characteristics, is exposed to the occurrence of natural phenomena; more than 90 million inhabitants live in risk areas, of which about 70% live in urban areas, 9.5% in semi-urban areas and the rest 20.5% live in rural areas" (p. 179). Article 27 of the CPEUM (2020) regulates urban law.

water (CESCR 2002) enumerates the requirements and obligations involved. At a regional level, the Inter-American Court of Human Rights recently spelled out the rights' legal foundations and content (IACtHR 2020). At a national level, the "very comprehensive" federal legal framework (Cancino et al. 2018, p. 112) and Mexican Constitution guarantee the right in a very similar way to the precepts raised in General Comment No. 15 (Cancino et al. 2018, p. 110):

> Everyone has the right to water access, disposal, and sanitation for personal and domestic consumption in a sufficient, healthy, acceptable, and affordable way. The State must guarantee the right, and the bases, supports, and modalities for the access and equitable and sustainable use of water resources, the Federation, federative entities and municipalities participation, and citizens' participation to achieve these ends will be regulated by law. (CPEUM 2020, Article 4)

The right to water is essential in the context of the scenarios that generate forced internal climate mobility. Regarding "development projects" related to the scenario of renewable energies, for example hydroelectric dams, people below the dam curtain can suffer from an acute water shortage. It can also apply in the slow-onset environmental degradation scenario. Severe periods of drought (due to scarcity and desertification) happen in Mexico, and climate change amplifies their severity and occurrence. We have already seen that drought affects agriculture and livestock in Mexico, putting at risk food security and bringing the probability of hunger. Furthermore, conflicts over access to water are already happening in Mexico (conflicts or violence caused by the scarcity of natural resources).

Distributive conflicts related to water are already occurring in Mexico. The conflict around the "Independence Aqueduct" in the State of Sonora (northwest of Mexico) illustrates this.[48] The project was to build an aqueduct to send water from the Yaqui River Basin to the Hermosillo city. It is an urban–rural conflict and, as is regularly the case in distributive conflicts, Indigenous Peoples were on the frontline of the socio-environmental conflict (in this case, the *Yaqui* Tribe). Accordingly, the case of the Independence Aqueduct also illustrates the differentiated approach we adopt in this research.

The National Human Rights Commission issued in 2012 a Recommendation (CNDH 2012) which recognizes the cultural dimension for water within the Tribe (para. 106). One of the commission's pivot of analysis is "the lack of public policies' implementation by the government of (*Sonora*) in order to avoid social conflict due to the perception of water scarcity, its inequitable supply and the construction of the 'Independence Aqueduct' for that purpose" (para. 65). The Commission also notes the tensions related to a "long-standing problem related to water access and supply, a circumstance that in recent years has gained relevance and has even generated a situation of tension among the inhabitants of that region" (par. 83). The Commission also mentions "polarization" regarding the aqueduct construction project. The

[48] For an overview of the judgments issued in this case, see Radonic (2017), who also very interestingly points out the ambivalence of the right to water: the "traditional" conception (the populations of the cities can have access to water—public authorities) *versus.* the Tribe's "territorial" concept of water ("territorial rights over water"—p. 156, or "right to water in their territory"—p. 157). This tension illustrates the limits of a rights-based and environmental rights approach mentioned earlier.

Commission highlights the State's obligation to prevent conflicts (paras. 91–92) and the essential role environmental procedural rights have for this purpose (paras. 103, 112–124). In the Recommendation, the Commission raises a series of requirements based, among others, on General Comment No. 15 (paragraphs 98–101) that details the right's scope beyond what the Recommendation mentions (it raises more extensive "basic obligations," CNDH 2012, para. 37).

General Comment No. 15 is also relevant regarding the three categories of migrants addressed in the present research, and more particularly, trapped climate migrants. General Comment No. 15 details the right to water and corresponding obligations. It distinguishes between the obligation to respect (States must "refrain from interfering directly or indirectly with the enjoyment of the right to water"), the obligation to protect (States must "prevent third parties from interfering with the right to water"), and the obligation to carry out or comply (States must "adopt appropriate legislative, administrative, budgetary, judicial, promotional and other measures to fully realize the right to water") (UN High Commissioner 2001, pp. 27–28). We have already addressed the first two obligations in the preceding paragraphs.

On the last type of obligation and trapped climate migrants General Opinion No. 15 provides the criteria for assessing the application of the rights. They are availability, quality, and accessibility (CESCR 2002, para. 12). The criteria apply in assessing trapped climate migrants' effective enjoyment of the right. They enjoy a particular protection, since they are a vulnerable group. Protection of vulnerable groups is firmly grounded in the General Opinion: "even in times of severe resource constraints, the vulnerable members of society must be protected by the adoption of relatively low-cost targeted programmes" (para. 13), together with the prohibition of de facto discrimination (para. 14), and the special obligation to "provide those who do not have sufficient means with the necessary water and water facilities" (para. 15). More specifically, the General Opinion requires a differentiated approach in the application of the positive obligations doctrine (obligation to perform or fulfill):

> Whereas the right to water applies to everyone, States parties should *give special attention to those individuals and groups who have traditionally faced difficulties in exercising this right*, including women, (…) minority groups, Indigenous Peoples (…) In particular, States parties should take steps to ensure that:
>
> (a) Women are not excluded from decision-making processes concerning water resources and entitlements. The disproportionate burden women bear in the collection of water should be alleviated;
>
> (b) *Rural and deprived urban areas* have access to properly maintained water facilities. Access to traditional water sources in rural areas should be protected from unlawful encroachment and pollution. Deprived urban areas, including informal human settlements, and homeless persons, should have access to properly maintained water facilities. *No household should be denied the right to water on the grounds of their housing or land status;*
>
> (c) *Indigenous Peoples*' access to water resources on their ancestral lands is protected from encroachment and unlawful pollution. States should provide resources for Indigenous Peoples to design, deliver and control their access to water;
>
> (d) (…) internally displaced persons and returnees have access to adequate water whether they stay in camps or in urban and rural areas. (…)

> (e) Groups facing difficulties with physical access to water, such as (…) *victims of natural disasters, persons living in disaster-prone areas, and those living in arid and semi-arid areas, or on small islands* are provided with safe and sufficient water. (ESCR Committee, 2002, para. 16, emphasis mine)

From the above, it results that trapped climate migrants enjoy special protection in terms of the right to water, either because of the areas they live in ("arid and semi-arid areas or on small islands," para. 16) or because of their economic vulnerability ("rural areas and disadvantaged urban areas"). Also, equity supports the differentiated approach to vulnerable groups:

> Any payment for water services has to be based on the principle of equity, ensuring that these services, whether privately or publicly provided, are affordable for all, *including socially disadvantaged groups*. Equity demands that poorer households should not be disproportionately burdened with water expenses as compared to richer households. (ESCR Committee, 2002, para. 27, emphasis mine).

2.5 Conclusion

We began this chapter with the territorial axis that is essential to proposing tools and guidelines for public policies in Mexico. We established and justified that the territory is an ideal unit of analysis to address the phenomenon. It allowed us to identify the challenges for the territories of origin, destination, and more particularly for cities and Indigenous Peoples. Identifying these challenges makes it possible to give flesh to the protection requirements, subsequent axis of the analysis grid.

We first demonstrated the protection requirements apply to the six climate internal forced mobility scenarios. It similarly applies to the three migrant categories. We also systematize the sources and scope of the protection requirements. We provide the protection requirements' content according to a rights-based and differentiated approach. The differentiated approach more particularly attends to three sectors of the population identified as vulnerable: Indigenous Peoples, people and communities which are economically and socially vulnerable, and women. We further complement this approach with environmental rights. They enable the addressing of the scenarios and categories applied in the research with more relevance. For example, the right to water enables trapped populations specific protection requirements.

Together, and despite the absence of a legal framework that explicitly addresses climate mobility, the differentiated, rights-based approach and environmental rights and related material and procedural rights make up a set of rights and obligations that apply to internal forced climate migration. The source of these rights is found in international, regional (Organization of American States), and national law. In our analysis of the national framework we only mention the Political Constitution of the United Mexican States. In the following two chapters, we perform a critical analysis of the Mexican political-legal framework. The protection requirements also guide the research carried out in the last chapter of the book. Based on the deficiencies identified in the Mexican legal framework and public policies, we identify the risks that flow from the regulation and public policies that could address internal climate migration in terms of protection requirements.

References

Adri N, Simon D (2018) A tale of two groups: focusing on the differential vulnerability of "climate-induced" and "non-climate-induced" migrants in Dhaka City. Climate Dev 10(4):321–336

Ambrosetti E, Petrillo ER (2016) Environmental disasters, migration and displacement. Insights and developments from L'Aquila's case. Environ Sci Policy 56:80–88

Anglés Hernández M (2014) Jurisprudencia interamericana: Acicate contra la discriminación y exclusión de pueblos originarios de México en relación con sus recursos naturales. Anu Mex de Derecho Int 14:261–299

Anglés Hernández M, Maldonado Madin MG (2020) La sequía como determinante del desplazamiento interno. Una mirada desde México. In: de Contreras MPP, Velázquez EO (eds) Migración forzada, derechos humanos y niñez. UNAM/IIJ, México, pp 111–130

Araya F, Faust KM, Kaminsky JA (2019) Public perceptions from hosting communities: the impact of displaced persons on critical infrastructure. Sustain Cities Soc 48:101508

Avcı D, Fernández-Salvador C (2016) Territorial dynamics and local resistance: two mining conflicts in Ecuador compared. Extr Ind Soc 3(4):912–921

Baldwin A (2016) Premediation and white affect: climate change and migration in critical perspective. Trans Inst Br Geogr 41(1):78–90

Baldwin A (2013) Racialisation and the figure of the climate-change migrant. Environ Plan A 45(6):1474–1490

Barkin D (2006) The new rurality: a framework for social struggle in the face of globalization. In: Ponencia presentada en la international conference on land, poverty, justice and development, pp 9–14

Becker SO, Ferrara A (2019) Consequences of forced migration: a survey of recent findings. Labour Econ

Bettini G, Andersson E (2014) Sand waves and human tides: exploring environmental myths on desertification and climate-induced migration. J Environ Dev 23(1):160–185

Birkenholtz T (2012) Network political ecology: method and theory in climate change vulnerability and adaptation research. Prog Hum Geogr 36(3):295–315

Blitz BK (2014) Location security and environmental-induced displacement: a case study of the Riverine Islands in Bangladesh. Refuge 29(2):63–74

Bradley M, Cohen R (2013) Disasters, Displacement And Protection: Challenges, Shortcomings And Ways Forward. In: Faist T, Schade J (eds) Disentangling migration and climate change. Springer, Dordrecht, pp 207–228

Burrows K, Pelupessy DC, Khoshnood K, Bell ML (2021) Environmental displacement and mental well-being in Banjarnegara, Indonesia. Environ Health Perspect 129(11). https://doi.org/10.1289/EHP9391

Cadena Iñiguez P, Camas Gómez R, López Báez W, Navarro Garza, H (2013) Practical and theoretical implications of the new rurality in Frailesca, Chiapas, Mexico. Rev Mex Ciencias Agrícolas 4(7), 1013-1026

Cancino MÀ, Dorantes–Díaz F, Anglès Hernandéz M, Coquis Velasco F, Rodríguez Rojas G (2018) Sistematización de normas en materia ambiental. Suprema Corte de la Nación, Serie: Derechos Humanos, 10

Castells-Quintana D, Lopez-Uribe MDP, McDermott TK (2021) Population displacement and urban conflict: global evidence from more than 3300 flood events. Universidad de los Andes, Colombia, Serie Documentos Cede, pp 2021–2026

Celorio M (2017) Desposesión de derechos humanos en el capitalismo contemporáneo. In: Estévez A, Vázquez D (eds) 9 razones para (des) confiar de las luchas por los derechos humanos. FLACSO, México, pp 263–291

Cissé G, McLeman R, Adams H, Aldunce P, Bowen K, Campbell-Lendrum D, Clayton S, Ebi KL, Hess J, Huang C, Liu Q, McGregor G, Semenza J, Tirado MC (2022) Health, wellbeing, and the changing structure of communities. In: Pörtner H-O, Roberts DC, Tignor M, Poloczanska ES, Mintenbeck K, Alegría A, Craig M, Langsdorf S, Löschke S, Möller V, Okem A, Rama B (eds) Climate change 2022: impacts, adaptation, and vulnerability. Contribution of working group II to the sixth assessment report of the intergovernmental panel on climate change. Cambridge University Press. https://www.ipcc.ch/report/ar6/wg2/

Clark N, Bettini G (2017) "Floods" of migrants, flows of care: between climate displacement and global care chains. Sociol Rev 65(2):36–54

Cohen R (2003) The role of protection in ending displacement. Forced Migr Rev 2003(17):21–23

Cohen IS, Spring ÚO, Padilla GD, Paredes JC, Inzunza Ibarra MA, López RL, Díaz JV (2013) Forced migration, climate change, mitigation and adaptive policies in Mexico: some functional relationships. Int Migr 51(4):53–72

Cohen R, Deng FM (2008) Orígenes y desafíos. RMF, GP10, pp 4–5

Comisión Nacional de Derechos Humanos (CNDH) (2011) Informe Especial sobre secuestro de migrantes en México

Comisión Nacional de Derechos Humanos (CNDH) (2012) Recomendación No. 37/2012 Sobre el caso del proyecto "Acueducto Independencia," en el Estado de Sonora

Comisión Nacional de Derechos Humanos (CNDH) (2018) Destaca CNDH gravedad del 65% de discriminación en México por origen étnico, y urge atender también pobreza, alimentación, vivienda, educación, salud y justicia para que las personas indígenas accedan a una vida digna. Comunicado de Prensa DGC/308/18. https://www.cndh.org.mx/sites/default/files/doc/Comunicados/2018/Com_2018_308.pdf. Accessed 30 Nov 2021

Committee on Economic, Social and Cultural Rights (CESCR) (2002) Substantive issues arising in the implementation of the international covenant on economic, social and cultural rights. General Comment No. 15: the right to water (Art.'s 11 and 12): 11/2000. E/C.12/2002/11

Comisión Económica para América Latina y el Caribe (CEPAL) (2018) Acuerdo Regional sobre el acceso a la información, la participación pública y el acceso a la justicia en asuntos ambientales en América Latina y el Caribe, adoptado el 4 de marzo de 2018

Constitución Política de los Estados Unidos Mexicanos (CPEUM) (2020) Last reform *Diario Oficial de la Nación*, 8 May 2020. http://www.diputados.gob.mx/LeyesBiblio/pdf_mov/Constitucion_Politica.pdf. Accessed 30 Nov 2021

Costa MM, Máñez KS, Paragay SH (2013) Adaptation to climate change under changing urban patterns: the climatic perspective of migration. In: Ruppel OC, Roschmann C, Ruppel-Schlichting K (eds) Climate change: international law and global governance. Nomos Verlagsgesellschaft mbH & Co. KG, pp 785–798

Cournil C (2010) Les défis du droit international pour protéger les "réfugiés climatiques:" réflexions sur les pistes actuellement proposées. In: Cournil C, Colard-Fabregoule C (eds) Changements climatiques et défis du droit. Bruylant, Brussels, pp 345–372

Cullen M (2020) Disaster, displacement and international law: legal protections in the context of a changing climate. Politics Gov 8(4):270–280

Décimo Octavo Tribunal Colegiado en Materia Administrativa del Primer Circuito (2013) Tesis I.18o.A.6 (10a), Registro: 2004891, Gaceta del Semanario Judicial de la Federación, Libro 26, noviembre de 2013a, Tomo II, p 1307

De Haas H (2010) Migration and development: a theoretical perspective 1. Int Migr Rev 44(1):227–264

Deng FM (2003) In the vacuum of sovereignty: the international challenge of internal displacement. Recuperado de: https://www.brookings.edu/articles/in-the-vacuum-of-sovereignty-the-international-challenge-of-internal-displacement/

Estévez A (2015) La crisis de derechos humanos y el dispositivo de administración del sufrimiento: necropolítica pública de víctimas, defensores y periodistas en México. El Cotidiano 194:7–17
Estévez A, Vázquez D (2017) 9 razones para (des) confiar de las luchas por los derechos humanos. FLACSO, Mexico
Fischer G, Shah M, Tubiello FN, Van Velhuizen H (2005) Socio-economic and climate change impacts on agriculture: an integrated assessment, 1990–2080. Philos Trans R Soc B: Biol Sci 360(1463):2067–2083
Francioni F (2010) International human rights in an environmental horizon. Eur J Int Law 21(1):41–55
Gamlen A, Bardsley DK, Wall J (2018) The evolution and impacts of Graeme Hugo's environmental migration research. Popul Environ 39(4):301–318
Garza-Guerra MT (2015) Derechos humanos y migración. In: Migrantes, transmigrantes, deportados y derechos humanos: un enfoque binacional, pp 231–243
Gemenne F (2011) Why the numbers don't add up: a review of estimates and predictions of people displaced by environmental changes. Glob Environ Chang 21:S41–S49
Gemenne F, Magnan A (2010) The other migrants, preparing for change. Environmental changes and migration in the Republic of Mauritius. IOM
Gemenne F, Blocher J (2017) How can migration serve adaptation to climate change? Challenges to fleshing out a policy ideal. Geogr J 183(4):336–347. https://doi.org/10.1111/geoj.12205
Global Protection Cluster Working Group (2007) Handbook for the protection of internally displaced persons. Global protection cluster working group. https://reliefweb.int/report/world/handbook-protection-internally-displaced-persons-enar#:~:text=The%20Handbook%20for%20the%20Protection,Geneva%20on%2024%20June%202010.&text=The%20Handbook%20also%20responds%20to,especially%20those%20on%20emergency%20deployments. Accessed 30 Nov 2021
Gómez GM, Mahecha OD (1998) Espacio, territorio y región: conceptos básicos para un proyecto nacional. Cuad Geogr: Rev Colomb Geogr 7(1–2):120–134
Gouritin A (2020) Towards a prevention-driven adaptation strategy, as applied to indigenous peoples' internal climate migration: some inputs based on a rights-based approach. In: Leal Filho W, Nagy GJ, Borga M, Muñoz PDC, Magnuszewski A (eds) Climate change, hazards and adaptation options. Springer International Publishing, Cham, pp 979–991
Gouritin A (2018) Human rights violations in the context of socio-environmental conflicts: wind farms in San Dionisio del Mar, Oaxaca (Mexico). In: Duyck S, Gouritin A (eds) The road from paris to sustainable development. Integrating human rights and gender equality into EU external policies. https://eu.boell.org/en/2018/06/12/road-paris-sustainable-development-integrating-human-rights-and-gender-equality-eu. Accessed 30 Nov 2021
Gouritin A (2016) EU environmental law, international environmental law, and human rights law: confirmation, complement and conflicts. In: The case of environmental responsibility. International environmental law series. Brill, Leiden
Gouritin A (2011) Potential liability of European States under the ECHR for failure to take appropriate measures with a view to adaptation to climate change. In: Faure M, Peeters M (eds) Climate change liability. Edward Elgar Publishing, Cheltenham, pp 134–164
Gouritin A, Aguilar A (2017) La adopción de la Declaración Americana sobre los Derechos de los Pueblos Indígenas: un análisis crítico desde el punto de vista de los derechos ambientales. Anu Mex Derecho Int 17:291–327
Haesbaert R (2011) El mito de la desterritorialización: del "fin de los territorios" a la multiterritorialidad. Siglo XXI, México
Hardoy J, Pandiella G (2009) Urban poverty and vulnerability to climate change in Latin America. Environ Urban 21(1):203–224
Hernández-Díaz J (2018) Avatars of community: the Zapotec migrants of the Zoogocho microregion. In: Robson JP, Klooster D, Hernández-Díaz J (eds) Communities surviving migration. Routledge, New York, pp 62–76

Hernández-Díaz J, Robson JP (2018) Migration dynamics and migrant organizing in rural Oaxaca. In: Robson JP, Klooster D, Hernández-Díaz J (eds) Communities surviving migration. Routledge, New York, pp 38–57

de Hidrocarburos L (2016) Diario Oficial de la Nación, Last reform 15 November 2016. http://www.diputados.gob.mx/LeyesBiblio/ref/lhidro/LHidro_orig_11ago14.pdf. Accessed 30 Nov 2021

High Commissioner for Human Rights (HCHR) (2001) El Derecho al Agua. Folleto informativo No 35. https://www.ohchr.org/documents/publications/factsheet35sp.pdf. Accessed 30 Nov 2021

High Commissioner for Human Rights (HCHR) (2010) Report of the representative of the secretary-general on the human rights of internally displaced persons, Walter Kälin Addendum framework on durable solutions for internally displaced persons. A/HRC/13/21/Add.4, p 1

Hodgkinson D, Young L (2013) In the face of looming catastrophe: a convention for climate change displaced persons. In: Gerrard MB, Wannier GE (eds) Threatened Island nations: legal implications of rising seas and a changing climate. Cambridge University Press, Nueva York, pp 299–336

Howden SM, Soussana JF, Tubiello FN, Chhetri N, Dunlop M, Meinke H (2007) Adapting agriculture to climate change. Proc Natl Acad Sci 104(50):19691–19696

Hugo G (2008) Migration, development and environment. Geneva: international organization for migration. https://publications.iom.int/books/mrs-no-35-migration-development-and-environment. Accessed 30 Nov 2021

Hunter LM, Kining S, Fussell E, King B, Rishworth A, Merdianoff A, Muttarak R, Riosmena F, Simon DH, Skop E, Van Den Hoek J (2021) Scales and sensitivities in climate vulnerability, displacement, and health. Popul Environ 43:61–81. https://doi.org/10.1007/s11111-021-00377-7

Ibarra Sarlat R (2021) Desplazados climáticos. Evolución de su reconocimiento y protección jurídica. UNAM, Mexico

Inter-American Commission on Human Rights (IAComHR) (2009) Indigenous and tribal people's rights over their ancestral lands and natural resources: norms and jurisprudence of the Inter-American human rights system, OEA/Ser.L/V/II, Doc. 56/09. https://www.oas.org/en/iachr/indigenous/docs/pdf/AncestralLands.pdf. Accessed 30 Nov 2021

Inter-American Commission on Human Rights (IAComHR) (2013) Derechos humanos de los migrantes y otras personas en el contexto de la movilidad humana en México, OEA/Ser.L/V/II. Doc 48/13. http://www.oas.org/es/cidh/migrantes/docs/pdf/informe-migrantes-mexico-2013.pdf. Accessed 30 Nov 2021

Inter-American Commission on Human Rights (IAComHR) (2015) Derechos Humanos de migrantes, refugiados, apátridas, víctimas de trata de personas y desplazados internos: normas y estándares del Sistema Interamericano de Derechos Humanos, OEA/Ser.L/V/II. Doc 46/15. http://www.oas.org/es/cidh/informes/pdfs/movilidadhumana.pdf. Accessed 30 Nov 2021

Inter-American Commission on Human Rights (IAComHR) (2018) Internal displacement in the Northern triangle of central America: public policy guidelines: approved by the inter-American commission on human rights on July 27, 2018, OAS/Ser.L/V/II. Doc. 101. https://www.oas.org/en/iachr/reports/pdfs/InternalDisplacement.pdf. Accessed 30 Nov 2021

Inter-American Court of Human Rights (IACtHR) (2001) Comunidad Mayagna (Sumo) Awas Tingni vs. Nicaragua. Merits, reparations and costs. Judgment of August 31, 2001. Serie C, No. 79

Inter-American Court of Human Rights. (IACtHR) (2005a) Comunidad indígena Yakye Axa vs. Paraguay. Merits, reparations and costs. Judgment of June 17, 2005. Serie C, No. 125

Inter-American Court of Human Rights (IACtHR) (2005b) Yatama vs. Nicaragua. Preliminary objections, merits, reparations and costs. Judgment of June 23, 2005b. Serie C, No. 127

Inter-American Court of Human Rights (IACtHR) (2005c) Comunidad Moiwana vs. Surinam. Preliminary objections, merits, reparations and costs. Judgment of June 15, 2005c. Serie C No. 124

Inter-American Court of Human Rights (IACtHR) (2006) Comunidad Indígena Sawhoyamaxa vs. Paraguay. Merits, reparations and costs. Judgment of March 29, 2006. Serie C No. 146

Inter-American Court of Human Rights (IACtHR) (2007) Pueblo Saramaka vs. Surinam. Preliminary objections, merits, reparations and costs. Judgment of November 28, 2007. Serie C, No. 172

Inter-American Court of Human Rights (IACtHR) (2008) Pueblo Saramaka vs. Surinam. Interpretation of the judgment on preliminary objections, merits, reparations and cost. Judgment of August 12, 2008. Serie C, No. 172

Inter-American Court of Human Rights (IACtHR) (2010a) Chitay Nech y otros vs. Guatemala. Preliminary objections, merits, reparations and costs. Judgment of May 25, 2010a. Serie C, No. 212

Inter-American Court of Human Rights (IACtHR) (2010b) Comunidad Indígena Xákmok Kásek vs. Paraguay. Merits, reparations and costs. Judgment of August, 24, 2010b. Serie C, No. 214

Inter-American Court of Human Rights (IACtHR) (2012a) Pueblo Indígena Kichwa de Sarayaku vs. Ecuador. Merits and Reparations. Judgment of June 27, 2012a. Serie C, No. 245

Inter-American Court of Human Rights (IACtHR) (2012b) Masacre de Santo Domingo vs. Colombia. Preliminary objections, merits, reparations and costs. Judgment of November 30, 2012b. Serie C, No. 259

Inter-American Court of Human Rights (IACtHR) (2013) Comunidades Afrodescendientes desplazadas de la Cuenca del Río Cacarica (operación génesis) vs. Colombia. Preliminary objections, merits, reparations and costs. Judgment of November 20, 2013. Serie C, No. 270

Inter-American Court of Human Rights (IACtHR) (2014) Pueblos Indígenas Kuna de Madungandí y Emberá de Bayano y sus miembros vs. Panamá. Preliminary objections, merits, reparations and costs. Judgment of October 14, 2014. Serie C, No. 284

Inter-American Court of Human Rights (IACtHR) (2015a) Comunidad Garífuna de Punta Piedra y sus miembros vs. Honduras. Preliminary objections, merits, reparations and costs. Judgment of October 8, 2015a. Serie C, núm. 304

Inter-American Court of Human Rights (IACtHR) (2015b) Comunidad Garífuna Triunfo de la Cruz y sus miembros vs. Honduras. Merits, reparations and costs. Judgment of October 8, 2015b. Serie C, núm. 305

Inter-American Court of Human Rights (IACtHR) (2015c) Pueblos Kaliña y Lokono vs. Surinam. Merits, reparations and costs. Judgment of November 25, 2015c. Serie C, No. 309

Inter-American Court of Human Rights (IACtHR) (2016a) Resolution. Solicitud de Medidas Provisionales respecto de Nicaragua. Asunto Pobladores de las Comunidades del Pueblo Indígena Miskitu de la Región Costa Caribe Norte. September 1, 2016. http://www.corteidh.or.cr/docs/medidas/miskitu_se_01.pdf. Accessed 30 Nov 2021

Inter-American Court of Human Rights (IACtHR) (2016b) Miembros de la Aldea Chichupac y comunidades vecinas del Municipio de Rabinal vs. Guatemala. Preliminary objections, merits, reparations and costs. Judgment of November 30, 2016b. Serie C No. 328

Inter-American Court of Human Rights (IACtHR) (2017) The environment and human rights (State obligations in relation to the environment in the context of the protection and guarantee of the rights to life and to personal integrity – interpretation and scope of Articles 4(1) and 5(1) of the American Convention on Human Rights). Advisory opinion OC-23/17 of November 15, 2017. Series A, No. 23

Inter-American Court of Human Rights (IACtHR) (2018) Pueblo Indígena Xucuru y sus miembros vs. Brasil. Preliminary objections, merits, reparations and costs. Judgment of February 5, 2018. Serie C, No. 346

Inter-American Court of Human Rights (IACtHR) (2020) Comunidades Indígenas Miembros de la Asociación Lhaka Honhat (Nuestra Tierra) vs. Argentina. Merits, Reparations and Costs. Judgment of February 6, 2020. Serie C, No. 400

International Labour Organization (ILO) (1989) International labour organization indigenous and tribal peoples convention 169, 27 June 1989. https://www.senado.gob.mx/comisiones/desarrollo_social/docs/marco/Convenio_169_PI.pdf. Accessed 30 Nov 2021

Kälin W (2001) How hard is soft law? The guiding principles on internal displacement and the need for a normative framework. Unpublished presentation given during the Ralph Bunche International Relations Roundtable Meeting, vol 19
Kälin W (2002) Comment on "when does displacement end?" The legal dimension. Washington: brookings. https://www.brookings.edu/research/comment-on-when-does-displacement-end-the-legal-dimension/. Accessed 30 Nov 2021
Kälin W (2008a) The climate change–displacement nexus. Washington: brookings. https://www.brookings.edu/on-the-record/the-climate-change-displacement-nexus/#:~:text=In%20the%20course%20of%20the,of%20recorded%20disasters%20has%20doubled. Accessed 30 Nov 2021
Kälin W (2008b) Guiding principles on internal displacement. Stud Transnatl Leg Policy 32:1–75
Kälin W (2019) International responses to climate-related migration. J Int Aff 73(1):255–260
Kalra N, Chander S, Pathak H, Aggarwal PK, Gupta NC, Sehgal M, Chakraborty D (2007) Impacts of climate change on agriculture. Outlook Agric 36(2):109–118
Kay C (2009) Estudios rurales en América Latina en el periodo de globalización neoliberal: una nueva ruralidad? Rev Mex Sociol 71(4):607–645
Kay C (2015) The agrarian question and the neoliberal rural transformation in Latin America. In: European review of Latin American and Caribbean Studies/Revista Europea de Estudios Latinoamericanos y del Caribe, pp 73–83
Kols A, Lewison D (1983) Migration, population growth, and development, Population Reports, M-7. Baltimore: John Hopkins University. Population Information Program, September-October
Kothari M (2007) Basic principles and guidelines on development-based evictions and displacement. Annex 1 of the report of the Special Rapporteur on adequate housing as a component of the right to an adequate standard of living A. HRC/4/18. https://www.ohchr.org/Documents/Issues/Housing/Guidelines_sp.pdf. Accessed 30 Nov 2021
Koubi V, Nguyen Q, Spilker G, Böhmelt T (2021) Environmental migrants and social-movement participation. J Peace Res 58(1):1–15
Lagos-Gallego M, Gutierrez-Segura JC, Lagos-Grisales GJ, Rodriguez-Morales AJ (2017) Post-traumatic stress disorder in internally displaced people of Colombia: an ecological study. Travel Med Infect Dis 16:41–45
Lombard M (2014) Constructing ordinary places: place-making in urban informal settlements in Mexico. Prog Plan 94:1–53
López RV (2007) Derecho global y desplazamiento interno: creación, uso y desaparición del desplazamiento forzado por la violencia en el derecho contemporáneo. Pontificia Universidad Javeriana
Marshall N (2016) Forced environmental migration: ethical considerations for emerging migration policy. Ethics Policy Environ 19(1):1–18
Martínez Quezada JÁ, Ramírez O, Sletza A (2015) Internalización de normas internacionales en materia de derechos de migrantes. El caso dee Chiapas. Norteamérica 10(1):191–213
Mayer B (2012) Sustainable development law on environmental migration: the story of an obelisk, a bag of marbles, and a tapestry. Environ Law Rev 14(2):111–133
McAdam J (2014) Creating new norms on climate change, natural disasters and displacement: international developments 2010–2013. Refug: Canada's J Refug 29(2):11–26
McGranahan G, Balk D, Anderson B (2007) The rising tide: assessing the risks of climate change and human settlements in low elevation coastal zones. Environ Urban 19(1):17–37
McMichael C, Barnett J, McMichael AJ (2012) An Ill wind? Climate change, migration, and health. Environ Health Perspect 120(5):646–654. https://doi.org/10.1289/EHP9391
Mooney E (2003) Bringing the end into sight for internally displaced persons. Forced Migr Rev 17:4–7
Morales Sánchez J (2019) Migración irregular y Derechos Humanos. Tirant Lo Blanch
Nelson GC, Rosegrant MW, Koo J, Robertson R, Sulser T, Zhu T, Ringler T, Msangi S, Palazzo A, Batka M, Magalhaes M, Valmonte-Santos R, Ewing M, Lee D (2009) Climate change: impact on agriculture and costs of adaptation, vol 21. Intl Food Policy Res Inst.

Obeng-Odoom F (2017). Unequal access to land and the current migration crisis. Land Use Policy 62:159–171
Okeowo D (2013) Examining the link: climate change. Environ Law Rev 15(4):273–289
Organization of American States (OAS) (1948a) Inter-American convention on the granting of civil rights to women, May 2, 1948a. http://www.oas.org/juridico/spanish/Tratados/a-45.html. Accessed 30 Nov 2021
Organization of American States (OAS) (1948b) Inter-American convention on the granting of political rights to women, May 2, 1948b. http://www.oas.org/es/cim/docs/CPRW%5BSP%5D.pdf. Accessed 30 Nov 2021
Organization of American States (OAS) (1969) American convention on human rights ("Pact of San José"), November 22, 1969. https://www.refworld.org.es/docid/57f767ff14.html. Accessed 30 Nov 2021
Organization of American States (OAS) (1988) Additional protocol to the American convention on human rights in the area of economic, social and cultural rights ("Protocol of San Salvador"), November 17, 1988. http://www.ordenjuridico.gob.mx/TratInt/Derechos%20Humanos/PI2.pdf. Accessed 30 Nov 2021
Organization of American States (OAS) (1994) Inter-American convention on the prevention, punishment, and eradication of violence against women ("Convention of Belem Do Para"), June 9, 1994. https://www.unhcr.org/47556c132.pdf. Accessed 30 Nov 2021
Organization of American States (OAS) (2007) Internally displaced persons. General assembly resolution. AG/RES. 2277 (XXXVII-O/07). August 13, 2007. https://www.oas.org/dil/esp/AG-RES_2277_XXXVII-O-07_spa.pdf. Accessed 30 Nov 2021
Organization of American States (OAS) (2011) Internally displaced persons. General assembly resolution. Resolución de la Asamblea General. AG/RES. 2667 (XLI-O/11). June 7, 2011. https://www.oas.org/dil/esp/AG-RES_2667_XLI-O-11_esp.pdf. Accessed 30 Nov 2021
Organization of American States (OAS) (2014) Internally displaced persons. Desplazados internos. General assembly resolution. AG/RES. 2850 (XLIV-O/14). June 4, 2014. http://www.oas.org/en/sla/docs/ag06712e04.pdf. Accessed 30 Nov 2021
Organization of American States (OAS) (2016) American declaration on the rights of indigenous peoples, AG/RES. 2888 (XLVI - O/16), June 14, 2016. https://www.oas.org/es/sadye/documentos/res-2888-16-es.pdf. Accessed 30 Nov 2021
Pascoe S (2015) Sailing the waves on our own: climate change migration, self-determination and the Carteret Islands. QUT l. Rev 15:72–85
Peguero Moreno M, Laura RDL (2019) El desplazamiento interno forzado en México: respuestas del estado y litigio estratégico. Tirant lo blanch, México
Petrova K (2021) Natural hazards, internal migration and protests in Bangladesh. J Peace Res 1–17
Piao S, Ciais P, Huang Y, Shen Z, Peng S, Li J, Zhou L, Liu H, Ma Y, Ding Y, Friedlingstein P, Liu C, Tan K, Yu Y, Zhang T, Fang J (2010) The impacts of climate change on water resources and agriculture in China. Nature 467(7311):43
Pisani E, Franceschetti G (2011) Territorial approaches for rural development in Latin America: a case study in Chile. Rev Fac Ciencias Agrar 43(1):201–218
Prieur M, Marguenaud JP, Monédiaire G, Bétaille J, Drobenko B, Gouguet JJ, Lavieille JM, Nadaud S, Roets D, Bouin F, Burgat F, Cournil C, Dinh V, Juste J, Lador Y, Mazzega P, Michelot A, Shelton D (2008) Projet de convention relative au statut international des déplacés environnementaux. Rev Eur Droit l'environnement 12(12):381–393. https://www.usherbrooke.ca/droit/fileadmin/sites/droit/documents/RDUS/volume_39/39-12-convention.pdf. Accessed 30 Nov 2021
Primer Tribunal Colegiado en Materia Administrativa del Primer Circuito (2017) Tesis IV.1o.A J/25 (10a), Registro: 2013745, *Gaceta del Semanario Judicial de la Federación*, Libro 39, February 2017, Tomo III, p 2112
Primer Tribunal Colegiado en Materia Administrativa del Primer Circuito (2013) Tesis XI.7o.A.T.10 K (10a), Registro: 2004606, *Gaceta del Semanario Judicial de la Federación*, Libro 24, September 2013b, Tomo III, p 2692

Radel C, Schmook B, Carte L, Mardero S (2018) Toward a political ecology of migration: land, labor migration, and climate change in northwestern Nicaragua. World Dev 108:263–273
Radonic L (2017) Through the aqueduct and the courts: an analysis of the human right to water and indigenous water rights in Northwestern Mexico. Geoforum 84:151–159
Raleigh C, Jordan L, Salehyan I (2008) Assessing the impact of climate change on migration and conflict. Paper commissioned by the world bank group for the social dimensions of climate change workshop, Washington, DC
Ramírez-Miranda C (2014) Critical reflections on the New Rurality and the rural territorial development approaches in Latin America. Agron Colomb 32(1):122–129
Rana Md. MP, Ilina IN (2021) Climate change and migration impacts on cities: lessons from Bangladesh. Environ Challs 5. https://doi.org/10.1016/j.envc.2021.100242
Reuveny R (2005) Environmental change, migration and conflict: theoretical analysis and empirical explorations. In: Human security and climate change: an international workshop (21–23 June 2005), vol 2
Reuveny R, Moore WH (2009) Does environmental degradation influence migration? Emigration to developed countries in the late 1980s and 1990s. Soc Sci Q 90(3):461–479
Rigaud KK, de Sherbinin A, Jones B, Bergmann J, Clement V, Ober K, Schewe J, Adamo S, McCusker B, Heuser S, Midgley A (2018) Groundswell: preparing for internal climate migration. World Bank, Washington D.C.
Roberts RE, Okanya O (2020) Measuring the socio–economic impact of forced evictions and illegal demolition; a comparative study between displaced and existing informal settlements. Soc Sci J 1–20
Robson JP, Klooster D (2018a) Communities surviving migration? The migration–community–environment nexus. Routledge. In: Robson JP, Klooster D, Hernández-Díaz J (eds) Communities surviving migration. Routledge, New York, pp 3–21
Robson JP, Klooster D (2018b) The changing landscapes of Indigenous Oaxaca. In: Robson JP, Klooster D, Hernández-Díaz J (eds) Communities surviving migration. Routledge, New York, pp 156–175
Romero Lankao P (2010) Water in Mexico City: what will climate change bring to its history of water-related hazards and vulnerabilities? Environ Urban 22(1):157–178
Rubio Díaz-Leal L (2017) Desplazamiento ambiental. Experiencia global, realidad mexicana. Comisión Mexicana de Defensa y Promoción de los Derechos Humanos, Heinrich Böll Stiftung México, Centroamérica y el Caribe. http://cmdpdh.org/project/desplazamiento-ambiental-experiencia-global-realidad-mexicana/. Accessed 30 Nov 2021
Ruiz SA, Carvajal M (2015) Hacia el desarrollo de políticas pública locales en movilidad humana, en el contexto de desastres naturales y cambio climático: El caso del Gobierno de la Provincia de Pichincha. Proyecto PRODEM-GIZ Migración y Asilo, Gobierno Autónomo Descentralizado de la Provincia de Pichincha (Quito, Ecuador)
Sánchez Mojica BE (2009) Cuando los derechos son la jaula. Trasplante rígido del soft law para la gestión del desplazamiento forzado. Estud Políticos 35:11–21
Schade J (2012) Les migrants des politiques climatiques: nouveaux défis face aux déplacements générés par le changement climatique. Cult Confls 88:85–110
Schütte S, Gemenne F, Zaman M, Flahault A, Depoux A (2018) Connecting planetary health, climate change, and migration. Lancet Planet Health 2(2):e58–e59
Scott M, Salamanca A (2020) A human rights-based approach to internal displacement in the context of disasters and climate change. Refug Surv Q 39(4):564–571
Segundo Tribunal Colegiado en Materia Administrativa del Primer Circuito (2007) Tesis I.4o.A.569 A, Registro: 173049, Gaceta del Semanario Judicial de la Federación, March 2007, Tomo XXV, p 1665
Secretaría de Medio Ambiente y Recursos Naturales (SEMARNAT), & Instituto Nacional de Ecología y Cambio Climático (INECC) (2018) Sexta Comunicación Nacional y Segundo Informe Bienal de Actualización ante la Convención Marco de las Naciones Unidas sobre el Cambio

Climático. https://cambioclimatico.gob.mx/sexta-comunicacion/introduccion.php. Accessed 30 Nov 2021

Séptimo Tribuna Colegiado en Materia Administrativa del Primer Circuito (2016a) Tesis I.7o.A J/6 (10a), Registro: 2012126, Gaceta del Semanario Judicial de la Federación, Libro 32, July 2016a, Tomo III, p 1801

Séptimo Tribunal Colegiado en Materia Administrativa del Primer Circuito (2016b) Tesis I.7o.A J/7 (10a), Registro: 2012127, *Gaceta del Semanario Judicial de la Federación*, Libro 32, July 2016b, Tomo III, p 1802

Skillington T (2015) Climate justice without freedom: assessing legal and political responses to climate change and forced migration. Eur J Soc Theory 18(3):288–307

Solutions D (2016) Peninsula principles on climate displacement within states (2013). http://displacementsolutions.org/wp-content/uploads/2014/12/Peninsula-Principles.pdf. Accessed 30 Nov 2021

Stojanov R, Rosengaertner S, de Sherbinin A, Nawrotzki R (2021) Climate mobility and development cooperation. Popul Environ 43(2):209–232. https://doi.org/10.1007/s11111-021-00387-5

Suprema Corte de Justicia de la Nación, Primera Sala (SCJN) (2014) Tesis 1a./ LXIX/2014 (10a), Registro: 2005624, *Gaceta del semanario Judicial de la Federación*, Libro 3, February 2014, Tomo I, p 648

Surrallés A, Hierro PG (eds) (2004) Tierra adentro: territorio indígena y percepción del entorno (No. 39). IWGIA critican los riesgos de despojo que conlleva esta concepción restrictiva

Tebas JM, García MDMH (2016) La incidencia del cambio climático en las migraciones y en la seguridad. Tiempo De Paz 120:44–55

Thompson S, Ballard M, Martin D (2014) Lake St. Martin First Nation community members' experiences of induced displacement: "we're like refugees." Refug: Canada's J Refug 29(2):75–86

Torres JM, Casey JA (2017) The centrality of social ties to climate migration and mental health. BMC Public Health 17(1):600

United Nations (UN) (1948) Universal declaration of human rights, December 10, 1948. https://www.un.org/es/documents/UDHRr/UDHRR_booklet_SP_web.pdf. Accessed 30 Nov 2021

United Nations (UN) (1952) United Nations convention on the political rights of women, December 20, 1952. https://www.oas.org/dil/esp/convencion_sobre_los_derechos_politicos_de_la_mujer.pdf. Accessed 30 Nov 2021

United Nations (UN) (1956) Supplementary convention on the abolition of slavery, the slave trade, and institutions and practices similar to slavery, September 7, 1956. http://www.ordenjuridico.gob.mx/TratInt/Derechos%20Humanos/D25.pdf. Accessed 30 Nov 2021

United Nations (UN) (1962) Convention on consent to marriage, minimum age for marriage and registration of marriages, December 10, 1962. http://www.ordenjuridico.gob.mx/TratInt/Derechos%20Humanos/D29.pdf. Accessed 30 Nov 2021

United Nations (UN) (1966a) International covenant on civil and political rights, December 16, 1966a. https://www.ohchr.org/SP/ProfessionalInterest/Pages/CCPR.aspx. Accessed 30 Nov 2021

United Nations (UN) (1966b) Pacto Internacional de Derechos Económicos, Sociales y Culturales, adoptado el 16 de diciembre de 1966b. https://www.ohchr.org/SP/ProfessionalInterest/Pages/CESCR.aspx. Accessed 30 Nov 2021

United Nations (UN) (1979) Convention on the elimination of all forms of discrimination against women, December 18, 1979. https://www.ohchr.org/sp/professionalinterest/pages/cedaw.aspx. Accessed 30 Nov 2021

United Nations (UN) (1992) Rio declaration on environment and development, August 12, 1992. https://www.un.org/spanish/esa/sustdev/agenda21/riodeclaration.htm. Accessed 30 Nov 2021

United Nations (UN) (1998a) Rome statute of the international criminal court. A/CONF.183/9, July 17, 1998a. 17 June 1998a. https://www.un.org/spanish/law/icc/statute/spanish/rome_statute(s).pdf. Accessed 30 Nov 2021

United Nations (UN) (1998b) Guiding principles on internal displacement. UN doc.E/CN.4/1998b/53/Add.2, February 11, 1998b. https://www.acnur.org/prot/prot_despl/5bff2c864/principios-rectores-de-los-desplazamientos-internos.html. Accessed 30 Nov 2021

United Nations (UN) (2005) Housing and property restitution for refugees and displaced persons principles, the resolution 2005/21 adopted on August 11th 2005. https://www.acnur.org/fileadmin/Documentos/Publicaciones/2008/6325.pdf. Accessed 30 Nov 2021

United Nations (UN) (2007) Declaration on the rights of indigenous peoples, A/RES/61/295. September 13, 2007. https://www.un.org/esa/socdev/unpfii/documents/DRIPS_es.pdf. Accessed 30 Nov 2021

United Nations (UN) (2020) Human rights and the environment. (A 3-Part Series of Nexus Dialogues). https://unemg.org/wp-content/uploads/2020/10/EMG_ND_HumanRightsEnv_Series_Integrated_Outcome_Report.pdf. Accessed 30 Nov 2021

Valle LM (2012) Apuntes para pensar el territorio desde una dimensión social. Ciências Sociais Unisinos 48(1):12–18

van der Vliet JA (2018) Climate refugees. A legal mapping exercise. In: Behrman S, Kent A (eds) Climate refugees. Routledge/GSE Research, London, pp 16–33

Vidal López R (2007) Derecho global y desplazamiento interno: creación, uso y desaparición del desplazamiento forzado por la violencia en el derecho contemporáneo. Pontificia Universidad Javeriana, Bogotá

Yates OE, Manuela S, Neef A, Groot S (2021) Reshaping ties to land: a systematic review of the psychosocial and cultural impacts of Pacific climate-related mobility. Clim Dev 1–18. https://doi.org/10.1080/17565529.2021.1911775

Zetter R (2009) The role of legal and normative frameworks for the protection of environmentally displaced people. In: Migration, environment and climate change: assessing the evidence, p 389

Zickgraf C (2018) Immobility. In: McLeman R, Gemenne F (eds) Routledge handbook of environmental displacement and migration. Routledge/GSE Research, London, pp 71–84

Chapter 3
Shortcomings of Legal Frameworks and Public Policies

Armelle Gouritin

In this chapter, we critically analyze the Mexican political-legal framework to determine whether or not it covers generically forced internal climate migration. In order to do this we draw on what we established in the two previous chapters: six scenarios that generate migration, the identification of three particularly vulnerable populations (Indigenous Peoples, women and people with limited economic resources), the definition and criteria that apply to the three categories of forced climate migrants, and the analytical grid (territory and protection requirements).

We subsequently discuss the limits of current academic research (1) and critically analyze the current legal framework in Mexico (2). Based on the shortcomings identified, we also consider another question: Do public policies consider the phenomenon even though legal frameworks fail to address it? Our examination of public policies concludes that this is indeed not the case. Public policies cover various aspects of internal climate mobilities, but do not address them as such (3). We do not analyze all of the laws in force in Mexico or relevant public policy instruments. Such analysis would be the subject of a book in itself. Instead, we select the standards and instruments that appear to be most relevant and have the most potential to address the phenomenon.

3.1 State-of-the-Art: The Shortcomings of Current Academic Knowledge

This chapter intends to fill a seemingly large void. So far, academics and international organizations have scarcely touched upon internal climate migrations in Mexico from the public policies and legal framework perspectives.

Studies that address forced internal migration in Mexico tend to focus on two mobility triggers: armed conflicts and situations of generalized violence (Velázquez Moreno 2017; Salazar Cruz 2014; Mondragón Market 2016; Mondragón Market 2018; Salazar and Alvarez Lobato 2018; Querales-Mendoza 2020). In addition,

A. Gouritin, *Climate Displacement in Mexico*,
https://doi.org/10.1007/978-3-031-10335-3_3

existing literature mainly aims to establish empirically international climate migrations with Mexico as a country of origin or transit to the United States of America (examples include, Papail 2002; Moctezuma 2002; Moraga 2013; Saldaña-Zorilla 2007; Schwartz and Notini 1994; Nawrotzki et al. 2013; Hunter et al. 2011; Feng et al. 2010).

Schwartz and Notini (1994) studied the relationship between desertification in Mexico and international migration. They identified poverty and economic inequality as essential factors in the decision to migrate from Mexico to the United States. They also found evidence that Mexican migration may be related to desertification, which occurs due to various factors (unsustainable cultivation techniques, logging, inefficient irrigation practices, mining, and urban expansion). Similarly, Hunteret al. (2011) investigated the link between changes in precipitation patterns and their influence on the impulse to migrate to the United States. They found migration can be an adaptation strategy to natural disasters for populations with a migratory tradition, suggesting that the reduction of natural environmental capital can increase the probability of migration. Feng et al. (2010) agree: they found a statistically significant link between climate change that affects yields in corn and wheat crops and migration rates to the United States. Considering climate change projections for 2080 (a temperature increase of 1–3 °C), climate change could drive between 1.4 and 6.7 million Mexican adults to migrate (between 2 and 10% of the Mexican population).

However, Nawrotzki et al. (2013) found less of a strong probability. They studied the link between changes in rainfall and migration to the United States between 2005 and 2009. They reported mixed results: the relationship between the environment and migration depends not only on climate scenarios but also on economic conditions in Mexico and the United States. Therefore, the forecast for future international migration from Mexico related to climate change is full of uncertainty.[1] Kniveton et al. (2008) studied rainfall patterns and their influence on migration in the state of Durango. They found a relationship between rainfall and migration to the United States, suggesting that, as more rainfall occurs, more migration is taking place. However, Kaivan Munshi found a negative correlation between rainfall and migration to the United States in the southwestern region of Mexico (Munshi 2003, cited in Kniveton et al. 2008, p. 3). Based on the above it appears that there is a relationship between precipitation and migration, but its nature is as yet unclear.

In contrast, few studies address internal climate migration in Mexico. They generally focus on the empirical link between climate change and internal migration (Albo and Ordaz 2011; Alscher 2010; Ruiz Meza 2012; Lupián and Campos 2015; Nawrotzki et al. 2015; Deheza 2013). In particular, Alscher (2010) analyzed the environmental factors that affect displacement in Chiapas and Tlaxcala. He found that, in Chiapas, many residents were already contemplating migrating due to environmental degradation (slow-onset phenomenon scenario), and two hurricanes in both 1998 and 2005 accelerated this decision. He also found age is a significant factor in the decision to migrate (young people had already migrated while the elderly

[1] Similarly, Riosmena et al. (2018) conclude with less categorical results than previous academic works regarding climate change-driven international migration (Mexico-United States).

preferred to stay in their communities). In Tlaxcala, soil erosion and disturbance to rainfall patterns were the environmental factors mentioned as factors to migration additional to economic considerations.

Even fewer studies analyze the political-legal framework that applies to the phenomenon (we identify Deheza 2013; Hardoy et al. 2014; Mercado Mondragón 2016; Ruiz Meza 2012; Pérez 2014). Ruiz Meza (2012) carried out an analysis in Chiapas and found a link between climate change, poverty, and migration along the southern Mexican border, where extreme hydrometeorological events have increased displacement. He also signals current climate change policies tend to focus on responding to emergencies, rather than reversing environmental degradation and social vulnerability and goes on to conclude that climate change policies and the design of programs should acknowledge international treaties on migration, labor, human rights, and food sovereignty.

Hardoy et al. (2014) lament that although a growing number of cities and municipal governments have adopted plans and programs on climate change adaptation, relatively few of those have made significant progress in the more complex task of implementing adaptation strategies. They recognize that Mexican leaders signed international agreements on the environment and climate change and adopted the General Law on Climate Change (LGCC 2020) and State Programs of Action against Climate Change. However, the authors also criticize the lack of specific processes for achieving climate change mitigation and adaptation goals. Additionally, they point out that Mexican public policies addressing migration and climate change are currently largely reactive rather than preventive.

In line with the above, Mercado Mondragón (2016) points out several deficiencies. He mentions the absence of public policies and a similar lack of specific definition of displacement (that is, not specifically climate related). He also signals that the phenomenon is still not officially recognized: it is classified as an economic migration process rather than a product of generalized violence or human rights violations, and public policies should be designed to serve the displaced population and protect them according to the framework of the Guiding Principles on Internally Displaced Persons (UN 1998).

Finally, Pérez (2014) reflects on why, contrary to Colombia, there is no national law to address internal displacement (not only climate displacement) in Mexico. She concludes that various public and social organizations do not coordinate and exchange information in order to identify and therefore partially address internal displacement in Mexico. The intervention of the authorities is elusive since the legal framework does not clearly define which institutions should address the phenomenon. Consequently, it is impossible to design and implement programs to address and reverse the phenomenon. Likewise, she mentions that the absence of institutional attribution generates a lack of information on the phenomenon.

To summarize the gaps in current knowledge on forced internal climate migration in Mexico, current studies focus primarily on international climate mobility (and, more specifically, migration from Mexico to the United States). Academics mainly dedicate themselves to establishing the link between generalized violence and forced internal mobility. This is understandable given the systemic violence that occurs in

Mexico. Violence was identified as a factor in all the cases of massive forced internal displacement in 2018 (11 491 persons). In most cases, the direct and primary cause of displacement was the violence of organized armed groups (6 156 persons) (CMDPDH 2019b, p. 20). Finally, the few studies that address the climate dimension of internal mobility in Mexico are mainly devoted to merely establishing it empirically. They do not consider the public policies or the legal framework that apply (or could apply) to the phenomenon. This is precisely what we intend to do in this chapter.

3.2 Deficiencies in the Mexican Legislative Framework

To determine whether the Mexican legal framework contemplates internal forced climate migrations and provides protection to people and communities affected we review the federal framework (Sect. 3.2.1), the Initiative for a General Law Internal Displacement (Sect. 3.2.2), and the two federate laws that address the phenomenon of forced internal migration (Sect. 3.2.3).

3.2.1 Critical Analysis of Mexican Norms at Federal Level

In this section, we identify the norms that have a link to forced internal climate migration and go on to classify them into three groups according to the recognition of the phenomenon and the protection afforded to populations which have been affected. We distinguish norms that ignore the phenomenon and the protection of affected people and communities (Sect. 3.2.1.1) and those that have the potential to address the phenomenon despite their literal content (Sect. 3.2.1.2). Eventually, we identify the norms that recognize the phenomenon and their limitations (Sect. 3.2.1.3).

Norms Which Ignore Forced Internal Climate Mobility

The three laws that contemplate migratory phenomena do not address internal mobility. The Law on Refugees, Complementary Protection, and Political Asylum (*Ley sobre Refugiados, Protección Complementaria y Asilo Político* 2014) does not address internal mobility. The definitions it provides in Article 2 limit its scope to people who come from other countries. The Migration Law (*Ley de Migración*, LM 2020) sheds much light on the principles of the Mexican migratory framework (Morales Vega 2012) and defines a "migrant" as being:

> the individual who for any reason leaves, transits, or arrives in a State's territory other than the State of his residence. (LM 2020, Article 3, Fr. XVII)

The Law does not cover internal migrants and instead focuses on international migration. Article 1 confirms this: the Law "aims to regulate the entry and exit of Mexicans and foreigners to (the Mexican) territory and the transit and stay of foreigners therein" (Article 1).

The General Population Law (*Ley General de Población*, LGP 2018) could address forced internal climate mobilities as its purpose is to "regulate the phenomena that affect the population in terms of their volume, structure, dynamics, and distribution in the national territory" (Article 1). Despite mentioning disasters (Article 3 Fr. XIII, sudden-onset disaster scenario, forced climate emergency migration category), it exclusively addresses international mobility (Articles 10, 76–84).

The General Law of Ecological Balance and Environmental Protection (*Ley General del Equilibrio Ecológico y la Protección al Ambiente*, LGEEPA 2018) is the Mexican framework Law on environmental matters. It defines climate change (in Article 3, Fr. V bis) and indicates the respective attributes of the Federation, states, and Municipalities regarding the formulation of climate change mitigation and adaptation actions and implementation (Article 5 Fr. XXI; Article 7 Fr XXI; and Article 8 Fr. XVI). It addresses various environmental issues, and specific Regulations derive from it (concerning e.g. ecological zoning, waste, environmental impact assessments, atmospheric emissions, protected natural areas, etc.). However, it does not elaborate on climate change issues in detail[2] and does not mention forced climate migrations.

Despite these shortcomings, it is an essential norm for climate forced mobilities from a preventive perspective. It states that human settlements should be avoided in areas where disasters due to the adverse impacts of climate change would generate a risk to populations (Article 23, Fr. X, sudden-onset disasters and forced relocation scenarios). It also regulates environmental protection, which is instrumental in preventing forced migration scenarios, either directly (e.g. the slow-onset environmental deterioration scenario) or indirectly (e.g. the conflicts or violence caused by the scarcity of natural resources scenario).

Like the LGEEPA, the Laws dedicated to diseases in animals and plants and pests (*Ley Federal de Sanidad Animal* 2018; *Ley Federal de Sanidad Vegetal* 2017) are relevant for this research for several reasons. They have a direct relationship with impacts resulting from climate change, as the causal link between climate change and the increase in diseases in animals, plants and pests has been firmly established (Oyhantçabal et al. 2011; Bulman and Lamberti 2011; Kuribreña et al. 2019; Hicke et al. 2022).[3] Diseases and pests can also generate forced internal climate migration scenarios: slow-onset phenomena (e.g. the disappearance of ecosystems caused by diseases or pests), areas designated unsuitable for human settlements (e.g. if an area becomes infected to such a degree that public authorities decide to relocate the people or communities), and conflicts generated by the scarcity of natural resources (e.g.

[2] The LGEEPA (2018) mentions climate change in Articles 2, Fr. V, 3, Fr. VBis; 5, Fr. XXI; 7, Fr. XXI; 8, Fr. XVI; 15, Fr. IV; 23, Fr. X; 39 and 41, and weather conditions in Articles 115 and 145, Fr. I. For its part, the Regulation of the General Law on Ecological Balance and Environmental Protection in matters of Ecological Planning (Regulation LGEEPA 2014) mentions climate change in Articles 12, Fr. IX, 23, Fr. III(b); 31, Fr. V; 43, Fr. III(e); and 44, Fr. III.

[3] On climate change and diseases in humans, see e.g. Williams et al. (2021).

access to healthy land, to an uncontaminated water source). However, these Laws do not mention human mobility.

Similar to the Laws mentioned above, the General Law for Waste Prevention and Comprehensive Management (*Ley General para la Prevención y Gestión Integral de los Residuos* 2018) and the National Water Law (Ley de Aguas Nacionales 2020) directly relate to the scenarios that generate internal climate mobility. They address water scarcity (slow-onset phenomena, conflicts and violence generated by water scarcity,[4] water scarcity motivated relocations), water-related environmental services (e.g. in the prevention of sudden-onset disasters),[5] the water-related disasters (articles 9, Fr. XLII, 22, 38, 84, sudden-onset disasters),[6] hydroelectric activity (scenario of development projects linked to renewable energies),[7] or the impact of waste on the environment (slow-onset phenomena, sinking islands and coasts).[8] Despite the above, the said Laws do not mention forced mobility.

The same applies to the General Law on Forests Sustainable Development (*Ley General de Desarrollo Forestal Sustentable*, LGDFS 2020). We have already mentioned the role of forests in the context of forced internal climate migration for the gradual deterioration of the environment (slow-onset phenomena), fires (sudden-onset disasters), conflicts related to access to resources, or sinking islands and coasts scenarios. Despite its importance, the Law does not differ from the other environmental laws. It does not mention mobility related to forest resources in the context of climate change. Similarly, the General Wildlife Law (*Ley General de Vida Silvestre*, LGVS 2018) is of the utmost importance for forced internal climate mobility scenarios. Its purpose is to "establish the Federal Government, States governments and Municipalities' attributions (…) regarding wildlife and its habitat' conservation and sustainable use" (Article 1). The purpose has implications for various scenarios (slow-onset phenomena, sudden-onset disasters, conflicts caused by access to scarce natural resources, sinking islands and coasts). However, it does not mention forced internal climate migration despite mentioning climate change in Article 3, Fr. XLIV.

For its part, the Sustainable Rural Development Law (*Ley de Desarrollo Rural Sustentable*, LDRS 2019) is another norm related to forced internal climate migration but does not go on to address it. In line with its purpose ("promoting the country's sustainable rural development, fostering an adequate environment", and "guaranteeing the State leadership and its role in promoting equity," Article 1), it is no surprise it provides elements related to internal climate migration scenarios. Climate change has a harsh impact on the agricultural sector and can generate forced mobility. Similar to the research presented in this book, the Law endorses a territorial axis: it focuses on rural territories, where populations greatly depend on the soil and

[4] See e.g. National Waters Law (2020), Article 7Bis, Fr. VI, Fr. XXI and XXII; 12Bis(6), Fr. XIV; and 29 Bis(5).

[5] See e.g. National Waters Law (2020), Article 3, Fr. XLIX.

[6] See also Regulations of the National Waters Law (2014), Articles 127–132.

[7] Regulations of the National Waters Law (2014), Articles 118–123.

[8] In Chap. 1, we saw that the sinking islands and coasts scenario is complex and mentioned some studies that link the scenario to poor waste management.

natural resources for their means of subsistence (that is, personal-consumption). The Law considers gradual environmental deterioration and the scarcity of natural elements in the context of rural development.[9] It also pays particular attention to access to water,[10] land use,[11] and disasters.[12] Accordingly, it addresses the slow-onset phenomena, violence caused by the scarcity of natural resources, and sudden-onset disaster scenarios.

Despite the above and various mentions of weather conditions,[13] it does not cover development projects linked to renewable energies, areas designated unfit for human settlement, or the sinking islands and coasts scenarios. Nor does it expressly address the internal migration phenomenon[14] or establish the protection requirements for those affected. The Law is a good candidate to address forced internal climate displacement. Poverty being a central axis of the Law,[15] enabling measures for the consideration of people and communities who can no longer live off their activities due to climate change and trapped migrants. In addition, the Law applies a differentiated approach (women and Indigenous Peoples).[16]

Finally, the General Law on Human Settlements, Land Planning, and Urban Development (*Ley General de Asentamientos Humanos, Ordenamiento Territorial y Desarrollo Urbano*, LGAH 2020) does not explicitly address the phenomenon. Neither does it cover all climate mobility scenarios, despite general provisions on the environmental impacts on human settlements and cities.[17] It has the potential to address

[9] LDRS (2019), Articles 1, 3, Fr. XIV, Fr. XI, XXVI and XXX, 4, 5, Fr. IV, 7, Fr. V, 9, 11, 15, Fr. IX, 22, Fr. IX, 27, Fr. X, 32, Fr. XI, 53, 55, Fr. VI, 56, Fr. IV and VIII, 57, 62, 71, Fr. V, 72, 87, 118, Fr VI, 136, Fr. V, 144, Fr. VIII, 154, Fr. V, 161, Fr. XI, 164, 170, 172-174, 176, 177, 183, Fr. VII, and 190, Fr. V.

[10] LDRS (2019), Articles 15, Fr. XIII, 32, Fr. III, 55, Fr. III, 88, 118, Fr. IV, 131, 136, Fr. V, 81–85, 168, 169, 171, and 190, Fr. III.

[11] LDRS (2019), Articles 15, Fr. XIII, 53, 55, Fr. V, 57, 71, Fr. V, 72, 88, 131, 136, Fr. V, 83–84, 164–167, 171, 173, and 190, Fr. IV.

[12] LDRS (2019), Articles 15, Fr. XVI, 53, 126, 131, 133, 154, Fr. VI, and 171.

[13] LDRS (2019), Articles 84, 116, 126, 129, 130, 134, 136, Fr. V, and 190, Fr. IV.

[14] In article 163, the Law mentions mobility:

> With the Mexican Council's participation, the Intersecretarial Commission will propose special programs to defend human rights and support for the migrant population and measures that address their attachment to their place of origin.

This article does not specify whether it refers to internal or international and voluntary or forced migration. We assume it refers to international migration (i.e. migration from rural Mexican territories to other countries) since the terminology used in Mexican legal norms refers to international migration when it mentions "migrant" (see e.g. the Migration Law).

[15] LDRS (2019), Articles 3, Fr. XX, 5, Fr. I, 8, 14, 15, Fr. VII, 80, 85, 116, 118, 146, 154, 155, 157, 159, 178, and 190, Fr. IV.

[16] LDRS (2019), Articles 15, Fr. X, 56, Fr. V, 29, 118, 144, Fr. IX, 154, 162, 174, and 175.

[17] It provides in Article 2 that "All persons, regardless of sex, race, ethnicity, age, physical disability, sexual orientation, are entitled to live in and enjoy cities and human settlements in sustainable, resilient, healthy, productive, equitable, fair, inclusive, democratic and safe conditions," which establishes the general framework for climate change impacts on the environment. Similarly, it

the slow-onset scenario. It defines human settlement as "the natural elements (…) that comprise it" (Article 3, Fr. IV) and conservation as an "action aimed at preserving areas with historical and cultural values, and protecting and maintaining the ecological balance in areas where environmental services can be identified" (Article 3, Fr. VIII). Furthermore, human settlement planning is "a public policy which goals are land rational occupation and use as a spatial basis for socioeconomic development and environmental preservation strategies" (Article 3, Fr. XXVI) which relies on the notions of environmental sustainability (Article 3, Fr. IX) and resilience (Article 3, Fr. VIII).

The Law also contemplates, to a certain extent, the scenario of conflicts generated by the scarcity of natural resources and, more specifically, the rural-urban conflicts generated by access to water:

> The Federation is responsible (...) To anticipate at the national level land necessities for urban Development and housing, considering water availability (...) and to regulate (...) the mechanisms to satisfy said necessities. (Article 8, Fr. VI)

In addition to the above, the Law refers to the sudden-onset disaster scenario. It defines risk as "[d]amage or probable losses on an affected agent, resulting from the interaction between its vulnerability and the presence of a disturbing agent" (Article 3, Fr. XXIX Bis), and Comprehensive Risk Management (*Gestión Integral de los Riesgos*) as the

> [a]ctions aimed at risk identification, analysis, evaluation, control and reduction, considering its multifactorial origin and a permanent elaborating process involving the three government levels and society sectors. It facilitates the implementation of public policies, strategies, and procedures that combat disasters' structural causes and strengthen the society's resilience or resistance capacities. It includes risks' identification and, where appropriate, their making process, forecasting, prevention, mitigation, preparation, assistance, recovery, and reconstruction. (Article 3, Fr. XXI)[18]

Likewise, the Law provides public policy elements on sudden-onset disasters and relocations decided by public authorities: "[a]ttention to emergency situations due to climate change and natural phenomena" and "[t]he delimitation of risk zones and the cushion and safeguard protection polygons establishment to guarantee person safety and the strategic national security facilities establishment" (Article 6, Fr. IX) are public utility causes that justify expropriation in the sense of article 27 of the Mexican Constitution (CPEUM 2020). In addition, the Federation must "[p]romote the development of tools to identify high-risk areas generated by disturbing phenomena of natural and anthropogenic origin in coordination with the Government Ministry (*Secretaría de Gobernación*)" (Article 8 Fr. XXIII).

Regarding expropriations, the Law regulates relevant aspects for the scenario of areas designated unfit for human settlement, such as the establishment of restricted areas (Articles 11, Fr. XVIII), irregular settlements (Articles 10, Fr. XI and 11,

provides a generic way that states must "[p]revent and avoid human settlements in high-risk areas following risk atlas and according to applicable legislation" (Article 10, Fr. XXIV).

[18] There are other references to risk in Articles 10, Fr. XXIV, and 11, Fr. XXIV.

Fr. XVI),[19] relocations (concerning "areas subject to hydrometeorological risks", Articles 64–66), and the availability of land (Article 11, Fr. XVII). The Law requires the consideration of people living in poverty and vulnerability (Article 11, Fr. XVII) regarding land availability. This is relevant as limited economic resources is one of the factors that generate trapped climate migrants.

In spite of the Law's potential to partially address four climate migration scenarios and definition of mobility in Article 3, Fr. XXV as the "capacity, ease and efficiency of transit or persons and goods' movement in the territory, prioritizing universal accessibility and its sustainability," it does not tackle the forced character of mobility. Instead, it establishes "universal accessibility," "closeness," and the "relationship between different urban activities" as public policy principles (Article 4, Fr. X). It does not provide a list of the rights which reflect the protection requirements for the populations forced into mobility for climate reasons,[20] with the notable exception of procedural rights (mentioned in the list of public policy principles):

> Democratic participation and transparency. Protect the right of all people to participate in the formulation, monitoring, and evaluation of policies, plans, and programs that determine cities and the territory's development. To achieve this, transparency and access to public information will be guaranteed following this Law's provisions and other applicable legislation on the matter. (Article 4, Fr. V)[21]

Standards that Have Potential Despite Their Literal Content

The Mexican Constitution (CPEUM 2020)[22] and the General Victims Law (*Ley General de Víctimas*, LGV 2017) could apply to forced internal climate migration, despite their limitations if we consider them literally. The Constitution considers expropriations (Article 27), which can relate to the areas designated unfit for human settlement. However, it does not mention internal displacement. Some authors go beyond this literal reading and argue that the Constitution protects forced internal migrants. Peguero and Rubio Diaz Leal (2019) point out that Mexico is a member of the Organization of American States (OAS), where the Inter-American Court of Human Rights serves as a jurisdictional body. The latter determined in a broad sense that internal displacement implies the right to move and reside freely (p. 51). Considering the Mexican Constitution guarantees this right (Article 11), "*ergo*, the

---

[19] In Chap. 2, we identified internal climate migration induces specific challenges for territories, including the irregular creation or development of settlements.

[20] The Law only mentions the requirement to "(p)rocure, promote, respect, protect, ensure, and take steps towards human rights related to land use and Urban Development's full realization" in Article 8, Fr. XXVI and human rights in Articles 10, Fr. III, and 11, Fr. XXII.

[21] In the same sense, regarding the attributes of federal authorities, see Articles 10, Fr. II, 11, Fr. XXII, 92–93, and 99–100.

[22] The constitutional reform (June 10, 2011) explicitly grounds human rights, established in international treaties, and to which Mexico is a party, to have the same hierarchical value as the fundamental rights guaranteed by the Constitution (CPEUM 2020, Article 1) and the *pro-persona* principle (the human rights interpretation towards broader protection—CPEUM, 2020, Articles 1, 11, 29 and 33).

Constitution implicitly contains protection against internal displacement" (Peguero and Rubio Diaz Leal 2019, p. 51).

We have some reservations about this interpretation. The Inter-American Court of Human Rights refers to internally displaced persons for any number of reasons, but it does not explicitly mention forced climate mobility. For the time being, the Court has not ruled on the recognition of climate migrants. Additionally, the protection is not explicit in the Constitution: it instead requires an interpretation of the Inter-American Court of Human Rights that relies on the right to free movement in the broadest sense. We therefore consider that the protection of internal climate migrants grounded in the Mexican Constitution remains elusive. Our analysis carried out in the following paragraphs confirms these reservations.

The General Victims Law (*Ley General de Víctimas*, LGV 2017) recognizes and guarantees the rights of "victims of crime and human rights violations" (Article 2, Fr. I). It recognizes that the people in a situation of internal displacement are exposed, as a group, to a greater risk of having their rights violated (Article 38), and Article 110 (amended on January 3 2017) regulates the recognition of the status of victimhood. It is a significant step towards recognising forced internal migrant.

However, the Law focuses solely on the victims of displacement due to violence or human rights violations. It establishes the classification of victims as the people who directly or indirectly "have suffered damage or impairment of their rights as a result of human rights violations or the commission of a crime" (Article 6, Fr. XIX). Consequently, the Law only applies *stricto* sensu to forced internal migrants if a human rights violation (e.g. to the right to water) or the committing of a crime (e.g. illegal logging) can be established. In the context of climate migration scenarios, it will not always be possible to establish such violations or evidence a crime as having been committed (e.g. it could hardly be demonstrated in the case of sinking islands or coasts), and climate internal migrants would not be considered victims.

Nevertheless, a jurisprudential development could open the door to a broader interpretation of "victim" that could extend to internal climate migrants. It concerns internal migration induced by organized crime. A recent Protection Resolution (*amparo*) specifies the scope of application of the Law regarding internal displacement driven by organized crime (Tenth Collegiate Court in Criminal Matters of the First Circuit, *Décimo Tribunal Colegiado en Materia Penal del Primer Circuito* 2018). It clarifies the obligations of the Executive Commission for the Assistance of Victims (*Comisión Ejecutiva de Atención a Víctimas*, CEAV). The Resolution establishes the CEAV's requirement to recognize and register a victim of organized violence a family victim of forced displacement in the state of Chihuahua (CMDPDH 2019a).

The Resolution is mostly relevant to our purposes. In this case it was organized violence which induced the internal forced displacement (not climate change). However, it endorses an expansive notion of internally displaced people which could, in turn, be extended to internal climate migrants:

> *the General Victims Law recognizes internal forced displacement and the multiple impacts it generates as human rights violations,* (...) (*internally displaced persons*) may not only be

> victims of other violations but also of crimes as a consequence of their particular vulnerability. (Tenth Collegiate Court in Criminal Matters of the First Circuit, 2018, p. 142, emphasis mine)
>
> This recognition is fundamental for the task for which the Executive Commission was created. *It has an obligation to guarantee victims' access to all their rights, which includes victims of internal displacement.* More particularly, this protection is fundamental given displaced persons' indeterminate specific legal status. They usually lack documentation, face difficulties in exercising the enjoyment of their rights (...) and have medical, psychological, legal, and social assistance necessities (...) since displacement usually implies employment, housing, and education loss. (Tenth Collegiate Court in Criminal Matters of the First Circuit, 2018, p. 145, emphasis mine)

Based on the above, and Article 110 of the Law, the Court granted constitutional protection to the displaced people. It ruled that the CEAV must acknowledge "its power to recognize and register internal forced displacement victims" (p. 155) and, where appropriate, "grant full reparation according to the General Victims Law" (p. 156). The broadening of the definition of victim to include internally displaced people could pave the way for subsequent decisions of the CEAV and judicial bodies to recognize the status as victims of internal forced climate migrants.

Limitations in the Norms that Recognize the Phenomenon

The General Law on Climate Change (LGCC 2020) adopted in 2012 is the key Mexican norm on climate change:

> Among other things, it defines the three levels of governmental responsibility and the coordination mechanisms between them; it transforms the current National Institute of Ecology (INE) into the National Institute on Ecology and Climate Change (*Instituto Nacional de Ecología y Cambio Climático*, INECC), turning it into a decentralized body of the federal administration; and establishes the National Climate Change System as a permanent communication, coordination and cooperation mechanism between the three levels of the government and the private and social sector. It also creates the Advisory Council to issue recommendations to the National Climate Change System to evaluate national policy on climate change and review compliance with the objectives, goals, and actions of the Special Climate Change Program on a regular and systematic basis. (…) It also established a national emissions registry, central to tackle climate change as it enables a national emissions diagnosis. It additionally set a climate change fund to gather and channel national and international public and private funds (...). (Ávila Akerberg 2012, pp. 187–188)[23]

It also provides central definitions of climate change (e.g. climate change, adaptation, mitigation, resilience, resistance, and vulnerability, Article 3). The Law is a priori the most appropriate norm to recognize forced internal migrations driven by climate

---

[23] Despite being a pioneering law (Ávila Akerberg 2012) and generally receiving positive comments (see e.g. Moraga and Meckievi 2016, pp. 36–38), Moraga and Meckievi (2016) provide an overview of the criticisms directed towards the Law (pp. 38–40). Among them, they mention mitigation measures are voluntary, "no public participation systems are defined," the Intersectoral Climate Change Commission does not have a clear mandate, the law does not establish a transversal coordination, the "law's wording" is more aspirational than directed towards the concrete or "hard" implementation, the "proliferation of regulatory instruments (plans, programs, regulations, etc.) generates potential duplications, contradictions and other problems of design policy that can negatively impact law's objectives effective implementation."

change. Since its 2016 reform, it mentions, for the first time in the Mexican regulatory framework, internal forced displacement induced by climate change (Articles 28, Fr. VII and 30, Fr. II). Article 28, Fr. VII reads:

> The federation must prepare a National Adaptation Policy within the framework of the National Climate Change System.
>
> In line with the scope of their attributions, the federation, federate entities and municipalities, must execute adaptation actions in the elaboration of policies, the National Strategy, the Special Climate Change Program, the National Adaptation Policy and programs in the following areas: (…) internal displacement of people caused by phenomena related to climate change.

This recognition is noteworthy. To date, it is the only mention of the phenomenon in the entire Mexican legal framework. However, the Law does not define "internal displacement of persons caused by phenomena related to climate change," set criteria, or establish the protection requirements for the people affected.

Finally, the General Civil Protection Law (*Ley General de Protección Civil*, LGPC 2018) is relevant to our purpose. It focuses on the prevention of risk and mitigation, including the risks generated by climate conditions. More specifically, it refers to the notion of disasters (defined in Article 2, Fr. XVI) in its notion of Comprehensive Risk Management (Article 2, Fr. XXVIII) (sudden-onset disasters scenario). The Law partially recognizes forced internal migration and the necessity to protect those affected. It defines the evacuee in a broad sense that corresponds to the forced internal migrant:

> [A] person who preventively and provisionally withdraws or is removed from his usual place of accommodation to guarantee his safety and survival because of a possible or certain emergency or disaster. (LGPC 2018, Article 2, Fr. XIX)

The Law mentions climate change explicitly: the knowledge surrounding climate change and adaptation to climate change is one of its priorities (Article 4, Fr. VII). Nevertheless, it does not draw a link between climate change and forced internal migrations ("victims and evacuees" in the Law).[24] The Law endorses population assistance in general but does not identify forced internal climate migrants as a particular group.

The Law establishes the prioritization of vulnerable social groups and that of those with scarce economic resources (Article 21). In other words, it endorses a differentiated approach. This is essential and resonates with the analytical grid contained within our research (a differentiated approach applied to Indigenous Peoples, women, and populations with limited economic resources). However, the said differentiated approach is limited. It does not provide a specific definition as to what should be considered a vulnerable group, neither does it provide the criteria for its identification. Similarly, it does not specify the content of the requirement to safeguard "the life, integrity and health of the population, as well as their property; the infrastructure (…) and the environment" (Article 2, Fr. XLIII) and does not clarify what a victim

[24] Regarding the slow-onset phenomena scenario, the LGPC Regulation (2018) links Integral Risk Management with environmental protection in Article 7, Fr. I[a]), and refers in Article 70, Fr. III to slow-onset phenomena (droughts, dry season and forest fires).

or evacuees' recovery and assistance is defined as in terms of rights. We found but a single mention of human rights. It refers to the actions of civil protection authorities (Article 5, Fr. VIII).

3.2.2 *Shortcomings of the 2020 Draft General Law to Prevent, Address, and Comprehensively Repair Internal Displacement*

As a background to the 2020 Draft General Law, Peguero and Rubio Díaz Leal (2019) refer to 11 legal initiatives related to internal forced displacement at federal level, between 1998 and 2016. None of these initiatives was successful. Against this background, the Mexican Chamber of Deputies' Government and Population Commission's approval of the General Law Project to Prevent, Address, and Integrally Repair Internal Displacement (*Proyecto de Ley General para Prevenir, Atender y Reparar Integralmente el Desplazamiento Interno*) on September 29 2020 is welcomed. The Draft Law (2020) is now before the Mexican Senate of the Mexican Republic.[25]

The Draft justifies the necessity to adopt a General Law. It mentions the invisibility of internal forced displacement (Draft Law 2020, III. Content of the Proposal, pp. 5–7), the necessity to involve the three government levels (federal, federate states and municipalities) (p. 15), and the gaps in the national framework regarding the recognition of forced internal migration.[26] The Draft provides definitions (Article 4), establishes the three government levels' obligations, powers and attributes (Articles 13–16, 36–38, 40–41, 43, 45–46, 48–50), creates a National Mechanism to prevent, address and generate durable solutions (Articles 20–32), creates the National Registry of Displaced Persons (Article 26, Fr. VIII (c) and XXVI), provides for the

---

[25] The remaining four subsequent stages for the Law to be finally adopted are: (i) Senate approval, (ii) Chamber of Deputies' approval, (iii) President of the Mexican Republic's promulgation and order to publish the corresponding Decree in the Official Gazette of the Federation (*Diario Oficial de la Federación*), and (iv) Publication in the Official Gazette of the Federation.

[26] Draft Law (2020), III. Content of the Proposal, (p. 9):

> Despite the presence and severe consequences of this phenomenon (*forced internal displacement*), our country does not possess the necessary national legal instruments to diagnose the problem, nor to face it from a comprehensive perspective based on the respect for integrity and individuals rights, reparation of damage and non-repetition of the facts. (…) Mexico has an institutional and regulatory vacuum to deal with the problem of internal forced displacement.

In the same sense, see the Initiative on Forced Displacement presented by Deputy Ulises García Soto of the Morena political party (Law Initiative 2019), Conceptual Framework, and CNDH (2016): "[t]he creation of a general norm that clearly and precisely determines which are the powers and obligations of each authority, be it a federal, state or municipal one is a necessity" (p. 131) and Velázquez Moreno (2017, pp. 76–77). *Contra*, see the "General Observations or Comments" of the Attorney General's Office on the Draft Law (2020), pp. 44–49: the General Directorate of Public Policies, Liaison and Inter-institutional Coordination of the Prosecutor's Office considers the Law duplicates the General Law of Victims (LGV 2017). Several Opinions to the Draft Law (2020) address the articulation of the law with the LGV.

development of a National Program to prevent, assist and generate durable solutions (Articles 17–18) and local programs (Article 19), an Early Warning System aimed at preventing forced internal displacement (Article 35), a National Assistance Fund (Articles 51–62), and the classification of internal displacement related crime in criminal law (Article 33, Transitory, Ninth). The analysis of the Draft reveals its potential and limitations when addressing forced internal climate migration.

Forced Internal Climate Migration Scenarios and the Categories of Migrants

The analysis of the Background, Postulates, and Articles of the draft that provide the definitions of forced internal displacement (Article 4, Fr VIII), internal forced displacement events (Articles 4, Fr. XII and 5), and people in situations of forced internal displacement (Article 4, Fr. XX), reveals the limitations of the recognition of forced internal displacement when compared to the internal forced climate migration concept applied in our research.

For the most part, the 2020 Draft Law partially covers the forced internal climate mobility scenarios. It contemplates four out of the six forced internal climate mobility scenarios: sudden-onset disaster (Article 5, Fr. VI), violence induced by the effects of climate change (scenario partially addressed in Article 5, Fr. II-V), relocations decided by the authorities (Article 42, which may correspond to cases of relocations decided by public authorities), and large-scale development projects linked to renewable energies (Article 5, Fr. IX). The Draft focuses on hydroelectric projects (Postulates of the Proposal, p. 7). The violence linked to the impact of climate change is also partially recognized. However, the gradual environmental deterioration (slow-onset phenomena) and sinking coasts and islands scenarios are not covered.[27] We subsequently analyze the climate mobility scenarios contained within the Draft in more detail.

The Draft mentions climate change in Article 5, Fr. VI. However, it explicitly associates climate change with sudden-onset disasters, disregarding other impacts: "The causes that generate internal forced displacement recognized by this Law include (…) (d)isasters associated to natural phenomena, caused by humans or climate change." This limited concept of climate change impacts does not allow all the scenarios to be covered or mentioned (at least literally). A more general reference to the impacts of climate change would have guaranteed the application of the Law to all its impacts which generate forced internal migration.

As for violence generated by the scarcity of natural resources, the Draft mentions territorial conflicts in the Postulates (p. 8). This mention resonates with the book's analytical grid. However, Article 5 (Fr. II-V) mentions agrarian, armed, communal, and property conflicts. It reflects an administrative approach to the delimitation of territorial conflicts rather than focusing on the objectives and rationale of the disputes

[27] The 2019 Legal Initiative did mention the sinking islands and coasts scenario: it mentionned "rising oceans" in its Conceptual Framework.

(access to land and natural resources). This limitation could be overcome by Article 5, Fr. XII, which generally mentions "other types of violence."

The Draft also disregards the gradual environmental deterioration scenario (slow-onset phenomena). The 2019 Initiative to the Draft Law shares this limitation.[28] One could argue that the scenario is covered when the deterioration can be legally defined as a violation of environmental rights as the Draft identifies the disregard of human rights as a cause that generates forced internal mobility (Article 5, Fr. IX). However, this argument is limited on several grounds.

In the previous chapter, we saw that environmental rights are not guaranteed on an equal basis: material and procedural environmental rights are widely recognized, but the autonomous right to a healthy environment (intrinsic dimension) is not. Consequently, the gradual environmental deterioration of its intrinsic value would hardly be considered as a cause of forced internal mobility. Similarly, the Draft establishes a violation of human rights as a cause of forced internal mobility. Interference with a right is not enough. The violation requirement seriously limits scope of application of environmental rights. We pointed out in the previous chapter that most of the rights through which environmental matters are addressed are relative (e.g. rights to health, private and family life, property). Unlike absolute rights, an interference with their enjoyment is not enough to conclude that the right has been violated. Interference is balanced against the legitimate objective that justifies it (e.g., a region's touristic development). Consequently, the application of this cause of migration to cover slow-onset phenomena is limited.

The Draft also omits the sinking islands and coasts scenario. The sudden-onset disaster scenario covered by the Draft tackles the sea-level rise impacts that generate such disasters (e.g. floods, Reidmiller et al. 2017). But the Draft does not cover the other impacts the sea-level rise induces (e.g. salinization of water, coastal erosion, Reidmiller et al. 2017). We might explain this limitation by the reliance of the Draft on the UN's Guiding Principles applicable to Forced Displacement (UN 1998) (pp. 9, 13)[29] since the UN Principles do not cover slow-onset phenomena.[30] The gap left by the UN Principles echoes in the Draft.

Finally, we also identify deficiencies in the Draft regarding the categories of migrants. The Draft mentions the forced nature of mobility (Articles 4, Fr. XX, 5) and therefore covers forced emergency migrants and climate migrant categories (although it does not provide details on the objective or subjective criteria which apply). However, it does not apply to trapped climate migrants as it requires movement (Article 4, Fr. XX: "escape or flee from their home or their usual place of residence").

---

[28] The scope of the 2019 Initiative's Conceptual Framework regarding the causes of forced internal displacement was broader than the Articles of the initiative. The Conceptual Framework mentioned three "contexts of expulsion" that could be related to the gradual environmental deterioration scenario (slow-onset phenomena): "anthropocentrism," "natural phenomena and cycles," and "natural and (or) man-made environmental disasters (… excessive agriculture and livestock)."

[29] In this regard, the Law Initiative (2019) was more distanced from those Principles (penultimate paragraph of the Conceptual Framework).

[30] See Chap. 2 of the book.

Protection Requirements

The Draft firmly recognizes the protection requirements of affected people. Article 2, Section I of the Draft mentions the protection requirement: it "aims to (...) (r)ecognize and guarantee the persons in situation of internal forced displacement' rights."[31] The Draft codifies the protection requirement in Title III ("On the Rights," Articles 6–12). Three situations are differentiated: the protection of people in a situation of internal forced displacement (Chapter I), contingency (Chapter II), and internal displacement (Chapter III). This codification reflects the protection requirements that we propose in terms of a human rights and a differentiated approach.

In addition to the attention paid to and protection against forced displacement (Article 6), the Draft provides an enunciative and non-limiting list of rights (Article 12). It is very exhaustive and essentially covers the rights systematized in the former chapter. The Draft explicitly mentions the right to access information, consultation, and "full participation" of "all the persons involved" in the context of large-scale development projects (Article 39). It is with no doubt a positive step towards a comprehensive rights-based approach. However, it does also refer to the "applicable regulations." We already saw that the legal framework on consultations and the participation of citizens in Mexico is limited in its design and implementation. The reference to the "applicable regulations" is therefore a limitation.

The Draft adopts a differentiated approach (background information p. 14, and Articles 3 and 4, Fr. IX). In the definition of the differentiated approach (Article 4, Fr. IX)[32] the Draft illustratively identifies criteria (i.e., the list does not exclude the recognition of other criteria)[33]: "sociodemographic, cultural, ethnic characteristics, gender, age, sexual orientation, gender identity, disability." It recognizes two of the three categories of individuals and communities which we study in particular: women (Articles 3, 5, 10, 12, 13, and 43) and Indigenous Peoples (Articles 5, 9, 10, 12, 13, and 43).

However, the Draft does not specifically mention characteristics of the third category (people with limited economic resources). Article 4, Fr. XXVII signals violence that causes economic damage or suffering. It differs from poverty and structural discrimination that do not result from a specific "act or omission."[34] In contrast, the 2019 Initiative did specify them. It mentioned "social and economic circumstances" and "a situation of vulnerability" in Article 4, Fr. IV. These specifications

[31] The protection requirement is the first object mentioned in the 2020 Draft. It was the second one in the 2019 Initiative.

[32] The Draft mentions a "differential" approach.

[33] It stems from Article 4, Fr. IX (emphasis mine):

> The perspective that government agencies and agencies (...) must apply to identify the particular needs of persons or groups of people in situations of forced internal displacement, according to their sociodemographic, cultural, ethnic, gender, age, sexual orientation, gender identity, disability, *among others*.

[34] The Draft Law (2020) mentions vulnerability in Article 4, Fr. VIII. However, it mentions the vulnerability of people "in a situation of displacement." This is not pre-existing vulnerability.

are of the utmost importance because this category of people is more inclined to become trapped migrants. We also established previously that pre-existing vulnerability to weather events or phenomena is an essential factor that drives mobility and its forced nature. The Law could easily overcome this deficiency with a definition of vulnerability that extends to pre-existing vulnerability in the Article that provides the definitions that apply for the purposes of the Law.

In summary, the Draft seeks to fill a legal void by recognizing the legal status of forced internally displaced people. The codification of the requirement for differentiated protection mainly corresponds to the requirement we identify in this book. However, we observe several limitations regarding climate mobility scenarios, the categories of forced internal migrants, and pre-existing vulnerability.

3.2.3 Critical Assessment of the Legislative Frameworks at State Level: Chiapas and Guerrero

Two state laws are, together with the 2020 Draft at the federal level, the norms that recognize forced internal migration, at least partially acknowledge the climate trigger, and set out the requirements it implies for human rights[35]. However, we do not elaborate on this Law in the chapter, since it ignores the climate trigger (the only trigger to internal displacement it recognizes is violence). The two state laws are (i) the Law for internal displacement prevention and attention in Chiapas (2012) and (ii) Law number 487 to prevent and address internal displacement in Guerrero (2014). Both apply partially to forced internal migrations driven by climate change and articulate the demand for the protection of those affected.

Limitations in the 2012 Law for the Prevention of Internal Displacement and the Attention Thereof in Chiapas

The first Law on forced internal migration in Mexico was adopted in Chiapas in 2012 ("Chiapas Law," 2012). This is of no surprise: Chiapas has a long history of forced internal migration. The state also has a high number of Indigenous Peoples (INEGI 2010]). The Law, in this case, has four purposes: to.

(1) "create the conceptual and guarantee framework regarding the rights of persons who, for various reasons, are forced to leave their habitual place of residence, defining what is considered an internally displaced person,"
(2) establish "internally displaced persons' rights,"

[35] A third state law recognizes internal displacement: the Sinaloa Law on internal displacement (Ley para Prevenir, Atender y Reparar Integralmente el Desplazamiento Forzado Interno, y se adiciona un Capítulo III, denominado "Desplazamiento Forzado Interno", que comprende el Artículo 175 Bis, al Título Quinto del Código Penal, ambas del Estado de Sinaloa 2021).

(3) mandate the creation of "the State Program for Internal Displacement Prevention and Attention," and
(4) "inter-institutional coordination through the establishment of the Comprehensive Attention to Internal Displacement State Council."

Article 2 of the Law provides:

> This law aims to establish the basis for the prevention of internal displacement, the assistance and implementation of durable solutions to overcome it, and provide a guarantee framework that attends and supports people in this situation.

The Law prohibits "arbitrary movements that force or oblige to leave the place of habitual residence" (Article 24). Article 25 defines such movement according to the cause that generates them. Several causes chime with the book's analytical framework. Armed conflicts (Fr. II) partially resonate with the conflicts or violence caused by the scarcity of natural resources. Development projects "on a large scale" "not justified by a higher or overriding public interest or do not seek to raise persons' human development level or combat poverty and population dispersion" (Fr. III) correspond to the development projects related to renewable energies. Disasters "unless the safety and health of the affected persons require their evacuation" (Fr. IV) reflect the sudden-onset disasters and relocations decided by public authorities.

Article 3 defines internally displaced persons:

> Internally displaced persons are those persons or groups of persons settled in the State of Chiapas who have been forced or compelled (*forzadas u obligadas*) to abandon, escape or flee from their place of habitual residence as a result of or to avoid the effects of an armed conflict, situations of generalized violence, human rights violations, or natural or man-made catastrophes, and who have not crossed the State territorial limits.

This definition takes up, almost verbatim, the United Nation's Guiding Principles on Internal Displacement (UN 1998). It attracts several comments when analyzed from the perspective of climate forced internal migration. These comments are similar to those formulated in the previous chapter regarding the Guiding Principles, but they are specified when analyzed in light of the context of Chiapas.

A first reading could indicate that the Law only applies scantly to forced internal climate migration. Most of the elements mentioned as mobility triggers, refer to violence (armed conflict, generalized violence and the violation of human rights). Natural or man-made disasters are a priori the only cause that resonates with forced internal climate migration scenarios, the sudden-onset disasters scenario.

However, from a closer reading it appears that the Law covers two other scenarios which occur in Chiapas: the conflicts or violence generated by the scarcity of natural resources (the numerous agrarian conflicts in Chiapas or the "development projects" illustrate this case, see CNDH 2018; Fund for Sustainable Development Goals 2013), and the sinking coasts or islands scenario (Bárcena Ibarra et al. 2020). Despite the above and similar to the limitations identified in the reach of the UN Guiding Principles identified in the previous chapter, the definition cannot be said to apply to the gradual degradation of the environment (slow-onset phenomena scenario).

Regarding the "guarantee framework that attends and supports" people in a situation of internal displacement (Chiapas Law 2012, Article 2), the Law provides the content of a human rights-based protection requirement. Displaced people enjoy human rights guaranteed at the international and national level (Article 4), the application of these cannot be discriminatory (Article 5), and a differentiated approach applies to the most vulnerable (Article 6). The Law mentions two of the most vulnerable groups of people identified in the book: women (Article 6) and Indigenous Peoples (Article 7). The Law does not explicitly mention people and communities with limited economic resources or in a situation of poverty. This is regrettable as we have seen that they are more prone to becoming "trapped," and poverty rates are very high in Chiapas (CONEVAL 2018).

Adding to the generic reference to human rights guaranteed at the international and national levels, the Law explicitly provides a list of guaranteed human rights: freedom, security, dignity and integrity, be it physical, moral or mental (Article 8), the "right to move freely and choose their place of residence in the terms the law provides" (Article 9), to enjoy "satisfactory living conditions, including the right to safety, health and hygiene" (Article 10), "to the identity and recognition of their legal personality" (Article 11), "to the protection of the law against arbitrary deprivation, appropriation, occupation or destruction of their properties and/or possessions, whether individual or collective, and where appropriate to the restitution or compensation of their violated rights regarding land, housing and property" (Article 12), "freedom of thought, conscience, religion or convictions, opinion and expression," "free choice of their work, in just and satisfactory conditions," "freedom of peaceful assembly and association," "to vote and be voted for positions of popular election" (Article 13), and the rights related to family life (Article 14).

The Law also indicates the three procedural rights essential in environmental matters: the right to access information (Article 28, Fr. I), to consultation and participation (Article 15), and to access justice (including the right to reparation, Article 16). The codification of these rights in the context of relocations (areas designated unfit for human settlement scenario) raises several points. Firstly, the right to consent (Article 28, Fr. II) establishes a veto for the populations affected's right to consultation:

> Except for exceptional situations or natural catastrophes, the competent authority must (…) [o]btain the free and informed consent of the population affected by the displacement. In the case of indigenous communities, their specific cultural and organizational needs must be addressed in the terms provided by this law.

Granting such a veto is of note as the international and national legal framework on the consultation of Indigenous Peoples essentially does not.

Secondly, Article 28 provides a series of specific guarantees in the case of relocations with the requirements to "involve the persons affected by displacement in the resettlement planning and management, especially women and female heads of family" (Fr. III), "provide displaced persons with adequate accommodation" (Fr. IV), and "carry out the displacement in satisfactory conditions of safety, food, health, and hygiene, without separating the members of the same family and without violating the rights to life, dignity and freedom of those affected" (Fr. V). These guarantees are essential, as relocations pose several challenges in terms of human rights and risks for the populations implicated in such movement (e.g., accentuated poverty).[36]

However, the rights guaranteed in Article 28 do not apply in case of "exceptional situations or natural catastrophes." It can limit the enjoyment of such rights in several scenarios: sudden-onset disasters, conflicts or violence caused by the scarcity of natural resources, and development projects related to renewable energies. It drastically reduces the application of the guarantees.

Finally, the Law suffers from other limitations. Academics criticize the Law because it needs to be supplemented with criminal law (Torrens 2013). Also, to date, the Law and State Programs Regulations have not been adopted. It limits the contribution of the law to protecting internal migrants (Velázquez Moreno 2017), including climate migrants.

Drawbacks in the 2014 Law Number 487 to Prevent and Address Internal Displacement in the State of Guerrero

The second Mexican state law that addresses forced internal displacement is Law number 487 to prevent and address internal displacement in Guerrero state (Guerrero Law 2014). The Considerations mention many internal displacements have taken place over "several decades" in Guerrero and have since accelerated. Among other factors described are, natural phenomena (the Considerations mention hurricanes) and socio-environmental vulnerability, development projects (and more particularly hydroelectric plants), "uses and customs," "religious intolerance," "intra-community problems, such as agrarian disputes and even armed conflicts" as well as insecurity.

The objectives of the Guerrero Law somewhat differ from the Chiapas Law, as Article 1 does not mention institutional aspects. However, it does mention the prevention of internal displacement, the protection of affected people, aid and assistance, the application of human rights (and humanitarian law), and a differentiated approach (directed more specifically to Indigenous Peoples, women, minors, the elderly, and the disabled). Articles 6, 7, 8, 9, and 11, Fr. III confirm the differentiated approach for these people.

The definition of internally displaced people (Article 2, Fr. V) does not differ from that of the Chiapas Law, so we refer to the comments we made in this regard. The Guerrero Law also provides specific definitions relevant to addressing forced internal climate migration. They depart from the definitions within Chiapas Law.

It defines the state of vulnerability as the "[l]evel of risk a family or individual faces to lose their life, their assets, properties, and their livelihood system in the event of a contingency. This level also corresponds to the degree of difficulty in recovering after such a catastrophe" (Article 2, Fr. VII). This definition does not refer to the pre-existing vulnerability we portrayed in previous chapters as an essential element to understanding the factors that drive climate mobility. The Law also defines resettlement as "the result of a new location or settlement" (Article 2, Fr. XIV).

[36] We see it in more detail in Chap. 5.

Finally, it defines the state of contingency as the "(risk) or event that can occur, especially if it is problematic and must be anticipated" (Article 2, Fr. IV). The Law's temporal application is not limited to the time the displacement occurs but also applies to risks. It is of the utmost importance in climate and environmental matters: unlike generalized violence or conflicts to access scarce natural resources, most of the climate change phenomena and impacts can be predicted or, at least, are well documented and foreseen (e.g. hurricanes, droughts, desertification, floods, ecosystems disappearance).

Like the Chiapas Law, the Guerrero Law prohibits "arbitrary displacements that force or oblige to leave a place of habitual residence" (Article 29). Article 3 defines them much as the Chiapas Law does. We will not repeat the comments made in this regard. Similarly, the Guerrero Law provides a list of rights of people affected (Article 35). The comments above on the procedural rights of the Chiapas Law apply in the same way to the Guerrero Law.

The Guerrero Law contains three mechanisms that the Chiapas Law does not foresee. Firstly, the displaced have the obligation and the duty to "accept the place or territory that the state may assign them" (Article 22, Fr. III). This obligation not only clashes with procedural rights (among them, nothing less than the veto of people affected). It can also lead to arbitrary relocations. This risk is all the more significant in a state where the same Law indicates the strong presence of organized crime and reports indicate strong corruption (INEGI 2020). Secondly, the displaced must register in the State Population Registry (Article 22). We can expect this mechanism contributes to increasing the visibility of those affected. Finally, the Law provides for the adoption of precautionary measures (Articles 9, 31, 32). Article 2, Fr. X provides their definition:

> They are actions dictated by judicial rulings to ensure that a particular right can be made effective in the case of litigation that recognizes the right's existence and legitimacy. Precautionary measures do not imply respect for a proceeding's right existence, but they do imply the adoption of judicial measures aimed at making effective the right that might be recognized.

Much like the registration mechanism, precautionary measures can be instrumental to the effective exercising of human rights in the context of forced internal migration, including those which are climate-driven. They are part of prevention in the Law and consistent of the definition of the contingency state mentioned above, which considers the events together with the risks that may occur.

Despite the notable advances of the Guerrero Law to address internal forced migration (or, from another perspective, adding to the limits indicated) and its existence since 2014, its implementation is questioned. For example, in November 2018, various notes in the media reported the complaints of civil society organizations. They wanted the Guerrero Government to declare a "humanitarian crisis due to the displacement of 6 000 persons driven by violence" (La otra realidad del desplazamiento interno en Guerrero 2019). They pointed out that there are no guidelines, state agendas, or, much less, political will to follow-up with a view to combating these problems from the human rights perspective. The regulation of the Law is also still yet to be adopted. This void jeopardizes the practical application of the Law.

In this section, we identified the shortcomings of the current legal framework to address forced climate internal mobility. These results lead us to reflect on another question: do public policies at a federal level address the phenomenon despite the restrictions of the legal framework? In the next section, we answer this question in the negative (Table 3.1).

3.3 Public Policy Instruments: Main Shortcomings

To determine whether or not public policies at a federal level address forced internal climate displacement, we analyze them according to their primary purpose: instruments that address climate change (Sect. 3.3.1), development and planning (Sect. 3.3.2), and those specifically aimed at addressing sudden-onset disasters (Sect. 3.3.3).

3.3.1 National Instruments That Address Climate Change

With the exception of the recent Special Climate Change Program 2021–2024 (PECC) (SEMARNAT 2021) and the 2020 Nationally Determined Contributions (NDC, SEMARNAT 2020), the instruments that focus on climate change impacts do not contemplate forced climate mobility. Consequently, they do not mention the mobility scenarios, categories of forced internal climate migrants, or a differentiated human rights approach to comply with the displaced' protection requirements. We subsequently analyze, in detail, the Special Climate Change Program 2021–2024 (PECC 2021–2024) (*Programa Especial de Cambio Climático 2021–2024*, SEMARNAT 2021), 2020 Nationally Determined Contributions (NDC, SEMARNAT 2020), National Climate Change Policy's Strategic Evaluation on subnational progress (*Evaluación Estratégica del avance subnacional de la Política Nacional de Cambio Climático*, INECC 2018), Inter-secretarial Commission on Climate Change's Internal Regulation (*Reglamento interno de la Comisión Intersecretarial de Cambio Climático*, Regulación CICC 2016), National Climate Change System's declarations (*Declaraciones del Sistema Nacional de Cambio Climático*, Declaración SINACC 2014, 2018), and the National Climate Change System's Regulation (*Reglamento del Sistema Nacional de Cambio Climático*, INECC 2015).

The Special Climate Change Program 2021–2024 (PECC) (SEMARNAT 2021) is undoubtedly a public policy instrument that stands out compared to all the instruments analyzed in this book. It explicitly mentions that climate change policy must follow a differentiated approach, including where directed at migrants:

Table 3.1 Attention to climate forced internal mobility: legal instruments' analysis

Instrument	Scenarios[d]						Internal forced climate mobility	Pre-existing socio-economic vulnerability	Human rights perspective		Trapped migrants
	Slow-onset	Sudden-onset	Sea-level rise	Relocations	Development projects	Violences			RBA	Differenciated	
Refugees, Complementary Protection and Political Asylum Law (2014)											
Migration Law (LM 2020)											
General Population Law (LGP 2018)		x									
Ecological Balance and Environmental Protection General Law (LGEEPA 2018)	x	x	x	x		x					
Animal Health Federal Law (2018)	x										
Vegetal Health Federal Law (2017)	x										

(continued)

Table 3.1 (continued)

Instrument	Scenarios[d]						Internal forced climate mobility	Pre-existing socio-economic vulnerability	Human rights perspective		Trapped migrants
	Slow-onset	Sudden-onset	Sea-level rise	Relocations	Development projects	Violences			RBA	Differenciated	
Waste Prevention and Management General Law (2018)	x	x	x			x					
National waters Law (2020)	x	x	x		x	x					
Forest Sustainable Development General Law (LGDFS 2020)	x	x	x			x					
Wild life General Law (LGVS 2018)	x	x	x	x		x					
Rural Sustainable Development Law (LDRS 2019)	x	x				x	x[a b c]	x		x	

(continued)

Table 3.1 (continued)

Instrument	Scenarios[d]						Internal forced climate mobility	Pre-existing socio-economic vulnerability	Human rights perspective		Trapped migrants
	Slow-onset	Sudden-onset	Sea-level rise	Relocations	Development projects	Violences			RBA	Differenciated	
Human Settlements, Land Management and Urban Development General Law (LGAH 2020)	x	x		x		x		x		x	
Mexican Constitution (CPEUM 2020)					x		x[b c]				
Victims General Law (LGV 2017)							x[b c]		x		
Climate Change General Law (LGCC 2020)	x	x	x	x	x		x[c]		x	x	
Civil Protection General Law (LGPC 2018)		x						x		x	

(continued)

Table 3.1 (continued)

Instrument	Scenarios[d]						Internal forced climate mobility	Pre-existing socio-economic vulnerability	Human rights perspective		Trapped migrants
	Slow-onset	Sudden-onset	Sea-level rise	Relocations	Development projects	Violences			RBA	Differenciated	
Draft General Law to Prevent, Attend and Comprehensively Repair Internal Displacement (2020)		x		x	x	x[c]	x		x	x	
Law for the Prevention and Attention of Internal Displacement in the State of Chiapas (2012)		x		x	x	x	x[b]	x	x[c]	x	
Law number 487 to Prevent and Address Internal Displacement in the State of Guerrero (2014)		x		x	x	x	x[b]	x	x[c]	x	

Note RBA: Rights-based approach
[a]Does not specify whether it is internal or international
[b]Does not specify whether it is induced by climate change
[c]With limitations
[d]Mentions to mobility causes are reported, but the instruments do not draw the link with the effect (i.e. mobility)

> human rights, the right to health, Indigenous Peoples, local communities, *migrants*, children, people with disabilities and people in vulnerable situations' rights and the right to development, as well as gender equality, the empowerment of women and intergenerational equity must be strictly respected when adopting measures to address climate change. (SEMARNAT 2021, p. 19, emphasis mine)

The mention of migrants' human rights is definitely a step towards establishing the link between migration and climate change. The PECC can be a first step towards recognizing and articulating climate mobilities in public policy. However, the PECC does not specify if this applies to internal or external migrants. Neither does it mention whether it deals with or includes migrations induced by climate change, or indeed, if it refers to migration entirely or mainly induced by other factors. This is regrettable more than five years following the General Law on Climate Change mentioned the climate change mobility issue.

Beyond this imprecise mention and literal analysis, the PECC clearly stands out from the other public policy instruments according to the analysis grid applied in this book. It integrates and emphasizes almost all of the elements of analysis identified in previous chapters. Firstly, it endorses the territorial axis of climate policies (see e.g. p. 10 regarding restoration). It specifies the requirement to consider local specificities (whether territorial or socioeconomic). Among many other examples, Specific Action 3.2.3 illustrates this point: "[t]o impulse the diversification of reliance on ecosystem services based on the territory and soil purpose through their inclusion in programs of ecological, land use and urban development, as well as tourism and agriculture" (See also SEMARNAT 2021, pp. 11, 22, and Strategic Priority 1.7).

Secondly, the 2021–2024 PECC clearly and firmly establishes that climate policy must articulate a rights-based approach. The fourth Priority Objective illustrates this requirement: "[r]einforce coordination mechanisms, financing and implementation (…) with the human rights approach." Similarly, Strategic Priority 2.1 reads: "[i]mpulse energy transition (…) with a rights-based approach to promote its generation and sustainable and inclusive use" (see also SEMARNAT 2021, pp. 11, 15, 19, 22, 23, 28, and 96). The PECC places a particular emphasis on participation (pp. 23: "social participation and key actors," 28, 86, and Strategic Priority 4.1). More specifically, the PECC portrays participation as a tool to reduce gaps (p. 31).

The human rights approach referred to here is differentiated and reflects the book's analytical grid. Among other examples, two Specific Actions illustrate the differentiated approach. Specific Action 1.1.8 reads: "[r]einforce coordination between levels of government levels and sectors to handle emergencies, *considering the population's different needs*" (emphasis mine), and Specific Action 4.6 0.7 specifies the necessity to "[i]nclude differentiated attention and gender criteria in the climate change policy so that indigenous peoples and communities, women, and young people participate, contribute, and have access to adaptation and mitigation processes under equal conditions and rights." The PECC also specifies that the differentiated approach applies to sudden-onset disasters (SEMARNAT 2021, p. 33).

The PECC adopts a differentiated approach to the three categories of particularly vulnerable populations addressed in the book: women (pp. 11, 15, 23, 28—in terms of participation, 31, 52, 98, and Specific Actions 4.6.7 and 4.7.8), Indigenous Peoples (pp. 11, 15, 23, 28—in terms of participation, 31, 52, 98, and Specific Actions 4.6.7 and 4.7.8), and individuals with limited economic resources. The attention to these people and communities in particular, stands out: poverty, marginalization, and pre-existing socioeconomic vulnerability is the object of the PECC's recurring focus (pp. 10, 11, 15, 23—it indicates that the analysis of pre-existing and future vulnerability is the adaptation cycle's first phase, 86, Priority Objective 1, Strategic Priority Strategy 1.1, and Specific Actions 1.1.9 and 4.2.9). The PECC very clearly expresses this p. 15: "socioeconomic conditions play a fundamental role in determining the degree of the population's vulnerability to climate change impacts. Factors derived from multidimensional inequalities, produced by economic and social development processes, can increase people's vulnerability" (references omitted).

The PECC does not mention all of the climate mobility scenarios. However, it does extensively cover sudden-onset disasters (e.g. p. 11) and slow-onset environmental deterioration. Regarding slow-onset environmental deterioration, the PECC stands out from other Mexican public policy instruments that do not clearly establish the link between climate change and slow-onset phenomena (see e.g. pp. 14–15, Strategic Priorities 1.3 and 3.1, and Priority Objective 3). It places particular emphasis on the slow-onset climate change impacts for water resources (pp. 10, 14, Strategic Priority 1.2), pollinators (p. 14), deforestation (p. 27), and the urgency to protect forests, jungles, and mangroves (pp. 73–74). It also mentions and details the extent of sea-level rise (p. 22) and affirms support for "clean energy" and projects related to renewable energy (p. 25, Strategic Priority 2.1). However, it does not mention the impact of development projects on the population (and more specifically, relocations). Finally, it does not mention violence and conflicts related to the scarcity of natural resources (but it does mention in Strategic Priority 1.2 a structural cause of conflicts related to access to water: equitable distribution) and areas designated unsuitable for human settlement (relocation).

Like the PECC (2021), the Mexican 2020 Nationally Determined Contributions (NDC, SEMARNAT 2020) is undoubtedly a step forward in recognizing and addressing internal forced climate migration.[37] More specifically, it almost entirely reflects the analysis matrix of the book. It mentions in detail the link between human rights and climate change (p. 9). The NDC specifies the differentiated human rights approach. It elaborates on a differentiated approach towards (among others) women, Indigenous Peoples, and groups "in situations of vulnerability" (p. 10). The NDC also refers to environmental procedural (p. 10) and material rights (e.g., water and a healthy environment—pp. 18–19).

It also endorses the fight against poverty and social inequalities as a central focus. Among other elements covered, it reports that "Mexican public policies have focused

[37] Greenpeace México initiated proceedings against the 2020 Mexican NDC regarding the mitigation part of the NDC. They succeeded to have the NDC suspended in a September 2nd Judgment until the Tribunal pronounces itself on the merits of the case (Tenth Collegiate Court in Administrative Matters of the First Circuit, *Décimo Tribunal Colegiado en Materia Administrativa del Primer Circuito*, 2021). At the time being it is not clear if the suspension concerns the whole NDC (and, hence, the reference to climate displacement in the NDC adaptation part) or only the NDC mitigation part. Greenpeace announced it would ask the Tribunal to clarify this aspect.

on poverty eradication," and the impacts of climate change are more significant for people in situations of "social, economic and environmental vulnerability" (p. 9). It specifically mentions the three particularly vulnerable groups identified in the book: women, Indigenous Peoples, and people with scarce economic resources.

It also mentions migrants (p. 9), but without specifying whether or not they are international or national migrants and whether climate change or other factors have driven their mobility. This lack of precision is not as regrettable as that of the PECC (2021): similar to the General Law on Climate Change (2020), the NDC explicitly mentions internal climate migrants (p. 16). The "Prevention and attention to negative impacts on the human population and in the territory" it foresees contains seven lines of action to be taken. Among them, line of action A7 points to forced internal climate migration ("identify and address the forced displacement of people due to the climate change adverse impacts"). The NDC presents it as a new line of action, with good reason. There is only one other explicit mention of the phenomenon in the entire Mexican political-legal framework. This is in the LGCC (2020), which does not foresee the adoption of measures for addressing such phenomena as the NDC does.

Furthermore, the NDC links forced internal climate migration with pre-existing socio-economic vulnerability (p. 15: "These are socio-economic conditions, coupled with the increase in hydrometeorological phenomena, such as floods and droughts, circumstances that have forced people to move from their homes and seek new opportunities in other territories"), and highlights Indigenous Peoples and women's vulnerability. The NDC also reflects the territorial approach adopted in the book (e.g., p. 10), and more specifically, in the context of climate mobility. It mentions two impacts of migration for the destination territories which we report in the book: conflicts and "competition for resources that can exacerbate pre-existing vulnerabilities" (p. 15).

The NDC mentions two explicit forced internal climate migration scenarios. Unlike the other political-legal instruments, the NDC establishes the link between migration, sudden-onset disasters (floods) and slow-onset phenomena (droughts). The NDC also mentions other impacts of climate change that trigger forced mobility but does not link them with the migratory phenomenon (similar to all the other political-legal Mexican instruments). It generally refers to the impacts of climate change on biological diversity (p. 18) and water scarcity (pp. 19–20). It also more specifically mentions changes in rainfall patterns (p. 16), deforestation (pp. 18–19), the overexploitation of natural resources (p. 16), pests and animal based diseases "facilitated and exacerbated by climate change" (p. 17), and impacts on marine ecosystems (p. 16, which includes the impact of rising sea-levels, but the NDC does explicitly state this). The NDC also alludes to islands impacted by climate change but, again, does not explicitly establish the link with sea-level rise (Action Line C6, p. 19).

The NDC does not, however, mention three significant mobility scenarios: conflicts, relocations, or development projects linked to renewable energies (despite foreseeing their protection against climate change impacts). Another limitation of the NDC is that it omits the trapped migrants' phenomenon.

Despite these limitations, the NDC should be greeted as an instrument that, together with the PECC (2021), constitutes a break with previous instruments. It explicitly recognizes the phenomenon, captures it as the object of a Line of Action (in theory, the adoption of measures to address it will follow), locates the phenomenon in a human rights framework, and articulates a differentiated approach towards Indigenous Peoples, women, and people and communities in pre-existing socio-economic vulnerability.

The National Climate Change Strategy (ENCC) (SEMARNAT 2013) does not mention forced internal migration, neither does it distinguish between the three categories of forced internal climate migrants. However, it does mention essential elements concerning the book's analytical grid. It points out the factors that generate climate mobility and establishes a differentiated approach towards women, Indigenous Peoples, and (to a lesser extent) those living in poverty.

In more detail, the ENCC considers sudden-onset disasters. The Glossary provides a series of relevant definitions: disaster, comprehensive risk management, risk reduction, and disaster risk. It also mentions the "attention to the population affected by disasters such as hurricanes and floods" (P1, Line of action P1.18), research on the matter (P3, Line of action P3.6), training the people responsible "for budget allocation" (P3, Line of action P3.17), and the reduction of vulnerability (P6.2; A1, A1.3, A1.5, and A1.12). In this respect, it defines vulnerability as the

> Level to which a system is susceptible or not capable of withstanding the adverse effects of Climate Change adverse effects, including climate variability and extreme events. Vulnerability is a function of the character, magnitude, and speed of climatic variation to which a system is exposed, its sensitivity, and its capacity to adapt. (Glossary)

In the same way, it links vulnerability reduction to poverty, a factor that generates trapped migrants. It explicitly mentions social vulnerability as an indication of vulnerability which determines the risk of disaster (at Fig. 4.1). It also provides the framework for considering poverty in a comprehensive definition of risk management.[38] Additionally, it prescribes a differentiated approach towards women (and to a lesser extent towards Indigenous Peoples) (5.2; P1, P1.7; P4, P4.7; 6.2; A1 and A1.11).

On the causes that generate forced mobility, the ENCC evokes sudden-onset disasters (P3, Line of Action P3.6; A1, A1.12), the slow-onset deterioration of the environment (6.3-4, A1, A1.7; A3), and relocations decided upon by public authorities (A1, A1.5). However, it does not indicate sea-level rise, the violence induced by

[38] Specifically,

> Based on the above elements, municipalities with high disaster risk were defined as high vulnerability and high risk of climatic events. The results show that, of the more than 2 440 municipalities in the country, 1 385 are in this category, and they concentrate 27 million inhabitants. Among them, 255 municipalities belong to the Crusade against Hunger, concentrating 10.4 million inhabitants. (SEMARNAT 2013, 6.1 Diagnosis of vulnerability to climate change).

See also Figs. 4.1 and 4.2, and the Atlas of water vulnerability in Mexico to climate change (AVHMCC), Arreguín Cortés et al. (2015). The AVHMCC incorporates six themes: health, education, housing, employment and income, and population (Soares et al. 2015).

the scarcity of natural resources amplified by climate change, or the mobility generated by development projects linked to renewable energies (it rather supports these projects—pp. 45–49).

In short, the ENCC partially addresses the causes identified within forced internal climate mobility and some essential aspects of it (poverty, differentiated approach). It falls short in articulating the protection requirement: it does not mention the rights of people or groups of people affected by mobility or the measures to guarantee effective human rights enjoyment.

The Mexican instruments that collect and disseminate information on the impacts of climate change do not endorse a differentiated protection approach. They do not cover the six forced internal climate mobility scenarios or the three forced internal climate migrant categories. However, they play an essential role: they provide the empirical data necessary to identify the areas where some of the causes that generate climate mobility occur, and identify the most vulnerable people and communities (including social vulnerability).

The Climate Change Information System website mentions it "integrates, updates and makes available to the public the statistical, geographical information and indicators generated and available in Mexico on topics such as: climate, greenhouse gas and compound emissions, mitigation projects, vulnerability, risks, population and biodiversity."[39] For its part, the National Atlas of Climate Change Vulnerability (ANVCC), developed by the National Institute of Ecology and Climate Change (INECC 2019), aims to be

> a structured and systematic set of maps that show territorial vulnerability to climate change, guide strategies' implementation within the adaptation process, and an input for decision-making in development planning in Mexico.[40]

It uses three vulnerability indexes: exposure, sensitivity, and adaptive capacity. The sensitivity and adaptive capacity indices can provide elements in order to propose a differentiated approach (women, Indigenous Peoples, and persons with limited economic resources). Of the six specific problems, three are directly related to the sudden-onset disaster scenario: human settlements' vulnerability to floods (pp. 32–35), landslides (pp. 56–75), and the vulnerability of livestock production to floods (pp. 176–211). On the gradual environmental deterioration scenario (slow-onset phenomena), it highlights livestock production (pp. 140–174) and foraging (pp. 104–139) and vulnerability to water stress. Beyond these specific problems, it also indicates, biodiversity loss and water resources (p. 15), pressure on natural resources (p. 119), deforestation (p. 184), and erosion (p. 117) among other elements.

The ANVCC Technical Sheets for Climate Change Vulnerability Evaluation dedicated to human settlements' vulnerability in the context of floods (SEMARNAT & INECC, n.d.(a)) and landslides (SEMARNAT & INECC, n.d.(b)) are also particularly instructive. Both refer to sudden-onset disasters and provide elements that echo the

---

[39] Climate Change Information System website. Available 30 November 2021, at http://www.gaia.inegi.org.mx/sicc/.

[40] ANVCC website. Available 30 November 2021, at https://www.cambioclimatico.gob.mx/atlas-nacional-de-vulnerabilidad-al-cambio-climatico/.

book's analytical grid. They signal the essential role of public policy instruments (e.g. the Municipal Risk Atlas, the Civil Protection Program, the Municipal Contingency Plan), and disaster declarations (SEMARNAT & INECC n.d.(c)). The Technical Sheet on landslides mentions the vulnerability related to settlements located in areas exposed to this risk. In this respect, we saw in Chap. 2 that the most economically vulnerable populations are more likely to settle in areas exposed to hazards generated by climate change (this also applies to the people who settle in these areas as a destination):

> Human settlements are usually located in areas not suitable for settlements because of population growth. They are most susceptible to extreme hydrometeorological events such as floods caused by frequent and intense rainfall. Therefore, it is necessary to evaluate where the population concentrates and if it is in areas susceptible to flooding. (SEMARNAT & INECC, n.d.(c), Criterion 2, variable 2.1, references omitted)

It explicitly mentions sea-level rise at a global level (p. 13). Even though the ANVCC does not detail this impact or disaggregate the effects of this scenario at a Mexican level, it indicates sea-level rise has the lowest impact when compared to other impacts (e.g. storms and severe climates, p. 15). Similarly, the ANVCC mentions rural/urban migration driven by the loss of soil fertility and its erosion (p. 117) and migration driven by dengue incidence (p. 84). However, it does not specify whether it is referring to internal or international mobility. Also, the ANVCC refers to poverty, but it does not mention it as a pre-existing vulnerability. It instead indicates poverty is a consequence of the impacts of climate change (pp. 117, 141). The ANVCC suffers from other limitations. It omits three climate mobility scenarios: the conflict related to the scarcity of natural resources amplified by climate change, human settlements potentially affected by public authorities' decision to relocate them, and development projects linked to renewable energies. Ultimately, it does not endorse a differentiated rights-based approach.

3.3.2 Federal Planning Instruments: Territorial and Ecological Development and Planning

Firstly, climate change is a significant issue in the National Development Plan 2019–2024 (*Plan Nacional de Desarrollo 2019–2024*, "the Plan") (SEGOB 2019). It is mentioned numerous times, as a "global and cross-border challenge" (p. 15), a risk (p. 161), a "risk and threat in matters of national security" (p. 62), and a multilateral foreign policy issue, among other things (p. 65). Moreover, the Plan establishes "vulnerability to climate change, resilience strengthening, and adaptation and mitigation capacities, especially if it impacts the most vulnerable populations or regions" as one of the criteria to which the Mexican Government "will adjust" (p. 38).

The plan claims its content has a territorial dimension (pp. 38, 186 and 188). However, it does not consider forced internal climate migrations generically. It instead focuses on international migration.[41] Despite this omission, it addresses essential aspects of forced internal climate migration. The Plan considers two scenarios that generate forced internal migration induced by climate change: slow-onset environmental deterioration and sudden-onset disasters.

It mentions the link between environmental deterioration, climate change, and mobility (p. 104):

> Environmental degradation affects people's quality of life, reduces their development possibilities, and contributes to land abandonment and migration. It has been caused mainly by resources' intensive use and the growing demand for energy to satisfy the population and productive sectors' necessities.

It does not, however, establish a link between slow-onset phenomena and mobility (see e.g. pp. 65, 104, 184, 185). It also mentions sudden-onset disasters on multiple occasions (and more specifically, extreme hydrometeorological phenomena—see e.g. p. 186). It raises the link between sudden-onset disasters and climate change (pp. 75, 114, 184–185) but, once again, it does not establish the link between sudden-onset disasters induced by climate change and forced internal mobility.

Departing from these shortcomings, the Plan endorses the protection requirements. It explicitly establishes the link between climate change and human rights (p. 107) and proposes a differentiated approach towards women (pp. 26, 34, 58, 90 and 180), including in the context of sudden-onset disasters (Strategy 2.7.1). Similarly, the Plan raises poverty as the pivotal issue that should guide the government's public policies (pp. 35, 52, 86–90, 114). This is essential as poverty is a factor that generates trapped internal climate migrants. In more general terms, the plan highlights that vulnerability amplifies the impact of climate change and, *vice-versa*, climate change aggravates pre-existing inequalities (pp. 38, 105, and 185).

Secondly, national instruments on territorial and ecological planning can, in theory, address forced internal climate migration. The object of the Program of General Ecological Planning of the Territory (*Programa de Ordenamiento Ecológico General del Territorio*, POEGT) (SEMARNAT 2012)[42] and the Sectorial Program for Agrarian, Territorial and Urban Development (*Programa Sectorial de Desarrollo Agrario, Territorial y Urbano*, PSDATU) (SEDATU 2020) grounds this potential.

For example, the General Law on Ecological Balance and Environmental Protection (LGEEPA 2018) establishes the POEGT's guidelines and indicates that the POEGT must consider aspects related to environmental deterioration (Articles 19 and 20) and the adequacy of human settlements with ecological conditions scenarios (Articles 19, Fr. II-IV, 20) (relevant for slow-onset phenomena and conflicts generated by scenarios stemming from the scarcity of natural resources). It also mentions "natural phenomena" (Article 19, Fr. III), which may refer to the sudden-onset disasters scenario. Another illustration is Article 28, Fr. VII of the General Law on Climate Change (LGCC) (2020). It links adaptation to climate change, ecological

---

[41] See SEGOB (2019), pp. 7, 14–16, 27, 34, 46, 64, 68–71, 126–127 and 204. *Contra*, see p. 176 regarding Indigenous Peoples and rural–urban mobility.

[42] On Mexican territorial ecological planning programs from a historical perspective, see Santana et al. (2015), and Rosete (2006).

land management, human settlement and urban development, and internal climate mobility:

> Within the scope of their competences, the federation, the federated entities and the municipalities, must execute adaptation actions in the elaboration of policies, the National Strategy, the Special Climate Change Program, the National Adaptation Policy and programs in the following areas: (…) Ecological planning of the territory, internal displacement of people caused by phenomena related to climate change, human settlements and urban development.

In the same sense, the Law portrays various aspects of the POEGT and PSDATU as adaptation actions to climate change in Article 29. It expressly mentions the "programs on human settlements and urban development" and "programs on development tourist" (Fr. XV and XVI). It also signals aspects directly associated with the forced internal climate mobility scenarios (slow-onset environmental deterioration and the conflicts generated by the scarcity of natural resources,[43] sudden-onset disasters—Fr. VII, and rising sea-levels).[44]

However, when analyzing these Programs, it turns out that the contributions are not as valuable as they may first appear. Firstly, academics indicate the limitations of the programs that can restrict their potential to address forced internal climate migration. Fernández and Bocco (2003) highlight civil society's participation and involvement is a prerequisite for the POEGT to "strengthen local capacities, those of the consultants who support the studies and those of the technical body of the communities that require the appropriation of these processes" (p. 21). In this regard, Sánchez Salazar et al. (2013) point out the program's deficiencies in terms of "social participation" (p. 39).[45] In another way, Wong-González (2009) argues that the Programs' sectoral nature is a limitation: "they cannot become substitutes for a comprehensive policy of sustainable regional-territorial development" (p. 14). To achieve greater efficiency and operability, they must "rethink" the territory to integrate "complexity, transdisciplinarity, diversity, virtuality and flexibility" (p. 33).

Similarly, Sánchez Salazar et al. (2013) point out a series of drawbacks and challenges. Among them are the deficient "planning culture at different government levels" (p. 30), the necessity to adopt a comprehensive and sustainable approach (p. 37), the need to consider programs as "decision-making benchmarks" (p. 37), the legal and institutional weaknesses (p. 37), and problems of vertical and transversal

---

[43] See e.g. the "strategic infrastructure in matters of water supply, health services and production and supply of energy" (Fr. XVIII), the "management, protection, conservation and restoration of ecosystems, forest resources and soils" (Fr. III), the "programs for the conservation and sustainable use of biodiversity" (Fr. XIII), the "use, rehabilitation or establishment of irrigation districts" (Fr. VIII), and the "establishment and conservation of protected natural areas and biological corridors" (Fr. X).

[44] It mentions the "conservation, sustainable use, rehabilitation of beaches, coasts, federal maritime, terrestrial zone, land reclaimed from the sea and any other deposit formed with maritime waters for tourist, industrial, agricultural, fishing, aquaculture or conservation" (Fr. IV), and the "establishment of population centers or human settlements, as well as in the actions of development, improvement, and conservation of the same" (Fr. II).

[45] In the same sense, see Martínez Flores et al. (2015), and Hernández Santana et al. (2015).

articulation and territorial Programs integration, which entails a "territorial problems' fragmentation" (pp. 37–38, 40).[46] Various Secretariats (Ministries) adopt the Programs. It can induce congruence problems regarding the objectives and priorities ("duality of approaches," Hernández Santana et al. 2015, p. 137). More specifically, Hernández Santana et al. (2015) mention the pending task of articulating the Management Programs with Risk Management Programs, with a particular emphasis on risks regarding several migration scenarios (sudden-onset disasters and sea-level rise). Finally, Sánchez Salazar et al. (2013) also point out that the Programs lack "political support" (p. 39), and the need to "deepen socio-environmental indicators' design" (p. 40).

Secondly, and more specifically regarding internal climate mobility, our analysis of the programs reveals their limited potential despite the explicit attention they dedicate to climate change. To begin with, neither the POEGT nor the PSDATU specifically mention forced internal mobility induced by climate change. The POEGT (SEMARNAT 2012) establishes climate change is an essential component of land use planning[47] and points out the link between physical and social vulnerability on the one hand and climate change adaptation on the other (Strategies 20 and 26). This is relevant as pre-existing vulnerability is one of the factors that generate climate mobilities. Additionally, the POEGT mentions the particularly vulnerable people and communities addressed in the research regarding forced climate mobility: women (Strategies 33, 37 and 39), Indigenous Peoples (Strategies 4, 7, 12, 20, 33 and 37), and people with limited economic resources (Strategies 24, 25, 31, 33, 34, 36 and 40).

Regarding the climate forced mobility scenarios, the POEGT points to sudden-onset disasters (Strategies 1, 25, 26 and 35) and makes an explicit reference to civil protection programs "for high-risk dams and diverse hydraulic infrastructure" (Strategy 26). It also mentions the link between tourism and climate change on the coast (Strategy 22), which is relevant for the sea-level rise scenario. The slow-onset phenomena scenario is widely covered (Strategies 1, 3, 14 and 26). For example, the POEGT refers to the "environmental functionality of the territory" (Strategy 26) and the impacts on agricultural production (Strategy 35). Likewise, it mentions the link between climate change and environmental conflicts (Introduction, Strategy 42, Annex 2). However, it does not mention relocations or mobility induced by development projects linked to renewable energies (despite supporting the use of renewable energies, see strategy 20).

The PSDATU (SEDATU 2020) mentions the rural/urban internal mobility phenomenon and its effects but attributes it to problems surrounding land tenure (pp. 10–11). Poverty and vulnerability are guidelines (p. 8; Planning Principle 10, p. 8; p. 13; pp. 18, 21, 24–27, 31, and specific actions 1.3.6, 1.3.8), and Indigenous Peoples and women receive particular attention (Planning Principles 3 and 9, p. 8; p. 13; Priority objective 1, p. 25; Priority strategy 1.5; Priority objective 2; and Priority strategy 2.4). Human rights and environmental protection, are a priori an

[46] In the same sense, see Ugalde (2010).

[47] See e.g. SEMARNAT (2012), Introduction.

important factor in the Program as it "defines Priority Objectives, Priority Strategies, and Specific Actions (...) from a perspective oriented to guarantee individual and collective human rights' enjoyment (...) and respect, attention, and protection of the environment" (p. 8).[48] However, a careful reading of the Program counteracts this assertion: the PSDATU addresses the scenarios that generate forced internal climate migration in a limited way.

It hardly touches upon environmental protection and the addressing of its gradual deterioration (slow-onset phenomena scenario): mention of the environment is superficial and not specific (e.g., an "environment of harmony, respect, and protection of the natural environment," p. 30). It does not detail, for example, the problems of access to water, ecosystem deterioration, or deforestation induced by urbanization and irregular settlements, nor climate change impacts as a factor which amplifies environmental deterioration.

Regarding the other scenarios that generate mobility, the PSDATU addresses socio-environmental conflicts (Planning Principle 13, p. 9 and p. 12, and Specific Actions 2.3.4 and 2.3.5) and mentions sudden-onset disasters (Specific actions 1.2.1, 1.2.2, 2.1.2; and Priority Strategies 3.2, 3.3) without detailing any specific measures to prevent them or address affected populations' particular cases (except for earthquakes). The only references to climate change are in the context of sudden-onset disasters (p. 33. In the same vein, see Specific Action 1.2.2 and Priority Strategy 4.3). The PSDATU mentions relocations, (Specific action 1.3.9 where it mentions "relocations") and more specifically, the "fair relocation of the population living in risk areas" (Specific action 3.2.5). Nevertheless, it does not specify what a "fair" relocation is (or how to achieve it), and it does not mention climate change in the context of relocations. Instead, it mentions access to strategic land (Specific action 1.3.9). It also refers to the violence generated by access to scarce natural resources, mentioning threats, forced eviction, and harassment to control lands (Specific action 4.2.1).

3.3.3 *The Specific Case of Federal Instruments that Address* **Sudden-Onset** *Disasters*

We distinguish between the instruments that (i) focus on prevention and reaction capacities and organization, (ii) generate empirical information to map risks and vulnerability, and (iii) those of a financial nature. Our purpose is not to provide a general analysis of attention to sudden-onset disasters in Mexico.[49] Instead, we analyze public policy instruments to determine whether they identify all of the three forced internal climate migrants categories (emergency forced climate migrants,

[48] In the same vein, see SEDATU (2020), Planning Principle 8, p. 8; pp. 13, 19, and Priority Goals, p. 20. Ibid., Planning Principles 3 and 9, p. 8; p. 13, and Priority objective 1, p. 25.

[49] For such studies, see among others Hernández et al. (2016), Aguilar and Villanueva (n.d.), Saldana-Zorrilla (2015), Audefroy et al. (2018), Estrada Díaz (2014), and OECD (2013).

forced climate migrants, and trapped migrants) and, also, whether they apply a differentiated approach towards women, Indigenous Peoples, and people with limited economic resources. These instruments apply to sudden-onset disasters. Therefore, we assume they also apply to the emergency forced climate migrants and forced climate migrant categories. We are yet to determine whether they apply to trapped internal migrants.

The General Law on Civil Protection (*Ley General de Protección Civil*, LGPC 2018) provides the framework for public policy instruments on sudden-onset disasters. The Law indicates government level attributions to reduce risks (Article 4, Fr. III). If "the emergency or disaster exceeds the municipality's response capacity or territorial demarcation, it will become the attribution of the corresponding federate entity according to the applicable legislation. If this is insufficient, the corresponding federal authorities will be informed." Local governments are the first level that responds to emergencies. The state and federal governments are involved only if the disaster or emergency escalates and exceeds local government capacities.

Concerning disaster response, the law indicates that special civil protection programs are

> the planning and operating instrument implemented ... in the context of a specific danger or risk derived from a disturbing agent in a given area or region, involving specific and vulnerable population groups, and which, due to their foreseeable characteristics, allow adequate planning time. (Article 38)

According to this definition, special civil protection programs seem to be an excellent tool for dealing with sudden-onset disasters. However, the elaboration of Special Programs and their implementation by the states, municipalities, and delegations are still pending. The information from the National Center for Disaster Prevention (*Centro Nacional de Prevención de Desastres*, CENAPRED) and the National Civil Protection System (*Sistema Nacional de Protección Civil*, SINAPROC) only contains guides for their preparation. Likewise, the information regarding states' civil protection systems shows they were mainly applied as a civil protection tool for public spectacles.

The National Civil Protection Program (*Programa Nacional de Protección Civil*, PNPC) (SEGOB 2014) establishes coordination strategies between government agencies and participating entities and their respective attributes to fulfill objectives, strategies, and lines of action. It recognizes the importance of approaches to preventive actions that reduce the risk of natural disasters supported with the knowledge of community vulnerabilities. However, it does not distinguish between the possible types of climate migrants to be dealt with.

The Plan for Assistance to the Civilian Population in Case of Disaster (*Plan de Auxilio a la Población Civil en Caso de Desastre*, Plan DN-III-E) (SEDENA 1966) originated from a presidential mandate in 1965. It is a "military operating instrument that establishes the general guidelines for the Mexican army and air force agencies to carry out relief activities for the population affected by disasters of natural or human origin" (SEDENA 2019a). It comprises of three phases: (a) prevention, (b) assistance (alerting, emergency plans, emergency coordination, damage assessment, security, search and rescue, strategic services, health and supply), and (c) recovery. It mentions activities that are relevant in the context of the internal climate migration scenarios (SEDENA 2019b), such as:

- Promoting military personnel's specialization and professionalization in environmental issues, giving conferences and courses on the subject,
- Participation in fighting forest fires,
- Applying the DN-III-E plan (*Plan for Assistance to the Civilian Population in Case of Disaster*), helping the population in cases of disasters anywhere in the national territory in order to contribute to the national effort to preserve people, their assets, and their environment, and
- Applying tree production and reforestation programs.

The Plan for Assistance to the Civilian Population in Case of Disaster is the main instrument to address the impacts of natural disasters on well-being (Constantino et al. 2011). However, there are doubts on the role of the Plan: repeated extreme events cannot be addressed "only through *ex-post* instruments or corrective strategies to assist the population; disasters are not the result of the occurrence of such types of phenomena's unpredictability, but rather of the privileged approach: build to rebuild" (Constantino et al. 2011, p. 25). Even when the Plan's application is formally foreseen in the context of preventive activities, their implementation probably does not have the same magnitude as in the emergency attention stages. Vera (1996) points out that the authorities confuse the disaster with the physical event so that the government's role occurs when the emergency occurs. As the months go by, it is then put to one side as new events occupy its attention. In the same way, Constantino et al. (2011) point out the "Mexican institutional design has focused more on reparation and assistance activities in the presence of extreme natural events than on the prevention of devastating effects of those" (p. 25).

Finally, two financial instruments addressed sudden-onset disasters. The Mexican Government extinguished the funds mentioned here (FONDEN and FOPREDEN) while writing these paragraphs (SEGOB 2020). However, we decided to maintain the paragraphs in the book as the limits indicated may inform the elaboration of new rules not adopted yet to replace or compensate for this termination (e.g., inclusion in other budgetary arrangements). The Natural Disaster Fund (*Fondo de Desastres Naturales*, FONDEN) was created in 1996 to allocate resources to reconstruct the damage suffered by a disturbing natural phenomenon (Morales 2017, p. 1). It provided economic resources to populations affected by extreme natural phenomena and the recovery of damaged public infrastructure. It was a useful tool for these disasters despite its reactive approach.

The General Law on Civil Protection's reform in 2003 created the Fund for Natural Disasters Prevention (*Fondo para la Prevención de Desastres Naturales*, FOPREDEN). It operated a shift of paradigm towards prevention:

> The federal government provides financial support to states and municipalities for disaster prevention programs through this fund (*FOPREDEN*) which complements FONDEN. Its creation in 2003 reflects a change in SEGOB's (*Government Ministry*) strategy from a

> reactive system to a proactive system. FOPREDEN co-finances federal, state and municipal projects in three main areas related to disaster prevention: risk assessment (e.g. Risk Atlas), risk mitigation or reduction (e.g. SAT or minor flood protection infrastructures), and strengthens prevention culture (e.g. educational materials). (Hernández Aguilar and Castillo Villanueva n.d., pp. 11–12)

The Funds' Rules and Guidelines reveal that they do not distinguish between internal climate migrant categories neither do they comprehensively articulate a differentiated approach as applied to women, Indigenous Peoples, and persons with limited resources. The Operational Rules of the Fund for the Prevention of Natural Disasters (*Reglas de Operación del Fondo para la Prevención de Desastres Naturales* 2010) emphasize the concept of vulnerability defined as "an affected system's susceptibility or propensity to suffer damage or loss in the presence of a disturbing natural phenomenon determined by its physical, social, economic and environmental conditions" (Article 6, Fr. XXXIX). It could ground a differentiated approach. However, it explicitly mentions women and Indigenous Peoples only in Annex 2 (Fr. IX, c) which concerns early warning systems and monitoring systems. Also, these Rules do not mention the case of trapped internal migrants. In addition, neither the Guidelines of the Fund for Emergency Attention (*Lineamientos del Fondo para la Atención de Emergencias*, FONDEN 2012) nor the Natural Disaster Fund's General Rules (*Reglas Generales del Fondo de Desastres Naturales* 2010) mention a differentiated approach according to individuals and communities affected and pre-existing vulnerability. Therefore, it is of no surprise that they do not specifically mention women or Indigenous Peoples and do not specifically address trapped internal migrants.

From the analysis conducted in this section, it can be seen that national public policies incorporate (or at least mention), in an incomplete manner, the characteristics of forced climate internal mobility: the causes that drive climate mobility, pre-existing vulnerability, and a differentiated rights-based approach. It is still necessary to address the phenomenon entirely and explicitly connect the dots with forced internal climate mobility (e.g., pre-existing vulnerability) (Table 3.2).

3.4 Conclusion

This chapter identified the current knowledge gap in terms of forced internal climate mobility in Mexico. We aim to at least partly fill this gap by analyzing the Mexican political-legal framework to determine whether or not it addresses the phenomenon.

Regarding the legal framework, we analyzed 20 regulations and identified their shortcomings. We can tabulate the result of our analysis, grouping the norms into four groups. Firstly, we found laws that ignore the phenomenon, the protection of affected individuals and communities and do not address the scenarios that generate the phenomenon. They are the Refugees Law, Complementary Protection and Political Asylum, the Migration Law, and the General Population Law.

Secondly, some laws ignore the phenomenon and the protection of the populations affected, but they do address the scenarios that generate the phenomenon. They

Table 3.2 Attention to climate forced internal mobility: public policy instruments' analysis

Instrument	Scenarios[d]						Internal forced climate mobility	Pre-existing socio-economic vulnerability	Human rights perspective		Trapped migrants
	Slow-onset	Sudden-onset	Sea-level rise	Relocations	Development projects	Violences			RBA	Differentiated	
Special Climate Change Program 2021–2024 (PECC) (SEMARNAT 2021)	x	x	x		x		x[a b c]	x	x	x	
Nationally Determined Contributions (NDC) (SEMARNAT 2020)	x	x	x[c]				x	x	x	x	
National Climate Change Strategy (ENCC) (SEMARNAT 2013)		x		x				x			
National Climate Change Vulnerability Atlas (ANVCC) (INECC 2019)	x	x					x[a c]	x			
ANVCC Technical Sheets (SEMARNAT & INECC, n.d.b) (SEMARNAT & INECC, n.d.c)		x	x[c]					x			

(continued)

Table 3.2 (continued)

Instrument	Scenarios[d]						Internal forced climate mobility	Pre-existing socio-economic vulnerability	Human rights perspective		Trapped migrants
	Slow-onset	Sudden-onset	Sea-level rise	Relocations	Development projects	Violences			RBA	Differentiated	
National Development Plan 2019–2024 (PND) (SEGOB 2019)	x	x						x	x	x [c]	
General Territory Ecological Planning Program (POEGT) (SEMARNAT 2012)	x	x				x		x		x	
Sectorial Program for Agrarian, Territorial and Urban Development (PSDATU) (SEGOB 2020)	x [c]	x		x		x	x [b c]		x		
National Civil Protection Program (PNPC) (SEGOB 2014)		x						x			

(continued)

Table 3.2 (continued)

Instrument	Scenarios[d]						Internal forced climate mobility	Pre-existing socio-economic vulnerability	Human rights perspective		Trapped migrants
	Slow-onset	Sudden-onset	Sea-level rise	Relocations	Development projects	Violences			RBA	Differentiated	
Assistance Plan for the Civilian Population in Case of Disasters (Plan DN-III-E) (SEDENA 1966)		x									
Natural Disasters Fund (FONDEN)		x									
National Natural Disasters Prevention Fund (FOPREDEN)		x								x[c]	
Rules of Operation of the Natural Disasters Prevention Fund (2010)		x								x[c]	
Guidelines of the Fund for Emergency Care FONDEN (2012)		x									

(continued)

Table 3.2 (continued)

Instrument	Scenarios[d]						Internal forced climate mobility	Pre-existing socio-economic vulnerability	Human rights perspective		Trapped migrants
	Slow-onset	Sudden-onset	Sea-level rise	Relocations	Development projects	Violences			RBA	Differentiated	
General Rules of the Natural Disasters Fund (2010)		x									

Note RBA: Rights-based approach
[a]Does not specify whether internal or international
[b]Does not specify whether or not induced by climate change
[c]With limitations
[d]Mentions to mobility causes are reported, but the instruments do not draw the link with the effect (i.e. mobility)

are the Ecological Balance and Environmental Protection General Law, the Animal Health Federal Law, the Plant Health Federal Law, the Waste Prevention and Integral Management General Law, the Wildlife General Law, the National Waters Law, the Sustainable Forest Development General Law, the Human Settlements, Territorial Planning Urban Development General Law, and the Sustainable Rural Development Law.

Thirdly, other norms could address the phenomenon despite their literal content: the Political Constitution of the United Mexican States and the Victims General Law. However, this potential is limited and depends on the interpretation provided by jurisdictional authorities.

In the fourth group, laws recognize the phenomenon *a minima*. The Climate Change General Law mentions the phenomenon and the scenarios that generate it but does not provide the internal climate migrants with a definition or criteria. In addition, the Civil Protection General Law is limited to the *sudden-onset* disasters scenario. Finally, the two laws adopted at a state level on forced internal mobility (Chiapas 2012; Guerrero 2014) partially apply to the phenomenon. Similarly, the 2020 Draft General Law to Comprehensively Prevent, Attend, and Repair Internal Displacement does not contemplate all forced internal climate mobility scenarios.

In short, we cannot conclude that the Mexican legal framework entirely addresses the phenomenon as we defined it and according to our analytical grid. However, we must nuance this conclusion. The scenarios that generate mobility are almost all covered. Furthermore, two laws contemplate poverty specifically, a factor that generates the phenomenon of trapped migrants (the Sustainable Rural Development Law and Human Settlements, Urban Development Territorial Order General Law).

Regarding public policy instruments at a federal level, we found restrictions similar to those identified in the legal framework: despite mentioning some essential characteristics of the phenomenon (e.g., vulnerability or a differentiated approach), public policies do not always apply a differentiated approach to all of the causes which drive climate mobility, and they do not address the case of trapped migrants. However, evoking some characteristics of the phenomenon is encouraging: it forms a solid foundation with which to address the phenomenon. What is, as yet, pending is mentioning and addressing the missing characteristics (e.g., all the causes that promote mobility—with a special attention given to rising sea-levels and gradual environmental deterioration which is generally omitted). There also remains a need to explicitly connect the dots between what is covered and forced internal climate mobility. In this respect, the 2020 Nationally Determined Contributions (SEMARNAT 2020) can be seen as a good step forward.

Finally, the reflections and conclusions of this chapter trigger another question: despite not being addressed by the legal and public policies framework generically, does the sub-national level fulfill the protection requirements? In the next chapter, we see that the answer is negative in the case of the State of Tabasco.

References

Albo A, Ordaz J (2011) Migración y Cambio Climático. El caso mexicano. Servicio de Estudios Económicos del grupo BBVA

Alscher S (2010) Environmental factors in Mexican migration: the cases of Chiapas and Tlaxcala. In: Environment, forced migration and social vulnerability. Springer, Berlin, Heidelberg, pp 171–185

Arreguín Cortés FI, López Pérez M, Rodríguez López O, Montero Martínez MJ (2015) Atlas de vulnerabilidad hídrica en México ante el cambio climático (AVHMCC). Instituto Mexicano de Tecnología del Agua, Jiutepec, Mor. https://www.atlasvulnerabilidad.inecc.gob.mx/page/fichas/ANVCC_LibroDigital.pdf. Accessed 30 Nov 2021

Audefroy JF, Padilla Lozoya R (Coord) (2018) Desastres asociados a fenómenos hidrometeorológicos. Instituto Mexicano de Tecnología del Agua, Jiutepec, Mor

Ávila Akerberg A (2012) Éxitos y fracasos de la legislación de cambio climático en América del Norte. Norteamérica 7(SPE):183–192

Bárcena Ibarra A, Samaniego J, Peres W, Alatorre JE (2020) La emergencia del cambio climático en América Latina y el Caribe: ¿seguimos esperando la catástrofe o pasamos a la acción? LC/PUB.2019/23-P. CEPAL. https://www.repositorio.cepal.org/handle/11362/45677. Accessed 30 Nov 2021

Bulman GM, Lamberti JC (2011) Parásitos y enfermedades parasitarias emergentes y reemergentes: Calentamiento global, cambio climático, transmisión y migración de especies. Evaluación de la participación del hombre. Sitio Argentino de Producción Animal. https://www.cutt.ly/0xJpU9E. Accessed 30 Nov 2021

Cámara de Diputados (Draft Law) (2020) Dictamen de la Comisión de Gobernación y Población, a la iniciativa con Proyecto de Decreto que expide la Ley General para Prevenir, Atender y Reparar Integralmente el Desplazamiento Interno, 29 September 2020. Dictámenes para declaratoria de publicidad. *Gaceta Parlamentaria*, Palacio Legislativo de San Lázaro, Núm. 5619-III, 29 September 2020. http://www.diputados.gob.mx/LeyesBiblio/iniclave/64/CD-LXIV-III-1P-186/02_dictamen_186_29sep20.pdf. Accessed 30 Nov 2021

Comisión Intersecretarial de Cambio Climático (CICC) (2016) Reglamento interno de la Comisión Intersecretarial de Cambio Climático. https://www.gob.mx/cms/uploads/attachment/file/261939/Reglamento_Interno_de_la_CICC_Modificado_el_8_de_diciembre_de_2016.pdf. Accessed 30 Nov 2021

Comisión Mexicana de Defensa y Promoción de los Derechos Humanos (CMDPDH) (2019a) Comunicado de la CMDPDH. Décimo Tribunal Colegiado en Materia Penal del Primer Circuito determina la facultad de la CEAV para realizar el reconocimiento y registro de víctimas de desplazamiento interno forzado. http://www.cmdpdh.org/2019a/03/decimo-tribunal-colegiado-en-materia-penal-del-primer-circuito-determina-la-facultad-de-la-ceav-para-realizar-el-recono cimiento-y-registro-de-victimas-de-desplazamiento-interno-forzado/. Accessed 30 Nov 2021

Comisión Mexicana de Defensa y Promoción de los Derechos humanos AC (CMDPDH) (2019b) Atención a víctimas. http://www.cmdpdh.org/temas/desplazamiento/atencion-a-vic timas. Accessed 30 Nov 2021

Comisión Nacional de Derechos Humanos (CNDH) (2018) Recomendación No. 90/2018 Sobre el caso de 971 personas víctimas de desplazamiento forzado interno en diversas comunidades del municipio de Chenalhó, Chiapas. https://www.cndh.org.mx/sites/default/files/documentos/2019-01/Rec_2018_090.pdf. Accessed 30 Nov 2021

Comisión Nacional de Derechos Humanos (CNDH) (2016) Informe Especial sobre Desplazamiento Interno Forzado en México. http://www.informe.cndh.org.mx/uploads/menu/15008/2016_IE_D esplazadosD.pdf. Accessed 30 Nov 2021

Consejo Nacional de Evaluación de la Política de Desarrollo Social (CONEVAL) (2018) Medición de la Pobreza, Chiapas, 2018. https://www.coneval.org.mx/coordinacion/entidades/Chiapas/Pag inas/Pobreza_2018.aspx. Accessed 30 Nov 2021

Constantino T, Roberto M, Dávila I, Hilda R (2011) Una aproximación a la vulnerabilidad y la resiliencia ante eventos hidrometeorológicos extremos en México. Política y Cultura 36:15–44

Constitución Política de los Estados Unidos Mexicanos (CPEUM) (2020) Last reform Diario Oficial de la Nación, 8 May 2020. http://www.diputados.gob.mx/LeyesBiblio/pdf_mov/Constitucion_Politica.pdf. Accessed 30 Nov 2021

Décimo Tribunal Colegiado en Materia Penal del Primer Circuito (2018) Revisión penal del décimo tribunal colegiado en materia penal del primer circuito, identificado como 208/2018

Décimo Tribunal Colegiado en Materia Administrativa del Primer Circuito (2021) Revisión adhesiva del Décimo Tribunal Colegiado en Materia Administrativa del Primer Circuito, identificado como 81/2021

Deheza E (2013) Cambio Climático, Migración y Seguridad. Política de buenas prácticas y opciones operacionales para México. United Services Institute. http://www.cenca.imta.mx/pdf/CC-Interim-report-Spanish-version.pdf. Accessed 30 Nov 2021

Estrada Díaz G (2014) Puesta en práctica de una política de desastres: los instrumentos de la gestión de riesgos en México. Bulletin De L'institut Français D'études Andines 43(3):611–632

Feng S, Krueger AB, Oppenheimer M (2010) Linkages among climate change, crop yields and Mexico–US cross-border migration. Proc Natl Acad Sci 107(32):14257–14262

Fernández GN, Bocco G (2003) El ordenamiento ecológico comunitario: una alternativa de planeación participativa en el contexto de la política ambiental de México. Gaceta Ecológica 68:9–22

Fondo para los Objetivos del desarrollo Sostenible (2013) Reporte Narrativo Final. Prevención de conflictos, desarrollo de acuerdos y construcción de la paz para desplazados en Chiapas, México. http://www.mdgfund.org/sites/default/files/Mexico%20-%20CPPB%20-%20Final%20Narrative%20Report.pdf. Accessed 30 Nov 2021

Hardoy J, Hernández I, Pacheco JA, Sierra G (2014) Institutionalizing climate change adaptation at municipal and state level in Chetumal and Quintana Roo Mexico. Environ Urban 26(1):69–85

Hernández Aguilar MLH, Castillo Villanueva Villanueva MLC (nd) El Sistema Nacional de Protección Civil en México y la Gestión de Riesgos en sus Estados y Municipios. https://www.academia.edu/33703465/EL_SISTEMA_NACIONAL_DE_PROTECCI%C3%93N_CIVIL_EN_M%C3%89XICO_Y_LA_GESTI%C3%93N_DE_RIESGOS_EN_SUS_ESTADOS_Y_MUNICIPIOS. Accessed 30 Nov 2021

Hernández Santana JR, Bollo Manent M, Méndez Linares AP, Espinosa Rodríguez LM (2015) Panorama contemporáneo del ordenamiento ecológico territorial en México Polígonos. Revista De Geografía 26:111–146

Hicke JA, Lucatello S, Mortsch LD, Dawson J, Domínguez Aguilar M, Enquist CAF, Gilmore EA, Gutzler DS, Harper S, Holsman K, Jewett EB, Kohler TA, Miller K (2022) North America. In: Climate change 2022: impacts, adaptation, and vulnerability. Contribution of Working Group II to the Sixth Assessment Report of the Intergovernmental Panel on Climate Change [Pörtner HO, Roberts DC, Tignor M, Poloczanska ES, Mintenbeck K, Alegría A, Craig M, Langsdorf S, Löschke S, Möller V, Okem A, Rama B (eds)]. Cambridge University Press. In Press. https://www.ipcc.ch/report/ar6/wg2/

Hunter LM, Murray S, Riosmena F (2011) Climatic variability and US migration from rural Mexico. University of Colorado, Institute of Behavioral Sciences, Boulder

Iniciativa que expide la Ley General sobre Desplazamiento Forzado Interno (Iniciativa de Ley) (2019) A cargo del Diputado Ulises García Soto, del grupo parlamentario de Morena, 19 February. http://www.sil.gobernacion.gob.mx/Archivos/Documentos/2019/04/asun_3841125_20190402_1550599387.pdf. Accessed 30 Nov 2021

Instituto Nacional de Ecología y Cambio Climático (INECC) (2019) Atlas Nacional de Vulnerabilidad al Cambio Climático México. 1st Edición (libro electrónico). INECC, México. https://www.atlasvulnerabilidad.inecc.gob.mx/page/fichas/ANVCC_LibroDigital.pdf. Accessed 30 Nov 2021

Instituto Nacional de Ecología y Cambio Climático (INECC) (2018) Evaluación Estratégica del avance subnacional de la Política Nacional de Cambio Climático. https://www.cambioclimatico.gob.mx/wp-content/uploads/2018/12/EVALUACION-ESTRATEGICA-AVANCE-SUBNACIONAL-PNCC.pdf. Accessed 30 Nov 2021

Instituto Nacional de Ecología y Cambio Climático (INECC) (2015) Reglamento del Sistema Nacional de Cambio Climático. https://www.gob.mx/inecc/documentos/reglamento-del-sistema-nacional-de-cambio-climatico. Accessed 30 Nov 2021

Instituto Nacional de Estadística y Geografía (INEGI) (2020) Encuesta Nacional de Calidad e Impacto Gubernamental (ENCIG) 2019. https://www.inegi.org.mx/contenidos/programas/encig/2019/doc/12_guerrero.pdf. Accessed 31 Mar 2022

Instituto Nacional de Estadística y Geografía (INEGI) (2010) Censos y conteo de población y vivienda. https://www.inegi.org.mx/temas/lengua/. Accessed 30 Nov 2021

Kniveton D, Schmidt-Verkerk K, Smith C, Black R (2008) Climate change and migration: improving methodologies to estimate flows. IOM

Kuribreña MA, Awada L, Mur L, Tizzani P (2019) Situación actual de la sanidad animal en el mundo: análisis de los eventos y las tendencias Organización Mundial de Sanidad Animal, Asamblea Mundial SG/2, 5 April. http://www.web.oie.int/downld/PROC2020/E_SANIDAD_ANIMAL.pdf. Accessed 30 Nov 2021

La otra realidad del desplazamiento interno en Guerrero (2019, 30 May) Síntesis de Guerrero. https://www.sintesisdeguerrero.com.mx/2019/05/30/la-otra-realidad-del-desplazamiento-forzado-interno-en-guerrero/. Accessed 30 Nov 2021

Ley de Aguas Nacionales (2020) Last reform Diario Oficial de la Federación, 6 January 2020. http://www.diputados.gob.mx/LeyesBiblio/ref/lan.htm. Accessed 30 Nov 2021

Ley de Desarrollo Rural Sustentable (LDRS, 2019) Last reform Diario Oficial de la Federación, 12 April 2019. http://www.diputados.gob.mx/LeyesBiblio/ref/ldrs.htm. Accessed 30 Nov 2021

Ley de Migración (LM) (2020) Last reform Diario Oficial de la Federación, 13 April 2020. http://www.diputados.gob.mx/LeyesBiblio/ref/lmigra.htm. Accessed 30 Nov 2021

Ley Federal de Sanidad Animal (2018) Last reform Diario Oficial de la Federación, 16 February 2018. http://www.diputados.gob.mx/LeyesBiblio/ref/lfsa.htm. Accessed 30 Nov 2021

Ley Federal de Sanidad Vegetal (2017) Last reform Diario Oficial de la Federación, 26 December 2017. http://www.diputados.gob.mx/LeyesBiblio/ref/lfsv.htm. Accessed 30 Nov 2021

Ley General de Asentamientos Humanos, Ordenamiento Territorial y Desarrollo Urbano (LGAH) (2020) Last reform Diario Oficial de la Federación, 6 January 2020. http://www.diputados.gob.mx/LeyesBiblio/ref/lgahotdu.htm. Accessed 30 Nov 2021

Ley General de Cambio Climático (LGCC) (2020) Last reform Diario Oficial de la Federación, 6 November 2020. http://www.diputados.gob.mx/LeyesBiblio/ref/lgcc.htm. Accessed 30 Nov 2021

Ley General de Desarrollo Forestal Sustentable (LGDFS) (2020) Last reform Diario Oficial de la Federación, 13 April 2020. http://www.diputados.gob.mx/LeyesBiblio/ref/lgdfs.htm. Accessed 30 Nov 2021

Ley General del Equilibrio Ecológico y la Protección al Ambiente (LGEEPA) (2018) Last reform Diario Oficial de la Federación, 5 June 2018. http://www.diputados.gob.mx/LeyesBiblio/ref/lgeepa.htm. Accessed 30 Nov 2021

Ley General para la Prevención y Gestión Integral de los Residuos (2018) Last reform Diario Oficial de la Federación, 19 January 2018. http://www.diputados.gob.mx/LeyesBiblio/ref/lgpgir.htm. Accessed 30 Nov 2021

Ley General de Población (LGP) (2018) Last reform Diario Oficial de la Federación, 12 July 2018. http://www.diputados.gob.mx/LeyesBiblio/ref/lgp.htm. Accessed 30 Nov 2021

Ley General de Protección Civil (LGPC) (2018) Last reform Diario Oficial de la Federación, 19 January 2018. http://www.diputados.gob.mx/LeyesBiblio/ref/lgpc.htm. Accessed 30 Nov 2021

Ley General de Víctimas (LGV) (2017) Last reform Diario Oficial de la Federación, 3 January 2017. http://www.diputados.gob.mx/LeyesBiblio/ref/lgv.htm. Accessed 30 Nov 2021

Ley General de Vida Silvestre (LGVS) (2018) Last reform Diario Oficial de la Federación, 19 January 2018. http://www.diputados.gob.mx/LeyesBiblio/ref/lgvs.htm. Accessed 30 Nov 2021

Ley para Prevenir, Atender y Reparar Integralmente el Desplazamiento Forzado Interno, y se adiciona un Capítulo III, denominado "Desplazamiento Forzado Interno", que comprende el Artículo 175 Bis, al Título Quinto del Código Penal, ambas del Estado de Sinaloa (2021) Last reform Periódico Oficial, 29 Oct 2021. https://gaceta.congresosinaloa.gob.mx:3001/pdfs/leyes/Ley_159.pdf. Accessed 27 Aug 2022

Ley sobre Refugiados, Protección Complementaria y Asilo Político (2014) Last reform Diario Oficial de la Federación, 30 October 2014. http://www.diputados.gob.mx/LeyesBiblio/ref/lrpcap.htm. Accessed 30 Nov 2021

Ley Número 487 para Prevenir y Atender el Desplazamiento Interno en el Estado de Guerrero (Ley de Guerrero) (2014) Periódico Oficial del Gobierno del Estado No. 58 Alcance I, 22 July 2014. https://www.vlex.com.mx/vid/ley-numero-487-prevenir-575287502. Accessed 30 Nov 2021

Ley para la Prevención y Atención del Desplazamiento Interno en el Estado de Chiapas (Ley de Chiapas) (2012) Periódico Oficial del Estado Número 355, 22 February 2012. https://www.congresochiapas.gob.mx/new/Info-Parlamentaria/LEY_0100.pdf?v=Ng. Accessed 30 Nov 2021

Lineamientos del Fondo para la Atención de Emergencias FONDEN (2012) Diario Oficial de la Federación, 3 July 2012. https://www.gob.mx/cms/uploads/attachment/file/492577/LINEAMIENTOS_2012_-3jul12-.pdf. Accessed 30 Nov 2021

Lupián LEO, Campos FJA (2015) Migración y cambio climático en México. Cimexus 10(1):35–51

Martínez Flores V, Romo Aguilar M, Córdova Bojórquez G (2015) Participación ciudadana y planeación del ordenamiento territorial en la frontera norte de México. Espiral (guadalajara) 22(64):189–220

Mercado Mondragón JM (2018) Pueblos indios y desplazamiento interno forzado. El camino recorrido para el establecimiento de una ley para el estado de Chiapas. Cuicuilco. Revista de ciencias antropológicas 25(73):133–163

Mercado Mondragón JM (2016) El desplazamiento interno forzado en México. El Cotidiano 200:181–192

Moctezuma L (2002) Los migrantes mexicanos en los Estados Unidos y la inversión productiva en México. Migraciones Internacionales 1(3):149–162

Moraga JFH (2013) Understanding different migrant selection patterns in rural and urban Mexico. J Dev Econ 103:182–201

Moraga P, Meckievi S (2016 Análisis comparativo de legislación de cambio climático. Centro de la Ciencia del Clima y la Resiliencia (CR)2

Morales CG (2017) Recursos Federales para la Atención de Desastres Naturales en México. http://www.bibliodigitalibd.senado.gob.mx/handle/123456789/3715. Accessed 30 Nov 2021

Morales Vega LG (2012) Categorías migratorias en México: Análisis a la Ley de Migración. Anuario Mexicano De Derecho Internacional 12:929–958

Munshi K (2003) Networks in the modern economy: Mexican migrants in the US labor market. Q J Econ 118(2):549–599

Nawrotzki RJ, Riosmena F, Hunter LM (2013) Do rainfall deficits predict U.S.-bound migration from rural Mexico? Evidence from the Mexican census. Popul Res Policy Rev 32(1):129–158

Nawrotzki RJ, Riosmena F, Hunter LM, Runfola DM (2015) Amplification or suppression: social networks and the climate change—migration association in rural Mexico. Glob Environ Chang 35:463–474

Organisation for Economic Co-operation and Development (OECD) (2013) Estudio de la OCDE sobre el Sistema Nacional de Protección Civil en México. OECD Publishing, Paris

Orozco Hernández ME, Campos Cámara BL, Velázquez Torres D (2016) Sociedad del Riesgo en México: Análisis y Perspectivas. Universidad de Quintana Roo, Quintana Roo. https://www.researchgate.net/publication/312309083_Sociedad_del_Riesgo_en_Mexico_Analisis_y_perspectivas. Accessed 30 Nov 2021

Oyhantçabal W, Vitale E, Lagarmilla P (2011) El cambio climático y su relación con las enfermedades animales y la producción animal. In: XV Congreso Latinoamericano de Buiatría/XXXIX Jornadas Uruguayas de Buiatría. Centro Médico Veterinario de Paysandú,

pp 169–177. http://www.centromedicoveterinariopaysandu.com/wp-content/uploads/2014/08/med.-amb.-lagarmilla-y-oyhanzabal-2011.pdf. Accessed 30 Nov 2021
Papail J (2002) De asalariado a empresario: la reinserción laboral de los migrantes internacionales en la región centro-occidente de México. Migraciones Internacionales 1(3):79–102
Peguero Moreno M, Rubio Díaz-Leal L (2019) El desplazamiento interno forzado en México: respuestas del estado y litigio estratégico. Tirant lo Blanch, Mexico
Pérez B (2014) El tema del desplazamiento interno en la agenda pública: una comparación de su situación en México y Colombia. El Cotidiano 183:7–16
Querales-Mendoza ME (2020) "No se pueden llevar a mi esposo:" desaparición forzada y desplazamiento interno forzado en el contexto de la guerra contra el narcotráfico en Michoacán (México). Historia y Sociedad 39:105–129
Reglamento de la Ley de las Aguas Nacionales (2014) *Last reform Diario Oficial de la Federación*, 25 August 2014. http://www.diputados.gob.mx/LeyesBiblio/regley/Reg_LAN_250814.pdf. Accessed 30 Nov 2021
Reglamento de la Ley General del Equilibrio Ecológico y la Protección al Ambiente en materia de Ordenamiento Ecológico (Reglamento LGEEPA) (2014) Last reform Diario Oficial de la Federación, 31 October 2014. http://www.diputados.gob.mx/LeyesBiblio/regley/Reg_LGEEPA_MOE_311014.pdf. Accessed 30 Nov 2021
Reglas de Operación del Fondo para la Prevención de Desastres Naturales (2010) Diario Oficial de la Federación, 23 December 2010. http://www.cenapred.unam.mx/es/documentosWeb/Fopreden/reglasdeoperacion.pdf. Accessed 30 Nov 2021
Reglas Generales del Fondo de Desastres Naturales (2010) Diario Oficial de la Federación, 3 December 2010. https://www.gob.mx/cms/uploads/attachment/file/492579/REGLAS_GENERALES_FONDEN_-3dic10-.pdf. Accessed 30 Nov 2021
Reidmiller DR, Avery CW, Easterling DR, Kunkel KE, Lewis KLM, Maycock TK, Stewart BC (2017) Fourth national climate assessment, Volume II: Impacts, risks, and adaptation in the United States US Global Change Research Program, Washington, DC
Riosmena F, Nawrotzki R, Hunter L (2018) Climate migration at the height and end of the Great Mexican Emigration Era. Popul Dev Rev 44(3):455–488
Rosete F (2006) Semblanza histórica del ordenamiento ecológico territorial en México. SEMARNAT-INE, México 53. http://www.xmlui/bitstream/handle/publicaciones/177/504_2006_Semblanza_historica_orden_ecologico_Mexico.pdf?. Accessed 30 Nov 2021
Ruiz Meza LE (2012) Cambio climático y migraciones laborales en la Frontera Sur de México. Revista Luna Azul (35)
Salazar Cruz LM (2014) Modalidades del desplazamiento interno forzado en México. Iztapalapa. Revista de ciencias sociales y humanidades 35(76):53–81
Salazar LM, Álvarez Lobato JA (2018) Violencia y desplazamientos forzados en México. Cuicuilco. Revista De Ciencias Antropológicas 25(73):19–37
Saldaña-Zorilla SO (2007) Socio-economic vulnerability to natural disasters in Mexico: rural poor, trade and public response, vol 92. United Nations Publications
Saldana-Zorrilla SO (2015) Assessment of disaster risk management in Mexico. Disaster Prev Manage 24(2):230–248
Sánchez Salazar MT, Casado Izquierdo JM, Bocco Verdinelli G (2013) La política de ordenamiento territorial en México: de la teoría a la práctica. Reflexiones sobre sus avances y retos a futuro. In: Sánchez M, Bocco G, Casado J (eds) La política de ordenamiento territorial en México: de la teoría a la práctica. UNAM, SEMARNAT & INECC, Mexico, pp 19–46
Schwartz ML, Notini J (1994) Desertification and migration: Mexico and the United States. US Commission on Immigration Reform, San Francisco, CA, USA
Secretaría de la Defensa Nacional (SEDENA) (2019a) Qué es el Plan DN-III-E? https://www.gob.mx/sedena/acciones-y-programas/que-es-el-plan-dn-iii-e. Accessed 30 Nov 2021
Secretaría de la Defensa Nacional (SEDENA) (2019b) Cambio Climático Plan DN-III-E. https://www.gob.mx/sedena/acciones-y-programas/cambio-climatico-plan-dn-iii-e. Accessed 30 Nov 2021

Secretaría de la Defensa Nacional (SEDENA) (1966) Plan de Auxilio a la Población Civil en Caso de Desastre (Plan DN-III-E). https://www.gob.mx/sedena/acciones-y-programas/plan-dn-iii-e. Accessed 30 Nov 2021

Secretaría de Desarrollo Agrario, Territorial y Urbano (SEDATU) (2020) Programa Sectorial de Desarrollo Agrario, Territorial y Urbano 2020–2024 (PSDATU). Diario Oficial de la Federación, 26 June 2020. https://www.gob.mx/sedatu/acciones-y-programas/programa-sectorial-de-desarrollo-agrario-territorial-y-urbano-2020-2024. Accessed 30 Nov 2021

Secretaría de Gobernación (SEGOB) (2020) por el que se reforman y derogan diversas disposiciones de la Ley para la Protección de Personas Defensoras de Derechos Humanos y Periodistas; de la Ley de Cooperación Internacional para el Desarrollo; de la Ley de Hidrocarburos; de la Ley de la Industria Eléctrica; de la Ley Federal de Presupuesto y Responsabilidad Hacendaria; de la Ley General de Protección Civil; de la Ley Orgánica de la Financiera Nacional de Desarrollo Agropecuario, Rural, Forestal y Pesquero; de la Ley de Ciencia y Tecnología; de la Ley Aduanera; de la Ley Reglamentaria del Servicio Ferroviario; de la Ley General de Cultura Física y Deporte; de la Ley Federal de Cinematografía; de la Ley Federal de Derechos; de la Ley del Fondo Mexicano del Petróleo para la Estabilización y el Desarrollo; de la Ley de Bioseguridad de Organismos Genéticamente Modificados; de la Ley General de Cambio Climático; de la Ley General de Víctimas y se abroga la Ley que crea el Fideicomiso que administrará el Fondo de Apoyo Social para Ex Trabajadores Migratorios Mexicanos. *Diario Oficial de la Federación*, 6 November 2020. https://www.dof.gob.mx/nota_detalle.php?codigo=5604411&fecha=06/11/2020. Accessed 30 Nov 2021

Secretaría de Gobernación (SEGOB) (2019) Plan Nacional de Desarrollo 2019—2024 (PND) Diario Oficial de la Federación, 12 July 2019. https://www.dof.gob.mx/nota_detalle.php?codigo=5565599&fecha=12/07/2019. Accessed 30 Nov 2021

Secretaría de Gobernación (SEGOB) (2014) Programa Nacional de Protección Civil 2014–2018 (PNPC). Diario Oficial de la Federación, 30 April 2014. http://www.dof.gob.mx/nota_detalle.php?codigo=5343076&fecha=30/04/2014. Accessed 30 Nov 2021

Secretaría de Medio Ambiente y Recursos Naturales (SEMARNAT) (2021) Programa Especial de Cambio Climático 2021–2024 (PECC). Diario Oficial de la Federación, 8 November 2021. https://www.dof.gob.mx/2021/SEMARNAT/SEMARNAT_081121_EV.pdf. Accessed 30 Nov 2021

Secretaría de Medio Ambiente y Recursos Naturales. (Semarnat) (2020) Contribución Determinada a nivel Nacional: México. Actualización 2020 (NDC). https://www.cambioclimatico.gob.mx/contribucion-determinada-a-nivel-nacional-actualizacion-2020/#:~:text=El%20presente%20documento%20constituye%20la,2%20y%201%2FCP. Accessed 30 Nov 2021

Secretaría de Medio Ambiente y Recursos Naturales (SEMARNAT) (2013) Estrategia Nacional del Cambio Climático (ENCC). Diario Oficial de la Federación, 3 June 2013. http://www.dof.gob.mx/nota_detalle.php?codigo=5301093&fecha=03/06/2013. Accessed 30 Nov 2021

Secretaría de Medio Ambiente y Recursos Naturales (SEMARNAT) (2012) Programa de Ordenamiento Ecológico General del Territorio (POEGT). Diario Oficial de la Federación, 7 September 2012. http://www.semarnat.gob.mx/archivosanteriores/temas/ordenamientoecologico/Documents/documentos_bitacora_oegt/dof_2012_09_07_poegt.pdf. Accessed 30 Nov 2021

Secretaría de Medio Ambiente y Recursos Naturales (SEMARNAT), & Instituto Nacional de Ecología y Cambio Climático (INECC) (n.d.(a)) Atlas Nacional de Vulnerabilidad al Cambio Climático, Ficha Técnica vulnerabilidad de los asentamientos humanos por inundaciones. https://www.atlasvulnerabilidad.inecc.gob.mx/page/fichas/FICHA_TECNICA_IAH_ANVCC.pdf. Accessed 30 Nov 2021

Secretaría de Medio Ambiente y Recursos Naturales (SEMARNAT), & Instituto Nacional de Ecología y Cambio Climático (INECC) (n.d.(b)) Atlas Nacional de Vulnerabilidad al Cambio Climático, Ficha Técnica vulnerabilidad de los asentamientos humanos por deslaves. https://www.atlasvulnerabilidad.inecc.gob.mx/page/fichas/FICHA_TECNICA_DAH_ANVCC.pdf. Accessed 30 Nov 2021

Sistema Nacional de Cambio Climático (SINACC) (2018) Declaración del Sistema Nacional de Cambio Climático. https://www.gob.mx/cms/uploads/attachment/file/345665/Minuta_I_Reuni_n_2018_del_SINACC.pdf. Accessed 30 Nov 2021

Sistema Nacional de Cambio Climático (SINACC) (2014) Declaración del Sistema Nacional de Cambio Climático. https://www.gob.mx/cms/uploads/attachment/file/40773/2015_declaracion_sinacc.pdf. Accessed 30 Nov 2021

Soares D, Romero R, López R (2015) Conceptualización de vulnerabilidad y construcción de un índice de vulnerabilidad social. In: Arreguín Cortés FI, López Pérez M, Rodríguez López O, Montero Martínez MJ (2021) Atlas de vulnerabilidad hídrica en México ante el cambio climático. Instituto Mexicano de Tecnología del Agua, Jiutepec, Mor, pp 9–39. https://www.atlasvulnerabilidad.inecc.gob.mx/page/fichas/ANVCC_LibroDigital.pdf. Accessed 30 Nov 2021

Torrens O (ed) (2013) El Desplazamiento interno forzado en México: un acercamiento para su reflexión y análisis. CIESAS

Ugalde V (2010) La coordinación institucional del ordenamiento territorial en México. Revista De Geografía Norte Grande 47:105–120

United Nations (UN) (1998) Principios Rectores De Los Desplazamientos Internos, UN doc.E/CN.4/1998/53/Add.2, 11 February 1998, Oficina de Coordinación de Asuntos Humanitarios de las Naciones Unidas. https://www.acnur.org/prot/prot_despl/5bff2c864/principios-rectores-de-los-desplazamientos-internos.html. Accessed 30 Nov 2021

Velázquez Moreno AL (2017) Desplazamiento interno por violencia en México. Causas, consecuencias y responsabilidades del Estado. CNDH, Mexico

Vera G (1996) Reseña de Estado, sociedad y gestión de los desastres en América Latina. In: Lavell AM (ed) Estado, sociedad y gestión de los desastres en América Latina: en busca del paradigma perdido. Soluciones Prácticas, Lima, pp 153–156

Williams PC, Bartlett AW, Howard-Jones A, McMullan B, Khatami A, Britton PN, Marais BJ (2021) Impact of climate change and biodiversity collapse on the global emergence and spread of infectious diseases. J Paediatr Child Health 57(11):1811–1818. https://doi.org/10.1111/jpc.15681

Wong González P (2009) Ordenamiento ecológico y ordenamiento territorial: retos para la gestión del desarrollo regional sustentable en el siglo XXI. Estudios sociales (Hermosillo, Son) 17(SPE):11–39

Chapter 4
Shortcomings at the Sub-national Level: Illustration with the State of Tabasco

Armelle Gouritin and José Antonio León Mella

This chapter aims to determine if, despite the shortcomings of the national political-legal framework (or, conversely, because of them), sub-national public policies address internal forced climate migrations. Shifting focus to a more sub-national level is all the more relevant as the effects of climate change and climate mobility scenarios occur, precisely at this level. In the same way, Ruiz and Carvajal (2015) justify that local levels should have the capacity to address the phenomenon within the framework of decentralized governance (pp. 44–54).

We focus on the state of Tabasco's public policies and analyze two case studies in this state. Several factors which justify this choice are, firstly, the Groundswell report of the World Bank (Rigaud et al. 2018) which identifies that Tabasco will be a territory of origin. Secondly, the number of natural disaster Declarations linked to climate change in Tabasco (14 Declarations between 2000 and 2020) (SEMARNAT n.d.), the Atlas on water vulnerability in Mexico in a climate change context (Arreguín Cortés et al. 2015) and the National Atlas on Climate Change Vulnerability in Mexico (INECC 2019) which confirms the acute nature of climate change vulnerability in Tabasco. Similarly, academics note that the effects of climate change significantly impact Tabasco. This is especially the case with extreme hydrometeorological phenomena and sea-level rise (Ruiz Liévano 2019; Ramos Reyes et al. 2016a, b; Rivera Hernández et al. 2016; Magaña 2012; Aguilar et al. 2020). Finally, the State Action Program against Climate Change (PEACC) (SERNAPAM 2011a) and the State Development Plan (PLED) (COPLADET 2019) confirm the vulnerability of Tabasco.

We first provide a critical overview of the public policy instruments of the state on climate change, territorial and ecological planning, the Development Plan, and risk management (Sect. 4.1). We then develop our analysis at the municipal level. Academics highlight the shortcomings of municipal instruments in Mexico in general and in the case of Tabasco, more acutely.

Ruiz Liévano (2019) mentions the omissions of Municipal Development Plans (Tabasco has 17 municipalities) regarding climate policies and their shortcomings

A. Gouritin, *Climate Displacement in Mexico*,
https://doi.org/10.1007/978-3-031-10335-3_4

as regards harmonization with the National Development Plan and the State Development Plan. On the municipal risk atlas, Ruiz Rivera et al. (2015) point out that, despite "an undeniable evolution in terms of the content and scope of the atlases, there is also an evident dissociation between the conceptual and normative content of these documents, and their real territorial effects in terms of reducing vulnerability or effective prevention of disaster risk" (pp. 159–160), inaccuracies about their content and use, "lack of clarity on the instances responsible for its implementation" (p. 146), and lack of sanctions in case of non-compliance. The authors conclude that they are more "informative than regulatory" (p. 160), and that it is.

> therefore important to draw attention to the need to strengthen institutional environments at the municipal level, which becomes identified as the weakest link in the Mexican National Civil Protection System. As long as the legislative and public administration spheres do not recognize the enormous importance of municipalities in territorial management and take measures to renew the legal and institutional architecture in favor of this government level, there will be a little possibility to apply the principles of complementarity and effectiveness (Ruiz Rivera et al. 2015, p. 160)

Against this background, our analysis confirms the limitations of municipal instruments in the case of climate mobility and the practical consequences for people and communities affected. We develop two case studies with a different approach: In the first case, we analyze public decisions related to a sudden-onset disaster event (Sect. 4.2). We tackle the 2007 floods in the Centro municipality with the Operational Guidelines on the Protection of Persons in Situations of Natural Disasters (Operational Guidelines) (IASC 2011) to determine if the protection requirements with a differentiated approach towards women were respected. In the second case, we analyze the public policies of the Centla municipality to find out if they meet the protection requirements with a differentiated approach towards Indigenous Peoples in the context of the sea-level rise scenario (Sect. 4.3). In both cases we conclude the protection requirements have not been fulfilled.

4.1 Sub-national Public Policy Instruments and Internal Forced Climate Mobility: A Critical Stance

This section does not develop a detailed analysis of Tabasco's public policies on climate change. That would exceed the scope of this book. We focus, instead, on a limited set of seven relevant instruments and identify that they do not mention climate mobility. We also find shortcomings regarding protection requirements (differentiated rights-based approach). However, they at least partially address climate mobility scenarios concerning the causes of forced mobility. We then analyze two climate policy instruments (Sect. 4.1.1), two territorial and ecological planning instruments (Sect. 4.1.2), the 2019–2024 State Development Plan (Sect. 4.1.3), and two risk management instruments (Sect. 4.1.4).

4.1.1 Climate Policy Instruments

The State Climate Change Action Program (*Programa Estatal Acción ante el Cambio Climático*, PEACC) (SERNAPAM 2011a) does not indicate climate change-driven human mobility or internal climate migrants categories. It does, however, mention the influence of social and economic factors in terms of "direct consequences of sea-rising levels" (p. 125). It could refer (among other concepts) to pre-existing social vulnerability, but it does not come with guidelines or indicators. This void is essential, as pre-existing vulnerability is a trigger for forced mobility, as it also is in the case of trapped migrants. Additionally, the PEACC partially covers the scenarios that generate climate mobility. It explains in much detail two of the scenarios: sea-level rise and sudden-onset disasters. It also covers, to a lesser extent, the slow-onset environmental deterioration scenario. The two case studies we detail in the subsequent sections concern the sudden-onset disaster and sea-level rise scenarios in line with the importance the PEACC attributes to these scenarios.

As to sea-level rise, it mainly frames this as a trigger to sudden-onset disasters or its amplification of them (e.g. floods, pp. 96, 108, 115, 124, and 288). The PEACC also mentions the destruction of land (p. 115) and the link between sea-level rise and the other scenarios that generate climate mobility. For example, it generates or amplifies slow-onset environmental deterioration (coastal erosion and floods—p. 124, the nexus between sea-level rise, soils and waters salinization, and its adverse effects for "biological productivity relationships," "biodiversity"—p. 125, and vice-versa—pp. 125–126).

Sea-level rise and its real-world impact are empirically established, quantified, projected, and mapped in Tabasco.[1] It is essential: it allows the identification of the people and communities most likely to have to migrate. The PEACC also empirically reports, quantifies, projects, and maps sudden-onset disasters (floods and extreme rains) and their impact.[2] It also establishes the link between the disasters themselves and gradual environmental deterioration (e.g., p. 111).

As for the slow-onset environmental degradation scenario on its own (i.e., not linked to sea-level rise or sudden-onset disasters), the PEACC establishes the link between environmental deterioration and climate change (pp. 14, 125). It explicitly mentions droughts (pp. 39 and 171), water availability shortages (p. 171), and detrimental impacts on the environmental services of ecosystems (p. 171).[3] However, departing from the sudden-onset disasters and sea-level rise scenarios, it does not go on to detail how climate change amplifies gradual environmental deterioration or

---

[1] SERNAPAM (2011a), pp. 129–131, 148–164. Of importance for attention to the migratory phenomenon, the PEACC provides maps of the impacted areas (Ibid., pp. 115–127). Sea-level rise and its effects are of the most significant importance for Tabasco: it has the highest percentage value of importance compared to the other covers (or layers) (p. 131).

[2] SERNAPAM (2011a), pp. 12, 95–114, 188–189.

[3] One might think Tabasco's environment is in such good shape that the gradual environmental deterioration scenario is superficial. This is not the case. Among others, the Climate Change Action Program SERNAPAM (2011a) indicates soil deterioration. See also Ramos-Reyes et al. (2016) and COPLADET (2019, pp. 287–288).

provide quantified empirical data, projections, or maps. Seemingly, the PEACC only indicates the relocation scenario as an action to "support the process of adaptation to extreme rains" (p. 188) without detailing the measures or guarantees surrounding them. Furthermore, it does not mention the scenario of violence generated by the scarcity of natural resource.

Regarding protection requirements, the PEACC is severely limited. It does not mention human rights or a differentiated rights-based approach applied to women or Indigenous Peoples.

The Tabasco Gender Agenda for Climate Change (*Agenda de Género ante el Cambio Climático para el Estado de Tabasco*, AGyCC)[4] (SERNAPAM 2011b) mentions flood-induced mobility ("migration to other places or asking for support from a relative to stay temporarily," p. 49) and hurricane-induced mobility (pp. 82, 86). Also, the Agenda mentions the construction of "Support Networks to attend families during contingency events and protect families displaced from their homes" as an adaptation action (p. 197). The Agenda explicitly draws a link between the impact of climate change and mobility without limiting the phenomenon to temporary mobility. However, it does not clearly distinguish between emergency climate migrants and climate migrants and does not mention trapped climate migrants.

The Agenda also mentions pre-existing vulnerability (among others, poverty and discrimination) and signals a differentiated human rights approach towards women in the context of climate change (pp. 113 and 203). However, it does not mention it in the context of forced internal climate mobility.

Despite the previous elements that depart from the instruments analyzed so far and address critical aspects of climate forced internal mobility, the Agenda attempts to tackle the causes that generate the phenomenon. It mentions slow-onset environmental deterioration (among them, it refers to water scarcity generated by droughts p. 70) and sudden-onset disasters (floods, hurricanes). This said, it does not mention sea-level rise,[5] relocations decided by the authorities, or conflicts caused by a scarcity of natural resources.

4.1.2 Territorial and Ecological Planning

Academics commonly signal the importance of state land use planning and Risk Atlas instruments and the resultant shortcomings (Frejomil et al. 2006; Santana et al. 2015; Ruiz Liévano 2019).[6] In this section, we focus on the shortcomings of Tabasco's instruments from the forced internal climate migration perspective.

---

[4] For an analysis of the Agenda and its adoption process, see Salazar Ramírez (2011), Tuñón Pablos and Molina Rosales (n.d.).

[5] The Agenda mentions sea-level rise p. 8. However, it does not explain it further or attribute specific effects to it.

[6] In the same sense, see the State Development Plan (PLED) (COPLADET 2019), p. 266.

The Sustainable Planning Program of the Territory of the State of Tabasco (*Programa de Ordenamiento Sustentable del Territorio del Estado de Tabasco*, POSTET) (Government of Tabasco 2018)[7] portrays climate change as a priority issue (e.g. pp. 10 and 113–115). However, it fails to mention climate mobility. It does mention internal migration and provides detailed data (e.g. p. 143) but does not specify what proportion corresponds to forced migration, nor does it indicate whether climate change and the six climate mobility scenarios play a role in these migrations. Similarly, it does not identify or distinguish between climate migrants from emergency climate migrants or trapped climate migrants. In addition, the POSTET provides a panorama of poverty in Tabasco[8] but does not link poverty with environmental degradation or the impact of climate change. As to the rights-based approach, it does not endorse a differentiated approach.

Concerning the causes that generate climate mobility, the Program mentions sudden-onset disasters (e.g. p. 50), slow-onset environmental deterioration (e.g. pp. 59, 111–112, and 116–119), and sea-level rise without disaggregating its impacts (e.g. p. 113). It also indicates conflicts related to access to land (p. 15) but does not specify how climate change impacts them. Finally, it does not evoke relocations decided by public authorities (although it identifies human settlements located in dangerous areas, e.g. p. 109) and development projects linked to renewable energies.

The Tabasco Regional Ecological Planning Program (*Programa de Ordenamiento Ecológico Regional del Estado de Tabasco*, POERET) (BIENESTAR 2019) does not mention mobility, displacement, or human migration induced by climate change. It only mentions the internal mobility of Tabasco residents to other states or abroad (p. 36) without specifying which factors induce this migration. Furthermore, it does not distinguish between the forced internal climate migrants categories. On climate mobility causes, the program mentions slow-onset environmental deterioration,[9] sudden-onset disasters (with an emphasis on floods),[10] sea-level rise (pp. 76–77, 128), support for renewable energies (including the generation of such energy, pp. 114, 125, and 131), and social conflicts (pp. 77 and 116). It also mentions the relocations decided upon by public authorities:

> if necessary, promote relocation mechanisms through consensus between the government and communities, avoiding generating controversies (p. 113).

---

[7] For an overview on territorial order in Tabasco see Campillo et al. (2018). On the Land Use Planning Programs' challenges and deficiencies at the state level, see Sánchez Salazar and Palacio Prieto (2004).

[8] E.g. POSTET (Government of Tabasco 2018), pp. 169–175; it points out p. 41 that "57.3% of the State's population is in a situation of poverty and 305,000 of them are in extreme poverty."

[9] The Environmental Agenda ranks topics in order of importance (BIENESTAR 2019, p. 11). The POERET places coastal erosion 1st, deforestation and land use change 4th, water pollution 5th, waste generation 8th, biodiversity loss 11th, conservation strategies' implementation deficiencies 13th, wetland fires 14th, monocultures 16th, air pollution 18th, restoration 19th, and water scarcity 20th.

[10] In the Environmental Agenda topics ranked in order of importance (BIENESTAR 2019, p. 11), the POERET ranks floods 2nd.

> The establishment of new human settlements is restricted in the UGA's (*Unidades de Gestión Ambiental, Environmental Management Units*) for hydrological protection (...). In the case of new flood protection infrastructure, the affected population's relocation will be considered (p. 129).

This mention is the most detailed that we have found considering all of the public policy instruments analyzed so far.

The program only partially covers essential aspects of the phenomenon. For example, vulnerability is essentially a "geographical" vulnerability to climate change impacts, not a pre-existing social or economic vulnerability amplified by these impacts (such as poverty, discrimination, fragile land tenure). The Program's articulation of poverty illustrates this point. It mentions quantified poverty in the state (p. 38). However, it does not link it to a pre-existing vulnerability that particularly exposes people and communities living in poverty to climate change impacts. Finally, the POERET does not mention a rights-based approach. It only refers to the right to a healthy and balanced environment (p. 6). It does not develop the content of this right or establish guidelines for action. Finally, it does not articulate a differentiated approach towards Indigenous Peoples and only superficially mentions a gender-based approach.[11]

4.1.3 State Development Plan 2019–2024

Climate change is of utmost importance in the State Development Plan (*Plan Estatal de Desarrollo*, PLED) (COPLADET 2019). For example, it is the object of one of the three transversal axes—"*Ordenamiento Territorial y Cambio Climático*," "Territorial Planning and Climate Change," pp. 287–296). Concerning forced internal climate mobility, the Plan mentions and aims to mitigate the rural/urban migration phenomenon (pp. 21, 196, 200–201, 203, 267, and 273–274) but does not detail the factors which drive it. It mentions only internal labor migration (p. 273). It does not rule out that migrating persons and communities qualify as forced internal climate migrants. In previous chapters, we mentioned the difficulty in identifying them. For example, if one asks migrants why they migrate, they may mention a job search. However, looking more closely, it may be that the lands they cultivated are no longer suitable because of slow-onset environmental deterioration amplified by climate change or sudden-onset disasters which an increase in frequency and intensity due to climate change. The Plan does not address the forced nature of internal mobility in itself. Therefore, it is no surprise that it does not elaborate on individuals and communities affected and their need for protection through a differentiated rights-based approach.[12] However, it does raise respect for human rights as a principle of "the policies, programs, projects and actions that make up the PLED 2019–2024"

[11] The only mentions of a gender-based approach refer to ecological strategies (BIENESTAR 2019, p. 11—organic gardens, p. 122—criteria for forest resources/investment, and p. 136—conservation).

[12] The PLED does it for international mobility. See e.g. COPLADET (2019), p. 217.

(p. 7). It also mentions environmental rights (p. 264) and the rights of Indigenous Peoples, women, and attention to vulnerable groups (including the poorest) in the context of environmental and climate policy (pp. 265–287).

As for pre-existing social and economic vulnerability and its amplification by the impact of climate change, it refers to "phenomena linked to climate change such as floods or droughts" that "affect to a greater extent those who do not have formal employment and who do not have (…) employee benefits, social security, or access to credit" (p. 203).[13] It also draws a link between poverty and internal migration (p. 203). It is essential for the reasons already mentioned: pre-existing vulnerability is a factor that generates climate mobility and trapped migrants.

On the scenarios that generate climate mobility, the Plan partially addresses the causes of internal climate mobility. It mentions slow-onset environmental deterioration (e.g. pp. 263, 274), sudden-onset disasters (e.g. pp. 55, 266, and 294), sea-level rise (pp. 266, 288, 291, and 294), relocations (pp. 266 and 279),[14] and supports the development of renewable energies (p. 172). Nevertheless, it does not mention violence or conflicts triggered by the scarcity of natural resources and access to land.[15]

4.1.4 *Risk Management Instruments*[16]

The Tabasco Atlas of danger due to natural phenomena (*Atlas de peligro por fenómenos naturales del estado de Tabasco*, Servicio Geológico Mexicano 2013) does not highlight forced climate change-driven mobility. However, it does mention in detail the risks induced by flood (a sudden-onset disaster). The danger is very present in Tabasco: it concerns about 90% of the territory (p. 87). The atlas distinguishes between three zones: high, medium, and low flood susceptibility zone. It also alludes to forced relocation (it recommends "relocating human settlements located in the lower parts," p. 124) and slow-onset environmental deterioration. It covers soil erosion, the link between environmental deterioration (e.g., deforestation) and floods, and details the adverse impacts of the oil industry on the environment. However,

[13] In the same sense, see COPLADET (2019), p. 266.

[14] The PLED (COPLADET 2019) does not mention forced relocations, but the "promotion of new participatory planning instruments" that can generate "the alternative of reducing settlements in conurbation spaces in risk areas" (p. 265).

[15] The factors indicated are "poverty, insufficient social development, (…) educational system deficiencies, institutional breakdown, the deterioration of social fabric, a crisis of civic values, addiction, dysfunctionalities and anachronisms of the legal framework, and (…) the persistence of long-standing inter-community, agrarian, and neighborhood conflicts" (COPLADET 2019, p. 47). As an exception to the above, the PLED refers to conflicts generated by access to land and the energy sector companies (p. 171).

[16] This section does not analyze the Tabasco Master Civil Protection Program 2013–2018 (Institute of Civil Protection of the State of Tabasco 2015) since it is not in force. The Program did not indicate forced climate mobility but institutes a differentiated rights approach towards women and Indigenous Peoples, and alludes to pre-existing social vulnerability.

it does not touch upon rising sea-levels, acts of violence induced by the scarcity of natural resources nor the migration generated by development projects linked to renewable energies. Finally, it does not hint at a protection approach based on human rights. Instead, it evokes the pre-existing vulnerability of people who are potentially affected and communities and points out that it exposes them to the risk of disaster (e.g. p. 55).

The Hydrometeorological Phenomena Contingency Plan (*Plan de Contingencias Fenómenos Hidrometeorológicos*, Instituto de Protección Civil del Estado de Tabasco 2016b) differs from the Atlas. It repeatedly refers to the link between climate change and hydrometeorological phenomena (e.g. p. 17). It focuses on the events that can generate sudden-onset disasters and does not mention the other climate mobility scenarios (e.g. slow-onset environmental deterioration, violence induced by scarce natural resources, or development projects linked to renewable energies). It nevertheless hints at relocation scenario (p. 8).

It does not explicitly refer to the mobility generated by hydrometeorological events. However, it does mention pre-existing vulnerability and its effects in terms of disasters (e.g. pp. 2, 20, 23), it also articulates a rights-based (pp. 23, 38, 58–59),[17] participatory (e.g. p. 22), and differentiated approach towards women (pp. 21, 23, 38, 51, 62, and 64). Unfortunately, it does not establish a differentiated approach towards Indigenous Peoples.

This section identifies a series of limitations at state level to address forced internal climate mobility. In the sections that follow, we find the municipal level instruments and measures adopted in the face of the 2007 floods and sea-level rise also suffer from restrictions.

4.2 The 2007 Flood and Women in the Centro Municipality

We now analyze a scenario of internal forced climate mobility: the 2007 floods (sudden-onset disaster) in the municipality of Centro, as regards the Operational Guidelines (IASC 2011) and a differentiated approach towards women. We present the analytical framework (Sect. 4.2.1), carry out a diagnosis of the municipality's instruments (Sect. 4.2.2), and report on the elements that question compliance with the guidelines in the case of the 2007 floods (Sect. 4.2.3).

[17] The plan refers to several needs that generate an echo with human rights in the context of contingency management: public services (drinking water, electrification, garbage collection, etc.), food production, distribution and supply, reactivation of social, productive, economic, and cultural life, public security (to "guarantee the integrity of the people, their heritage, and the environment"), health, essential infrastructure "to guarantee the rights of the population and promote development," employment creation and self-employment, and biodiversity conservation and the resilience of the maintenance of ecosystems (Instituto de Protección Civil del Estado de Tabasco 2016b, pp. 58–60).

4.2.1 Presentation of the 2007 Flood and 2011 Operational Guidelines

The 2007 floods in Tabasco is our case-study for several reasons. Tabasco has a long history of floods. Population concentrates in flood plains, the lower Grijalva hydrological system (Arreguín et al. 2014) and the presence of the two largest Mexican river in the State (Grijalva and Usumacinta) (Gordillo and Pablos 2016, p. 126) explain the flood patterns in Tabasco that generate internal migration (Gordillo and Pablos 2016, p. 127). Additionally, sectors of the population identified as particularly vulnerable to forced internal climate migration live in Tabasco.

There are many Indigenous Peoples (Ayapaneco Peoples, Chontal de Tabasco, Ch'olCh'ol, Nahua, Tseltal, and Zoque INPI 2019), and in 2018, 53.6% of the state population was living in a situation of poverty (the percentage at the national level was 41.9%), and 12.3% of the population was in a situation of extreme poverty (the national average was 7.4%) (CONEVAL 2019). Finally, according to the State Institute for Women, women in Tabasco are particularly and severely affected by the floods (731 354 women suffered the adverse effects of torrential rains), with damage to their houses and public infrastructure (CEPAL, CENAPRED and SEPLAN 2012).

In 2007 (October 29–November 1), the atypical rain event generated a flood that affected around 1,500,000 people and 123,386 homes (CENAPRED n.d.). It is the largest natural disaster on record at national level due to flooding (Trejo et al. 2009). Trejo et al. (2009, p. 165) describe the chain of events that brought it about (p. 165). They mention excessive rainfall throughout October, saturated soil, very intensive rainfall in a the short period of October 29–November 1, an increase in rivers levels, excess spillway of the *Peñitas* dam as of October 29, and slow drainage towards the sea.[18]

The analysis of the case-study focuses on the municipality of Centro (Villahermosa is Tabasco's capital, hereafter "Centro"). Most of the state's population is concentrated in Centro (more than half in Villahermosa, SERNAPAM 2011a).[19] The Municipality's Risk Atlas (Ayuntamiento Constitucional de Centro 2015) indicates regular flooding, environmental degradation (a factor which leads floods to become disasters), and rural–urban migration which has generated many irregular settlements. The 2007 floods affected a considerable proportion of the Tabasco territory, but Centro was most severely affected (Ayuntamiento Constitucional de Centro 2015).

Massive mobility took place in the aftermath of the 2007 floods. The mobility mainly occurred within the same locality (Gordillo and Pablos 2016). This data illustrates that the destination territory of climate mobility can, in fact, be very close to the territory of origin. Accordingly, the definition of climate migrants based on rights triggers the application of the protection requirements (risks of human rights

---

[18] For another explanation of the causes of the flood, see Perevochtchikova and de la Torre (2010).

[19] 684,847 people live in the municipality of Centro (353,577 live in the urban area, and 331,270 live in the rural area) (*Plan Municipal de Desarrollo 2018–2021 del Municipio de Centro*, Centro Ayuntamiento 2018).

violations) in the numerous cases where a territorial criterion would not because of the lack of difference between the origin and destination territory).

This section determines whether or not Centro's public policy instruments endorse internal climate migrants' protection requirements and women's needs (Sect. 4.2.2). We also more particularly identify whether protection requirements and women's needs were respected in the 2007 floods (Sect. 4.2.3). We use the Operational Guidelines for the Protection of People in Situations of Natural Disasters (Operational Guidelines) issued by the Inter-Agency Standing Committee (IASC 2011) as a frame of reference. They do not bind states.[20] Instead, they provide tools applicable to the scenario of sudden-onset disasters (p. 9). They protect specific groups of people, including women. We select 20 Guidelines which have the potential to address them. They cover the protection requirements once mobility occurs, i.e. they do not address the prevention of the events that generate it (see Table 4.1 below).[21]

4.2.2 *Diagnosis of the Centro Municipality's Instruments*

The Climate Action Plan of Centro (*Plan de Acción Climática del municipio de Centro*, PACMUN) (Cabildo del Municipio Centro 2015) aims to "guide municipal public policies on mitigation and adaptation to the effects of climate change." It does not support compliance with the Operational Guidelines and mainly focuses on mitigation. Concerning adaptation, it mentions the extreme hydrometeorological events in Centro (including floods) and focuses mainly on the resultant impacts for productive sectors (in particular agriculture). It does not mention the adaptation of human settlements to sudden-onset disasters linked to climate change. *A fortiori*, it does not mention the migration induced by these disasters nor the particular situation of women.

In contrast, the Civil Protection Regulation for the Municipality of Centro (*Reglamento de Protección Civil para el Municipio de Centro*, Ayuntamiento Constitucional de Centro Villahermosa 2003)[22] provides a solid base to comply with the Guidelines. Article 4 establishes a participatory approach regarding civil protection

[20] The Operational Guidelines (IASC 2011) indicate that international humanitarian law, as a branch of law applicable in armed conflict situations, is not applicable in natural disaster situations. The Operational Guidelines' primary purpose is to "focus on what operational standards humanitarian actors may be guided by in order to implement a rights-based approach to humanitarian action in the context of natural disasters" (p. 9).

[21] The reader can refer to Table 4.1 in Chap. 2 to pin-point the legal grounds for the protection requirements that correspond with the Operational Guidelines.

[22] We could not find Centro Municipal Civil Protection Program (*Programa Municipal de Protección Civil de Centro*) on the Internet. Despite sending several emails, we were not able to access it. It might not yet have been yet. As of January 31, 2019, it had apparently not yet been adopted (Ayuntamiento Constitucional de Centro 2019).

Table 4.1 Protection specificities for vulnerable groups in the operational guidelines—women

Guideline
No discrimination (I.1)
Participation and consultation (I.3)
Protection activities according to identified needs (I.8)
Lifesaving measures (A.1.1)
Protection against violence (A.4.1)
Protection against gender-based violence (A.4.2)
Protection against trafficking, child labor, contemporary forms of slavery (A.4.3)
Security in host families and communities, or in collective shelters (A.5.2)
Access to and adequate provision of humanitarian goods and services (B.1.1, B.1.2)
Specific gender-based roles in humanitarian action (B.1.4)
Inclusion of women in the planning, design, and implementation of food distribution (B.2.1)
Security in access to sanitation facilities in camps and collective shelters (B.2.2)
Adequate shelter that meets specific needs (B.2.3)
Special attention to women's health needs (B.2.5)
Access to education under equal conditions (B.2.6)
Assistance in claiming property and acquiring titles in own name (C.1.5)
Consultation and participation in planning and implementation of shelter and housing programs (C.2.3)
Access to livelihoods, professional training, working conditions (C.3.2)
Equal access to documentation issued in own name (D.1.1)
Opinion re. Disaster Response (D.4.1)

Source Own elaboration, based on IASC (2011)

(see more particularly fractions I, IV, V, VII, and VII).[23] Although the article does not explicitly reflect a gender-based approach, it reflects on the Operational Guidelines concerning (i) participation and consultation, (ii) the inclusion of women in food distribution planning, design, and implementation, and (iii) consultation and participation in the planning and implementation of shelter and housing programs. Participation ensures that women's specific needs reflected in the other Guidelines are taken into account. The regulation also provides for the protection of people and their assets (Art. 9). It echoes the protection of life, physical integrity, and health of people exposed to imminent risks guideline. The regulation mentions measures as regards medical care, accommodation, food, recreation, and public services restoration (Art. 55). However, it does not specify the content of these measures in detail, nor does it foresee a differentiated approach.

[23] In the same sense, see Reglamento de Protección Civil para el Municipio de Centro, Ayuntamiento Constitucional de Centro Villahermosa (2003), Art. 32–36.

Several lines of action of the Municipal Development Plan 2018–2021 (*Plan Municipal de Desarrollo 2018–2021 del Municipio de Centro*, the Plan) (Ayuntamiento de Centro 2018) are relevant concerning the sudden-onset disasters scenario and, more specifically, floods linked to climate change. Regarding land-use planning, a Line of action aims to prevent disasters from occurring and, where this fails mitigate their effects (Line of action 7.1.1.1.1). The Plan also provides for better compliance with environmental regulations (Lines of action 7.3.1.1.1–7.3.1.1.5). Some Lines of action aim to improve the environment of the municipality. More specifically, they address reforestation and the recovery of deteriorated ecosystems. This is essential as it allows the mitigating of the effects of extreme hydrometeorological events, Lines of action 7.4.1.1.2, 7.4.1.1.3, and 7.5.1.1.2).

Also, the "Resilient Municipality" Program (Program 7.9) establishes Plans of action which refer to sudden-onset disasters to prevent, assist, and respond to emergencies and contingencies (Strategy 7.9.1.1, Lines of action 7.9.1.1.1–7.9.1.1.3), more specifically in the context of floods (Lines of action 1.3.1.1.1–1.3.1.1.4). However, it does not institute a differentiated approach towards women. The Plan's Operational Guidelines provide elements regarding pre-existing vulnerability. They mention poverty at the municipal level and provide specific lines of action for vulnerable groups (within which women can be found, Lines of action 3.3.1.1.1–3.4.1.1.5). They also provide a Line of action on women's employment (Line of action 5.3.1.1.1). However, it is not designed to apply in the specific context of sudden-onset disasters. They also touch upon violence against women. However, they do not specify if it increases with a sudden-onset disaster or how to address it if that is the case. Furthermore, the Plan expressly adopts a differentiated approach towards women (Lines of action 3.5.1.1.1–3.5.1.1.3) to "[d]iffuse women's rights, facilitating access to services before the competent authorities" (Strategy 3.5.1.1). The Plan also foresees a line of action on training the indigenous population in their mother tongue "with a gender and human rights perspective" (Line of action 3.6.1.1.3).

The Plan also establishes Lines of action that can support compliance with the Guidelines without adopting a differentiated approach or locating these actions in a health disaster context[24] (Guidelines for the protection of life, physical integrity, and health of the persons exposed to imminent risks, with particular attention to the health needs of women) and services provided by the civil registry (Guidelines for assistance in claiming property and acquiring titles in their own name, and access to documentation issued in own name under equal conditions, Program 4.4).

Centro's Risk Atlas (*Atlas de Riesgos del Municipio Centro*, Ayuntamiento Constitucional de Centro 2015) aims to provide empirical data related to floods, and this data is to inform the decision-making process. We do not find many public policy elements that echo the Operational Guidelines. The Atlas does not support the Guidelines on consultation and participation (the General Recommendations part reflects a top-down approach). People and communities affected (and especially women)'s

[24] Ibid., Program 3.11. Access to health services in the municipality is particularly worrying: in 2010, "less than 70% of the municipal population" had access to a health institution (Ayuntamiento de Centro 2018, p 81).

consultation and participation are not central. Furthermore, the Atlas merely provides a list of minimum equipment for temporary shelters.[25] It does not endorse any element to address women's specific needs about health, education and training, protection against violence, and access to livelihoods and documentation.

Finally, the Urban Development Program Villahermosa and Centro's Metropolitan Centers (*Programa de Desarrollo Urbano del Centro de Población de la Ciudad de Villahermosa y Centros Metropolitanos del Municipio de Centro*, Ayuntamiento Constitucional de Centro 2008) brings little from the guidelines and their specific elements in order to achieve compliance. It also does not institute a differentiated approach. Similarly, it does not provide the empirical elements to determine to what extent temporary shelters have the necessary equipment to comply with the Guidelines' requirements (e.g., health or hygiene).[26] Lastly, it does not address violence against women (including in shelters) or the lack of women's participation or consultation. On participation and consultation, it mentions the importance of citizen's participation, indicates a list of organizations, and establishes "consultation forums" (p. 589), but does not detail how the participatory process should be organized or the frequency of the consultation.

In the subsequent section, we rely on the Operational Guidelines to identify whether the protection requirements were met and, more specifically, focus on women. We find the measures adopted concerning the 2007 floods have neither overcome nor offset the national, state, and municipal framework deficiencies.

4.2.3 Case Study: The 2007 Flood

We now analyze the implementation of the Guidelines and the application of measures adopted to address the 2007 flood. We identify a very limited number of sources to identify and compare quantitatively and qualitatively the measures with the 20 Operational Guidelines selected and mentioned above (IASC 2011). Such a limited number of sources might in itself reflect a limited interest in the differentiated approach to women forced to migrate because of climate impacts. The sources being a documentary (Discovery Channel 2008), journalistic notes, a report of the Boell Foundation (Jungehülsing 2012), two reports from the UN Office for the Coordination of Humanitarian Affairs (OCHA 2007a, b), and a few academic articles. Hernández Hernández's doctoral thesis on the 2007 floods (2011) has also been

[25] Ayuntamiento Constitucional de Centro (2015), p. 219:

> The minimum equipment of the temporary shelters is: cistern with sufficient capacity (...), toilets and showers in adequate quantity, kitchen, (...) electricity supply, among other things. Supposing the establishment does not have the necessary equipment but is classified as a shelter, it should be provided with portable equipment such as kitchens, bathrooms, water purification plants, electric power plants, etc.

[26] The data it provides on drainage, piped water, electricity, and roofs do not include shelters.

particularly useful. We assume no measure was adopted when none of the sources mentions a measure that could support compliance with the Guidelines' requirements, but cannot confirm it as such. The sources at our disposal do not allow us to do so. Such statements would require fieldwork.[27]

The Operational Guidelines are grouped below in increasing order according to their compliance: (i) the measures adopted incorporate a differentiated approach towards women, (ii) we did not find information to determine that the measures met the Operational Guidelines for migrants in general (and, more specifically, for women), and (iii) the available information points to possible, severe shortages for migrants in general, and women in particular.

We found only one case of measures that effectively considered the prioritization of women in applying an Operational Directive: assistance in claiming property. The relocation program addressed the population that lost their home due to the flood granted houses to relocate them outside Villahermosa. Tabasco's former governor stated, "the houses will be attributed to the mothers. They show affection for raising their children" (Hernández Hernández 2011, p. 92). However, Jungehülsing (2012) signals a series of deficiencies with the housing programs: expensive public transport that hampers women from reaching workplaces from the relocation sites, lack of infrastructure in the relocation sites (e.g. clinics, childcare).

In our review of the available information, we found no mention of measures adopted to comply with 7 of the 20 guidelines applicable to women as a vulnerable group. For the reasons stated above, we cannot claim they are omissions. If that were the case, the human rights supported by the Operational Guidelines would be at risk of being violated during the emergency. They are:

- Protection against gender-based violence,
- Protection against trafficking, child labor, contemporary forms of slavery,
- Specific gender-based roles in humanitarian action,
- The inclusion of women in food distribution planning, design, and implementation,
- Security in access to sanitation facilities in camps and collective shelters,
- Special attention to women's health needs, and.
- Access to livelihoods and professional training.

We now turn to the cases where bibliographic evidence indicates the adopted measures do not comply with Operational Guidelines. In these cases, the sources consulted point out deficiencies. A first series of deficiencies refers to internal migrants and applies regardless of gender. Regarding the non-discrimination requirement, discriminatory behaviors were evidenced between the affected and unaffected populations and between government workers in charge of delivering goods:

> Those who lived in the shelter are perceived and stigmatized as persons accustomed to receiving everything from the government and lacking instruction to live in decent housing under the gaze of those who did not live in this place, while the latter, or those who do not

[27] Among other factors, observed through extensive interviews with people involved in flood responses.

share their trajectories similar to the rest of the group, are viewed with suspicion by the former. (Hernández Hernández 2011, p. 109)

This was also in evidence after the emergency and during the relocation process: "the surrounding towns do not look favorably at the relocated persons' settlements... because they perceive the (relocated persons) are treated in a paternalistic way. Everything is given to them, they are used to receiving everything from the government" (Hernández Hernández 2011, p. 116). In addition, the state government authorities acted discriminatorily. For example, "the (relocated persons settlements)' administrator mentions that because they lived for almost ten months in a shelter, these persons got used to having free food and they thought that the government had to maintain them" (Hernández Hernández 2011, p. 116). More particularly, Hernández Hernández (2011) highlights re-victimization process of the forced migrants.

Firstly, some works to prevent future floods, undertaken in the aftermath of the 2007 flood, have been criticized. They are portrayed as white elephants to justify using public resources and focusing on safeguarding the capital's integrity to the detriment of "second-rate" populations. Governmental preference to safeguard the capital population over the rest of the state's population resembles a discriminatory attitude. For example, the works to protect Villahermosa flooded the municipalities of Nacajuca, Jalpa de Méndez, Cunduacán and Centla (Megainundación 2007... 2014).

Secondly, the decision-making process on emergency care and evacuation to shelters adopted a top-down approach concerning participation and consultation (Hernández Hernández 2011, p. 78). The concentration of the decision-making process at the federal and state governments levels without consulting migrants was the common denominator in light of the analysis of public policies in response to emergencies. The migrants who left their homes for security reasons were not consulted on the destination site and the governmental guarantees regarding security conditions of the temporary shelters. Gordillo and Pablos (2016) point out the programs' verticality and the absence of local participation.

Thirdly, adopted measures were also deficient in terms of the prioritization of protection activities according to the needs assessment. The state government evaluated the situation of people located in areas with high vulnerability to flooding and the installation of shelters in the city. The evaluation is congruent and necessary given the magnitude of the emergency. However, it did not consider particularities such as the protection of vulnerable groups, including internal forced migrants in the decision-making process, adequate mechanisms for the distribution of products and services and access to information. The deficiencies generated dissatisfaction among the affected population.[28]

Fourthly, access to education under equal conditions also suffered deficiencies. We did not find evidence that education was a priority and that concrete actions have been adopted in that regard. On the contrary, according to journalistic reports in 2014 (seven years after the flood), there were still schools affected by the flood. The

[28] CEPAL, CENAPRED and SEPLAN (2012) report on government decisions to prevent floods after the emergency at p. 47.

situation induced demonstrations (e.g., groups of women demonstrated and blocked roads in Villahermosa) (Desaloja policía … 2014).

Fifthly, the sources consulted mention severe deficiencies in consultation and participation regarding shelters and the planning and implementation of housing programs. In contrast with the Operational Guidelines, public authorities did not consult migrants. This is particularly true for the relocation stage, where "relocated families' consultation and participation did not occur" (Hernández Hernández 2011, p. 112). No reference was found on migrants (and more specifically women) being able to participate in the decision-making process before, during, or after the emergency.

Sixthly, the measures adopted did not comply with the access under equal conditions to documentation issued in the names of women affected. Two years after the delivery of the houses for relocated people, they still did not "legally belong to their occupant" who did not have "any document that accredit[ed] them as future owners" (Hernández Hernández 2011, p. 98). This deficiency induces another one: the difficulty surrounding women's access to the loans necessary to begin remunerated activities. Banks require a deed to grant a loan. Without deeds, women cannot start businesses and generate incomes (Jungehülsing 2012).

Another set of deficiencies refer to measures adopted by the authorities to protect climate migrants without applying a differentiated approach to women. Regarding primarily the protection of life, physical integrity, and health of people exposed to imminent risks. In addition to the State Secretariat of Health's participation during the emergency, brigades of 54 Cuban doctors, epidemiologists, obstetricians, pediatricians supported sanitation and the establishment of sanitary fences (Llega a Tabasco … 2007). The organization Doctors without Borders also provided medical care and logistics during the floods (Discovery Channel 2008). The sources available do not mention abnormal outbreaks of disease when the population took refuge in temporary shelters. They also fail to mention a differentiated protection approach to protecting women.

Secondly, on the protection of women against violence (including in the camps and collective centers during and after the emergency), we found few references to mandatory coexistence rules of in temporary shelters (Hernández Hernández 2011). However, they were general and did not particularly consider the protection of women. The deficiency is all the more significant as security conditions were not optimal. The insecurity regarding migrants' personal property occurred despite the deployment of elements of the Secretariat of National Defense (*Secretaría de la Defensa Nacional*, SEDENA) to support rescue and security activities. For example, a journalistic note of November 5, 2007, indicates that 54 people were detained for acts of robbery (Tabasco: suman 54 detenidos … 2007). We could not find any reference to public security efforts with particular attention to women's safety and assistance.

Thirdly, access to humanitarian goods and services and adequate provision of them was also deficient. Hernández Hernández (2011) mentions that the concentration of goods and services was one of the main problems during the 2007 flood,[29] and resource hoarding and poor distribution among the people affected. She explains the situation by the "centralization of the resources in Quinta Grijalva" (where the then Governor lived); "affected persons had to stand in long lines to be able to get support, whether in food, clothing or footwear" (p. 83). Hernández Hernández (2011) provides another example. The state government implemented a policy to grant vouchers to purchase and replace household appliances for an amount of 10 000 Mexican pesos (around 500 USD). The last 15 000 vouchers were delivered in May 2008, seven months after the flood occurred. The amount of money granted for the replacement of material losses may be questionable (it seems insignificant compared to the affected population's economic losses). The provision of economic support more than half a year after the flood is even more questionable. Other than this, we could not identify the application of measures specifically devoted to women.

Fourthly and regarding adequate shelters that meet specific needs, the first shelters began to be installed in Villahermosa on October 28, 2007. The establishment of temporary shelters was one of the urgent measures given the number of internally forced climate displaced persons. Temporary shelters were set in Centenario Park (Parque Centenario, with a capacity of 4 000 people), in Atasta Recreation Center (centro recreativo de Atasta, with a capacity of 3,000 people), and in Milenio Square (Plaza Milenio) (Hernández Hernández 2011). We are not sure they provided the essential services to meet women' specific needs. The people who stayed in the shelter for an extended time endured their conditions, hoping they would later be assigned a home (Hernández Hernández 2011). We found only one reference to the shelters meeting women's needs. It relates to pregnant women. The UN Office for the Coordination of Humanitarian Affairs (OCHA) reports specific shelters "for pregnant women in the last trimester" (OCHA 2007a, b). This is critical given women's specific vulnerability in a temporary or permanent relocation process and living in shelters. The literature indicates women's differing vulnerability compared to men for their exposure to.

> greater risks of violence, abuse and sexual coercion, the inability to negotiate safe sex, unwanted pregnancies, the spread of sexually transmitted infections (...), including the human immunodeficiency virus (HIV), prostitution (...) or, in the worst case, death. (Ramírez et al. 2012, p. 119)

---

[29] Hernández Hernández (2011) points out at p. 61 that

> A week after the flood, some shelters no longer had sufficient supplies, neither human nor material, due to the number of people sheltering ... Some people worried about food being rationed, unhealthy bathrooms, and saturation. Among some people affected, there was restlessness and uncertainty; the feeling that food was beginning to be scarce became real even when the media announced the largest food concentration in Quinta Grijalva.

The fifth case concerns opinion on disaster response. A disaster perception study in Tabasco has consulted residents over 18 years of age, affected by recurring natural disasters. The disaster most frequently mentioned in the perception study was flooding, with a qualitative assessment of "very important" and a degree of attention assigned to "little attention" (López and Garrido n.d.). The floods they have suffered through have a high degree of importance, but the attention of the authorities responsible has not been of the same magnitude. Women assigned higher importance to floods than men. In other words, perceived vulnerability in the occurrence of a natural disaster was higher for women than for men. Even though both men and women consider the government has paid little attention to floods, the share of federal resources assigned to the state to address the floods has been significant. For example, Tabasco received 32.5 billion pesos in 2007 (Tabasco, tragedia... 2007) (the equivalent of all of the resources assigned to the Secretariat of National Defense for security activities in the country).

In this section, we identified a series of limitations of Centro's instruments and measures to address the sudden-onset disaster scenario and the 2007 flood. In the subsequent section, we identify other limitations regarding forced mobility induced by another scenario, that being sea-level rise.

4.3 Sea-Level Rise Scenario and Chontal People in the Centla Municipality

We continue with the analysis of the instruments adopted at municipal level, focusing on the sea-level rise scenario and Indigenous Peoples. We present the analysis framework (Sect. 4.3.1) and analyze the instruments adopted by the municipality of Centla to address the disaggregated impacts of sea-level rise (Sect. 4.3.2).

4.3.1 Analysis Framework

We present the disaggregation of the impact of sea-level rise (Sect. 4.3.1.1). Then, we present the requirement for protection of Indigenous Peoples (Sect. 4.3.1.2). These requirements will allow us to evaluate municipal instruments. Lastly, we present the municipality of Centla, the object of our analysis and the Chontal People to which we direct our attention (Sect. 4.3.1.3).

Breaking Down the Impact of Sea-Level Rise

The sea-level rise scenario applies to Mexican islands and coasts. The country has 11,122 km of coastline, of which 7,828 km correspond to the Pacific Ocean and

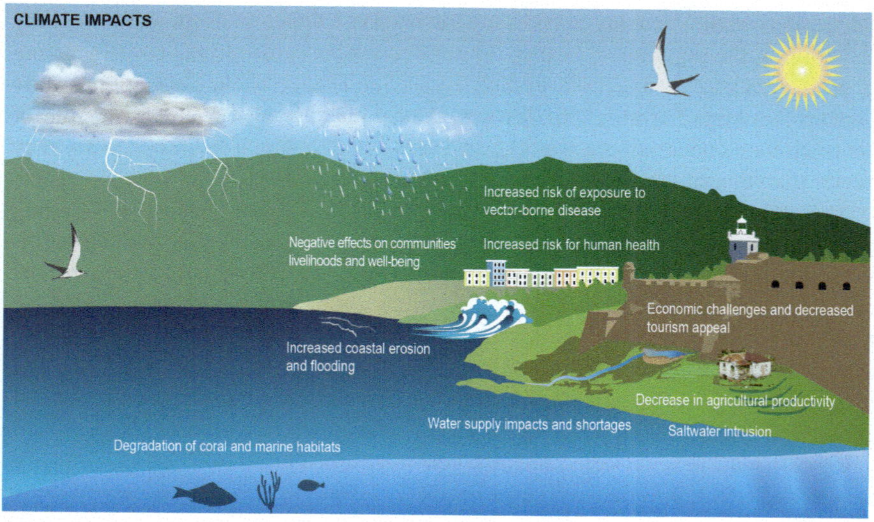

Fig. 4.1 Risks for coastal systems induced by climate change. *Source* Reidmiller et al. (2017, p. 818)

3,294 km to the Gulf of Mexico and the Caribbean Sea (Statistical 2014). It is, therefore, essential to analyze the vulnerability of human settlements located in this territorial strip. Sea-level impacts are highly complex. We start with their disaggregation and relationship to the mobility scenarios to diagnose the policy instruments of the Centla municipality.

Sea-level rise (considering an increase in temperature due to climate change of between 1.5 and 2 °C) generates the following impacts: (i) coastal flooding and impacts on populations, (ii) damage to infrastructure and assets (high confidence), (iii) water stress (medium confidence), (iv) risks to marine ecosystems (high confidence), (v) risks to critical sectors (medium confidence),[30] and (vi) changes in the salinity of coastal groundwater (high confidence) (Hoegh Guldberg et al. 2018) which can cause, in the worst case, land loss. Figure 4.1 below presents the risks to coastal systems due to climate change.

Turning to the scenarios that generate forced internal climate mobility, coastal floods (sudden-onset disasters) can impact the populations settled in coastal areas and forcibly induce migration due to repeated flooding or land loss generated by erosion (Botello et al. 2010). Coastal floods can also generate the trapped migrant phenomenon if the people and communities affected cannot migrate. For example, in a case study to assess the vulnerability of a coastal municipality in the north of the state of Veracruz, González Turrubiates (2010) found a high municipal vulnerability in terms of the number of homes and populations affected in the event of floods,

[30] In this regard, Field et al. (2014) indicate that critical sectors refer to diverse infrastructures such as water supply, health and emergency services, pipelines, electrical networks, and transportation systems.

and "resilience deficiencies considering the response capacity of the authorities and the action of the population ... they do not know who to turn to in case of disaster" (p. 602). In a study carried out in the City of San Francisco de Campeche, Posada Vanegas and Vega Serratos (2010) found that in the case of a 0.59-m sea-level rise due to climate change, there would be an area flooded which would be three times greater than that of a scenario without sea-level rise (p. 618).

Also, saline intrusion can generate water stress (Yáñez-Arancibia and Day 2010), reducing water availability and forcing people and communities affected to migrate. Water stress (slow-onset phenomena scenario) has consequences on physical, environmental, and human receptors (e.g., less water available for human use, crops, and freshwater ecosystems). For example, Ospina Noreña et al. (2010) signal trends of decreasing freshwater availability in the hydrological basins of the Gulf of Mexico that could have "important implications for the natural vegetation cover and its consequent effects on the different existing biotic–physical elements" (p. 83).

Risks in marine ecosystems can also generate forced migration. Yáñez-Arancibia and Day (2010) argue that wetlands on the coastline (e.g., mangroves) can move inland if sea-level rises. If no land is available, however, they cannot. Consequently, ecosystems can disappear due to not finding a territory to settle in or the adequate conditions in which to persist. In turn, their disappearance leaves the coastal area exposed to storms and floods (p. 17). The link is made with the sudden-onset disasters scenario. Also, climate change can alter and damage marine ecosystems such as coral reefs (e.g. bleaching). The death of these ecosystems can also damage the economic activities that depend on them, the exchange of nutrients, and biomass production (Padilla Souza et al. 2010). The loss of coastal-marine ecosystems implies a decrease in biomass generation that will generate lower fish yields (which could put at risk the self-sufficiency of communities that depend on these catches) and loss of income from tourism and trade. Consequently, damage to marine ecosystems relates to several climate migration scenarios: slow-onset phenomena and conflicts or violence caused by the scarcity of natural resources.

Risks in critical sectors include, damage to port infrastructure and marine transport, landslides affecting road infrastructure, and telecommunications. Treviño Trujillo et al. (2010) mention various cases on coastal highways in Mexico. The construction of artificial reefs on the Campeche coastal highway as well as protection structures and "best routes" on the Tamaulipas coastal highway are the options chosen to protect them from sand obstructions and wave induced landslides. These protection mechanisms are adaptive strategies. However, if they are not effective, the level of vulnerability of the population increases. They also become isolated. In turn, the entry of aid and the mobility of the population from critical areas is difficult or impossible and can generate trapped migrants. This impact assimilates the sudden-onset disasters scenario.

On changes in the salinity of coastal groundwater, de la Lanza Espino et al. (2010) carried out a case study in Marismas Nacionales (a Ramsar site in the Sinaloa and Nayarit coasts). They identified the progression of saline intrusion generated by, among other things, increased tides because of channel openings, a lower contribution from two rivers in the region, and dams that decreased in water flow. They found flood levels drop and mangroves are damaged. The saline intrusion also impacts the availability of water for the populations located in the coastal zones and the ecosystem thereof. This impact is related to the scenarios of slow-onset phenomena and conflicts or violence caused by the scarcity of natural resources.

Finally, sea-level rise can generate land loss because of erosion and submergence (Wong et al. 2014). We consider it to be the most drastic impact as it involves the definitive loss of territory. It can generate internal forced climate migrants and, at the same time, be related to the scenarios of slow-onset phenomena and areas designated unsuitable for human settlements.

The Rights of Indigenous Peoples at Risk of Being Violated

Mapping the rights of Indigenous Peoples at risk of being violated due to sea-level rise impacts will allow us to conduct the case study in the Centla municipality. These rights vary according to the consequences of rising sea-levels (from gradual environmental degradation—slow-onset phenomena scenario, to coastal floods—rapid-onset disasters scenario) and Indigenous Peoples' differentiated rights at risk of being violated range from the right to access means of subsistence to the conservation and improvement of their habitat, protection against forced displacement (and where appropriate, protection during migration), to the right to ownership regarding their lands, territories, and resources.

International (ILO 1989; United Nations 1998, 2007), regional (OAS 2016), and national (CPEUM 2020) standards firmly ground the rights of Indigenous Peoples. The Operational Guidelines (IASC 2011) provide a section with 16 specific protection guidelines for Indigenous Peoples. They are non-discrimination (I.1), participation and consultation (I.3), protection activities that must be prioritized according to the needs assessment (I.8), the protection of the life, physical integrity, and health of people exposed to imminent risk (A.1.1), protection against violence (including in the camps and collective centers during and after the emergency) (A.4.1), security and protection in the camps and collective centers (A. 5.2), treatment of mortal remains, taking into account ethnic and religious aspects (A.6.4), access to humanitarian goods and services and their adequate provision (B.1.1 and B.1.2), adequate shelter that takes into account culture (B.2.3), equal access to education considering culture (B.2.6), the protection of abandoned property (C.1.2), respect for traditional claims to land titles and ownership (C.1.6), the consultation and participation in the planning and implementation of shelter and housing programs (C.2.3), ensuring that training programs do not reinforce the economic marginalization of such groups (C.3.1), opinion on disaster response (D.4.1), and assistance that takes into account the culture and right to exercise religious faith and cultural traditions (D.4.2 and D.4.3).

Several instruments adopted at federal level could also establish guidelines for protecting the rights of Indigenous Peoples in the context of forced internal climatic migration. However, through analysis, we conclude they do not. Firstly, one of the specific objectives of the National Program for Indigenous Peoples 2018–2024

(*Programa Nacional de los Pueblos Indígenas 2018–2024*, INPI 2018) is to guarantee the "lands, territories and natural resources' defense, conservation and use" (p. 25). Likewise, it considers within its strategy 1.3 the "attention to the migrant indigenous population" and the elaboration and implementation of a "special plan for the migrant indigenous population" (p. 29). Another Line of action of the program is promoting measures for the adaptation to and active mitigation of climate change (p. 31). However, it does not delve into their content. The line of action most closely linked to our subject establishes that "support and accompaniment" must be provided to the "indigenous and Afro-American population in a situation of forced displacement" (Line of action 8.2.3, p. 49) through poorly defined inter-institutional strategies. In short, the Program does not provide concrete lines of action to evaluate responses to the disaggregated effects of sea-level rise in municipal instruments.

The Law of the National Institute of Indigenous Peoples (*Ley del Instituto Nacional de los Pueblos Indígenas*, LINPI 2018) establishes, as one of its objectives, the proposal, promotion, and implementation of the measures required to guarantee the fulfillment of Indigenous and Afro-American Peoples' rights and the generation and maintenance of sustainable lifestyles, facing the consequences of climate change (Article XVI). There is no further reference regarding climate change and climate mobilities. In the first six articles, it mentions the rights of some Indigenous Peoples already guaranteed by other regulations (self-determination, autonomy, political participation and representation, lands and the defense and conservation of natural resources). However, most of the 29 articles that make up this law focus on delimiting the attributions and objectives of the National Institute of Indigenous Peoples (*Instituto Nacional de los Pueblos Indígenas*). Neither is it, therefore, an essential instrument for our analysis.

Finally, the Project for Attention to Displaced Indigenous People (*Proyecto para la Atención a Indígenas Desplazados*, Comisión Nacional para el desarrollo de los Pueblos Indígenas 2016) was deactivated when restructuring the then National Commission for the development of Indigenous Peoples (*Comisión Nacional para el desarrollo de los Pueblos Indígenas*, currently the National Indigenous Peoples Institute). This Program aimed to protect the indigenous population from forced displacement (without specifically considering displacement driven by climate change). It contemplated four types of support to Indigenous Peoples that can inform their needs when in a situation of internal mobility: the acquisition of lots for housing, materials for housing construction, farmland, and inputs for productive activities (with maximum amounts of 80,000 Mexican pesos—around 4,000 USD—per household).

Introducing the Case Study: Centla Municipality and Pueblo Chontal

The sea-level rise scenario will particularly affect the Ayapaneco, Chontal de Tabasco, and Nahuas Indigenous Peoples in Tabasco. It results from comparing the projections of areas to be flooded by a 1-m sea-level rise (Botello et al. 2010, p. 342) with the distribution of Indigenous Peoples in Tabasco. In the case study, we consider the Chontal de Tabasco people because they are closest to the coastal

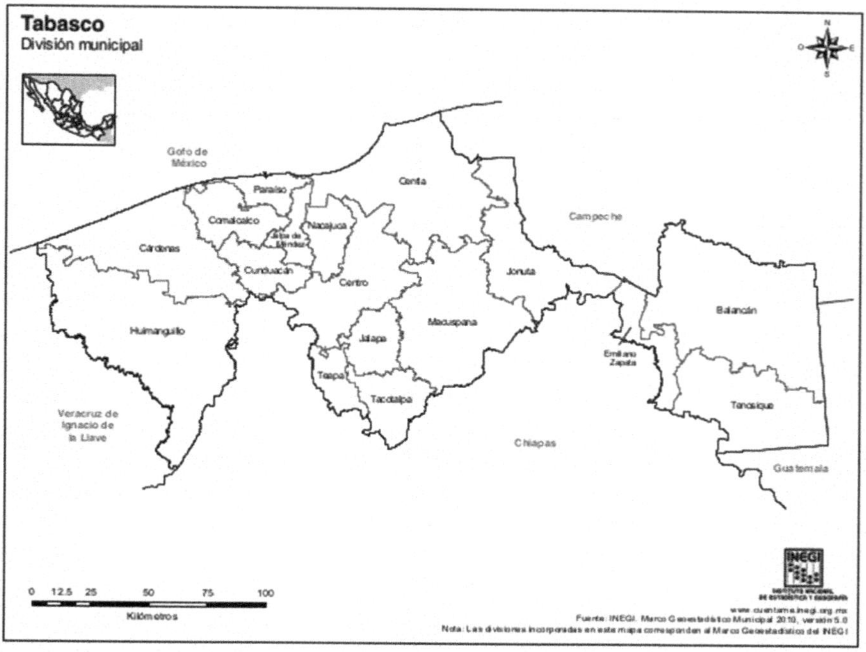

Map 4.1 Map of Tabasco. Municipal Division. *Source* INEGI (2018)

area and constitute the most significant number of people with indigenous identity exposed to rising sea levels.

The Chontal People reside mainly in six of the 17 Tabasco municipalities: Centla, Centro, Jonuta, Macuspana, Jalpa de Méndez and Nacajuca (INPI 2019). We chose to focus on the Centla Municipality since it is close to the sea with the greatest presence of Chontal People (INPI 2019). Centla is located on Map 4.1.

The main economic activities of the Chontal People in Centla are agriculture, livestock, and fishing. Due to Centla's high rates of marginalization, the economic development of the people has been complex (Ayuntamiento Constitucional de Centla 2019). As of 2015, 45.4% of Tabasco's population was living in a situation of poverty—1,085,114 inhabitants). The percentage rises to 59.7% in Centla (70,819 inhabitants) (CONEVAL 2015). This data points out the possibility that the impact of rising sea-level may force the Chontal People to migrate. However, they may well not have the necessary resources to do so. They would become trapped forced climate migrants. Also, Centla's public infrastructure and municipal services are deficient (Ayuntamiento Constitucional de Centla 2019, p. 33). The deficiencies could have significant consequences when it comes to the attention needed to the impacts of rising sea-levels on the entire population of the municipality and, more particularly, on the Chontal People.

As to Centla environment, it is worth mentioning the "Pantanos de Centla" Ramsar wetland. The wetland is of great biological importance (Ramsar Convention Secretariat 2014). Also, the environment in Centla is already degraded. The Atlas of Disaster Risks in the face of hydrometeorological phenomena of Centla (*Atlas de Riesgos de desastres ante fenómenos hidrometeorológicos del municipio de Centla*, Coordinación de la Unidad de Protección Civil Municipal n.d.) mentions deforestation and damage and the destruction of the mangroves (Characterization, Vegetation). Similarly, the Municipal Development Plan (*Plan Municipal de Desarrollo*, Ayuntamiento Constitucional de Centla 2019) signals "high pollution and environmental deterioration," the "irrational use" of natural resources and that "some flora and fauna resources are threatened with extinction" while "some species have already become extinct" (p. 70). Contamination by hydrocarbons (Ayuntamiento Constitucional de Centla 2019, p. 70) and contamination attributed to oil sector activities are also pointed out, and more particularly the activities of Petróleos Mexicanos (PEMEX) (Coordinación de la Unidad de Protección Civil Municipal n.d., p. 4). In this context, the activity of the *Dos Bocas* refinery in the Paraíso Municipality raises concerns (the refinery is under construction). It could amplify the contamination of neighboring municipalities (slow-onset deterioration scenario) and generate a sudden-onset scenario in the event of an accident. Furthermore, the sea-level rise phenomenon in the coastal area where the refinery is being built could be another factor that generates such an accident.

4.3.2 Shortcomings of Municipal Instruments

We analyze Centla's public policy instruments according to each one of disaggregated effects of sea-level rise. We first assess the importance of each effect in Centla and if the instruments address them and reflect a differentiated approach to Indigenous Peoples. We analyze five instruments: the Centla Municipal Development Plan 2018–2021 (*Plan Municipal de Desarrollo 2018–2021*, Ayuntamiento Constitucional de Centla 2019), Atlas of dangers due to natural phenomena in the State of Tabasco (*Atlas de peligros por fenómenos naturales en el Estado de Tabasco*, Servicio Geológico Mexicano 2013), Centla Atlas of Disaster Risks in the face of Hydrometeorological phenomena (*Atlas de Riesgos de desastres ante fenómenos hidrometeorológicos del Municipio de Centla*, Coordinación de la Unidad de Protección Civil Municipal n.d.), Tabasco Risk Atlas (*Atlas de Riesgos de Tabasco*, Seguridad, CNPC and CENAPRED 2017), and the study Pilot Program of Adaptation to Climate Change for the Pantanos subregion (*Programa Piloto de Adaptación ante el Cambio Climático para la Subregión Pantanos*, ECOSUR and SERNAPAM 2014).[31]

[31] The Atlas of Disaster Risks in the event of hydrometeorological phenomena (Coordinación de la Unidad de Protección Civil Municipal n.d.) mentions other municipal instruments: Emergent Program Against Floods (*Programa Emergente Contra Inundaciones*), Program of Risks Prevention and Protection (*Programa de Prevención y Protección de Riesgos*), and Municipal Contingency Plan

We can formulate general comments on the instruments before going into more detail in our analysis. No instrument provides quantitative data on mobility induced by the impacts of climate change (nor *a fortiori* by the sea rise scenario). Also, the study Pilot Program for Adaptation to Climate Change for the Pantanos Subregion (ECOSUR and SERNAPAM 2014) is the only document that mentions sea-level rise (p. 34). When we identify public policy elements that address the disaggregated effects of the sea-level rise in other instruments, the link with sea-level rise is not explicit.

In addition, the Centla Development Plan (Ayuntamiento Constitucional de Centla 2019), Atlas of Disaster Risks in the event of hydrometeorological phenomena (Coordinación de la Unidad de Protección Civil Municipal n.d.) and the study Pilot Program of Adaptation to Climate Change for the Pantanos subregion (ECOSUR and SERNAPAM 2014) refer to pre-existing vulnerability, a key element in the generation of forced mobility. It is worth mentioning, as it is almost absent from state-level instruments and only scarcely appears in federal-level instruments.

Likewise, with the exception of the Municipal Development Plan (Ayuntamiento Constitucional de Centla 2019), no instrument indicates the specific situation of Indigenous Peoples (be it their pre-existing specific vulnerability, vulnerability when facing the impacts of climate change, or their close relationship to their land). The Centla Municipal Development Plan is the only instrument that mentions Indigenous Peoples will have preference in "all government programs" concerning social development (p. 69). However, when detailing public policies, it does not articulate the implications of public policies for this population (nor their impact on public policies—e.g., through their traditional knowledge). The Development Plan explicitly carries out a differentiated approach towards people with disabilities (Objective 2.1), the elderly (Objective 2.2), women (Objective 2.4), and young people (Objective 2.5). However, there is no such specific approach for Indigenous Peoples. Based on this deficiency, we felt the measures aimed at addressing sea-level rise impacts in Centla were not going to reflect a differentiated protection approach. The subsequent analysis confirms our suspicion.

We distinguish between the seal-level impacts that

(i) are not relevant in Centla and for the Chontal People (water stress), and those for which the municipal instruments
(ii) do not address the sea-level rise impact (land loss),
(iii) address the sea-level rise impact but do not endorse a differentiated approach (coastal flooding, marine ecosystems losses), and
(iv) address the sea-level rise impact and endorse a very limited differentiated approach (changes in coastal soils and the salinity of groundwater, risks in critical sectors).

(*Plan Municipal de Contingencia*). Unfortunately, we were unable to find them on the internet. Neither did the municipal authorities we contacted by phone and e-mail provide them to us (in some cases, they were unaware of the existence of such instruments).

(i) **Water stress in Centla is not related to sea-level rise**

On water stress and the Chontal People, water availability per capita in the hydrological region Frontera Sur (where Centla is) is very high (18 852 m^3/inhabitant/year) (CONAGUA 2016). It is the hydrological region with the highest water availability in Mexico (CONAGUA 2016, p. 14). As of 2018, the Centla aquifer had groundwater availability of 832.711.527 m^3 per year for new concessions (CONAGUA 2018, p. 9). Despite water availability, the information presented in Table 4.2 reflects deficiencies in the coverage of drinking water and municipal sewerage services in Centla.

The percentage of the Centla population with drinking water service is meager compared to the state average. The disparity is even more significant if we consider the rural population (13.8%). Drainage coverage at the municipal and state level is high, but the sewerage system is practically nonexistent at the rural level. In the same sense, the Municipal Development Plan 2018–2021 (Ayuntamiento Constitucional de Centla 2019) indicates that only 33.30% of the homes in Centla have piped water (p. 23).

In this context, we infer that the stress to water of the Centla rural and indigenous communities (including the Chontal peoples) relates more to infrastructure deficiencies than to climate change impacts. It also points to an essential climate mobility aspect: the importance of local context as well as the social and institutional context.

(ii) **Land loss: the impact of sea-level rise not addressed**

Land loss due to rising sea levels can generate forced climate migrants who (furthermore and logically) cannot return to their place of origin as they would no longer exist. Coastal erosion has already caused the loss of houses in Tabasco (Devora erosión litoral tabasqueño n.d.). The Chontal People settled on the coastline are at risk of losing their lands and territories. Despite this risk, we found no references to protecting lands at risk of being lost or the people and communities at risk of losing their lands and territories in the Centla Municipal Development Plan 2018–2021 (Ayuntamiento Constitucional de Centla 2019).

(iii) **Sea-level rise impacts addressed without a differentiated approach**

First and concerning coastal flooding, the Atlas of Dangers due to Natural Phenomena in the State of Tabasco (Servicio Geológico Mexicano 2013) provides data that indicates a risk of river flooding in coastal localities near

Table 4.2 Coverage of drinking water, drainage and sewerage in the municipality of Centla, Tabasco

Municipality	Population	Coverage (%)								
		Drinking water			Sewer system			Sewerage		
		Rural	Urban	Total	Rural	Urban	Total	Rural	Urban	Total
Centla	92 755	13.8	52.0	30.4	78.1	97.1	86.3	0.6	28.5	12.7
State total	1 989 969	58.4	87.8	74.4	88.0	97.8	93.4	8.5	71.5	42.8

Source CONAGUA (2009, p. 728)

bodies of water and the risk of marine flooding to most of Centla beaches. The Tabasco Risk Atlas (Seguridad, CNPC and CENAPRED 2017) confirms these risks. Municipal civil protection also confirms the vulnerability of coastal communities to flooding (Ayuntamiento Constitucional de Centla 2019). For example, in 2019, in Centla 600 people were affected by a storm surge in two localities, 200 by the Usumacinta River flooding and 2 499 by the Grijalva River influx (Ayuntamiento Constitucional de Centla 2019). To attend to this population, the state civil protection Institute reports 67 temporary shelters in Centla, with the capacity to receive 15 180 people (Instituto de Protección Civil del Estado de Tabasco 2016a). The Municipal Development Plan (Ayuntamiento Constitucional de Centla 2019) does not mention quite as many temporary shelters: it mentions just 22 of them (p. 45).

Against this background, the Chontal People are particularly exposed to those risks. Furthermore, climate change is expected to increase the frequency and magnitude of coastal flooding (Ayuntamiento Constitucional de Centla 2019, p. 42). The number of people affected will probably increase, generating emergency forced climate migrations and trapped migrants given the Chontal People's socio-economic preexisting vulnerability and isolation. The Centla Atlas of Disaster Risks to Hydrometeorological Phenomena (Coordinación de la Unidad de Protección Civil Municipal n.d.) confirms the risk of trapped population:

> Those who have this perception (*the high possibility of river flooding*) accept the possibility of flooding with a particular frequency in exchange for continuing to live on their property or where their resources allow. Relocating is not considered a viable alternative, mainly for its high costs. (Diagnosis of the knowledge and sensitivity of community and population leaders regarding the risks existing in the locality and the need to prevent their possible consequences).

The knowledge of the Indigenous Peoples in flood management could limit the magnitude of the risks. For example, Rivera Ospina and Ramos Muñoz (2019, p. 252) and the Atlas of Disaster Risks in the event of hydrometeorological phenomena (Coordinación de la Unidad de Protección Civil Municipal n.d.) mention adaptation strategies already applied by communities exposed to flooding (e.g., the construction of homes on high ground or in the form of stilt houses). However, this knowledge cannot be relied upon to fully evade the reported risks.

As regards the risk to the Chontal People of having to forcefully migrate or being trapped by coastal flood events, we identify the shortcomings of Centla's public policy instruments. The Centla Development Plan 2018–2021 (Ayuntamiento Constitucional de Centla 2019) foresees a series of relevant lines of action for floods that can limit the risks to buildings in danger zones (Lines of action 2.9.1.11–12, 2.10.1.2, 2.10.1.4–5, 2.10.1.8, 2.10.1.12, and 3.1.3.6). It also recognizes various gaps: there is, for example, no action plan to prevent future disasters, no population map to help families "living in low-lying areas," no municipal risk atlas, neither is there a Contingency Disaster

Program (pp. 86–87). Regarding the rights of Indigenous Peoples, no plan of action mentions their specific needs or reflects the protection requirements and the specificities codified in the Operational Guidelines.

The Centla Atlas of Disaster Risks to Hydrometeorological Phenomena (Unidad de Protección Civil Municipal n.d.) is devoted almost exclusively to flood risks. It performs a diagnosis of risks and the deficiencies of current measures (most notably as regards budget). However, it does not establish the link between sea-level rise caused by climate change and floods nor does it mention that the measures must reflect a differentiated approach towards Indigenous Peoples.

Secondly, marine ecosystems include various types of wetlands that provide services to ecosystems. The "Ramsar Convention on Wetlands" website on wetlands of global importance describes the following attributes:

> Wetlands provide us with water, they protect us from floods, droughts and other disasters, they provide food and livelihoods to millions of people, they support rich biodiversity, and they store more carbon than any other ecosystem. (Ramsar Convention Secretariat 2019)

In the same sense, Guzmán Amaya et al. (2010) point out the importance of wetlands to the provision of food. The loss of 1 hectare of mangrove is equivalent to the loss of production of more than 700 tons of shrimp per year. Risks to marine ecosystems involve their physical loss and the loss of the services to the ecosystems they provide. It hinders the capacities of the populations that settle there to survive economically.

Wetlands suffer deterioration in Mexico.[32] Losses in wetland ecosystems have been increasing, even though they provide environmental services and their deterioration generates economic losses, and the creation of protected areas for conservation (such as Protected Natural Areas and areas designated as Ramsar sites).[33] At a national level, 62.1% of wetland coverage has been lost, and in Tabasco, the percentage is around 60% (Landgrave and Moreno-Casasola 2012, p. 27). Erosive processes documented by Hernández Santana et al. (2008) have more specifically generated the destruction of mangrove ecosystems in Centla (Hernández Santana et al. (2008), p. 17). With information

[32] The National Institute of Statistics and Geography (INEGI 2013) indicates that the economic effort registered in favor of the environment during 2013 was 5.1 times less than the damage caused.

[33] At a national level, the National Commission for Protected Natural Areas (CONANP) is the designated authority to manage Protected Natural Areas and "to apply the guidelines derived from the agreements and commitments adopted by the Ramsar Convention on Wetlands of International Importance" (CONANP 2019). The *Pantano de Centla* is within the ANP (category of Biosphere Reserve—with decree date of August 6, 1992). It has a management program (September 26, 2000), and is a Ramsar site (entry date of June 22, 1995). At a national level, the National Wetlands Policy defines strategic lines and policies to conserve wetlands. It involves different government sectors, with the National Water Commission (*Comisión Nacional del Agua*—CONAGUA) and National Commission on Protected Natural Areas (*Comisión Nacional de Áreas Naturales Protegidas*—CONANP) acting as leading institutions (SEMARNAT 2017).

from the National Institute of Statistics and Geography (*Instituto Nacional de Estadística y Geografía*—INEGI) (2015) we identify 21 localities settled on the Centla coast (4 271 people) who could be directly affected. Academic studies confirm it. For example, Rivera Ospina and Ramos Muñoz (2019) found that women in the Quintin Arauz community have had to diversify their productive activities.

The Indigenous Peoples in Centla have settled preferentially in floodplain and coastal zones. These ecosystems provide them with the resources to survive. The gradual loss of the marine ecosystem induced by sea-level rise could induce their forced mobility as they directly depend on accessing the ecosystem and associated resources (slow-onset scenario). Additionally, scarcity within the ecosystem could generate significant conflict and violence (violence scenario).

The Centla Municipal Development Plan (Ayuntamiento Constitucional de Centla 2019) mentions several points of action aimed at protecting marine ecosystems at a local level. For example, line of action 2.14.17 refers to "programs that prevent the coast, rivers and lagoons deterioration to repopulate and preserve marine and aquaculture species,"[34] and line of action 3.1.2.13 foresees the creation of a "manatee sanctuary." However, they do not articulate a differentiated approach to Indigenous Peoples (e.g., determining which species are essential to them, either for their economic development or for their worldview).

(iv) **Impacts of sea-level rise addressed with a very limited differentiated approach**

First and regarding changes in coastal soils and groundwater' salinity, ECOSUR and SERNAPAM (2014, p. 9) report on the salinity of soils in Tabasco and Mérida (2002) points out that changes in coastal groundwater salinity or saline intrusion occur when.

> a hydraulic load of the freshwater present in the aquifer and containing seawater advance is removed … by water extraction, aquifer recharge, coastal modifications (anthropic and natural), and sea level. (p. 841)

In the same way, Aguilar (2012) evaluated the effect of saline intrusion in 19 home gardens distributed throughout the Tabasco municipalities of Paraíso, Jalpa de Méndez, and Comalcalco. He found high salinities in orchards near the Mecoacán lagoon. Also, some tree species have a greater ability to regulate exposure to the impacts of salinity that generate a loss of the soil's agricultural productivity. He also mentions that salinity is not limited to coastal areas and can enter rivers and lagoons, affecting inland sites (p. 36).

[34] Similarly, regarding the municipal agenda for climate change (Strategy 3.1.2 "Implementing a municipal agenda for climate change") but without specifically referring to marine ecosystems, Line of action 3.1.2.5 provides for the protection, conservation, and restoration of the municipality's ecosystems and their natural resources, Line of action 3.1.2.6 refers to restoration and sanitation programs for affected ecological zones, and Line of action 3.1.3.11 establishes the need to "implement closed programs to preserve and repopulate endangered species.".

Accordingly, the Chontal People may be forcefully migrating due to changes in coastal soils and groundwater salinity, the decrease in soil productivity, and the availability of freshwater (together with the lack of adequate infrastructure for its provision identified in the water stress case). This relates to two scenarios that generate internal forced climate migrations: slow-onset phenomena and conflicts and violence generated by the scarcity of natural resources.

Turning to Centla's public policy instruments, we did not find any mention of objectives, strategies, or lines of action to address the problem of water salinity explicitly. However, we did find several Lines of action in the Municipal Development Plan (Ayuntamiento Constitucional de Centla 2019) which are relevant in the context of soil salinity. They aim to improve soil quality or support for the populations as regards the production losses they suffer. Among them, the preparation of "research studies for the optimal use of the land establishing policies for reforestation of the same" (Line of action 2.13.1.9),[35] and the development of "training, advisory, technical and operational support programs required by indigenous populations and marginalized groups in the municipality, to carry out productive projects or social development actions" (Line of action 2.12.1.16). It is the only mention of Indigenous Peoples. Therefore, we conclude Centla's instruments do not endorse a differentiated approach.

Secondly, regarding the road infrastructure, Ramos Reyes et al. (2016a, b) demonstrated that coastal vulnerability induced by sea-level rise, impacts the highways in Centla. Damage to the roads increases the indigenous population's vulnerability: it can cause temporary isolation and make it difficult for goods and people to enter and leave the zone where they settle. It would then, in turn, generate a particular group of forced internal climate migrants: trapped forced climate migrants.[36]

The road infrastructure closest to the coastline in the municipality of Centla goes from Puerto Ceiba (municipality of Paraíso, Tabasco) to Ciudad del Carmen, Campeche, passing through the towns of Cuauhtémoc, Vicente Guerrero, and Frontera, all in Centla. From this highway, there are secondary roads to coastal towns such as Carlos A. Madrazo, Emiliano Zapata, Pico de Oro, Miramar Beach, El Privilegio, El Milagro, El Bosque, Nuevo Centla, Barra de San Pedro, and Nuevo Campechito (INEGI 2017). Due to their proximity to the coastline, sea-level rise could adversely affect these road sections. From the bibliographic review carried out, we found no references regarding damage or potential damage to these road infrastructures. It does not mean damage has not happened or cannot happen. Instead, to date, they have not been investigated, and no projections have been realized.

In terms of public policies, the creation and maintenance of the road network is the responsibility of different government levels: the federal government administers toll-free federal highways and networks, the state government is responsible for state highways, state government feeder networks and rural network, while the

[35] Reforestation also appears in line of action 2.1.2.7 on the municipal agenda for climate change.

[36] The potential of Indigenous Peoples to be trapped because of damage to the access roads adds to another factor: their social marginalization. They may not have the necessary economic resources to move (a car, fuel to fill the tank etc.).

municipal government manages local feeder networks, rural networks and improves road surfaces (Junta Estatal de Caminos 2019).

The capacity of the state to provide adequate infrastructure is apparently low, as the maintenance of the roads and repair is insufficient. The Tabasco 2019–2024 State Development Plan indicates that from 2013 to 2016, the road network increased 334 km (3.27%). However, its conservation (state government's responsibility) in 2018 was: 20% in good conditions, 48% in normal conditions, and 32% in poor conditions (Gobierno de Tabasco 2019, p. 270). Consequently, the current conditions of the roads already favor the generation of trapped forced climate migrants, and the sea-level rise will further impact roads already damaged.

In this context, the Centla Municipal Development Plan (Ayuntamiento Constitucional de Centla 2019) mentions in the promotion of public paving works (Line of action 2.11.1.2), implementing actions to improve the communication routes of rural area (Line of action 2.11.1.4), and a program for repaving roads (Line of action 2.11.1.5). Of these lines of action, the improvement of communication routes in the rural area (where the Chontal People concentrate) stands out. Beyond that, no action addresses Indigenous Peoples and, more specifically, the needs of the Chontal People.

In short, the instruments in place in Centla are grossly deficient in addressing the sea-level rise impacts the Chontal People will suffer. They do not address all of the sea-level rise impacts. When they do address the impacts, they do not implement a differentiated approach to Indigenous Peoples, or it is, at best, very superficial. In turn, these deficiencies not only increase the Chontal People's exposure to forced climate internal mobility. They also increase their risk of being trapped.

4.4 Conclusion

In this chapter, we aimed to determine whether the limitations and legal and public policy shortcomings at a federal level are also evident at a sub-national level. We chose the Tabasco state to conduct our analysis and found that the instruments it has in place also suffer from shortcomings. Similarly, when we directed our attention to the municipal level with a differentiated approach towards women and Indigenous Peoples and focused on two climate mobility scenarios, we found several further limitations. It is worth noting that we found limitations in the two cases analyzed despite their differences. One case refers to a specific case sudden-onset disaster (2007 floods), and the other one relates to the sea-level rise disaggregated impacts regarding both contingency and prevention.

Consequently, we can conclude from this chapter and the previous one that there is no solid regulatory and public policy framework to address this phenomenon at a federal, state, or municipal level. This answer to the initial question of the research justifies the need to adopt (or at least adapt) the legal frameworks and public policies to ensure compliance with the protection requirements for the people and communities affected. In this sense, there remains the need to reflect on the potential guidelines to include such legal and public policy frameworks. The following chapter expands on this topic.

References

Aguilar CMZ, Velázquez MA, Manzanilla AV (2020) Proyecciones de Cambio Climático para la Zona Centro de Villahermosa, Tabasco. Kuxulkab 26(54):21–26

Aguilar AA (2012) Efecto de la intrusión salina en huertos familiares en la costa de Tabasco, México. Tesis de Doctorado, El Colegio de la Frontera Sur

Arreguín FI, Rubio H, Domínguez R, Luna FD (2014) Análisis de las inundaciones en la planicie tabasqueña en el periodo 1995–2010. Tecnología y Ciencias Del Agua 5(3):05–32

Arreguín Cortés FI, López Pérez M, Rodríguez López O, Montero Martínez MJ (2015) Atlas de vulnerabilidad hídrica en México ante el cambio climático, Jiutepec, Mor.: Instituto Mexicano de Tecnología del Agua. Available 30 Nov 2021, at https://atlasvulnerabilidad.inecc.gob.mx/page/fichas/ANVCC_LibroDigital.pdf

Ayuntamiento Constitucional de Centla (2019) Plan municipal de desarrollo 2018–2021. Available 30 Nov 2021, at https://centla.gob.mx/index.php/plan-de-desarrollo-municipal/

Ayuntamiento Constitucional de Centro Villahermosa (2003) Reglamento de Protección Civil para el Municipio de Centro. Suplemento B al Periódico Oficial Núm. 6311, 1 March 2003. Available 30 Nov 2021, at http://www.ordenjuridico.gob.mx/Estatal/TABASCO/Municipios/Centro/11REG.pdf

Ayuntamiento Constitucional de Centro (2008) Programa de Desarrollo Urbano del Centro de Población de la Ciudad de Villahermosa y Centros Metropolitanos del Municipio de Centro, Tabasco. Available 30 Nov 2021, at http://www.ordenjuridico.gob.mx/Estatal/TABASCO/Municipios/Centro/1PROGRAMA.pdf

Ayuntamiento Constitucional de Centro (2015) Atlas de Riesgos del Municipio de Centro, Tabasco. Available 30 Nov 2021, at https://transparencia.villahermosa.gob.mx/images/Documentos_transparencia/Informacion_de_Interes/Instituto_Municipal_de_Planeaci%C3%B3n_y_Desarrollo_Urbano/AtlasDeRiesgosCentro2015.pdf

Ayuntamiento Constitucional de Centro (2019) Coordinación de transparencia acceso la información pública; Villahermosa, Tabasco, 8 de febrero de 2019. Solicitud de información: "Programa Interno de Protección Civil durante el trienio 2016–20," "Programa Interno de Protección Civil del Trienio 2018–2021". Available 30 Nov 2021, at https://transparencia.villahermosa.gob.mx/doctos/concluido/2019/ac.disponibilidad141-19-21-R.pdf

Ayuntamiento de Centro (2018) Plan Municipal de desarrollo 2018–2021. Available 30 Nov 2021, at https://www.villahermosa.gob.mx/doctos/2019/agosto/PMD_2019.pdf

Botello AV, Fragoso V, Galaviz R, Luiscoed J (2010) Vulnerabilidad de las zonas costeras mexicanas ante el cambio climático. Gobierno del Estado de Tabasco, Semarnat-IINE, UNAM-ICMYL, Universidad Autónoma de Campeche, Campeche

Cabildo del Municipio Centro (2015) Plan de Acción Climática del municipio Centro, Tabasco (PACMUN). Available 30 Nov 2021, at https://transparencia.villahermosa.gob.mx/images/Documentos_transparencia/articulo-10-fraccion-i/inciso-q/ano-2015-16/4to%20trimestre/Acta_Sesion_50_4totrim_15.pdf

Campillo LMG, López HMD, Torres RAC, Conde EDCS, Ordoñez EJM (2018) Políticas del ordenamiento ecológico de Tabasco. Kuxulkab' 24(48):37–42

Centro Nacional de Prevención de Desastres (CENAPRED). (n.d.). Base de datos sobre el impacto socioeconómico de los daños y pérdidas ocasionados por los desastres en México. Available 30 Nov 2021, at http://www.atlasnacionalderiesgos.gob.mx/archivo/descargas.html

Colegio de la Frontera Sur (ECOSUR), & Secretaria de Recursos Naturales y Protección Ambiental del Estado de Tabasco (SERNAPAM) (2014) Programa Piloto de Adaptación ante el Cambio Climático para la Subregión Pantanos. Available 30 Nov 2021, at https://www.researchgate.net/publication/274007053_Programa_Piloto_de_Adaptacion_ante_el_Cambio_Climatico_para_la_Subregion_Pantanos

Comisión Económica para América Latina y el Caribe (CEPAL), Centro Nacional de Prevención de Desastres (CENAPRED) & Secretaría Técnica de Planeación y Evaluación (SEPLAN) (2012) Impacto socioeconómico de las inundaciones registradas en el estado de Tabasco de septiembre a noviembre de 2011. Available 30 Nov 2021, at https://www.cepal.org/es/publicaciones/25923-impacto-socioeconomico-inundaciones-registradas-estado-tabasco-septiembre

Comisión Nacional para el Desarrollo de los Pueblos Indígenas (2006) Proyecto para la Atención a Indígenas desplazados. Lineamientos. Diario Oficial de la Federación, 15 June 2016. Available 30 Nov 2021, at https://www.dof.gob.mx/nota_detalle.php?codigo=4911544&fecha=15/06/2006

Comisión Nacional de Áreas Naturales Protegidas (CONANP) (2019) Humedales de México. Available 30 Nov 2021, at https://www.gob.mx/conanp/documentos/manual-operativo-del-comite-nacional-de-humedales

Comisión Nacional del Agua (CONAGUA) (2009) III.5.e Agua potable y saneamiento. Available 30 Nov 2021, at https://www.gob.mx/cms/uploads/attachment/file/111728/III.5.E.-_AGUA_POTABLE_Y_SANEAMIENTO.pdf

Comisión Nacional del Agua (CONAGUA) (2016) Atlas del agua en México. Available 30 November 2021 at http://201.116.60.25/publicaciones/AAM_2016.pdf

Comisión Nacional del Agua (CONAGUA) (2018) Actualización de la disponibilidad media anual de agua en el acuífero Centla (2704), Estado de Tabasco. Available 30 Nov 2021, at https://sigagis.conagua.gob.mx/gas1/Edos_Acuiferos_18/tabasco/DR_2704.pdf

Comité de Planeación para el Desarrollo del estado de Tabasco (COPLADET) (2019) Plan Estatal de Desarrollo 2019–2024 (PLED). Available 30 Nov 2021, at https://tabasco.gob.mx/sites/default/files/users/planeacion_spf/PLED%202019-2024.pdf

Consejo Nacional de Evaluación de la Política de Desarrollo Social (CONEVAL) (2019) Evolución de la pobreza y pobreza extrema nacional y en entidades federativas, 2008, 2010, 2012, 2014, 2016 y 2018. Available 30 Nov 2021, at https://www.coneval.org.mx/Medicion/MP/Paginas/Pobreza-2018.aspx

Consejo Nacional de Evaluación de la Política de Desarrollo Social (CONEVAL) (2015) Pobreza a nivel municipio 2010 y 2015. Available 30 Nov 2021, at https://www.coneval.org.mx/Medicion/Paginas/Pobreza-municipal.aspx

Constitución Política de los Estados Unidos Mexicanos (CPEUM) (2020) Last reform *Diario Oficial de la Nación*, 8 May 2020. Available 30 Nov 2021, at http://www.diputados.gob.mx/LeyesBiblio/pdf_mov/Constitucion_Politica.pdf

Coordinación de la Unidad de Protección Civil Municipal (n.d.) Atlas de Riesgos de desastres ante fenómenos hidrometeorológicos del Municipio de Centla. Available 30 Nov 2021, at https://transparencia.centla.gob.mx/images/doc/2016/Art76/XLIX/Atlas-de-Riesgos-del-Municipio-de-Centla.pdf

Desaloja policía a mujeres que exigen reparar escuela en Tabasco (2014) Available 30 Nov 2021, at https://www.proceso.com.mx/nacional/estados/2014/9/8/desaloja-policia-mujeres-que-exigen-reparar-escuela-en-tabasco-137010.html

Devora erosión litoral tabasqueño (n.d.) Novedades de Tabasco. Available 30 Nov 2021, at https://novedadesdetabasco.com.mx/2016/04/22/devora-erosion-litoral-tabasqueno/

Discovery Channel (2008) Tabasco bajo el agua (Documental). Available 30 Nov 2021, at https://www.youtube.com/watch?v=_OUp7zYQxrI

Estadístico A (2014) Geográfico de los Estados Unidos Mexicanos 2013. INEGI, México

Field CB, Barros V, Dokken DJ, Mach KJ, Mastrandrea MD, Bilir TE, Chatterjee M (2014) Impacts, Adaptation, and Vulnerability. Part A: Global and Sectoral Aspects, 1132. Global and Sectoral Aspects, Cambridge, UK

Frejomil EP, Sánchez-Crispín A, Izquierdo JC (2006) Las divergencias socioeconómicas territoriales como alternativa metodológica del ordenamiento territorial en México. Geographicalia 49:157–175

Gobierno de Tabasco (2018) Programa de Ordenamiento Sustentable del Territorio del Estado de México (POSTET). Available 30 Nov 2021, at http://periodicos.tabasco.gob.mx/media/periodicos/7920B1.pdf

Gobierno de Tabasco (2019) Informe de la gestión gubernamental 2019. Available 30 Nov 2021, at https://informe.tabasco.gob.mx/pdf/informe_gobierno/informe_gobierno_tabasco.pdf

González Turrubiates DM (2010) Valoración y gestión del riesgo por inundaciones en municipios costeros (Una aproximación con base en indicadores). In: Rivera E, Azuz I, Alpuche L, Villalobos GJ (eds) Cambio Climático en México un Enfoque Costero-Marino. Universidad Autónoma de Campeche, Cetys-Universidad, Gobierno del Estado de Campeche, Campeche, pp. 591–606. Available 30 Nov 2021, at https://www.researchgate.net/profile/Leticia_Alpuche_Gual/publication/267510460_Cambio_Climatico_en_Mexico_un_enfoque_costero_y_marino/links/59193a924585152e19a24a6f/Cambio-Climatico-en-Mexico-un-enfoque-costero-y-marino.pdf#page=99

Gordillo GDCÁ, Pablos ET (2016) Vulnerabilidad social de la población desplazada ambiental por las inundaciones de 2007 en Tabasco (México). Cuadernos De Geografía-Revista Colombiana De Geografía 25(1):123–138

Guzmán Amaya P, Rojas Carrillo PM, Morales García GZ, Espino Barr E (2010) Retos para el sector pesquero y acuícola ante el cambio climático. In: Botello AV, Fragoso V, Galaviz R, Luiscoed J (eds) Vulnerabilidad de las zonas costeras mexicanas ante el cambio climático. Gobierno del Estado de Tabasco, Semarnat-IINE, UNAM-ICMYL, Universidad Autónoma de Campeche, Campeche, pp 113–164

Hernández Hernández M (2011) Inundación, reubicación y cotidianidad. El caso de Villahermosa, Tabasco, 2007 (Tesis de doctorado, tesis de maestría en Antropología Social, México, Centro de Investigaciones y Estudios Superiores en Antropología Social, Unidad Golfo). Available 30 Nov 2021, at https://ciesas.repositorioinstitucional.mx/jspui/bitstream/1015/534/1/TE%20H.H.%202011%20Ma%20Magdalena%20Hernandez%20Hernandez.pdf

Hernández Santana JR, Ortiz Pérez MA, Méndez Linares AP, Gama Campillo L (2008) Morfodinámica de la línea de costa del estado de Tabasco, México: tendencias desde la segunda mitad del siglo XX hasta el presente. Investigaciones Geográficas 65:7–21

Hoegh-Guldberg O, Jacob D, Taylor M, Bindi M, Brown S, Camilloni I, Diedhiou A, Djalante R, Ebi K, Engelbrecht F, Zhou G, Joel G, Hijioka Y, Mehrotra S, Payne A, Seneviratne S, Thomas A, Warren R, Halim S, Guillén Bolaños T (2018) Impacts of 1.5 °C global warming on natural and human systems. In: Masson-Delmotte V, Zhai P, Pörtner H-O, Roberts D, Skea J, Shukla PR, Pirani A, Moufouma-Okia W, Péan C, Pidcock R, Connors S, Matthews JBR, Chen Y, Zhou X, Gomis MI, Lonnoy E, Maycock T, Tignor M, Waterfield T (eds) Global Warming of 1.5°C. An IPCC Special Report on the impacts of global warming of 1.5°C above pre-industrial levels and related global greenhouse gas emission pathways, in the context of strengthening the global response to the threat of climate change, sustainable development, and efforts to eradicate poverty. IPCC, Ginebra, pp 175–311. Available 30 Nov 2021, at https://www.ipcc.ch/site/assets/uploads/sites/2/2019/02/SR15_Chapter3_Low_Res.pdf

Instituto Nacional de Estadística y Geografía (INEGI) (2015) Panorama sociodemográfico de Tabasco 2015. Available 30 Nov 2021, at http://internet.contenidos.inegi.org.mx/contenidos/Productos/prod_serv/contenidos/espanol/bvinegi/productos/nueva_estruc/inter_censal/panorama/702825082390.pdf

Instituto de Protección Civil del Estado de Tabasco (2016a) Refugios temporales. Available 30 Nov 2021, at https://tabasco.gob.mx/sites/default/files/users/ipcettabasco/menu_Refugios_Temporales_2016a.pdf

Instituto de Protección Civil del Estado de Tabasco (2016b) Plan de Contingencias Fenómenos Hidrometeorológicos. Available 30 Nov 2021, at https://tabasco.gob.mx/sites/default/files/users/ipcettabasco/menu_planesyprogramas_5_fenomenos_hidrometeorologico_2016b.pdf

Inter-Agency Standing Committee (IASC) (2011) IASC Operational Guidelines on the Protection of Persons in Situations of Natural Disasters. Brookings-Bern Project on Internal Displacement. Available 30 Nov 2021, at https://www.cndh.org.mx/sites/all/doc/OtrosDocumentos/Doc_2017_045.pdf

International Labour Organization (ILO) (1989) Convenio 169 Pueblos Indígenas y Tribales en Países Independientes, adopted 27 June 1989. Available 30 Nov 2021, at https://www.senado.gob.mx/comisiones/desarrollo_social/docs/marco/Convenio_169_PI.pdf

Instituto Nacional de Ecología y Cambio Climático (INECC) (2019) Atlas Nacional de Vulnerabilidad al Cambio Climático México. 1ª. Edición (libro electrónico). INECC, Mexico. Available 30 Nov 2021, at https://atlasvulnerabilidad.inecc.gob.mx/page/fichas/ANVCC_LibroDigital.pdf

Instituto Nacional de Estadística y Geografía (INEGI) (2017) Carta topográfica 1:250 000 Frontera E1505. Available 30 Nov 2021, at http://internet.contenidos.inegi.org.mx/contenidos/Productos/prod_serv/contenidos/espanol/bvinegi/productos/geografia/historia/topo/1_250_000/889463531968_geo.pdf

Instituto Nacional de Estadística y Geografía (INEGI) (2018) Mapa de Tabasco. División Municipal. Available 30 Nov 2021, at http://www.cuentame.inegi.org.mx/mapas/pdf/entidades/div_municipal/tabascompios.pdf

Instituto Nacional de Estadística y Geográfica (INEGI) (2013) Sistema de Cuentas Nacionales de México: cuentas económicas y ecológicas de México 2013: preliminar: año base 2008. Available 30 Nov 2021, at http://internet.contenidos.inegi.org.mx/contenidos/productos/prod_serv/contenidos/espanol/bvinegi/productos/nueva_estruc/702825068752.pdf

Instituto Nacional de los Pueblos Indígenas (INPI) (2019) Atlas de los Pueblos Indígenas de México. Available 30 Nov 2021, at http://atlas.inpi.gob.mx/?page_id=247

Instituto Nacional de los Pueblos Indígenas (INPI) (2018) Programa Nacional de el 19 de noviembre de 2020. Available 30 Nov 2021, at https://www.gob.mx/cms/uploads/attachment/file/423227/Programa-Nacional-de-los-Pueblos-Indigenas-2018-2024.pdf

Instituto de Protección Civil del Estado de Tabasco (2015) Programa Maestro de Protección Civil 2013–2018 del Gobierno de Tabasco. Available 30 Nov 2021, at https://tabasco.gob.mx/sites/default/files/users/setabasco/16-PROGRAMAMAESTRODEPROTECCIONCIVIL.pdf

Jungehülsing J (2012) Gender relations and women's vulnerability to climate change: contribution from an adaptation policy in the state of tabasco toward greater gender equality; the reconstruction and reactivation program to transform Tabasco. Available 3 May 2021 at https://reliefweb.int/sites/reliefweb.int/files/resources/estudio_tabasco_ingles.pdf

Junta Estatal de Caminos (2019) Evaluación de consistencia y resultados en materia de diseño. Programa presupuestario: k008 Carreteras para el año 2016. Available 30 Nov 2021, at https://tabasco.gob.mx/sites/default/files/users/jectabasco/CyRmD_K008_13102017_VFinal.pdf

Landgrave R, Moreno Casasola P (2012) Evaluación cuantitativa de la pérdida de humedales en México. Investigación Ambiental Ciencia y Política Pública 4(1):19–35

de la Lanza Espino G, Gómez Rojas JC, Blanco y Correa M, Flores Verdugo, Hernández Pulido S (2010) Vulnerabilidad costera: caso de estudio del sistema de humedales Marismas Nacionales. In: Rivera-Arriaga E, Azuz-Adeath I, Alpuche-Gual L, Villalobos-Zapata GJ (eds) Cambio climático en México: un enfoque costero y marino (pp. 205–230). Universidad Autónoma de Campeche, Cetys-Universidad, Gobierno del Estado de Campeche, Campeche. *Cambio climático en México*, 205. Available 30 Nov 2021, at https://www.researchgate.net/profile/Leticia_Alpuche_Gual/publication/267510460_Cambio_Climatico_en_Mexico_un_enfoque_costero_y_marino/links/591710980f7e9b70f49f4adb/Cambio-Climatico-en-Mexico-un-enfoque-costero-y-marino.pdf#page=231

Ley del Instituto Nacional de los Pueblos Indígenas (LINPI) (2018) Última reforma *Diario Oficial de la Federación*, 4 December 2018. Available 30 Nov 2021, at http://www.diputados.gob.mx/LeyesBiblio/ref/linpi.htm

Llega a Tabasco una brigada de médicos de cuba (2007) Proceso. Available 30 Nov 2021, at https://www.proceso.com.mx/nacional/2007/11/9/llega-tabasco-una-brigada-de-medicos-de-cuba-38317.html

López A, Garrido J (n.d.) Percepción de la importancia de los desastres y atención que reciben los afectados en Tabasco. Educación Ambiental, Cambio Climático y Sustentabilidad, 107

Magaña V (2012) Guía metodológica para la evaluación de la vulnerabilidad ante cambio climático. Instituto Nacional de Ecología y Cambio Climático (INECC) y Programa de las Naciones Unidas para el Desarrollo (PNUD), México

Megainundación de 2007 sumergió a Tabasco en más miseria e impunidad: Codehutab (2014) Proceso. Available 30 Nov 2021, at https://www.proceso.com.mx/nacional/estados/2014/10/27/megainundacion-de-2007-sumergio-tabasco-en-mas-miseria-impunidad-codehutab-138994.html

Mérida LA (2002) La Evolución de la Intrusión Salina en el Acuífero Marplatense. Ejemplo De Una Gestión Sustentable 10:841–850

Office for the Coordination of Humanitarian Affairs (OCHA) (2007a) Mexico: Tabasco and Chiapas Floods OCHA Situation Report No. 4. Available 30 Nov 2021, at https://reliefweb.int/report/mexico/mexico-tabasco-and-chiapas-floods-ocha-situation-report-no-4

Office for the Coordination of Humanitarian Affairs (OCHA) (2007b) Mexico: Tabasco and Chiapas Floods OCHA Situation Report No. 7. Available 30 Nov 2021, at https://reliefweb.int/report/mexico/mexico-tabasco-and-chiapas-floods-ocha-situation-report-no-7

Organization of American States (OAS) (2016) Declaración Americana sobre los Derechos de los Pueblos Indígenas, adopted 14 June 2016, AG/RES. 2888 (XLVI—O/16). Available 30 Nov 2021, at https://www.oas.org/es/sadye/documentos/res-2888-16-es.pdf

Ospina Noreña JE, Sánchez Torres Esqueda G, Conde Álvarez CC (2010) Impactos del cambio climático en las regiones hidrológicas del Golfo de México. In: Rivera-Arriaga E, Azuz-Adeath I, Alpuche-Gual L, Villalobos-Zapata GJ (eds) Cambio climático en México: un enfoque costero y marino.: Universidad Autónoma de Campeche, Cetys-Universidad, Gobierno del Estado de Campeche, Campeche, pp 73–87. Available 30 Nov 2021, at https://www.researchgate.net/profile/Leticia_Alpuche_Gual/publication/267510460_Cambio_Climatico_en_Mexico_un_enfoque_costero_y_marino/links/59193a924585152e19a24a6f/Cambio-Climatico-en-Mexico-un-enfoque-costero-y-marino.pdf#page=99

Padilla Souza C, Alafita-Vásquez H, Andreu-Montalvo E (2010) Factores de riesgo para los arrecifes coralinos y sus mecanismos de respuesta ante los efectos del cambio climático global. In: Yáñez-Arancibia A (eds) Impactos del Cambio Climático Sobre la Zona Costera. Instituto Nacional de Ecología INE-SEMARNAT and Instituto de Ecología AC INECOL, Mexico, pp 181–203

Perevochtchikova M, de la Torre JLL (2010) Causas de un desastre: Inundaciones del 2007 en Tabasco, México. J Lat Am Geogr 73–98

Posada Vanegas G, Vega Serratos BE (2010) Evaluación de zonas inundables para la ciudad de San Francisco de Campeche. In: Rivera E, Azuz I, Alpuche L, Villalobos GJ (eds) Cambio Climático en México un Enfoque Costero-Marino. Universidad Autónoma de Campeche, Cetys-Universidad, Gobierno del Estado de Campeche, Campeche, pp 607–622. Available 30 November 2021, at https://www.researchgate.net/profile/Leticia_Alpuche_Gual/publication/267510460_Cambio_Climatico_en_Mexico_un_enfoque_costero_y_marino/links/59193a924585152e19a24a6f/Cambio-Climatico-en-Mexico-un-enfoque-costero-y-marino.pdf#page=99

Ramírez DK, Nazar A, Zapata E, Estrada E, Salvatierra B (2012) Vulnerabilidad, derechos sexuales y reproductivos de mujeres centroamericanas residentes en dos comunidades de la zona rural de Tapachula, Chiapas, Mexico. Papeles De Población 18(72):113–146

Ramos Reyes R, Gama Campillo LM, Núñez Gómez JC, Sánchez Hernández R, Hernández Trejo H, Ruíz Álvarez O (2016a) Adaptación del modelo de vulnerabilidad costera en el litoral tabasqueño ante el cambio climático. Revista Mexicana De Ciencias Agrícolas 7(SPE13):2551–2563

Ramos Reyes R, Zavala-Cruz J, Gama Campillo LM, Pech Pool D, Ortiz Pérez MA (2016b) Indicadores geomorfológicos para evaluar la vulnerabilidad por inundación ante el ascenso del nivel del mar debido al cambio climático en la costa de Tabasco y Campeche, México. Boletín De La Sociedad Geológica Mexicana 68(3):581–598

Ramos-Reyes R, Sánchez-Hernández R, Gama-Campillo LM (2016) Análisis de cambios de uso del suelo en el municipio costero de Comalcalco, Tabasco, Mexico. Ecosistemas y Recursos Agropecuarios 3(8):151–160
Ramsar Convention Secretariat (2014) Ficha informativa de los humedales de Ramsar. Available 30 Nov 2021, at https://rsis.ramsar.org/RISapp/files/RISrep/MX733RIS.pdf
Ramsar Convention Secretariat (2019) The global wetland Outlook. Available 30 Nov 2021, at https://proesteros.org/en/global-wetland-outlook/
Reidmiller DR, Avery CW, Easterling DR, Kunkel KE, Lewis KLM, Maycock TK, Stewart BC (2017) Fourth national climate assessment, Volume II: Impacts, risks, and adaptation in the United States. US Global Change Research Program, Washington, DC
Rigaud KK, de Sherbinin A, Jones B, Bergmann J, Clement V, Ober K, Schewe J, Adamo S, McCusker B, Heuser S, Midgley A (2018) Groundswell: preparing for internal climate migration. World Bank, Washington D.C.
Rivera Hernández B, Aceves Navarro LA, Juárez López JF, Méndez Adorno JM, Ramos Álvarez C (2016) Evidencias del cambio climático en el estado de Tabasco durante el periodo 1961–2010. Revista Mexicana De Ciencias Agrícolas 7(SPE14):2645–2656
Rivera Ospina MB, Ramos Muñoz DE (2019) Mujeres en la Reserva de la Biósfera Pantanos de Centla, México: Género, adaptación y variabilidad climática. CIAIQ2019 3:239–248
Ruiz SA, Carvajal M (2015) Hacia el desarrollo de políticas pública locales en movilidad humana, en el contexto de desastres naturales y cambio climático: El caso del Gobierno de la Provinica de Pichincha. Proyecto PRODEM-GIZ Migración y Asilo, Gobierno Autónomo Descentralizado de la Provincia de Pichincha (Quito, Ecuador)
Ruiz Liévano SR (2019) Desafíos del cambio climático en tabasco, México. Barataria: revista castellano-manchega de ciencias sociales (26):207–224
Ruiz Rivera N, Casado Izquierdo JM, Sánchez Salazar MT (2015) Los Atlas de Riesgo municipales en México como instrumentos de ordenamiento territorial. Investigaciones Geográficas 88:146–162
Salazar Ramírez H (2011) Cambio climático y género: una visión desde lo local. Secretaría de Recursos Naturales y Protección Ambiental. Available 30 Nov 2021, at https://www.nofrackingmexico.org/wp-content/uploads/2019/06/Genero-y-CC-Tabasco-publicaci%C3%B3n29-nov.pdf
Sánchez Salazar MT, Palacio Prieto JL (2004) La experiencia mexicana en la elaboración de los Programas Estatales de Ordenamiento Territorial: Diagnóstico, problemática y perspectivas desde el punto de vista de la participación del Instituto de Geografía de la UNAM. Investigaciones Geográficas 53:75–97
Santana JRH, Manent MB, Linares APM, Rodríguez LME (2015) Panorama contemporáneo del ordenamiento ecológico territorial en México. Polígonos. Revista De Geografía 26:111–146
Secretaría de Bienestar, Sustentabilidad y Cambio Climático del Estado de Tabasco (BIENESTAR) (2019) Programa de Ordenamiento Ecológico Regional del Estado de Tabasco (POERET). Available 30 Nov 2021 (4 parts), at https://tabasco.gob.mx/sites/default/files/users/sbstabasco/POERET_2019-1.pdf; https://tabasco.gob.mx/sites/default/files/users/sbstabasco/POERET_2019-2.pdf; https://tabasco.gob.mx/sites/default/files/users/sbstabasco/POERET_2019-3.pdf; https://tabasco.gob.mx/sites/default/files/users/sbstabasco/POERET_2019-4.pdf
Secretaría de Medio Ambiente y Recursos Naturales (SEMARNAT) (n.d.) Declaratorias de desastre natural publicadas en el Diario Oficial de la Federación y número de municipios señalados por entidad federativa y tipo de fenómeno. Available 30 Nov 2021, at http://dgeiawf.semarnat.gob.mx:8080/ibi_apps/WFServlet?IBIF_ex=D1_DESASTRE00_06&IBIC_user=dgeia_mce&IBIC_pass=dgeia_mce&NOMBREENTIDAD=*&NOMBREANIO=*
Secretaría de Medio Ambiente y Recursos Naturales (SEMARNAT) (2017) Política nacional de humedales. Available 30 Nov 2021, at https://agua.org.mx/wp-content/uploads/2017/07/politica-nacional-de-humedales.pdf
Secretaria de Recursos Naturales y Protección Ambiental del Estado de Tabasco (SERNAPAM) (2011a) Programa Estatal de Acción ante el Cambio Climático. SERNAPAM, México. Available

30 Nov 2021, at https://www.researchgate.net/publication/308905598_Programa_Estatal_ante_el_Cambio_Climatico_de_TabascoMexico
Secretaria de Recursos Naturales y Protección Ambiental del Estado de Tabasco (SERNAPAM) (2011b) Agenda de Género ante el Cambio Climático para el Estado de Tabasco (Agenda GyCC). Available 30 Nov 2021, at https://www.gob.mx/cms/uploads/attachment/file/316925/Agenda_G_nero_y_Cambio_Clim_tico.pdf
Secretaría de Seguridad y Protección Ciudadana (Seguridad), Coordinación Nacional de Protección Civil (CNPC), & Centro Nacional de Prevención de Desastres (CENAPRED) (2017) Atlas de Riesgos. Tabasco. Available 30 Nov 2021, at http://www.atlasnacionalderiesgos.gob.mx/app/Estados/VisorTabasco/
Servicio Geológico Mexicano (2013) Atlas de peligro por fenómenos naturales del estado de Tabasco. Informe Técnico. Available 30 Nov 2021, at http://rmgir.proyectomesoamerica.org/PDFAtlasEstatales/TABASCO_2013.pdf
Tabasco, tragedia de la corrupción (2007). Proceso. Available 30 Nov 2021, at https://www.proceso.com.mx/nacional/2007/11/3/tabasco-tragedia-de-la-corrupcion-38206.html
Tabasco: suman 54 detenidos por actos de pillaje (2007) *El Universal*. Available 30 Nov 2021, at https://archivo.eluniversal.com.mx/notas/459240.html
Trejo FR, Cortés GS, Fernández JB (2009) La inundación de 2007 en Tabasco, México: evolución de niveles de agua. Tecnología y Ciencias Del Agua 24(4):159–166
Treviño Trujillo J, Rolon Aguilar JC, Pichardo Ramírez R (2010) Las vías de comunicación y el transporte ante las amenazas del cambio climático y su impacto en el desarrollo regional de México. In: Rivera-Arriaga E, Azuz-Adeath I, Alpuche-Gual L, Villalobos-Zapata GJ, Cambio climático en México: un enfoque costero y marino. Universidad Autónoma de Campeche, Cetys-Universidad, Gobierno del Estado de Campeche, Campeche, pp 407–420. Available 30 Nov 2021, at https://www.researchgate.net/profile/Leticia_Alpuche_Gual/publication/267510460_Cambio_Climatico_en_Mexico_un_enfoque_costero_y_marino/links/59193a924585152e19a24a6f/Cambio-Climatico-en-Mexico-un-enfoque-costero-y-marino.pdf#page=433
Tuñón Pablos E, Molina Rosales D (n.d.) Elaborando La Agenda de Género ante el Cambio Climático desde lo local: el caso de Tabasco. Secretaria de Recursos Naturales y Protección Ambiental del Estado de Tabasco. Available 30 Nov 2021, at https://www.americalatinagenera.org/es/documentos/2010/cop16/agenda-genero-cc.pdf
United Nations (UN) (1998) Principios Rectores De Los Desplazamientos Internos, UN doc.E/CN.4/1998/53/Add.2, 11 February 1998, Oficina de Coordinación de Asuntos Humanitarios de las Naciones Unidas. Available 30 Nov 2021, at https://www.acnur.org/prot/prot_despl/5bff2c864/principios-rectores-de-los-desplazamientos-internos.html
United Nations (UN) (2007) Declaración de las Naciones Unidas sobre los derechos de los pueblos indígenas, adopted 13 September 2007, A/RES/61/295. Available 30 Nov 2021, at https://www.un.org/esa/socdev/unpfii/documents/DRIPS_es.pdf
Wong PP, Losada IJ, Gattuso JP, Hinkel J, Khattabi A, McInnes KL, Saito Y, Sallenger A (2014) Coastal systems and low-lying areas. In: Field CB, Barros VR, Dokken DJ, Mach KJ, Mastrandrea MD, Bilir TE, Chatterjee M, Ebi KL, Estrada YO, Genova RC, Girma B, Kissel ES, Levy AN, MacCracken S, Mastrandrea PR, White LL (2014) Climate Change 2014: Impacts, Adaptation, and Vulnerability. Part A: Global and Sectoral Aspects. Contribution of Working Group II to the Fifth Assessment Report of the Intergovernmental Panel on Climate Change. Cambridge University Press, Cambridge and New-York, pp 361–409. Available 30 Nov 2021, at https://www.ipcc.ch/site/assets/uploads/2018/02/WGIIAR5-Chap5_FINAL.pdf
Yáñez-Arancibia A, Day JW (2010) La zona costera frente al cambio climático: vulnerabilidad de un sistema biocomplejo e implicaciones en el manejo costero. In: Yáñez-Arancibia A (eds) Impactos del Cambio Climático Sobre la Zona Costera. Instituto Nacional de Ecología INE-SEMARNAT and Instituto de Ecología AC INECOL, Mexico, pp 12–35

Chapter 5
Addressing the Phenomenon in Mexico: Risks from the Protection Requirements Perspective

Armelle Gouritin

In previous chapters, we established that Mexican law and public policies do not satisfactorily address internal forced climate mobility. With this void in mind, the phenomenon's breadth, and the risks posed to migrants regarding protection requirements, we now reflect on the measures put in place by Mexican authorities to build up the public policies that would address this phenomenon.[1] Jakobsson (2021) portrays forced climate mobility as a "creeping crisis." As a creeping crisis, key factors "work against politicians' motivations to tackle this increasingly worrisome creeping crisis" (p. 144). If Mexican authorities were to dismiss this qualification and address climate displacement, a number of risks would arise. We expose them in this chapter.

In order to critically analyze the measures researched by academics and found in the reports of international organizations, we present them and determine their potential and risks in Mexico from the perspective of the territory and protection requirements. We then look into the "mobility" and climate dimensions. The "mobility" dimensions are critically analyzed while also acknowledging the need to recognize the phenomenon (1) and forced relocation (2). We then assess the climate dimension taking into account preventative and reactive adaptation measures (3). We ground our critical analysis on academic literature, existing reports, and within the context of Mexico (informed, among others, by the knowledge generated in previous chapters).

---

[1] The objective of this chapter is not to propose the content of said public policies but rather the guidelines that would inform their formation. On the content of policies, see the Brookings Institution national twelve-point framework (Kälin 2004). They are (i) Prevention, (ii) Raise National Awareness of the Problem, (iii) Data Collection, (iv) Training on the Rights of IDPs, (v) A National Legal Framework Upholding the Rights of IDPs, (vi) A National Policy or Plan of Action on Internal Displacement, (vii) A National Institutional Focal Point for IDPs, (viii) A Role for National Human Rights Institutions, (iv) Participation by IDPs in Decision-Making, (x) Durable Solutions, (xi) Adequate Resources, (xii) Cooperation with International and Regional Organizations.

A. Gouritin, *Climate Displacement in Mexico*,
https://doi.org/10.1007/978-3-031-10335-3_5

5.1 Recognizing Forced Internal Climate Migrants

5.1.1 The Necessary Recognition

We have already seen that the current legal framework in Mexico does not generically recognize the internal mobility phenomenon. *A fortiori*, it does not recognize forced internal climate migrants at all. Public policies could have been made to address the phenomenon in substance in spite of this said void. However, we have found that public policies do not even attempt to do so. As an initial step, recognizing the phenomenon of internal mobility and its challenges is crucial towards addressing the phenomenon. Cruz (2014) expressed it very clearly, identifying the impact of this lack of attention:

> Considering (...) there is no official and institutional recognition of the displaced population in Mexico, (there are no) public policies for protection, attention, defense, return, resettlement and integration after it (Internal Forced Displacement—IFD), reparation and full restitution of their previous conditions, internally displaced persons are made invisible, so the information is segmented and exploratory. (…)
>
> We can affirm that IFD occurs in Mexico but is not recorded. (…) Official instances assist the displaced population in the modalities of "shelter," "help" and "immediate attention" (which are insufficient since permanent or monitored support cannot be implemented as (...) displacement does not exist institutionally) (…). Against this context, (...) there is no systematic and/or centralized registry or observation (...), nor recognition that displacements occur permanently. Consequently, the phenomenon's complexity has not been accurately valued. (pp. 57–59)[2]

Recognizing the phenomenon would make it possible to comply with the protection requirements and the obligations of the corresponding State concerning the entire mobility process (transit, settlement in transit and destination territories, return, relocation). Regarding planning, it would also allow the necessary *ex-ante* mobility planning to take place regarding destination territories and go on to integrate compliance with the protection requirements. It would, for example, consider quantitative studies that project a substantial migration from rural areas to urban areas due to increased temperatures and low rainfall.[3] *Ex-ante* planning would consider the public problems (e.g., the challenges we previously identified for urban territories induced by the increase in urban population).

A group of academics and NGOs is currently highlighting the need to recognize forced internal mobilities in general (i.e. not specifically climate mobilities). This

[2] In the same sense, see IAComHR (2013, p. 182) and CNDH (2016, pp. 191–192).

[3] Among other studies that establish this link in Mexico, see Nawrotzki et al. (2017).

movement is gaining more visibility and traction.[4] Mexico could draw on the recognition currently being seen in other countries[5] and the recommendations made by several international organizations (IDMC and Brookings-LSE 2013).

5.1.2 Risks Associated with the Recognition

Adopting a law or public policies to recognize the phenomenon and protect forced internal migrants would have to be consultative. Citizens should be consulted from the outset (including the design of the process itself) to the last stages (including monitoring and implementation). Were this not to take place, laws and policies would run the risk of resembling the State's "suffering management" ("*administración del sufrimiento*") and "public necropolitics" ("*necropolíticas públicas*"). This criticism is particularly valid in Mexico. The "suffering management" would then conceal the necessity to address the root causes which bring about the phenomenon in the first place. We have already seen that they do not exclusively refer to the climate and the environment. Poverty, exclusion, marginalization, and deficiencies in terms of public infrastructures are a variety of factors that participate in promoting climate mobility. The suffering management could allow for the erasure of the State's responsibility to address them.

In addition, a law or public policy could be implemented to define the phenomenon too restrictively. This would leave certain people or communities without protection. We found, for example, that this is indeed the case for the people and communities displaced by slow-onset phenomena in the Mexican Draft Law on Forced Internal Displacement (2020). However, too extensive a definition could generate difficulties in implementation (e.g. regarding the articulation with the policies that apply to voluntary migrants).

Also of concern is the legal effect of such recognition. Is the recognition declarative or constitutive? Peguero Moreno and Rubio Díaz Leal (2019) signal that according to the Inter-American Court of Human Rights, "the recognition and registration of a person as a victim of internal displacement is declarative but not constitutive" (pp. 51–52). They contrast this position with the Mexican General Victims Law (LGV 2017). According to the authors, this law supports the constitutive nature of the recognition. We agree and would add that the constitutive character also contradicts victims of

[4] This movement generally addresses forced internal mobility driven by violence and addresses the environmental (climate) dimension in a residual way. See, for example, the Mexican Commission for the Defense and Promotion of Human Rights (CMDPDH) (http://cmdpdh.org/temas/desplazamiento/). This visibility contrasts with forced internal mobility "invisibility." Domínguez (2014, p. 15) according to the author, this invisibility due to not crossing borders. On the invisibility of internal displacement in Mexico, see also Peguero Moreno and Rubio Díaz Leal (2019, p. 20). They refer to internal mobilities generated by violence.

[5] An updated 2018 list of countries that adopted laws or policies is available on the Global Protection Cluster Working Group website (http://www.globalprotectioncluster.org/global-database-on-idp-laws-and-policies/).

forced internal mobility as the holders of rights and the corresponding obligation of the State to respect those rights and guarantee their effective enjoyment regardless of the recognition of forced internal migration. These rights and obligations pre-exist the recognition. The recognition does not trigger the application of these rights and obligations. Accordingly, encapsulating recognition as a constitutive act does not seem appropriate.

In the first chapter, we argued that the definition should avoid the strict following of the UN Guiding Principles on Forced Internal Displacement (UN 1998) as doubts arise regarding their application to all climate mobility scenarios and, more specifically, the slow-onset phenomenon scenario. It is a weakness we also identified in the Mexican Draft Law on Forced Internal Displacement (2020). We also point out that the definition and the criteria that would apply would have to be both objective and subjective. They are elements that would have to be incorporated into law and (or) public policies.

Finally, recognizing the phenomenon and forced internal climate migrants should endorse the protection requirements fully. This recognition should be sufficiently flexible to enable a differentiated approach and enable all facets of the phenomenon to fall within the law or the scope of application of policies.

5.2 Facilitating and Organizing the Relocation of Migrants

5.2.1 Relocation and Climate Mobility Scenarios

The State can facilitate (organize) the relocation of individuals and communities. This measure has the very real potential to be successfully applied to forced climate internal mobility. It applies directly to the scenarios of designated areas unfit for human settlement and "development projects" related to renewable energy. The link is direct as it is constitutive of these scenarios (it distinguishes these scenarios from the others). It does not mean that it does not apply to other scenarios. We have already stated that the scenarios are not hermetic or static. A given context can trigger another scenario (e.g., from the slow-onset phenomenon scenario to the scenario of designated areas unfit for human settlement). The sudden-onset disaster scenario may also imply the facilitation of relocation (e.g., the State facilitates relocation in the context of a hurricane). Finally, the State can facilitate the relocation of trapped populations.

Several countries have already applied this measure in the context of mobilities induced by climate change and, more generally, in the framework of environmental degradation or development projects. Many academics identify the potential of the measure (Nishimura 2015; de Sherbinin et al. 2011), which is also prioritized at international level. The Cancun Agreements (UN 2010) illustrate this preference.[6] Within

[6] Decision 1 / CP.16 Cancun Agreements: result of the work of the Ad Hoc Working Group on long-term cooperation under the Convention, p. 14(f).

the framework of the book, we cannot report an analysis of all of the academic literature and reports on relocations in the context of development projects, environmental degradation, or that induced by climate change. Instead, in the following section, we summarize the risks, potential, and main weaknesses identified by academic literature and reports from international institutions based on those previous experiences. They fall within the two axes of the analytical grid (territory and protection requirements). We also reflect on the relevance of these risks in the Mexican context.

5.2.2 *Risks Associated with Forced Relocation*

Academics and various reports identify a series of risks that arise in the context of forced relocation (Johnson 2018; Schade 2013; Miller 2020; Mycoo et al. 2022; Cissé et al. 2022). These risks also depend on the specificities of each case (Vigil 2018, p. 374). In the case of "development projects,"

> The displacement impacts of development projects are massive. Experience has shown that most people, who experience relocation, are unable to recover their previous standards of living and end up worse off than before. Generally, the worst affected happen to be among the poorest. They lose their homes, lands, livelihoods and social networks, remain inadequately compensated, and are forcibly moved to unfamiliar places. On the lives of tribal people and women, in particular, displacement is known to have produced some of the worst impoverishing effects. A sudden increase in impoverishment is indeed the most visible impact on the lives of people, displaced by development projects. (Mathur 2015, p. 120, references omitted)

On the territorial dimension, Cernea (2000) mentions the characteristics of the risks which have a strong territorial dimension. Based on these, a "risks and reconstruction model" is constituted for their reversal—see Box 5.1).

Box 5.1 Risks and Reconstruction Model

Risk characteristics: Landlessness, Joblessness, Homelessness, Marginalization, Increased morbidity and mortality, Food insecurity, Loss of access to common property, Social disarticulation

Reconstruction: From landlessness to land-based reestablishment, From joblessness to reemployment, From homelessness to house reconstruction, From social disarticulation to community reconstruction, From marginalization to social inclusion, From expropriation to restoration of community assets, From food insecurity to adequate nutrition, From increased morbidity to better health care

Source Cernea (2000, pp. 1572–1576).

Mathur (2015) also identifies the factors that avoid the risks: a relocation policy and relocation planning. In the same sense, the Sendai Framework for Disaster Risk Reduction (UN 2015) mentions the formulation of "public policies (...) aimed at addressing the issues of prevention or relocation, where possible, of human settlements in disaster risk-prone zones, subject to national law and legal systems" (para. 27(k)) and the strengthening of "the capacity of local authorities to evacuate persons living in disaster-prone areas" (para. 33(m)).

However, these policies and the related planning suffer from shortcomings. Regarding public policies, Mathur (2015, pp. 122–123) mentions the gap between the expected (promise) and the realization (performance). Experiences in other countries also highlight the shortcomings in terms of implementation:

> Too often, the relocation of affected people has been seen as the end of the resettlement exercise, not rebuilding their livelihoods in the new place. But what is truly appalling is that relocation has been essentially carried out through official coercion. (...) But planners prefer to identify and develop resettlement sites all by themselves without consulting the people who have to resettle there. (...) Equally critical is failure to consult the host populations in planning resettlement. This neglect can prove counterproductive, as the potential for resettler–host conflict is always present in these situations. (Mathur 2015, p. 123, references omitted)[7]

The conditions of the place of relocation (or destination territory) are also crucial. Based on previous experiences, it appears this place may often not be suitable from an environmental and climate point of view. Consequently, it is prone to reproducing the same set of conditions which led to people and communities relocating in the first place (De Sherbinin et al. 2011; Schade 2013; Tan 2018). Finally, academics and international reports signal the problems related to the use of force, inadequate compensation (Schade 2013), and discrimination:

> The likelihood that authorities will prefer relocation to more expensive in situ adaptation options such as flood protection or a comprehensive water management system is particularly high in the case of socially vulnerable groups that have no voice to in situ alternatives. Relocations might thus turn out to be a kind of 'second-class adaptation' for marginalized segments of a population. (Schade 2013, p. 194)

Regarding the protection requirements, almost all of the authors who have studied the relocation phenomenon mention compliance with human rights requirements can avoid the risks mentioned above (Domínguez 2014; Johnson 2018; Franck 2018; Bronen 2018; Marter-Kenyon 2020). They particularly highlight procedural environmental rights (Schade 2013, pp. 200–202; Jacobs and Almeida 2020), the necessity to consult and include people affected (Martin and Warner 2012; Franck 2018; Kieval 2020; Dodman et al. 2022; Cissé et al. 2022; Mycoo et al. 2022; Hicke et al. 2022; Schipper et al. 2022)[8] and the need to respect the right to access information (Gharbaoui 2018, p. 311).

---

[7] In the same sense, see Raleigh et al. (2008, pp. 25–28) and Schade (2012, p. 102).

[8] Bronen (2018, pp. 397–398) emphasizes the impossibility for effective procedural rights to be out in place (participation and informed consent) to prevent the risks mentioned above in the context of development projects financed by the World Bank. Instead, Robin Bronen emphasizes self-determination as a tool to allow the empowerment of communities and to make decisions regarding

On the consultation of people and communities affected, they report examples of "successful" relocations (Gharbaoui 2018, pp. 313–314), where the interaction was successful (Bronen 2018). In a case-study in Alaska, Bronen (2018) reports interactions between public authorities NGOs, and tribal institutions. In the case reported, she also mentions Guiding Principles based on the tribe's way of life. These Principles guided the relocation of the tribe and were integrated into the relocation plan (which guides the intervention at federal and state level in this case) (p. 391). However, academics mostly formulate criticism in similar cases. They highlight the shortcomings of the existing practices regarding procedural rights and mention cases or systems structurally designed according to a top-down approach instead of a bottom-up one (Tan 2018).

Academics also describe cases where the communities proactively and voluntarily plan their relocation, but with adverse outcomes. In their study of the Jean Charles Island tribe (Louisiana), Maldonado and Peterson (2018) reveal the "obstacles, challenges, and opportunities" that arise (pp. 189–199). They mention the lack of availability of land and budget as one of those obstacles (pp. 295–296) and highlight the difficulty of proactively planning the relocation of a community by itself. In the case of Indigenous Peoples, the collective dimension of their rights is critical. The authors mention culture, the "sense of autonomy" and community values (p. 294). However, in the second chapter, we saw that the differentiated approach where applied to Indigenous Peoples entails much more. Without repeating what we have already written and detailed, it is also about the spiritual dimension and the mere survival of Indigenous Peoples.[9] Despite this reservation, the consequences indicated by the authors are fully valid in terms of the risks associated with relocation policies which do not take into account the collective dimension as regards the Indigenous Peoples affected:

> Without one's community, the process to relocate individually is colonizing in and of itself and works to reinstate the forced assimilation policies. It is about more than just relocating people from place to another, as if the same model could be used in any geographical or cultural context. Resettlement should be just that – re-establishing an intact community settlement that includes the key physical, economic, cultural, social, and spiritual infrastructure that enable the community to thrive in such a way as they determine, in a location that the community chooses and makes sense to them. It should envision a community that can thrive and resemble the positive attributes of the dispersed community, so as to reunite those who have been displaced. (p. 294)[10]

Beyond the relevance and need to apply a differentiated approach to Indigenous Peoples, another crucial aspect which we have already mentioned several times is that

---

their relocation. Also, Ajibade (2019) analyzes the retreat of urban populations in Lagos (Nigeria) and reports that "retreat was stated in climate change policy but in practice only the urban poor were forcibly removed from waterfront areas and in their place new urban development projects are being constructed." (p. 299). He calls for a holistic approach that builds on "a legislative approval, extensive consultation with affected stakeholders, cost–benefit analysis, and a thoughtfully planned relocation process" (p. 313).

[9] In the same sense, McDonnell (2021) signals the importance of Indigenous Peoples' customary tenure arrangements.

[10] In the same sense, see Bronen (2018, pp. 396–397).

the differentiated approach does not apply equally to all Indigenous Peoples. Instead, its application to relocation policies and planning has to consider the specificities of each group of Indigenous People (e.g., the characteristics concerning their territory, beliefs, etc.).

In short, the risks mentioned above are significant because relocations can be broadly applied in the context of forced internal climate mobility. Previous relocation experiences in other countries are generally very negative (de Sherbinin et al. 2011).[11] Due to all the risks involved in the relocation process, it is considered the last resort, where others (mitigation, prevention, and adaptation) can no longer be applied (Gharbaoui 2018; Bronen 2018; O'Neill et al. 2022). Some authors portray relocation as an adaptation measure. For example, Robin Bronen (2018) writes: "The combination of repeated extreme weather events, ongoing and accelerated rates of environmental change such as erosion, and the loss of protective coastal barriers is forcing coastal communities to choose to relocate as a long-term adaptation strategy" (p. 389). We disagree and present our arguments in Sect. 5.3.

5.2.3 *Very High Risks in Mexico*

The context of Mexico presents specific challenges regarding the risks mentioned in the preceding lines. Firstly, they concern the territory, the hidden objectives of the state and the subsequent risk of land grabbing. Secondly, they refer to compliance with the individuals affected and community-based rights.

The risk that States will use relocation to pursue objectives that have nothing to do with protecting the people or communities affected, and instead pursue unofficial objectives[12] resonates with the risk of land grabbing. Some authors report cases where individuals and communities affective have suspected the hidden aim of the relocation was to evict them so that companies could exploit the natural resources present in the lands where they resided. For example, in the case of relocations in Louisiana,

> Some residents perceived the government wanting residents to relocate so the oil industry could have free range over the area without interference. Individuals from local tribes grew up either with the personal experience or with the stories of family and friends having had their land confiscated. (Maldonado and Peterson 2018, p. 293)

Vigil (2018) addresses the risk of land grabbing in the context of climate change in detail. To define the phenomenon, Vigil refers to the Tirana Declaration (International Land Coalition 2011) which identifies five alternative criteria:

---

[11] These authors mention similar experiences in Vietnam, Mozambique, Alaska, Mongolia, Papua New Guinea. In the first chapter (Sect. 1.2.4) we mentioned experiences in China, Canada, Sri Lanka, New Zealand, and the Maldives. On the Chinese experience, see Tan (2018).

[12] E.g., in Chap. 1 (Sect. 1.2) we reported on the government of the Maldives which pursued the objective to gather populations split on various islands to rationalize the economic management of the islands.

> [A]cquisitions or concessions that are one or more of the following:
>
> (i) in violation of human rights, particularly the equal rights of women;
>
> (ii) not based on free, prior and informed consent of the affected land-users;
>
> (iii) not based on a thorough assessment, or are in disregard of social, economic and environmental impacts, including the way they are gendered;
>
> (iv) not based on transparent contracts that specify clear and binding commitments about activities, employment and benefits sharing, and;
>
> (v) not based on effective democratic planning, independent oversight and meaningful participation. (Commitment 4)

Land grabbing is not always illegal: "[a]lthough some transactions are illegal, most land is acquired or rented legally" (Vázquez García 2017). To identify legal land grabbing it is crucial to look at the scope and adverse consequences of sales and rentals agreements. In this book, one of the obvious adverse effects is the forced displacement of people and communities.

Vigil (2018) mentions some factors that favor the recent surge in land grabbing. Among them is the need for renewable energy and mitigation imperatives. Climate change has become a justification for green grabbing:

> [C]limate change has reinvigorated the trend (*resource expropriationsin the name of the environment*) with novel discourses around climate mitigation being deployed as justifiers. (…) Today, on top of biodiversity conservation and eco-tourism, land-intensive climate mitigation policies such as biofuel production and forest carbon projects are becoming more and more prominent. (pp. 372–373)

Within the framework of the scenarios that generate climate migration, land grabbing can occur directly in two cases. Firstly, it can be a consequence of "development project" scenarios related to renewable energies (e.g., in the case of construction and operation of infrastructures that generate "clean energy"—hydroelectric complexes, biofuels production, etc.). Public authorities can size lands to carry out these projects and initially relocate the local population on to these available lands (the relocation decided by—or facilitated by—public authorities). Public authorities can also relocate the people or communities not initially located on the seized lands but affected by development projects (e.g., water scarcity generated by the project using all available water). In other words, a grabbing following a grabbing. Grabbing can also occur more indirectly in the context of conflicts or violence caused by the scarcity of natural resources. In these cases, land grabbing generates conflict or violence. In turn, this generates mobility. The case of development projects is particularly delicate: in theory, these projects aim to meet mitigation or adaptation goals (e.g. in the case of forest management). The literature calls this phenomenon "green grabbing:"

> "Green grabbing" for the purpose of conservation, tourism and carbon sequestration has been identified in Guatemala, Colombia and Mexico. It is usually portrayed as an environment and community-friendly alternative because it does not involve the whole alienation of land. However, it often leads to changing patterns of resource management which are experienced by locals as land dispossession and enclosure (Vázquez García 2017, section "Land grabbing: extension, scale, purpose and novelty," references omitted)

Finally, green grabbing also occurs for a particularly vulnerable category of persons identified in this book: trapped populations. Green grabbing can deprive individuals or communities of the economic resources they need to migrate, consequently generating trapped migrants. In the same sense, Vigil (2018) mentions the in-situ displacement phenomenon: "[l]acking the necessary assets to make a move, many of those who lose land become 'involuntarily immobile'" (p. 377).

The risk of land grabbing is currently high in Mexico. It can result from an act of the state (public authorities) or organized crime. It is worth underlining the importance of land grabbing by acts of organized crime, as it is already an established reality in Mexico (Witness 2018; Gasparello 2018, p. 78). Neo-extractivism is also part of the Mexican context. It generates a series of its own difficulties.

Academics report on the scarcity of land available and suitable for agricultural activities (Gay et al. 2006, p. 284). The scarcity of land available can also be an obstacle to relocation (suitable alternative lands regarding the relocation of Indigenous Peoples). Neo-extractivism amplifies the difficulty (e.g., more than 25% of the national territory is allocated to mining activities, Tetreault 2015). Also, the neo-extractivism model generates ways to deprive Indigenous Peoples of their lands in favor of extractive projects. For example, the 2013 energy reform established a legal servitude mechanism in favor of investors in cases of conflict with landowners if mediation fails (Hydrocarbons Law 2016, Article 109). Similarly, the Agrarian Law was modified several times to legally facilitate the transition from *ejidal* lands regimes (a common use and tenure regime) to private property (Vázquez 2013). Referring to the oil, mining, and electricity sectors, Vázquez García (2017) points out that.

> According to the Movement for Food Sovereignty and Defense of Land, Water, Natural Resources and Territory, the Energy Reform could affect 60 % of *ejidos* and agrarian communities, with around 200 conflicts already occurring in the country. (Section "Land grabbing in Mexico," emphasis in the text, references omitted)

Finally, we reiterate, as mentioned in the first chapter that: structural violence against defenders of Indigenous Peoples' rights and the environment is a characteristic of the Mexican context. More specifically, the risks in terms of the violation of procedural environmental rights are very real. Previous experience of environmental relocation support this:

> [I]n those decades (2000–2010), a series of relocations were carried out (also called by the government "resettlements" ["*repoblaciones*"]). They were meant to be both preventive measures (…) for marginalized, dispersed, and vulnerable communities in high-risk environmental areas and durable ex-post solutions for people displaced by disasters. The cases of relocations in Chiapas and Tabasco illustrate the implementation of new post-disaster prevention and care policies. They reflect the absence of a focus on human security and the affected people's human rights. Although the political discourse mentions them as a comprehensive civil protection policy, the analysis of the cases demonstrates the lack of planning, consultation, and participation of the "benefited" communities (...) and, even worse, in many cases, the relocation areas have represented new risks and threats for the displaced. (Rubio Díaz-Leal 2017, p. 78)

Ultimately, violations of procedural environmental rights in the context of relocations can be understood if we consider the structural and systematic deficiencies of those rights in Mexico against a background of neo-extractivism, corruption,[13] and impunity.[14] From the differentiated approach perspective, the Mexican context is even more worrying, particularly for the rights of Indigenous Peoples. We have already mentioned their particular vulnerabilities regarding (among other issues) the maintenance of their structures, worldview, and specific economic vulnerability. In acknowledgement of this, academics and both national and international organizations already document the systematic violation of Indigenous Peoples' procedural environmental rights. The situation of women is also very problematic. In the first chapter, we highlighted their particular exposure to the risks forced mobility brings about. The relocation measure would induce additional risks as their rights to land and role in negotiations are fragile (Vázquez García 2017).

In short, relocation facilitation (organization) measures would be very problematic in Mexico. We can expect the violation of the rights of individuals and communities affected. This would coincide with Michael Cernea's risk model mentioned above. In Mexico, numerous factors point to the high risk of dispossession and land grabbing officially motivated by the very safety of populations forcibly displaced by the impacts of climate change. However, they would be motivated by legal and illegal authorities seeking to control lands. They would violate the rights of people and communities affected and, indeed, the requirement to protect the environment in a context of total impunity.

Consequently, many crucial safeguards would be needed to frame this measure. They would prevent violent socio-environmental conflict as "the safeguards put into place for the benefit of local populations have resulted not from the good intentions of investors, but rather from social organization and mobilization" (Vázquez García 2017, in the section "Land acquisitions: a development tool for the rural poor?," references omitted). Safeguards could also break the cycle of damage suffered by relocated people and communities (Sánchez 2011).

---

[13] Mexico is not the only country where corruption can involve the violation of human rights in the context of disasters. Voigt and Thorntorn (2015) point out the numerous corruption cases related to Hurricane Katrina (United States—New Orleans) and their consequences for human rights.

[14] For an assessment of Mexico in terms of state weakness, see the "Fragile State Index" produced by The Fund for Peace. In 2019 (and since 2006, the first indexation), Mexico was classified as being in the "warning" category (there are four categories: "alert," "warning," "stable," and "sustainable"). Within the framework of the factors that we indicate, the reader can refer to the "state legitimacy," "Human Rights," "refugees and IDPs," and "security apparatus" indicators. Except for the "state legitimacy" indicator, all indicators indicate a worsening over the years. The Index is available at https://fragilestatesindex.org/country-data/. On impunity, see among others Witness (2018, pp. 19–20).

5.3 Preventive and Reactive Adaptation Measures

We can place adaptation measures into two categories: in-situ preventive adaptation (*ex-ante* mobility) and ex-situ reactive adaptation (*ex-post* mobility) (Martin and Warner 2012). With *ex-post* (ex-situ) adaptation, the focus is on populations living outside their place of origin. With *ex-ante* (in-situ) adaptation, the measure is expressed in terms of resilience. In this section, we successively analyze preventive and reactive adaptation measures.

5.3.1 Resilience Strengthening to Prevent Mobility

The prevention of climate change-driven mobility is not neutral. It entails (or reveals) positions or preconceptions in favor of or against mobility (Sánchez Mojica 2009; Evans and Reid 2016).[15] Public policies must rely on data and models that project the extent of the phenomenon and its specificities to prevent forced internal climate mobility. For example, what are the origin and destination territories? Who are the people or communities more exposed to forced internal mobility for climate reasons?

In previous chapters, we established the difficulties related to projections and the current knowledge regarding Mexico. The phenomenon has been foreseen and is already taking place. It is different to other migratory phenomena induced by largely unpredictable factors (e.g., a civil war, a financial crisis, or an earthquake). In this sense, public policies should endorse prevention as a central goal. The Sendai Framework for Disaster Risk Reduction (UN 2015) illustrates this. Prevention is its primary objective (para. 17).[16] Prevention is the strategy most commonly used by people, communities, and public authorities (Gemenne et al. 2013). We now focus on a concept that guides public policies to prevent the phenomenon addressed: resilience. It is "the capacity of a socioecological system to maintain its normal

---

[15] A growing number of academics signal and criticize such positions and their inherent limits. For example, Lietaer and Duran-Delacre (2021) denounce the dominant "sedentary bias" (p. 17) as opposed to the "minority counter-discourse" that "emphasises freedom of movement" (p. 17). Similarly, Ayeb-Karlsson et al. (2018) identify the risks linked to the "trapped" concept. They mention it can reflect or induce resettlement is the panacea while such panacea is not neutral. They distinguish between trapped persons and identify risks:

> In situations where immobility is involuntary and people willingly self-identify as 'trapped', assisted migration similar to that initiated when a refugee is offered 'refuge' in a safe state may be welcomed. However, where immobility is voluntary, it will represent an imposition into the lives of people who do not want to leave their homes. Climate policy recommending resettlement and relocation must be approached in a manner that reflects the incredibly complex and sensitive nature of the process and acknowledges the power and prejudices that may underlie its use. (pp. 569–570, references omitted).

[16] In the same sense, see the Inter-American Court of Human Rights' case-law (mentioned in IAComHR (2015) para. 247).

functioning without undergoing significant change when it experiences perturbations" (McLeman 2018, p. 323). Taken in context of the current book, a measure that implements a resilience approach aims to strengthen resilience and reduce vulnerability to the impacts of climate change in order to prevent climate change-driven forced internal mobilities.

Resilience has become a mantra in academic circles and international institutions.[17] Resilience is presented as an "ideal," "the fundamental property that peoples and individuals around the world must possess in order to demonstrate their capacities to live with danger" (Evans and Reid 2016, p. 26). For example, resilience is the entry point in climate migration discussions within the UN Framework Convention on Climate Change. Adaptation is the perspective with which to address climate migration (Article 14[f] of the Cancun Agreement UN 2010), and resilience plays an essential role in adaptation (e.g., Article 11 of the same Agreement). Adaptation and resilience are also pivotal in Mexican climate change public policies and legal instruments. The General Law on Climate Change (LGCC 2020) defines resilience as the "[n]atural or social systems' capacity to recover or withstand climate change impacts" (Article 3, Fr. XXXV), and vulnerability is the.

> [l]evel to which a system is susceptible or is not capable of withstanding the adverse effects of Climate Change, including climate variability and extreme events. Vulnerability is a function of the character, magnitude, and speed of the climatic variation to which a system is exposed, its sensitivity, and its capacity to adapt. (Article 3, Fr. XLII)

The General Law on Climate Change also establishes the centrality of resilience in adaptation to climate change: "the national policy of adaptation to climate change will be based on instruments of diagnosis, planning, measurement, monitoring, reporting, verification and evaluation, and will have the following objectives: (…) Strengthening natural and human systems' resilience and resistance" (Article 27, Fr. II).[18]

Despite its status as cure-all, resilience as a guiding concept to strengthen adaptation and prevent forced climate (and environmental) mobilities raises concerns. Some are conceptual, and others refer to the practical application of the concept. On the conceptual side, resilience as a solution gives itself to "the security idea itself and its possibility by taking in the necessity to be exposed to all kinds of dangers as a means to live well" (Evans and Reid 2016, p. 26). We can break down this criticism into several arguments from an environmental rights point of view.

If the resilience mantra shifts responsibility from States and public authorities towards the actual or potential victims of the impact of climate change, environmental protection deficiencies (regarding environmental law's content and the implementation of the environmental rights perspective) can become overshadowed or lose

---

[17] See e.g. McCarney and Kent (2020).

[18] In the same sense, see LGCC (2020), Articles 101, Fr. II and 30, Fr. XVIII. Reducing vulnerability is also an objective (see e.g. LGCC (2020), Article 2, Fr. IV; and SEMARNAT (2013), section 6.2). The Special Climate Change Program 2020–2024 (SEMARNAT (2021) also establishes resilience is a central axis of climate policies (pp. 11, 15, 86, Priority Objective 1, Specific Action 1.3.6, and Priority Strategy 4.7).

precedence. Consequently, it would divert attention away from the current shortcomings of the regulations surrounding environmental protection and addressing climate change in terms of environmental rights (e.g., the right to water and its quantitative and qualitative aspects, health, biodiversity protection, food). It would also depart from the Sendai Framework for Disaster Risk Reduction (UN 2015), which recognizes the importance of compliance with environmental law and environmental rights (paragraph 27[d]). This risk is of crucial importance in Mexico. It would reinforce the many shortcomings concerning the regulation of environmental protection and environmental rights standards. Academics commonly report on these deficiencies (Ugalde 2010; Delgado Ramos et al. 2015).

Furthermore, Evans and Reid (2016) quite rightly point out that academics tend to confuse resilience and resistance. The authors argue that resilience not only absorbs resistance but extinguishes it (p. 66). In terms of of this book, this confusion is far from neutral. Resilience as the capacity to face the impacts of climate change in order to prevent forcefully mobility would shift responsibility for the causes and impacts of climate change induced by the action or inaction of States and public authorities to the populations themselves. It also has the potential to make resistance disappear. In other words, resilience absorbs co-opts resistance. It implies renouncing security and taking insecurity for granted: "[t]o be in resilience, the subject must deny any belief in the possibility of being safe" (Evans and Reid 2016, p. 70), "[t]he construction of subjects in resilience involves deliberately disabling the habits, tendencies and political capacities of peoples and replacing them with others that are adaptive" (Evans and Reid 2016, p. 140).

The same authors point out that resilience exists within a context where the challenges related to "critical moments" (the Anthropocene) do not question economic liberalism or "liberal power regimes" (p. 34).[19] Instead, "[t]he change involves the adaptation of *specific* populations' *particular* behaviors in such a way that the fundamental principles of liberalism' survival is not questioned in the face of accelerated global change" (p. 34, emphasis in the text). In the case of Mexico, it can be identified, among other regulation, in the 2013 energy reform. It emphasizes fuel supply continuity and does not question consumption patterns.[20] In the context of forced internal climate mobility, public policies aimed at prevention around the concept of resilience, address "certain behaviors" to reduce the probability of forced mobility. They focus on "particular populations:" those most exposed to the impact of climate change (populations identified as most vulnerable) and do not question the general population's adaptive capacity or behavior.

Finally, a resilience approach can "depoliticize" the more structural context that people and communities susceptible to forced internal migration face. For example, measures to increase the resilience of some people or communities do not question the root cause of limited resilience (e.g. poverty). They instead accept them as fact,

---

[19] In the same sense, see Stensrud (2016).

[20] For criticism, see CEMDA (n.d.).

considered an immutable reality and a working basis[21] (Evans and Reid 2016, pp. 64–65).[22] This criticism echoes the criticisms already mentioned regarding human rights: the state "manages the suffering" in "public necropolitics."[23] In the same sense, Evans and Reid (2016) argue "resilience discourses are not about local empowerment, but rather a governance technology" (p. 126). This criticism is valid within Mexican policy making. Climate policies do not address the root causes of this vulnerability (e.g., poverty) with concrete lines of action: "it is necessary to attend priority issues with strategies that help empower groups economically and socially vulnerable and address the underlying vulnerability causes, including gender, inequality, and marginalization" (Delgado Ramos et al. 2015, p. 139). In the same vein, the National Institute of Ecology and Climate Change (INECC 2018) points out: "state and municipal authorities report multiple adaptation actions, such as reforestation, protection of natural areas, without giving an account of how they reduce vulnerability and without considering other factors that can exacerbate it, such as poverty" (p. 16). In this respect, the Special Climate Change Program 2021–2024 (PECC) (SEMARNAT 2021) departs from previous policy instruments. It identifies the root causes of vulnerability and firmly grounds attention to those as a priority.

Against this background, a differentiated rights-based approach could overcome the associated risks. Human rights tools and mechanisms can question territorial policies and the acceptance or imposition of insecurity by design. They can also empower the populations susceptible to forced migration so that they are not the objects of public policies aimed at increasing their resilience and are instead their subjects and agents. Procedural and material environmental rights can articulate this potential through strategic litigation (and more particularly, the proportionality analysis in the litigation process) and their implementation.

Public policies on resilience indicate that this potential exists but also evidence serious implementational problems. More concretely, applying measures based on the concept of resilience as an adaptation measure is highly problematic in the context of Mexico. Firstly, these measures must be designed according to local needs and particularities (Campos et al. 2014).[24] Academics themselves should also take this requirement into account. This would prevent the formulation of dubious general statements. For example, Martin and Warner (2012, p. 157) affirm that many people in Mexico have the necessary assets and network to migrate. The authors do not, however, provide a comparison. It is speculative to suggest, for example, that Indigenous Peoples who are already highly vulnerable economically have all the economic means or networks necessary to migrate. Rather, they are precisely among the people and communities most likely to be trapped. This is also true of Argentina (see IACtHR 2020, para. 285). Put simply, we doubt that making statements about the capacity to

[21] In the same vein, see Kelman (2014).

[22] *Contra*, Surugu and Chutab (2020) identify the potential of transformational adaptation to address the roots of limited resilience. The authors provide references to academic works that reflect this potential.

[23] In the same sense, Evans and Reid (2016) mention the "suffering politicization" (p. 223).

[24] In the same sense, see Spring (2019).

migrate or adapt at State level is valid given the high disparities that can exist within a country. This also resonates with the requirements we identified in Chap. 2 with the concept of territory and the General Law on Climate Change (LGCC 2020) which affirms the need to consider local needs and specificities.[25]

The requirement to guarantee the participation of people and communities affected complements the requirement to consider the local context in the design and implementation of in-situ adaptation measures.[26] It points to environmental procedural rights (access to information, participation, and access to justice) violations and questions who gets to define resilience: the populations subject to that definition, public authorities, experts, or academics? For resilience not to become a vehicle for dominant powers (partly responsible for climate change and its impacts that drive the forced nature of mobility), populations must themselves define resilience and identify the best measures to increase it.[27] A decisive criterion is how these people and communities perceive the environment (i.e., their worldview and ancestral knowledge), including if (or perhaps especially if) it does not fit with the dominant utilitarian and anthropocentric approach to the environment.

In the context of Mexico, we have already pointed out the need for improvement of environmental procedural rights. There is also an abundance of existing literature on the governance of in-situ adaptation, including in the context of Mexico (Schroth et al. 2009; Galicia et al. 2015; Calliari et al. 2019). According to the literature dedicated to adaptation in Mexico, the practice is, thus far, far from complying with these two requirements related to the territory and environmental rights. In their analysis of case studies and the national framework, most of the authors signal the inadequacy of public strategies and measures in terms of implementation at a local level (Muñoz Sevilla et al. 2016), the requirements for participation, and taking into account local context (Delgado Ramos et al. 2015; Muñoz Sevilla et al. 2016; Rueda Abad et al. 2017).

Detailing or providing a list of those studies would be beyond the scope and purpose of the book. We instead provide a few examples as illustration. Vázquez (2017) analyzed adaptation measures in Tabasco and, applying a political ecology approach, identified contradictions in public measures and conflicts of interest between the various actors present in the Gulf of Mexico. The author emphasizes the need to consider the local context, local needs and requirements, and how they identify problems and potential solutions. The author singles out the failings of public adaptation strategies. In the same sense and focusing on the participation of women and the necessity to consider their specific needs and situation (including at the local level), academics point out substantial deficiencies (Delgado Ramos et al. 2015). Bojórquez and Aguilar (2015) The authors point out significant gaps regarding participation in a case study in Coahuila. Sánchez et al. (2013) also mention shortcomings

[25] LGCC (2020), Articles 27, Fr., II; 30, Fr. I, IV; 63, Fr. III; 72, Fr. II; 77, Fr. IV; and 101, Fr. I, IV.

[26] LGCC (2020), Articles 7, Fr. III and Fr. XII, 8, Fr. IV, XI, 26, Fr. VII, 57, Fr. III, 60, 68, and 109–110; National Climate Change Strategy (2016), Sects. 2.3.2, 3.4, 5.1, P4.3, A1.13, A2.3, A3.16 and 8; and United Nations (2015), paras. 19(d), 24(i), 27(e)(h), 30(j), 33(l), and 36(a).

[27] In the same sense (but more generally), see Ortiz Espejel and Azuara García (2019).

in acknowledging local characteristics in an empirical analysis in Michoacán: "general solutions that do not respond to complex local realities keep on being proposed" (p. 344).

Not all studies conclude negatively in terms of considering the local context and participation. Several studies report relatively successful experiences in, for example, Tabasco (Giannini et al. 2017), Tlaxcala (Galicia et al. 2015), Puebla (Martínez-Corona 2012), and Veracruz (Maldonado González and González Gaudiano 2013).

5.3.2 Celebrating Forced Mobility as a Successful Reactive Adaptation[28]

International Organizations tend to view climate migration as a successful adaptation to climate change, and most academics seem to have the same understanding. Scholars argue that migration can strengthen resilience to climate change (e.g. through remittances—Black, Kniveton, and Schmidt-Verkerk 2011). They also argue that "migration is a form of adaptation, not a failure to adapt" (Geddes and Jordan 2012, p. 1030). Migration is "part of the solution rather than the problem" (Tacoli 2009, p. 514). These academics generally try to escape the security narrative associated with climate migration (a narrative the author also rejects):

> Knowing that displacement will occur presents an opportunity to plan responses and create broad reaching adaptive measures. (…) the discourse must shift from its focus on climate change migrants as a national security threat to discussion of migration as an adaptive mechanism. (Nishimura 2015, p. 20)[29]

However, other academics criticize this position. They argue, for example, that this is not a neutral position and instead reveals skewed perspectives, pressures, and manifestations of power (Cannon and Müller-Mahn 2010; Felli and Castree 2012; Eriksen, Nightingale and Eakin 2015; Baldwin 2016; Ayeb-Karlsson, Smith and Kniveton 2018). Some authors have already established the normative weight of the discourse that favors ex-situ adaptation (Gebauer and Doevenspeck 2015). In the same vein, Bettini, Nash and Gioli (2017) criticize such a framing for removing "structural inequalities" from climate migration understanding. Others argue celebrating climate mobility as a successful ex-situ adaptation depoliticizes the issues and ignores the social dimension of the phenomenon, a dimension that needs to be addressed (Kelman 2014; Methmann and Oels 2015; Offner and Marlowe 2021). In this sense, climate forced mobility can be framed as "maladaptation" (Caretta et al. 2022, p. 119; Bezner Kerr et al. 2022; Schipper et al. 2022; O'Neill et al. 2022).

---

[28] This section is based on Gouritin (2020).

[29] Few academic papers mention security in such a way that it departs from the "classic" definition of national security. For example, the broad definition can encompass cultural security, potentially threatened by climate change impacts and climate mobility (e.g. cultural security in terms of world views). See e.g. Krupocin and Krupocin (2020).

Additionally, in another publication, we argue that internal forced climate migration requires preventive actions when Indigenous Peoples have a particular link with their lands and territories. The rights of Indigenous Peoples ground this requirement (Gouritin 2020). Empirical data also questions the portrayal of ex-situ forced internal migration as a successful adaptation. It contradicts the value of this type of migration for those who "stayed behind." Adding to the empirical studies that identified a loss of knowledge and workforce in the territories of origin, scholars identified exacerbated risks of poverty (Jacobson, Crevello, Chea and Jarihani 2019). In case studies in Cambodia, "Migration causes labour shortages and welfare issues, but does not necessarily improve food security. This and climate trends lead us to argue that migration may be maladaptive over the long term, resulting in a climate-induced poverty trap" (Jacobson et al. 2019, p. 101).

From the perspective of the adaptation of destination territories in the face of climate mobilities, we can signal the requirement to also apply a bottom-up approach. This entails the consultation and participation of migrants themselves, and considering the particular situation of a destination territory from the perspective of said migrants. Similarly, Nishimura (2015) states "regional and community specific vulnerabilities should (...) be addressed" (p. 21).[30] We have already elaborated on the consultation and participation of potential migrants. Academic literature and reports dedicated to ex-situ climate mobilities have yet to investigate greatly into the specificities of destination territories from a migrant protection perspective. We were able to find very few references to these aspects in terms of public policies. Tacoli (2009) argues for the adoption of measures "to support a more decentralized pattern of urbanization and industrialization" aimed at reducing the number of climate and environmental migrants living in risk areas (p. 519) and the provision of public services and public infrastructure (p. 522). More generally, the author points out the public policies necessary regarding development and poverty reduction (p. 522).

In the context of the limited knowledge on climate mobilities as ex-situ adaptation regarding destination territories and the perspectives of migrants, decision-makers can rely on academics who dedicate themselves to migratory issues without limiting their work to climate mobilities. Within the limits of this book, we cannot report fully on this. Instead, we signal the source of information and the necessity to rely on it to address climate mobilities and destination territories. Decision-makers can also rely on the rights-based approach, specifically on material (e.g. the right to water, food) and procedural environmental rights (access to information, participation, and access to justice).

In short, if we can conceive of ex-situ internal forced climate migration as a form of adaptation, we can hardly articulate it as the generic solution public policies should be built on. Qualifying it as a successful solution or adaptation depends on each particular mobility case. It depends on the local context and the characteristics of the population involved. Consequently, in-situ adaptation and the necessity to take into account the local context are essential.

[30] In the same sense on the importance of local participation, see Wolff (2021).

Despite the above, we can expect Mexican public policies will not be limited to prevention. A preventive approach focused on resilience cannot entirely prevent forced internal climate mobilities. This is true of most climate change impacts: we are already experiencing them today. Consequently, Mexican public policies will most probably also portray the mobility phenomenon as an ex-situ adaptation. Accordingly, the Mexican political agenda should prioritize solving the implementation deficiencies of adaptation strategies mentioned in the previous section. We do not wish to say that Mexico is unlike other countries in not having or implementing an effective or timely policy in terms of adaptation strategies. All States, globally (and their different governmental levels), face obstacles and difficulties with climate change public policies.

5.4 Initial Thoughts on How to Overcome the Risks and Articulate the Solutions

Considering the risks and potential solutions identified throughout this chapter, we now reflect on how to articulate the measures needed within Mexico to address internal forced climate migration. We focus on an approach that could guide the design and implementation of said measures: the Socio-Ecological Systems approach.[31] We present herein initial thoughts rather than fully developed recommendations.

Sustainability Sciences focus on the interactions between society and the environment and are the product of scientific and technological progress. They aim to understand the relationship between society and nature and articulate social and natural sciences (Gibson, Ostrom and Ahn 2000; Watson 2005). The integration of both fields of knowledge also involves transforming the evaluation of the changes involved in terms of climate change and natural origin phenomena worldwide. It is necessary to consider social actors and their political interactions when articulating social sciences and biophysics. Evaluations can no longer only be quantitative. It is also necessary to integrate qualitative evaluations in order to reflect the complexity of social processes. All in all, the primary objective of sustainability sciences is to improve the quality of life of populations in the face of environmental deterioration (Pacheco-Vega in press).

The Socio-Ecological System concept builds on the above. The concept encompasses all types of social relationships between human beings and the environment and any symbolic or material benefit obtained from natural resources (Liehr et al. 2017). It builds on the interdependence between social and biophysical systems (Pérez-Soba and Dwyer 2016; Balvanera, Astier, Gurri and Zermeño-Hernández 2017).

It also has great potential to overcome the risks identified in the chapter and articulate the solutions. For example, it proposes focusing on the causes of natural disasters

---

[31] This reflection builds on and is a modified and shorter version of Velázquez Melo et al. (2021).

and not only describing them (Lerner and Charli-Joseph 2018). This is important to address the root causes of migration scenarios. Biogeophysical phenomena are not disasters in themselves. They are a threat that can turn into a disaster if the population impacted does not have the necessary tools to develop its capacity for resilience. Failure in the social system explains why the threat exceeds the capacity to cope of the population and institutions. With this taken into account, the Socio-Ecological Systems approach allows the understanding of the causes of disasters and the formulation of appropriate guidelines as preventative measures.

In the previous chapters, we referred to pre-existing vulnerability and the distribution of both access to natural resources and the cost of environmental deterioration and disasters (some authors refer to asymmetries). According to the Socio-Ecological Systems approach, if there is an unbalanced relationship between the actors within the social system, the severity of the disaster is increased. The institutions dedicated to the governance of a social system tend to focus on decisions related to managing risk (for example, awareness-raising regarding the early warning systems, or what to do in the event of an earthquake). However, what guarantees a decreased vulnerability is the stability between social class and the equitable distribution of resources that reduces vulnerability in the poorest sectors (Runfola, Romero-Lankao, Jiang, Hunter, Nawrotzki and Sanchez 2015; Scovronick, Lloyd and Kovats 2015). Thus, even though poverty is generally portrayed as a cause of vulnerability, it is instead caused by the development of social processes that do not allow the correct distribution of resources and by insufficient knowledge of environmental changes (Chaffin and Gunderson 2016; Ching 2018; Klein et al. 2005; Moser 2010). Not knowing the environmental processes implies resistance to it and, therefore, increases the severity of a disaster.

It is also necessary to consider the negative impacts of industrial processes on the environment (Graham 2010; Joseluis 2008; Vogel and Henstra 2015). The extraction of resources and numerous industrial processes have modified the environment and participated in creating the threats that tend to become disasters (Bocco, Priego and Cotler 2010; Boin't et al. 2005; Long and Rice 2018; Handmer and Dovers 2007; Singh 2015; Von Lucke 2018). Against this background, public policies must generate well-being in the population and the conserve of the environment as a matter of urgency, that is, to guarantee the stable operation of the Socio-Ecological System (Rebollar Plata and Oliva Aguilar 2011, p. 81).

In addition, the Socio-Ecological Systems approach does not provide the tools to counteract the correlation of forces that cross institutional and legal entities. It does, however, provide two requirements that could influence habits and behaviors of the population through social institutions and then bring about an active role for such populations in state institutions: participation and interdisciplinarity.

Regarding participation, the approach endorses the orientation of public policies towards active participation and collaboration between the population, the public sector, the private sector, and the government (Rebollar Plata and Oliva Aguilar 2011). In previous chapters, we saw that each region and population is vulnerable to site and context-specific issues (such as the Chontal indigenous population, whose geographic location places them at risk).[32] The impacts caused by human use and practices on the environment affect the same society (Young, Berkhout, Gallopin, Janssen, Ostrom and van der Leeuw 2006). Each social system has specific practices, knowledge, uses, and customs concerning its natural environment. Thus, the requirement to integrate participation into the making of public policies allows States to address the requirement for local integration when making public policies, which, together with the territorial perspective, also allows for the consideration not only of the biological aspects but all the territorial dimensions (e.g., the social, historical and spiritual dimensions).

Concerning the interdisciplinarity requirement, all Socio-Ecological Systems approach elements involve substantial complexity, and collective action implies local knowledge and interdisciplinarity. Useful analysis cannot rely on a single aspect of the region to achieve successful resource management. All of them must be made known (Pacheco-Vega 2005, 2006, 2014). The Socio-Ecological Systems approach integrates different disciplines and knowledge around a social system that operates together with a biophysical system (Thiel, Adamseged and Baake 2015).

Further to this it fully articulates the necessity to consider the local dimension. Elinor Ostrom's substantial contribution in this field was the acknowledgement of a need for articulation between institutional arrangements, formal and informal rules. She also articulated them with the ecosystems of the communities within a local context by integrating biophysical systems (Ostrom 2010) in the analytical framework of the Institutional Analysis and Development (IAD). She included local knowledge in the concepts guiding social institutions. Thoroughly knowledge of a region is therefore essential for resource management. Her study of how institutions should act within this perspective allowed her to demonstrate that resources can be managed locally and without the need for privatization. Unlike her contemporaries, Elinor Ostrom believed this should be carried out from homes, businesses, or small spaces. For her, locals can make much better use of common resources than government authorities (Ostrom 2005; Ostrom et al. 1988). Institutional robustness responds to the monitoring and resolution of conflict that institutions dedicated to local management must be able to articulate (Pacheco-Vega 2006; Pacheco-Vega and Basurto 2008). The success of these institutions is entirely context-specific. That is why it is necessary to have a full understanding of the context, something that will only be achieved if the local population is involved in what Elinor Ostrom calls "spaces of action" where the actors' different points of view converge (Anderies and Janssen 2013; Basurto 2013; Merino Pérez 2005).

The approach also departs from "traditional" conceptions regarding resilience. It is, furthermore, possible to evaluate resilience according to the changes that can happen without generating alterations (Rubinos and Anderies 2020). The objective is to achieve a significant resistance to natural origin disturbances: they would not generate chaos for the Socio-Ecological System. In the same vein, the analysis of resilience within Socio-Ecological Systems analyzes non-linear processes (Gain, Giupponi, Renaud and Vafeidis 2020). It is essential for Sustainability Sciences.

[32] In the same sense, see Hoffmann, Šedová and Vinke (2021).

They aim to build a new qualitative interpretation of the processes within Socio-Ecological Systems which depart from the former linear interpretation of sustainable development (Meine 2020).

Nevertheless, this approach also has its limitations. If a restricted notion of communities is adopted, these populations must belong to a specific community to be heard according to the precepts of the Socio-Ecological Systems approach (Anderies and Janssen 2013; Kamperman and Biesbroek 2017; Luna 2014; Oki and Quiocho 2020). When they do not belong to a community, the Socio-Ecological Systems approach will not support the differentiated human rights approach. The specificities of a person in terms of risk of human rights violations and specific links with the territory would not be acknowledged. The limit is, therefore, critical: it hampers the approach from solving socio-environmental problems (such as poverty) with regards to those people. For the Socio-Ecological Systems approach to guide public policies that would address gender related issues and serve people with limited economic resources, it should go beyond the *stricto* sensu notion of community and cover individual actors that can form *lato* sensu communities. This issue involves further research on the scope of the Socio-Ecological Systems approach.

5.5 Conclusion

We critically analyzed the public policy tools potentially applied to address internal forced climate migration and identified risks from a theoretical point of view within the context of Mexico. We also identified how to mitigate these risks and determined the analytical research axes (territory and protection requirements) are most relevant: they point to possible risks and signal possible solutions or prevention tools for said risks.

In our analysis, we identified a series of risks implied by the recognition of migrants. The process of the adoption of legal norms or public policies would not be consultative and adhere too closely to the United Nations Guiding Principles. We also addressed forced relocations. They pose significant risks in terms of human rights violations and their territorial dimension. We lastly discussed the measures designed to address the climate perspective of the phenomenon.

We first approached the measures to strengthen the in-situ adaptation of populations and expressed strong doubts regarding resilience as a concept and the objectives and foundations of public policies. Not only can resilience hamper the political dimension of the phenomenon from being addressed, it can also be a vehicle for human rights violations.

Regarding the measures aimed at strengthening the concrete application adaptation in-situ, current understanding points to severe deficiencies in terms of territorial dimensions (by not taking into account the local context) and human rights (by failing to firmly place those strategies in a procedural environmental rights approach).

Secondly, we analyzed the measures aimed at strengthening the ex-situ adaptation. Similar to what we found regarding resilience, we identified the risks of depoliticizing the phenomenon and ignoring its underlying causes. The risks entailed are those of negating the social, local, and political dimensions thereof and violating procedural environmental rights (right of participation/consultation).

Another problem identified throughout this chapter (and mentioned in previous chapters) is deficiencies of implementation in Mexico. Public policies and human rights implementation deficiencies generate risk and, at the same time, severely reduce their potential to prevent the risks mentioned above.

We then closed the chapter with a general reflection on the potential of the Socio-Ecological Systems approach to guide future public policies addressing internal climate forced migration in Mexico and to avoid the risks as well as articulate the potential solutions mentioned above.

References

Ajibade I (2019) Planned retreat in Global South megacities: disentangling policy, practice, and environmental justice. Clim Change 157(2):299–317. https://doi.org/10.1007/s10584-019-02535-1

Anderies JM, Janssen MA (2013) Robustness of social-ecological systems: implications for public policy. Policy Stud J 41(3):513–536

Ayeb-Karlsson S, Smith CD, Kniveton D (2018) A discursive review of the textual use of "trapped" in environmental migration studies: the conceptual birth and troubled teenage years of trapped populations. Ambio 47(5):557–573. https://doi.org/10.1007/s13280-017-1007-6

Baldwin A (2016) Premediation and white affect: climate change and migration in critical perspective. Trans Inst Br Geogr 41(1):78–90

Balvanera P, Astier M, Gurri F, Zermeño-Hernández I (2017) Resiliencia, vulnerabilidad y sustentabilidad de sistemas socio ecológicos en México. Rev Mex Biodivers 88:141–149

Basurto X (2013) Linking multi-level governance to local common-pool resource theory using fuzzy-set qualitative comparative analysis: insights from twenty years of biodiversity conservation in Costa Rica. Glob Environ Chang 23(3):573–587

Bettini G, Nash SL, Gioli G (2017) One step forward, two steps back? The fading contours of (in)justice in competing discourses on climate migration. Geogr J 183(4):348–358. https://doi.org/10.1111/geoj.12192

Bezner Kerr R, Hasegawa T, Lasco R, Bhatt I, Deryng D, Farrell A, Gurney-Smith H, Ju H, Lluch-Cota S, Meza F, Nelson G, Neufeldt H, Thornton P (2022) Food, fibre, and other ecosystem products. In: Climate change 2022: impacts, adaptation, and vulnerability; In: Pörtner H-O, Roberts DC, Tignor M, Poloczanska ES, Mintenbeck K, Alegría A, Craig M, Langsdorf S, Löschke S, Möller V, Okem A, Rama B (eds) Contribution of working group ii to the sixth assessment report of the intergovernmental panel on climate change. Cambridge University Press. In Press. https://www.ipcc.ch/report/ar6/wg2/

Black R, Kniveton D, Schmidt-Verkerk K (2011) Migration and climate change: towards an integrated assessment of sensitivity. Environ Plan A 43(2):431–450

Bocco G, Priego A, Cotler H (2010) The contribution of physical geography to environmental public policy development in Mexico. Singap J Trop Geogr 31(2):215–223

Boin A, 't Hart P, Stern E, Sundelius B (2005) The politics of crisis management: public leadership under pressure. Cambridge University Press, Cambridge, MA, USA

Bojórquez GC, Aguilar MDLR (2015) Gobernanza climática: actores sociales en la mitigación y adaptación en el estado de Coahuila, México. Nóesis. Revista de Ciencias Sociales y Humanidades 24:129–146

Bronen RT (2018) Climate-induced community relocations. In: McLeman R, Gemenne F (eds) Routledge handbook of environmental displacement and migration. Routledge/GSE Research, London, pp 388–402

Calliari E, Michetti M, Farnia L, Ramieri E (2019) A network approach for moving from planning to implementation in climate change adaptation: evidence from southern Mexico. Environ Sci Policy 93:146–157

Campos M, McCall MK, González-Puente M (2014) Land-users' perceptions and adaptations to climate change in Mexico and Spain: commonalities across cultural and geographical contexts. Reg Environ Change 14(2):811–823

Cannon T, Müller-Mahn D (2010) Vulnerability, resilience and development discourses in context of climate change. Nat Hazards 55(3):621–635

Cámara de Diputados (Draft Law) (2020) Dictamen de la Comisión de Gobernación y Población, a la iniciativa con Proyecto de Decreto que expide la Ley General para Prevenir, Atender y Reparar Integralmente el Desplazamiento Interno, 29 de septiembre de 2020. Dictámenes para declaratoria de publicidad. *Gaceta Parlamentaria*, Palacio Legislativo de San Lázaro, Núm. 5619-III, 29 September 2020. http://www.diputados.gob.mx/LeyesBiblio/iniclave/64/CD-LXIV-III-1P-186/02_dictamen_186_29sep20.pdf. Accessed 30 Nov 2021

Caretta MA, Mukherji A, Arfanuzzaman M, Betts RA, Gelfan A, Hirabayashi Y, Lissner TK, Liu J, Lopez Gunn E, Morgan R, Mwanga S, Supratid S (2022) Water. In: Pörtner H-O, Roberts DC, Tignor M, Poloczanska ES, Mintenbeck K, Alegría A, Craig M, Langsdorf S, Löschke S, Möller V, Okem A, Rama B (eds) Climate change 2022: impacts, adaptation, and vulnerability. Contribution of working group II to the sixth assessment report of the intergovernmental panel on climate change. Cambridge University Press. In Press. https://www.ipcc.ch/report/ar6/wg2/

Centro Mexicano de Derecho Ambiental A. C. (CEMDA) (n.d.) Postura ante la Reforma Energética. https://www.cemda.org.mx/postura-ante-la-reforma-energetica/. Accessed 30 Nov 2021

Cernea MM (2000) Risks, safeguards and reconstruction: a model for population displacement and resettlement. Econ Polit Wkly 3659–3678

Chaffin BC, Gunderson LH (2016) Emergence, institutionalization and renewal: rhythms of adaptive governance in complex social-ecological systems. J Environ Manag 165:81–87

Ching L (2018) The paradox of social resilience: explaining delays in water infrastructure provision in Kathmandu. Water Altern 11(1):61–85

Cissé G, McLeman R, Adams H, Aldunce P, Bowen K, Campbell-Lendrum D, Clayton S, Ebi KL, Hess J, Huang C, Liu Q, McGregor G, Semenza J, Tirado MC (2022) Health, wellbeing, and the changing structure of communities. In: Pörtner H-O, Roberts DC, Tignor M, Poloczanska ES, Mintenbeck K, Alegría A, Craig M, Langsdorf S, Löschke S, Möller V, Okem A, Rama B (eds) Climate change 2022: impacts, adaptation, and vulnerability. Contribution of working group II to the sixth assessment report of the intergovernmental panel on climate change. Cambridge University Press. In Press. https://www.ipcc.ch/report/ar6/wg2/

Comisión Nacional de Derechos Humanos (CNDH) (2016) Informe Especial sobre Desplazamiento Interno Forzado (DIF) en México. http://informe.cndh.org.mx/uploads/menu/15008/2016_IE_DesplazadosD.pdf. Accessed 30 Nov 2021

Cruz LMS (2014) Tres dimensiones del desplazamiento interno forzado en México. Iztapalapa Revista De Ciencias Sociales y Humanidades 35(76):53–81

Delgado Ramos GC, De Luca Zuria A, Vázquez Zentella V (2015) Adaptación y mitigación urbana del Cambio Climático en México. UNAM, Mexico

Dodman D, Hayward B, Pelling M, Castan Broto V, Chow W, Chu E, Dawson R, Khirfan L, McPhearson T, Prakash A, Zheng Y, Ziervogel G (2022) Cities, settlements and key infrastructure. In: Pörtner H-O, Roberts DC, Tignor M, Poloczanska ES, Mintenbeck K, Alegría A, Craig M, Langsdorf S, Löschke S, Möller V, Okem A, Rama B (eds) Climate change 2022: impacts, adaptation, and vulnerability. Contribution of working group II to the sixth assessment report of the intergovernmental panel on climate change. Cambridge University Press. In Press. https://www.ipcc.ch/report/ar6/wg2/

Domínguez JC (2014) Desplazamiento forzado por proyectos de desarrollo: retos para la cooperación internacional en América Latina. Instituto de Investigaciones Dr. José María Luis Mora

Eriksen SH, Nightingale AJ, Eakin H (2015) Reframing adaptation: the political nature of climate change adaptation. Glob Environ Chang 35:523–533

Evans B, Reid J (2016) Una vida en resiliencia: el arte de vivir en peligro. Fondo de Cultura Económica, Mexico

Felli R, Castree N (2012) Neoliberalising adaptation to environmental change: foresight or foreclosure? Environ Plan A 44(1):1–4

Franck M (2018) UNHCR's perspectives and activities on displacement in the context of climate change. In: McLeman R, Gemenne F (eds) Routledge handbook of environmental displacement and migration. Routledge/GSE Research, London, pp 403–407

Gain AK, Giupponi C, Renaud FG, Vafeidis AT (2020) Sustainability of complex social-ecological systems: methods, tools, and approaches. Reg Environ Change 20(3):3–6

Galicia L, Gómez-Mendoza L, Magaña V (2015) Climate change impacts and adaptation strategies in temperate forests in Central Mexico: a participatory approach. Mitig Adapt Strat Glob Change 20(1):21–42

Gasparello G (2018) Análisis del conflicto y de la violencia en Cherán, Michoacán. Relaciones. Estudios De Historia y Sociedad 39(155):77–112

Gay C, Estrada F, Conde C, Eakin H, Villers L (2006) Potential impacts of climate change on agriculture: a case of study of coffee production in Veracruz, Mexico. Climatic Change 79(3–4):259–288, 284

Gebauer C, Doevenspeck M (2015) Adaptation to climate change and resettlement in Rwanda. Area 47(1):97–104

Geddes A, Jordan A (2012) Migration as adaptation? Exploring the scope for coordinating environmental and migration policies in the European Union. Eviron Plann C Gov Policy 30(6):1029–1044

Gemenne F, Gray C, Bilsborrow R (2013) Environmental influences on human migration in rural Ecuador. Demography 50(4):1217–1241

Gharbaoui DJ (2018) Social and cultural dimensions of environment-related mobility and planned relocations in the South Pacific. In: McLeman R, Gemenne F (eds) Routledge handbook of environmental displacement and migration. Routledge/GSE Research, London, pp 300–319

Giannini V, Ramieri E, Chávez MC, del Prado Gasca KS, Martínez KP (2017) A participatory process to design climate change adaptation measures for the Carmen-Pajonal-Machona lagoon system in Mexico. In: Leal Filho W, Keenan J (eds) Climate change adaptation in North America. Springer International Publishing, Cham, pp 27–49

Gibson CC, Ostrom E, Ahn TK (2000) The concept of scale and the human dimensions of global change: a survey. Ecol Econ 32(2):217–239

Gouritin A (2020) Towards a prevention-driven adaptation strategy, as applied to indigenous peoples' internal climate migration: some inputs based on a rights-based approach. In: Leal Filho W, Nagy GJ, Borga M, Muñoz PDC, Magnuszewski A (eds) Climate change, hazards and adaptation options. Springer International Publishing, Cham, pp. 979–991

Graham S (ed) (2010) Disrupted cities: when infrastructure fails. Routledge, New York and London

Handmer J, Dovers S (2007) Handbook of disaster & emergency policies & institutions. Earthscan, London, UK, Sterling, VA, USA

Hicke JA, Lucatello S, Mortsch LD, Dawson J, Domínguez Aguilar M, Enquist CAF, Gilmore EA, Gutzler DS, Harper S, Holsman K, Jewett EB, Kohler TA, Miller K (2022) North America. In: Pörtner H-O, Roberts DC, Tignor M, Poloczanska ES, Mintenbeck K, Alegría A, Craig M, Langsdorf S, Löschke S, Möller V, Okem A, Rama B (eds) Climate change 2022: impacts, adaptation, and vulnerability. Contribution of working group II to the sixth assessment report of the intergovernmental panel on climate change. Cambridge University Press. In Press. https://www.ipcc.ch/report/ar6/wg2/

Hoffmann R, Šedová B, Vinke K (2021) Improving the evidence base: a methodological review of the quantitative climate migration literature. Glob Environ Chang 71:102367. https://doi.org/10.1016/j.gloenvcha.2021.102367

Inter-American Commission on Human Rights (IAComHR) (2013) Derechos Humanos de los migrantes y otras personas en el contexto de la movilidad humana en México, OEA/Ser.L/V/II. Doc 48/13. http://www.oas.org/es/cidh/migrantes/docs/pdf/informe-migrantes-mexico-2013.pdf. Accessed 30 Nov 2021

Inter-American Commission on Human Rights (IAComHR) (2015) Derechos Humanos de migrantes, refugiados, apátridas, víctimas de trata de personas y desplazados internos: normas y estándares del Sistema Interamericano de Derechos Humanos, OEA/Ser.L/V/II. Doc 46/15. https://www.oas.org/es/cidh/informes/pdfs/MovilidadHumana.pdf. Accessed 30 Nov 2021

Inter-American Court of Human Rights (IACtHR) (2020) Caso comunidades indígenas miembros de la Asociación Lhaka Honhat (nuestra tierra) vs. Argentina. Fondo, Reparaciones y Costas. Judgment 6 February 2020. Serie C No. 400

Internal Displacement Monitoring Centre (IDMC), Brookings-LSE (2013) National instruments on internal displacement. A guide to their development. IDMC - Norwegian refugee council, brookings-LSE project on internal displacement. https://www.ohchr.org/Documents/Issues/IDPersons/Law-and-policymaking_Guide-2013.pdf. Accessed 30 Nov 2021

International Land Coalition (2011) Securing land access for the poor in times of intensified natural resources competition. ILC. Tirana. https://www.landcoalition.org/en/about-ilc/governance/assemblydeclarations/. Accessed 30 Nov 2021

Instituto Nacional de Ecología y Cambio climático (INECC) (2018) Evaluación Estratégica del Avance Subnacional de la Política Nacional de Cambio Climático. https://www.gob.mx/cms/uploads/attachment/file/418239/EVALUACION_ESTRATEGICA_AVANCE_SUBNACIONAL_PNCC.pdf. Accessed 30 Nov 2021

Jacobs C, Almeida B (2020) Government-led resettlement after natural disasters as a durable solution? The case of cyclone idai. Refug Surv Q 1–9

Jacobson C, Crevello S, Chea C, Jarihani B (2019) When is migration a maladaptive response to climate change? Reg Environ Change 19(1):101–112

Jakobsson E (2021) Political attention in a creeping crisis: the case of climate change and migration. In: Boin A, Ekengren M, Rhinard M (eds) Understanding the creeping crisis. Palgrave, Macmillan, pp 131–147. https://doi.org/10.1007/978-3-030-70692-0

Johnson CA (2018) Climate, migration and displacement. In: McLeman R, Gemenne F (eds) Routledge handbook of environmental displacement and migration. Routledge/GSE Research, London, pp 342–355

Joseluis S (2008) Cambio climático y desarrollo en América Latina y el Caribe: una reseña. CEPAL, Santiago de Chile

Kälin W (2004) Addressing internal displacement: a framework for national responsibility. Brookings Institution–U of Bern Project on Internal Displacement, Washington, DC. http://www.globalprotectioncluster.org/_assets/files/tools_and_guidance/protection_of_idps/20050401_nrframework-EN.pdf. Accessed 30 Nov 2021

Kamperman H, Biesbroek R (2017) Measuring progress on climate change adaptation policy by Dutch water boards. Water Resour Manag 31(14):4557–4570

Kelman I (2014) No change from climate change: vulnerability and small island developing states. Geogr J 180(2):120–129

Kieval M (2020) On thin ice: exploring solutions for climate-induced displacement in the face of disappearing permafrost. Arctic Yearbook 2020

Klein RJT, Schipper ELF, Dessai S (2005) Integrating mitigation and adaptation into climate and development policy: three research questions. Environ Sci Policy 8(6):579–588

Krupocin D, Krupocin J (2020) The impact of climate change on cultural security. J Strat Secur 13(4):1–28

Lerner A, Charli-Joseph L (2018) Marcos teóricos para el estudio de los sistemas socio-ecológicos y las ciencias de la sostenibilidad. In: Ávila Foucat VS, Perevochtchikova M (eds) Sistemas socio-ecológicos: Marcos analíticos y estudios de caso en Oaxaca, México. UNAM-IIEs, Ciudad de México, pp 149–184

Ley de Hidrocarburos (2016) Last reform Diario Oficial de la Nación, 20 May 2021. http://www.diputados.gob.mx/LeyesBiblio/pdf/LHidro_200521.pdf. Accessed 30 Nov 2021

Ley General de Cambio Climático (LGCC) (2020) Last reform *Diario Oficial de la Federación*, 6 November 2020. http://www.diputados.gob.mx/LeyesBiblio/ref/lgcc.htm. Accessed 30 Nov 2021

Ley General de Víctimas (LGV) (2017) Last reform *Diario Oficial de la Federación*, 3 January 2017. http://www.diputados.gob.mx/LeyesBiblio/ref/lgv.htm. Accessed 30 Nov 2021

Liehr S, Röhrig J, Mehring M, Kluge T (2017) How the Social-Ecological Systems Concept Can Guide Transdisciplinary Research and Implementation: Addressing Water Challenges in Central Northern Namibia. Sustainability 9(7):1109–1128

Lietaer S, Durand-Delacre D (2021) Situating "migration as adaptation" discourse and appraising its relevance to Senegal's development sector. Environ Sci Policy 126:11–22. https://doi.org/10.1016/j.envsci.2021.09.008

Long J, Rice JL (2018) From sustainable urbanism to climate urbanism. Urban Stud 56(5):992–1008

Luna F (2014) "Vulnerability", an interesting concept for public health: the case of older persons. Public Health Ethics 7(2):180–194

Maldonado JK, Peterson KA (2018) A community-based model for resettlement. In: McLeman R, Gemenne F (eds) Routledge handbook of environmental displacement and migration. Routledge/GSE Research, Londres, pp 289–299

Maldonado González AL, González Gaudiano ÉJ (2013) De la resiliencia comunitaria a la ciudadanía ambiental: El caso de tres localidades en Veracruz, Mexico. Rev Integr Educ 6(3):14–28

Marter-Kenyon J (2020) Origins and functions of climate-related relocation: an analytical review. Anthropocene Rev 7(2):159–188

Martin S, Warner K (2012) Climate change, migration, and development. In: Omelaniuk I (ed) Global perspectives on migration and development. Springer, Dordrecht/OIM, pp 153–172

Martínez-Corona B (2012) Género, participación social, percepción ambiental y remediación ante desastres naturales en una localidad indígena, Cuetzalan, Puebla. Ra Ximhai 8(1):113–126

Mathur HM (2015) Climate change and displacement: learning from resettlement in the development context. Soc Chang 45(1):118–130

McCarney R, Kent J (2020) Forced displacement and climate change: time for global governance. Int J 75(4):652–661

McDonnell S (2021) The importance of attention to customary tenure solutions: slow onset risks and the limits of Vanuatu's climate change and resettlement policy. Curr Opin Environ Sustain 50:281–288. https://doi.org/10.1016/j.cosust.2021.06.008

McLeman R (2018) Thresholds in climate migration. Popul Environ 39(4):319–338

Meine C (2020) From the land to socio-ecological systems: the continuing influence of Aldo Leopold. Socio-Ecol Pract Res 2(1):31–38

Merino Pérez L (2005) El desarrollo institucional de esquemas de pago por servicios ambientales. Gac Ecológica 74:29–42

Methmann C, Oels A (2015) From "fearing" to "empowering" climate refugees: governing climate-induced migration in the name of resilience. Secur Dialogue 46(1):51–68

Miller F (2020) Exploring the consequences of climate-related displacement for just resilience in Vietnam. Urban Stud 57(7):1570–1587

Moser SC (2010) Now more than ever: the need for more societally relevant research on vulnerability and adaptation to climate change. Appl Geogr 30(4):464–474
Muñoz Sevilla N P, Azuz-Adeath I, Le Bail M (2016) La Política Nacional de Adaptación ante el Acuerdo de París. In: Rueda Abad J, Gay García C, Quitana Solórzano F (eds) *21 visiones de la COP21*. UNAM, Mexico, pp 173–185
Mycoo M, Wairiu M, Campbell D, Duvat V, Golbuu Y, Maharaj S, Nalau J, Nunn P, Pinnegar J, Warrick O (2022) Small Islands. In: Pörtner H-O, Roberts DC, Tignor M, Poloczanska ES, Mintenbeck K, Alegría A, Craig M, Langsdorf S, Löschke S, Möller V, Okem A, Rama B (eds) Climate change 2022: impacts, adaptation, and vulnerability. Contribution of working group II to the sixth assessment report of the intergovernmental panel on climate change. Cambridge University Press. In Press. https://www.ipcc.ch/report/ar6/wg2/
Nawrotzki RJ, DeWaard J, Bakhtsiyarava M, Ha JT (2017) Climate shocks and rural-urban migration in Mexico: exploring nonlinearities and thresholds. Clim Change 140(2):243–258
Nishimura L (2015) "Climate change migrants:" impediments to a protection framework and the need to incorporate migration into climate change adaptation strategies. Int J Refug Law 27(1):107–134
Offner S, Marlowe J (2021) Reconceptualising climate-induced displacement in the context of terminological uncertainty. Environ Hazards 1–16
Oki T, Quiocho RE (2020) Economically challenged and water scarce: identification of global populations most vulnerable to water crises. Int J Water Resour Dev 36(2–3):416–428
O'Neill B, van Aalst M, Zaiton Ibrahim Z, Berrang Ford L, Bhadwal S, Buhaug H, Diaz D, Frieler K, Garschagen M, Magnan A, Midgley G, Mirzabaev A, Thomas A, Warren R (2022) Key risks across sectors and regions. In: Pörtner H-O, Roberts DC, Tignor M, Poloczanska ES, Mintenbeck K, Alegría A, Craig M, Langsdorf S, Löschke S, Möller V, Okem A, Rama B (eds) Climate change 2022: impacts, adaptation, and vulnerability. Contribution of working group II to the sixth assessment report of the intergovernmental panel on climate change. Cambridge University Press. In Press. https://www.ipcc.ch/report/ar6/wg2/
Ortiz Espejel B, Azuara García G (2019) Retos y desafíos del cambio climático en el siglo XXI: una perspectiva biocultural y de gobernanza. In: Rueda Abad JC (eds) Aún estamos a tempo para el 1.5 C? Voces y Visiones sobre el Reporte Especial del IPCC. UNAM, Mexico, pp 247–260
Ostrom E (2005) Understanding Institutional Diversity. Princeton University Press, Princeton, New Jersey and Oxford, UK
Ostrom E (2010) Response: the institutional analysis and development framework and the commons. Cornell Law Rev 95(4):807–816
Ostrom V, Feeny D, Picht H (1988) Rethinking institutional analysis and development. International Center For Economic, San Francisco
Pacheco-Vega R (2014) The impact of Elinor Ostrom's scholarship on commons governance in Mexico. An overview. Policy Matters 1(19):23–33
Pacheco-Vega R, Basurto F (2008) Instituciones en el saneamiento de aguas residuales: reglas formales e informales en el Consejo de Cuenca Lerma-Chapala. Rev Mex Sociol 70(1):87–109
Pacheco-Vega R (in press) La gobernanza policéntrica de mitigación y adaptación al cambio climático en México en el contexto de la arquitectura global de política climática. In: Solorio I (ed) México ante la encrucijada de la gobernanza climática. UNAM, Ciudad de México
Pacheco-Vega R (2005) Institutional analysis within the Lerma-Chapala Region: new challenges for watershed management. In: Unu-Inweh (Ed.) UNU-INWEH/UNESCO-MAB-IHP international workshop "Water and ecosystems: water resources management in diverse ecosystems and providing for human needs." Fort Collins, CO: US Geological Survey (USGS)
Pacheco-Vega R (2006) Instituciones para el manejo del agua: Una visión analítica de los consejos de cuenca. Derecho Ambiental y Ecología (Febrero-Marzo), pp 52–55
Peguero Moreno M, Rubio Díaz-Leal L (2019) El desplazamiento interno forzado en México: respuestas del estado y litigio estratégico. Tirant lo blanch, Mexico

Pérez-Soba M, Dwyer J (2016) The social- ecological system concept, DG agri workshop. Wageningen University. http://ec.europa.eu/information_society/newsroom/image/document/2017-1/3-4-_socio-ecological_concept_-_marta_perez-soba_and_janet_dwyer_41127.pdf. Accessed 30 Nov 2021

Raleigh C, Jordan L, Salehyan I (2008) Assessing the impact of climate change on migration and conflict. In: Paper commissioned by the world bank group for the social dimensions of climate change workshop, Washington, DC

Rebollar Plata M, Oliva Aguilar V (2011) Medio ambiente y la construcción del riesgo. In: Pérez Campuzano E, Valderrábano Almegua M (eds) Medio ambiente, sociedad y políticas ambientales en México contemporáneo: una revisión interdisciplinaria. Universidad Autónoma de Guerrero/Porrúa, Ciudad de México, pp 69–81

Rubinos C, Anderies JM (2020) Integrating collapse theories to understand socio-ecological systems resilience. Environ Res Lett 15(7):1–13

Rubio Díaz-Leal L (2017) Desplazamiento ambiental. Experiencia global, realidad mexicana. Comisión Mexicana de Defensa y Promoción de los Derechos Humanos, Heinrich Böll Stiftung México, Centroamérica y el Caribe. http://cmdpdh.org/temas/desplazamiento/wp-content/uploads/2019/04/cmdpdh-desplazamiento-ambiental-laura-rubio-2018.pdf. Accessed 30 Nov 2021

Rueda Abad JC, Gay García C, Ortiz Espejel B (2017) Gobernanza Climática en México. Aportes para la consolidación estructural de la participación ciudadana en la política climática nacional. Volumen I. Diagnóstico. UNAM, Mexico

Runfola DM, Romero-Lankao P, Jiang L, Hunter LM, Nawrotzki R, Sanchez L (2015) The influence of internal migration on exposure to extreme weather events in Mexico. Soc Nat Resour 29(6):750–754

Sánchez GMF (2011) La reubicación de poblaciones como estrategia de ordenamiento territorial. Revista Geográfica De América Central 2:1–18

Sánchez MC, Herrador D, Valdés CM, McCall MK (2013) Estrategias de adaptación al cambio climático en dos comunidades rurales de México y El Salvador. BAGE (61)

Sánchez Mojica BE (2009) Cuando los derechos son la jaula. Trasplante rígido del soft law para la gestión del desplazamiento forzado. Estud Políticos 35:11–21

Schade J (2012) Les migrants des politiques climatiques: nouveaux défis face aux déplacements générés par le changement climatique. Cult Conflits 88:85–110

Schade J (2013) Climate change and planned relocation: risks and a proposal for safeguards. In: Faist J, Schade J (eds) Disentangling migration and climate change. Springer, Dordrecht, pp 183–206

Schipper ELF, Rev A, Preston BL, Carr ER, Eriksen SH, Fernandez-Carril LR, Glavovic B, Hilmi NJM, Ley D, Mukerji R, Muylaert de Araujo MS, Perez R, Rose SK, Singh PK (2022) Climate resilient development pathways. In: Pörtner H-O, Roberts DC, Tignor M, Poloczanska ES, Mintenbeck K, Alegría A, Craig M, Langsdorf S, Löschke S, Möller V, Okem A, Rama B (eds) Climate change 2022: impacts, adaptation, and vulnerability. Contribution of working group II to the sixth assessment report of the intergovernmental panel on climate change. Cambridge University Press. In Press. https://www.ipcc.ch/report/ar6/wg2/

Schroth G, Laderach P, Dempewolf J, Philpott S, Haggar J, Eakin H, Castilleros T, García Moreno J, Soto Pinto L, Hernández R, Eitzinger A (2009) Towards a climate change adaptation strategy for coffee communities and ecosystems in the Sierra Madre de Chiapas, Mexico. Mitig Adapt Strat Glob Change 14(7):605–625

Secretaría de Medio Ambiente y Recursos Naturales (SEMARNAT) (2013) Estrategia Nacional del Cambio Climático (ENCC). *Diario Oficial de la Federación*, 3 de junio de 2013. http://www.dof.gob.mx/nota_detalle.php?codigo=5301093&fecha=03/06/2013. Accessed 30 Nov 2021

Secretaría de Medio Ambiente y Recursos Naturales (SEMARNAT) (2021) Programa Especial de Cambio Climático 2020–2024. Copy with the autor

Scovronick N, Lloyd SJ, Kovats RS (2015) Climate and health in informal urban settlements. Environ Urban 27(2):657–678

De Sherbinin A, Castro M, Gemenne F, Cernea MM, Adamo S, Fearnside PM, Krieger G, Lahmani S, Oliver-Smith A, Pankhurst A, Scudder T, Singer B, Tan Y, Wannier G, Boncour P, Ehrhart C, Hugo G, Pandey P, Shi G (2011) Preparing for resettlement associated with climate change. Science 334(6055):456–457

Singh RB (ed) (2015) Urban development challenges, risks and resilience in Asian mega cities. Springer, Tokyo

Spring ÚO (2019) Climate change and its impact on vulnerable people. In: Spring ÚO (ed) Úrsula Oswald Spring: pioneer on gender, peace, development, environment, food and water. Springer, Cham, pp 481–506

Stensrud AB (2016) Harvesting water for the future: reciprocity and environmental justice in the politics of climate change in Peru. Lat Am Perspect 43(4):56–72

Surugu JIM, Chutab DN (2020) Going beyond incrementalism: climate projects' beneficiaries' perspective on what could be described as transformational adaptation. Int J Rural Manag 1–20

Tacoli C (2009) Crisis or adaptation? Migration and climate change in a context of high mobility. Environ Urban 21(2):513–525

Tan YL (2018) Environmental stressors and population mobility in China. In: McLeman R, Gemenne F (eds) Routledge handbook of environmental displacement and migration. Routledge/GSE Research, Londres, pp 238–256

Tetreault, D. V. (2015). México: La ecología política de la minería. In H.Y Veltmeyer & J. Petras, *El neoextractivismo. ¿Un modelo posneoliberal de desarrollo o el imperialismo del siglo XXI?* (253–280). México: Crítica

Thiel A, Adamseged ME, Baake C (2015) Evaluating an instrument for institutional crafting: How Ostrom's social-ecological systems framework is applied. Environ Sci Policy 53:152–164

Ugalde V (2010) La aplicación del derecho ambiental. In: Graizbord B, Lezama JL (eds) Los grandes problemas de México, Medio Ambiente. El Colegio de México, pp 227–256

United Nations (UN) (1998) Principios Rectores De Los Desplazamientos Internos, UN doc.E/CN.4/1998/53/Add.2, 11 February 1998, Oficina de Coordinación de Asuntos Humanitarios de las Naciones Unidas. https://www.acnur.org/prot/prot_despl/5bff2c864/principios-rectores-de-los-desplazamientos-internos.html.Accessed 30 Nov 2021

United Nations (UN) (2010) Convención Marco sobre el Cambio Climático. Informe de la Conferencia de las Partes sobre su 16o período de sesiones, celebrado en Cancún del 29 de noviembre al 10 de diciembre de 2010. Decisión 1/CP.16 Acuerdos de Cancún: resultado de la labor del Grupo de Trabajo Especial sobre la cooperación a largo plazo en el marco de la Convención FCCC/CP/2010/7/Add.1. https://unfccc.int/resource/docs/2010/cop16/spa/07a01s.pdf. Accessed 30 Nov 2021

United Nations (UN) (2015) Marco de Sendai para la Reducción del Riesgo de Desastres 2015–2030. https://www.unisdr.org/files/43291_spanishsendaiframeworkfordisasterri.pdf. Accessed 30 Nov 2021

Vázquez JLP (2013) Mercado de tierras y propiedad social: una discusión actual. Anales De Antropología 47(2):9–38

Vázquez LM (2017) Implementation challenges of climate change adaptation initiatives in coastal lagoon communities in the Gulf of Mexico. Maritime Stud 16(1):14–30

Vázquez García V (2017) El acaparamiento de tierras en México: extensión, escala, propósito y novedad. Revista Mexicana De Ciencias Forestales 8(44):10–28

Velázquez Melo T, Pacheco-Vega R, Gouritin A (2021) Lineamientos de atención al fenómeno: potencial del enfoque de Sistemas Socio- Ecológicos. In: Gouritin A (ed) Migrantes Climáticos en México. Flacso, Mexico, pp 289–326

Vigil SD (2018) Green grabbing-induced displacement. In: McLeman R, Gemenne F (eds) Routledge handbook of environmental displacement and migration. Routledge/GSE Research, London, pp 370–387

Vogel B, Henstra D (2015) Studying local climate adaptation: a heuristic research framework for comparative policy analysis. Glob Environ Chang 31:110–120

Voigt L, Thornton WE (2015) Disaster-related human rights violations and corruption: a 10-year review of post–Hurricane Katrina New Orleans. Am Behav Sci 59(10):1292–1313

Von Lucke F (2018) Linking climate change and security in Mexico: explorations into an attempted securitisation in the Global South. J Int Relat Dev 21(2):415–441

Watson RT (2005) Turning science into policy: challenges and experiences from the science-policy interface. Philos Trans R Soc Lond B Biol Sci 360(1454):471–477

Witness G (2018) A Qué Precio? Negocios irresponsables y el asesinato de personas defensoras de la tierra y del medio ambiente en 2017. https://www.business-humanrights.org/sites/default/files/documents/Defenders_report_spanish-7%20%282%29.pdf. Accessed 30 Nov 2021

Wolff E (2021) The promise of a "people-centred" approach to floods: types of participation in the global literature of citizen science and community-based flood risk reduction in the context of the Sendai Framework. Prog Disaster Sci 10:100171. https://doi.org/10.1016/j.pdisas.2021.100164

Young OR, Berkhout F, Gallopin GC, Janssen MA, Ostrom E, van der Leeuw S (2006) The globalization of socio-ecological systems: an agenda for scientific research. Glob Environ Chang 16(3):304–316

General Conclusion

This manuscript presents the results of an investigation whose main objectives were to provide answers to the following questions: In Mexico, does the legal framework and current public policies address the phenomenon of forced internal climate mobility? If not, what are the potential elements of an effective legal framework and public policies to address the phenomenon?

When approaching the phenomenon, we felt it was complex and entailed numerous tensions that could generate other questions throughout the research process. We were not wrong. The crossover between climate change (per se crossed by empirical and theoretical tensions and debates) and mobility (which also generates empirical and normative debates—e.g., on the security narrative) is, without doubt, a fertile field, intellectually challenging, and intrinsically humane in its aims. The challenges are numerous: among others, avoiding racism or paternalistic biases, nor reproducing and reinforcing dominant paradigms without questioning them.

The first stage of the investigation (Chaps. 1 and 2) aims to reveal these tensions by establishing the phenomenon's theoretical and contextual framework. Although this research is qualitative, we report the first stage empirical data which allows us to contemplate the scale of the phenomenon in Mexico: according to a 2018 World Bank report, by 2050 and in the most pessimistic scenario, up to 3.1 million people will be internal forced climate migrants in Mexico. They will represent 11% of internal migrants in Mexico. This empirical data is accompanied by a typology or taxonomy that allows the disaggregating of the phenomenon to answer the questions that guide the research. Based on the academic literature, we modify the proposed typologies and work with six scenarios: sudden-onset disasters, slow-onset phenomena, "sinking" islands and coasts, conflicts or violence caused by a scarcity of natural resources, areas designated unfit for human settlement, and "development projects" related to renewable energies. We also identified three particularly vulnerable populations who received special attention throughout this work: Indigenous Peoples, women, and people with limited economic resources.

A. Gouritin, *Climate Displacement in Mexico*,
https://doi.org/10.1007/978-3-031-10335-3

The next step consisted of proposing definitions and criteria, a necessary exercise in any scientific work (and also when adopting a legal framework and public policies). This exercise revealed lines of tension between skeptics and alarmists. In spite of these differences of opinion, we propose and justify definitions, three categories of forced internal migrants (emergency forced climate migrants, forced climate migrants, and trapped forced climate migrants), and related identification criteria. In proposing and justifying these, another tension emerged between a subjective or an objective approach. We chose to combine these approaches and do not rule out a subjective one since it allows us to consider people affected and the agency of communities and not place them in the role of a passive victim.

With these six scenarios, the identification and selection of three particularly vulnerable populations and the proportion of definitions, the three categories of migrant and corresponding criteria, it was possible to develop the analytical grid further. It was composed of two facets based on territory as well as protection requirements. We place our territory-based approach in a broad territorial context which encompasses the symbolic, economic, social, cultural, and historical dimensions of the territory. We justify this and identify the challenges inherent to the phenomenon in terms of territories. As to the demand for protection, we adopt a rights-based approach despite the criticism that this provokes. More specifically, we adopt a differentiated rights-based approach complemented by environmental rights. After justifying this approach and determining its potential in the research context, we visualize the content of the rights and obligations in several tables.

In the first two chapters, it emerged that numerous tensions cross climate mobilities. A first series of discussion concerns the empirical identification of the phenomenon. Environmentalists tend to place great empirical importance on this ("waves of climate migrants"). Conversely, specialists in migratory phenomena tend to deny the importance of the climate trigger. This throws up another tension: that of identifying the climate dimension in the migratory phenomenon. Many factors drive and can be said to explain it. Accordingly, identifying the climate trigger and its relative importance is delicate. Another tension refers to the interplay between the environmental and climate dimensions: is the climate dimension a "sub-part" of the environmental dimension? Or does the climate dimension amplify the environmental migration triggers? Another tension relates to the call for the protection of migrants: should it be regulated according to the obligations of a State or based on the rights of migrants?

It is also worth mentioning the tension between a bottom-up or top-down approach: the latter entails the risk of racist biases, political objectives that have nothing to do with the effects of climate change in themselves and identifying the phenomenon as a window of opportunity for land-grabbing or authoritarian land-planning, and further, runs the risk of failing to address the underlying problems (among which poverty is the most pressing). A bottom-up approach would, conversely, make it possible to give full importance to the agency of migrants and avoid the risk of perceiving them as passive agents. The last tension that we mention here (the book addresses more) is the perception surrounding the phenomenon: can forced climate mobility be perceived as a successful adaptation to climate change or, on the other hand, as a failed adaptation?

In the second stage research, we analyze the political-legal framework of Mexico to determine whether it addresses the phenomenon (Chaps. 3 and 4). We find the political-legal framework at the federal, state (Tabasco), and municipal levels (Tabasco municipalities of Centro and Centla) to suffer from various gaps and shortcomings, leaving internal climate migrants unprotected.

Taking into account the shortcomings of the Mexican political-legal framework, the third stage is dedicated to possible responses to address the phenomenon with an approach to protection firmly rooted in the territory (Chap. 5). Starting with a critical analysis of the tools Mexico could potentially adopt, we focus on four tools: the recognition of the legal framework of internal climate forced migrants, forced relocations, preventive adaptation measures (in situ adaptation organized around the concept of resiliency) and reactive adaptation measures (ex situ adaptation, mobility considered a successful adaptation). We identify a large number of particularly acute risks in the Mexican context. Two are recurring: violating the rights of people and communities affected to participate in identifying problems and solutions, and not taking the local context into account. We then identify the potential of the Socio-Ecological Systems approach and limits to guide public policies aimed at addressing the phenomenon.

Completing this project is by no means a conclusion of the investigation and work surrounding this phenomenon. We are very aware of its limits and, if possible, we would like to continue to dig deeper and cast out net wider into aspects that we were unable to develop due to the limits of a one-year research project. Among these potential areas of focus are, the task of further validating the analytical grid remains alongside the findings with fieldwork, carrying out a comparative analysis of the legal frameworks and public policies, developing the case study of the four-forced internal climate mobility scenarios not addressed in case-studies, and widening the reflections on a regional scale (for example, starting with Central America). A further pending task (and no less important) is the disseminating of the results of the research among legislators and public authorities in order for the research to have an impact beyond academic circles.

The authors, December 2021.